Integrated Pest Management in Agricultural Crops and Forest Ecosystems

Integrated Pest Management in Agricultural Crops and Forest Ecosystems

Collection Editors

Nickolas G. Kavallieratos
Maria C. Boukouvala

MDPI

Basel • Beijing • Wuhan • Barcelona • Belgrade • Novi Sad • Cluj • Manchester

Collection Editors

Nickolas G. Kavallieratos
Department of Crop Science
Agricultural University
of Athens
Athens
Greece

Maria C. Boukouvala
Department of Crop Science
Agricultural University
of Athens
Athens
Greece

Editorial Office
MDPI AG
Grosspeteranlage 5
4052 Basel, Switzerland

This is a reprint of the Topical Collection, published open access by the journal *Insects* (ISSN 2075-4450), freely accessible at: https://www.mdpi.com/journal/insects/topical_collections/Agricultural_Forest.

For citation purposes, cite each article independently as indicated on the article page online and as indicated below:

Lastname, A.A.; Lastname, B.B. Article Title. *Journal Name* **Year**, *Volume Number*, Page Range.

ISBN 978-3-7258-4767-9 (Hbk)
ISBN 978-3-7258-4768-6 (PDF)
https://doi.org/10.3390/books978-3-7258-4768-6

Cover image courtesy of Maria C. Boukouvala

Contents

About the Editors

Nickolas G. Kavallieratos

Nickolas G. Kavallieratos is Professor of Agricultural Entomology and Acarology at the Agricultural University of Athens (AUA). During 2002–2014, he was a researcher at Benaki Phytopathological Institute (BPI), serving in all scientific ranks and administration positions (Head of the Laboratory of Agricultural Entomology, Head of the Department of Entomology and Agricultural Zoology, and Deputy Director) at BPI. He has also worked at several universities and institutes in Europe, North America, Africa, and Asia. He studies morphology, morphometry, systematics, zoogeography, phylogeny, genetics and behavior of the Aphidiinae (Braconidae) parasitoids of aphids, providing descriptions of new species in science, redescriptions of species and genera, revisions of genera and reviews around the world (Europe, Asia, Africa, North and South America) since 1994. The overall outcome of his research deals with the generation of tools that enable entomologists to identify Aphidiinae parasitoids with ease in different types of habitats, host plants, and host aphids. Since 2000, he has also been working on the biology, ecology, demography, behavior, and management of stored-product pests (insects and mites) with the use of natural (diatomaceous earths and other inert dusts), biological agents (entomopathogenic fungi, entomopathogenic nematodes, bacteria), chemical insecticides, and novel chemical or natural compounds (essential oils, botanical molecules, nanoemulsions, microemulsions) that exhibit insecticidal properties. Prof. Kavallieratos has published >360 papers in peer-reviewed journals and coordinated > 40 projects.

Maria C. Boukouvala

Maria C. Boukouvala is a Post Doctoral Researcher in the Laboratory of Agricultural Zoology and Entomology at the Agricultural University of Athens (AUA). She has been involved in the investigation of new insecticidal substances for the protection of stored grains, such as essential oils and their nanoemulsions, which are environmentally friendly natural products, with chemical insecticides that exhibit low toxicity to mammals and non-target organisms, with active substances based on the metabolites of microorganisms, and with the use of entomopathogenic nematodes and fungi as biological control agents. Dr Boukouvala has worked in several national and EU projects. Dr Boukouvala has obtained international laboratory experience from her participation in eight Short-Term Scientific Missions (STSMs) funded by the European Union, of a total duration of >1 year. Dr Boukouvala has long academic experience at the AUA since she officially undertook the teaching responsibility as Academic Scholar for several courses between 2021 and 2024. Her academic experience extended to the international level, given that she has delivered numerous lectures to several universities and institutes (University of Pisa, Poznan University of Life Sciences, and Institute of Dendrology—Polish Academy of Sciences). Dr Boukouvala has published >88 papers in international journals.

Preface

The target Topical Collection aimed to shed light on novel alternative paths to manage dangerous arthropod species under field conditions in diverse ecosystems (i.e., crops, forests, and urban or suburban areas) and controlled environments. The remarkable number of the 18 published papers around the globe (Japan, Chile, Germany, Belgium, Czech Republic, Austria, France, USA, Poland, China, Greece, Vietnam, Italy, Spain, Mexico, Romania, Hungary, Slovenia, South Africa) evaluated the potential of the following: using biological control agents, the enhancement of natural biological control, the use of biopesticides, compounds that are produced by insects, novel attractants, semiochemicals and agricultural practices for the management of lepidopterans, coleopterans, hemipterans, thysanopterans, and dipterans. The published important findings introduce valuable tools in the frame of IPM and, simultaneously, open new areas of research on the same studied organisms or in other noxious species that threaten the health of plants and the overall balance within different ecosystems.

Nickolas G. Kavallieratos and Maria C. Boukouvala
Collection Editors

insects

MDPI

Article

Population Fitness of *Eupeodes corollae* Fabricius (Diptera: Syrphidae) Feeding on Different Species of Aphids

Shanshan Jiang [1], Hui Li [1], Limei He [2] and Kongming Wu [1,*]

[1] State Key Laboratory for Biology of Plant Diseases and Insect Pests, Institute of Plant Protection, Chinese Academy of Agricultural Sciences, Beijing 100193, China; jiangss97@163.com (S.J.); lihuilh521@163.com (H.L.)

[2] Institute of Urban Agriculture, Chinese Academy of Agricultural Sciences, Chengdu 610200, China; helimei91@163.com

* Correspondence: wukongming@caas.cn

Simple Summary: As a candidate for controlling aphid populations, *Eupeodes corollae* Fabricius also performs a pollination function which is critical to agricultural systems. We evaluated the population life tables and flight performance of *E. corollae* fed on three prey species (*Aphis craccivora* Koch, *Myzus persicae* Sulzer and *Megoura japonica* Matsumura) to select suitable aphid species for keeping *E. corollae* indoors. The results showed that *E. corollae* completed development and reproduction on these three aphid species while achieving the shortest generation time, the maximum fecundity, and the highest intrinsic natural growth rate and flight ability on *M. japonica*. Our study indicated that *M. japonica* is the most suitable prey for *E. corollae*, providing a basis for utilizing the ecological service function of the hoverfly.

Abstract: *Eupeodes corollae* Fabricius, as one of the most common beneficial predatory insects in agricultural ecosystems, provides pollination and biological control services that help improve crop yield and maintain biodiversity. However, systematic research is needed on the species of aphids used for propagation. To develop highly fit populations of the important insect predator and crop pollinator, *E. corollae*, for research and commercial use, further research is needed to develop the most nutritious diet and efficient propagation methods. Here, the fitness of *E. corollae* was assessed in the laboratory after larvae were fed an aphid diet of *Aphis craccivora* Koch, *Myzus persicae* Sulzer or *Megoura japonica* Matsumura. The larval survival rate on *M. japonica* was significantly lower than on *A. craccivora* and *M. persicae*. The developmental duration for larvae (7.6 d) and pupae (6.9 d) was longest on *A. craccivora*. The pupal emergence rate on *A. craccivora* (98.0%) was significantly higher than on the other two, and lowest (64.7%) on *M. japonica*. On *A. craccivora*, *M. persicae*, and *M. japonica*, respectively, the generation time was 24.85 d, 23.12 d and 21.05 d; the value for the intrinsic rate of natural increase was 0.19, 0.20, and 0.21; and the value for the finite rate of increase was 1.21, 1.22, and 1.23. For flight variables, *E. corollae* attained the fastest velocity and longest distance and duration on *M. japonica*. The *M. japonica* diet, thus, provided the shortest generation time, the highest intrinsic rate of natural increase and finite rate of increase, the maximum fecundity and the greatest flight ability. Thus, to improve the survival rate of *E. corollae* larvae, *A. craccivora* or *M. persicae* can be used to feed newly hatched larvae, and *M. japonica* can be used for second- and third-instar larvae. These results provide a theoretical basis for feeding *E. corollae* and optimizing its ecosystem services.

Keywords: *Eupeodes corollae*; fecundity; life history; flight ability; aphid species

1. Introduction

Syrphidae, also known as flower flies or hoverflies, comprise almost 6000 species in 200 genera [1,2] and are found all over the world except for some remote islands in Antarctica and the Pacific Ocean (such as Hawaii) [3,4]. In China, 465 species in 80 genera of Syrphidae are known [5]. Hoverfly larvae may be saprophages, carnivores or other types of feeders [6–8]. About one third of hoverfly species have carnivorous larvae that prey on whiteflies, thrips and aphids. Thus, the larvae of these species are natural enemies that might be used to control aphids in agroecosystems [9]. The third-instar larvae of *E. corollae* consume about 60 cabbage aphids per day per individual and 300 cabbage aphids in total [10]. In addition, most hoverfly adults visit flowers and are the second most important pollinators after bees. They are even more effective than bees at pollinating individual plants such as *Paphiopedilum dianthum* [11–13]. Their pollination efficacy, however, varies depending on the crop species, season and location [14]. For example, hoverfly pollination increased fennel yield by 104.9% [15] and parsley seed yield nearly 3-fold compared with a control group not pollinated by hoverflies [16].

Eupeodes corollae Fabricius (Diptera: Syrphidae), a common hoverfly species worldwide, is a promising candidate for aphid control and pollination [17]. The larvae are carnivorous and feed on a variety of aphid species and small lepidopteran larvae [18,19]. In one study, larvae consumed 64.7 wheat aphid individuals in a field, providing a control rate of 74.8% [20]. When 8-day-old larvae of *E. corollae* were released at a benefit-to-harm ratio of 1:10, the number of aphids decreased by 85.9% after 8 days [21]. As pollinators, *E. corollae* adults can significantly increase crop yields. The fruit yield and seed set of sweet peppers pollinated by *E. corollae* increased, respectively, by 390% and 395% compared to the control group not pollinated by hoverflies [22].

For the study and development of hoverflies for agricultural use, however, a large supply of high-quality insects at different stages is needed. The few studies on large scale propagation have focused on aspects of adult nutrition, prey species and host plants. The larvae of *Episyrphus balteatus* (De Geer) and *Eupeodes bucculatus* (Rondani) that fed on an artificial diet mainly composed of drone honeybee brood powder (DP) grew and developed normally, but did not lay eggs as adults [23,24]. At present, hoverflies are all reared using aphids as food; however, this process has many problems such as self-mutilation by larvae, obligate diapause of adults, and mating difficulty [25]. In addition, there is a lack of research on various species of aphids and their effects on fitness variables of *E. corollae*.

Hoverflies are excellent candidates to control aphid populations due to their high predation and reproductive capacity [26,27], but the development of solid or liquid diets has been difficult, because *E. corollae* larvae do not eat dead aphids [28]. Here, we used the major aphid pests *Aphis craccivora* Koch, *Myzus persicae* Sulzer and *Megoura japonica* Matsumura (Homoptera: Aphididae) [29], which readily reproduce in greenhouses [21], as diets for *E. corollae* in the laboratory. We then assessed larval developmental, reproductive and flight variables for *E. corollae* to select the best nutritional source as a theoretical basis for rearing *E. corollae* indoors. This work will advance research on syrphids.

2. Materials and Methods

2.1. Aphid Collection and Culture

Aphids (*A. craccivora*, *M. persicae* and *M. japonica*) were collected from the experimental field at the Langfang Experimental Station of Agricultural Sciences in Hebei Province, China (CAAS; 39°30′29′′ N, 116°36′8′′ E) in 2016. *Aphis craccivora* and *M. japonica* were reared on bean plantlets, and *M. persicae* were reared on pea plantlets until more than 30 generations in the greenhouse at 25 ± 1 °C, 50 ± 5% RH and 16:8 (L:D) h. All plantlets were grown in nutrient soil and vermiculite in plastic boxes (50 × 40 × 18 cm).

2.2. Hoverfly Collection and Culture

In June 2018, 25 adults of *E. corollae* (♀:♂= 15:10) were also collected from the field at the Langfang Experimental Station of Agricultural Sciences (CAAS; 39°30′29′′ N, 116°36′8′′ E). Adults were fed with a mixture of pollen (rape: corn = 3:1) and 10% *v/v* honey water in nylon gauze cages (30 × 40 × 50 cm). Broad bean plantlets infested with mixed aphids (*A. craccivora*:*M. persicae*:*M. japonica* ≈ 1:1:1) were placed in the cages for laying eggs. Larvae were fed on *A. craccivora* on bean plantlets until 10 consecutive generations in laboratory at 25 ± 1 °C, 50 ± 5% RH and 16:8 (L:D) h.

2.3. Life Table Study for E. corollae

Newly hatched larvae of *E. corollae* were raised with *A. craccivora*, *M. persicae* or *M. japonica*, individually, in petri dishes (3.5 cm diameter × 1 cm height), with 100 aphids added daily until pupation. On the third day after pupation, each pupa was weighed on an electronic scale (Sartorius CPA225D; Suzhou Sainz Instrument Co., Ltd., Suzhou, China). The development (whether they molted) and survival of *E. corollae* larvae were observed and recorded every day with a stereo microscope (XTL-165-VT, Phenix Optical Co., Ltd., Shangrao, China) until they pupated or died. Each diet treatment consisted of 30 eggs or newly hatched larvae and was replicated three times. Larval survival rate, pupation rate, and emergence rate were calculated as Larval survival rate = No. of mature larvae/No. of tested larvae × 100; Pupation rate = No. of pupae/No. of mature larvae × 100; and Emergence rate = No. of adult/No. of pupae × 100.

Emerging adults (♀:♂= 1:1) were reared in 20 × 30 × 45 cm cages (200-mesh nylon) with broad bean plantlets infested with mixed aphids (*A. craccivora*:*M. persicae*:*M. japonica* ≈ 1:1:1), and also with a mixture of pollen (rape:corn = 3:1), and 10% honey water provided daily. *Eupeodes corollae* eggs on broad bean plantlets were counted and transferred to a petri dish (9 cm diameter × 1 cm height) to observe hatching. The plantlets were replaced every day with fresh ones until *E. corollae* adults died.

The data on developmental and reproductive variables, generation time (T), finite rate of increase (λ), intrinsic rate of natural increase (r), and net reproductive rate (R_0) of the experimental populations of *E. corollae* were calculated according to the following formulas [30,31]:

$$l_x = \sum_{j=1}^{m} S_{xj}$$

$$m_x = \frac{\sum_{j=1}^{m} S_{xj} f_{xj}}{\sum_{j=1}^{m} S_{xj}}$$

$$\sum_{x=0}^{\infty} l_x m_x e^{-r(x+1)} = 1$$

$$R_0 = \sum_{x=0}^{\infty} l_x m_x$$

$$T = \frac{\ln R_0}{r}$$

$$\lambda = e^r$$

where x is the time interval in days (d); l_x is the survival rate of *E. corollae* from egg to x days old; m is the number of stages; S_{xj} is the survival rate of *E. corollae* from egg development to x days old and developmental stage j; f_{xj} is the age-specific fecundity at age x; and m_x is the average population fecundity from egg to x days old.

2.4. Flight Tests for E. corollae

Flight variables (duration, distance, velocity) of 5-day-old adults were measured using a flight mill (FXMD-24-USB, Jiaduo Science Industry and Trade Co., Ltd., Hebi, China) as described previously [32,33]. The mesothorax was attached to the arm of a flight mill using 502 glue (Shenzhen Jinsan second Adhesive Co., Ltd., Shenzhen, China). The mill was kept in a climate chamber (MGC-450HP, Shanghai Yiheng Scientific Instrument Co., Ltd., Shanghai, China) at 25 ± 1 °C with 50 ± 5% RH. Flights started at approximately 20:00 and ran for 10 h in the dark. More than 30 individuals were successfully tested for each prey treatment.

2.5. Data Analysis

Differences in the flight ability variables and pupal mass were analyzed using non-parametric Kruskal–Wallis test followed by a Bonferroni-adjusted significance test for pairwise comparison at the 0.05 significance level. Differences in larval survival rate, pupation rate, emergence rate and hatching rate of eggs were determined using a one-way analysis of variance (ANOVA) followed by Tukey's honestly significant difference (HSD), with proportional data first arcsine square-root-transformed to meet the assumptions of normality and heteroscedasticity. Differences in the duration of developmental stages between genders were determined using a Student's *t*-test, and differences in survival curves of *E. corollae* after feeding on different aphids were analyzed using a log rank test. A paired bootstrap test with 100,000 replications was used for the precise estimation of the mean and standard error among the demographic parameters and developmental duration of the 3 diet treatments. All tests were performed in the program SPSS version 25 (IBM, Armonk, NY, USA), except life table variables, which were calculated and differentially analyzed in TWO-SEX-MSChart (Chi 2019) and plotted in OriginPro 2021 (OriginLab Corporation, Northampton, MA, USA).

3. Results

3.1. Development of *E. corollae* Fed on Different Aphids

All the *E. corollae* larvae that fed on the three aphid species completed development, but the larval and pupal stages varied, and development differed between sexes in all prey treatments (Table 1). More specifically, there were significant differences in the duration of the first instar, second instar, third instar and larval stage (first–third instar) between the different aphid treatments. The duration of larvae (7.6 d) and pupae (6.9 d) after feeding on *A. craccivora* was significantly longer than for those fed on *M. persicae* or *M. japonica*. The first instar (1.4 d) was the shortest on *M. persicae*. The second instar (1.8 d), third instar (3.0 d) and pupal (6.1 d) stages were the shortest on *M. japonica*. The duration of the combined larva–adult stage and the adult lifespan of *E. corollae* did not differ significantly among on the different aphid diets. On *A. craccivora*, the duration of first-instar larvae ($t = 2.816$, $df = 62$, $p = 0.007$), larval stage (first–third instar) ($t = 2.398$, $df = 62$, $p = 0.020$), pupae ($t = 2.667$, $df = 62$, $p = 0.012$), adult ($t = 4.172$, $df = 62$, $p < 0.001$) and larva–adult ($t = 4.728$, $df = 62$, $p < 0.001$) of females were significantly longer than those of males. On *M. persicae*, the duration of first instar ($t = 2.703$, $df = 39$, $p = 0.014$), pupae ($t = 6.658$, $df = 39$, $p < 0.001$), adult ($t = 3.162$, $df = 39$, $p = 0.003$) and larva–adult ($t = 3.792$, $df = 39$, $p = 0.001$) of females were significantly longer than those of males. On *M. japonica*, the first instar ($t = -3.135$, $df = 25$, $p = 0.005$) of females was significantly shorter than that of males.

Table 1. Stage durations for *E. corollae* after larvae fed on different aphid species.

Stage	Sex	Stage Duration (d)					
		A. craccivora	*n*	*M. persicae*	*n*	*M. japonica*	*n*
Egg	♀+ ♂	2.0 ± 0.00 a	90	2.0 ± 0.00 a	90	2.0 ± 0.00 a	90
	♀	2.0 ± 0.00 A	31	2.0 ± 0.00 A	20	2.0 ± 0.00 A	13
	♂	2.0 ± 0.00 A	33	2.0 ± 0.00 A	21	2.0 ± 0.00 A	14
1st instar	♀+ ♂	2.0 ± 0.06 a	75	1.4 ± 0.06 b	90	1.8 ± 0.09 a	55
	♀	2.1 ± 0.06 A	31	1.5 ± 0.18 A	20	1.5 ± 0.18 B	13
	♂	1.8 ± 0.09 B	33	1.0 ± 0.00 B	21	2.2 ± 0.11 A	14
2nd instar	♀+ ♂	2.1 ± 0.03 a	72	1.9 ± 0.06 b	76	1.8 ± 0.07 b	49
	♀	2.1 ± 0.04 A	31	1.7 ± 0.14 A	20	1.8 ± 0.15 A	13
	♂	2.1 ± 0.04 A	33	1.8 ± 0.09 A	21	1.6 ± 0.13 A	14
3rd instar	♀+ ♂	3.5 ± 0.07 a	65	3.6 ± 0.13 a	55	3.0 ± 0.13 b	41
	♀	3.6 ± 0.09 A	31	3.1 ± 0.08 A	20	2.8 ± 0.10 A	13
	♂	3.5 ± 0.10 A	33	3.4 ± 0.13 A	21	3.0 ± 0.23 A	14

Table 1. *Cont.*

Stage	Sex	Stage Duration (d)					
		A. craccivora	*n*	*M. persicae*	*n*	*M. japonica*	*n*
Larval stage (1st–3rd instar)	♀+ ♂	7.6 ± 0.09 a	65	6.7 ± 0.16 b	55	6.7 ± 0.16 b	41
	♀	7.8 ± 0.10 A	31	6.4 ± 0.11 A	20	6.2 ± 0.20 A	13
	♂	7.4 ± 0.14 B	33	6.1 ± 0.16 A	21	6.8 ± 0.23 A	14
Pupae	♀+ ♂	6.9 ± 0.04 a	64	6.3 ± 0.07 b	41	6.1 ± 0.07 b	27
	♀	7.0 ± 0.00 A	31	6.7 ± 0.10 A	20	6.1 ± 0.10 A	13
	♂	6.8 ± 0.07 B	33	6.0 ± 0.00 B	21	6.1 ± 0.10 A	14
Adult	♀+ ♂	11.8 ± 0.69 a	64	12.5 ± 0.84 a	41	12.4 ± 0.77 a	27
	♀	14.5 ± 1.11 A	31	14.9 ± 1.25 A	20	13.3 ± 0.91 A	13
	♂	9.3 ± 0.55 B	33	10.1 ± 0.89 B	21	11.6 ± 1.21 A	14
Egg–adult	♀+ ♂	28.3 ± 0.70 a	64	27.1 ± 0.88 a	41	27.1 ± 0.83 a	27
	♀	31.3 ± 1.07 A	31	30.0 ± 1.26 A	20	27.7 ± 1.08 A	13
	♂	25.5 ± 0.59 B	33	24.3 ± 0.86 B	21	26.6 ± 1.26 A	14

Values are mean ± SE. Different lowercase letters on the same row mean significant differences (paired bootstrap test, $p < 0.05$). Means for male and female at the same developmental stage in same column followed by different capital letters differed significantly (Student's *t*-test, $p < 0.05$).

The aphid species greatly affected the larval survival rate ($F_{2,6} = 12.022$, $p = 0.008$), pupal emergence ($F_{2,6} = 14.853$, $p = 0.005$), and pupal mass (Kruskal–Wallis, $H = 26.607$, $p < 0.001$) of *E. corollae* (Table 2). The larval survival rate of *E. corollae* on *M. persicae* (84.4%) was significantly higher than that on *M. japonica* (54.4%), but the pupal emergence rate was significantly higher on *A. craccivora* (98.0%) than on *M. persicae* (73.9%) and *M. japonica* (64.7%).

Table 2. Variables and pupal mass of *E. corollae* after larvae fed on different aphid species.

Variable	*A. craccivora*	*n*	*M. persicae*	*n*	*M. japonica*	*n*
Larval survival rate (%)	80.0 ± 5.77 a	3	84.4 ± 2.94 a	3	54.4 ± 4.84 b	3
Pupation rate (%)	89.9 ± 5.52 a	3	72.8 ± 6.78 a	3	83.1 ± 7.62 a	3
Emergence rate (%)	98.0 ± 1.96 a	3	73.9 ± 6.29 b	3	64.7 ± 4.05 b	3
Hatching rate of egg (%)	90.0 ± 5.09 a	3	88.89 ± 4.01 a	3	82.2 ± 6.76 a	3
Mass (g)						
♀	0.0564 ± 0.0133 A	31	0.0317 ± 0.0009 A	20	0.0315 ± 0.0006 A	13
♂	0.0332 ± 0.0005 A	33	0.0307 ± 0.0009 A	21	0.0309 ± 0.0009 A	14
♀+ ♂	0.0444 ± 0.0066 a	64	0.0312 ± 0.0006 b	41	0.0312 ± 0.0005 b	27

Values are mean ± SE. Different lowercase letters on the same row reflect significant differences (one-way ANOVA, Tukey's HSD, $p < 0.05$). Different capital letters in the same column indicate significant differences (Student's *t*-test, $p < 0.05$) in pupal mass between male and female fed on the same aphid.

The aphid species also significantly affected the generation time (T) of *E. corollae.* On *A. craccivora*, T was the longest, and the finite rate of increase (λ) and intrinsic rate of natural increase (r) were the lowest. On *M. japonica*, however, T was the shortest, and λ and r were the highest. The order of population growth from highest to lowest on the three aphid diets was *M. japonica* > *M. persicae* > *A. craccivora* (Table 3).

Table 3. Life table variables for *E. corollae* after larvae fed on different aphid species.

Prey	*A. craccivora*	*M. persicae*	*M. japonica*
T	24.85 ± 0.34 a	23.12 ± 0.41 b	21.05 ± 0.33 c
λ	1.2052 ± 0.0089 a	1.2167 ± 0.0137 a	1.2335 ± 0.0182 a
r	0.1866 ± 0.0074 a	0.1961 ± 0.0113 a	0.2098 ± 0.0149 a
R_0	105.38 ± 19.65 a	96.39 ± 24.90 a	86.16 ± 24.42 a

Values are mean ± SE. Different lowercase letters on the same row reflect significant differences (paired bootstrap test, $p < 0.05$).

The survival rate of *E. corollae* larvae, pupae and adults was the lowest on *M. japonica* (0.52 for egg development to second-instar larvae, 0.51 for third-instar larvae, 0.41 for pupae and 0.15 for adults) (Figure 1). The survival rate was highest for egg development to second-instar larvae (0.88) and third-instar larvae (0.84) on the *M. persicae* diet, but highest on *A. craccivora* for egg development to pupae (0.72) and adult (0.35).

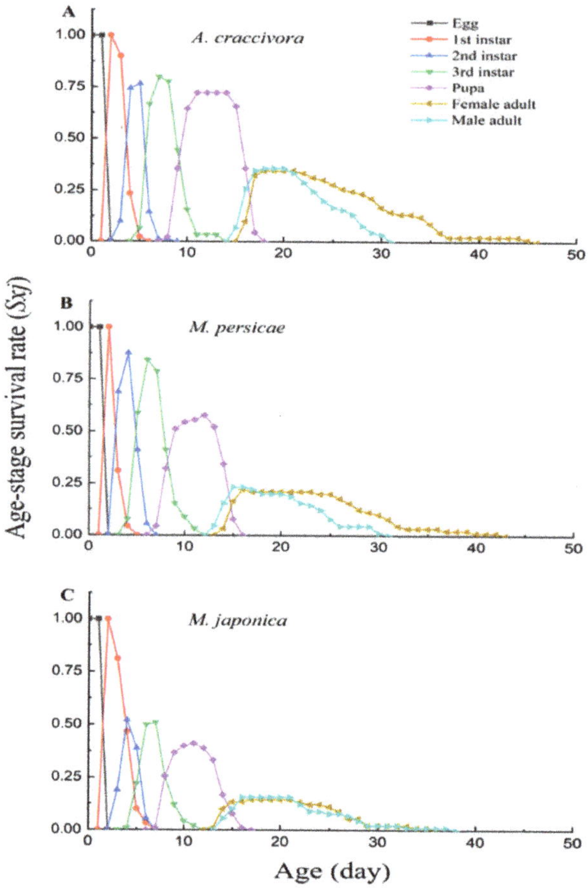

Figure 1. Survival rate curves for *E. corollae* after larvae fed on *A. craccivora* (**A**), *M. persicae* (**B**) and *M. japonica* (**C**) in the laboratory.

The l_x curves for *E. corollae* also differed significantly among the three aphid diets ($x^2 = 26.594$, $df = 2$, $p < 0.001$). For the *M. japonica* diet, the l_x curve of *E. corollae* decreased by 0.7 from day 4 to day 16, which was a faster decrease than for *A. craccivora* and *M. persicae*. From day 20 to day 30 (adult stage), the l_x curve of *E. corollae* that fed on *A. craccivora* decreased rapidly, indicating a higher mortality rate for adults. The peak spawning day for females was day 23 on *A. craccivora*, day 21 on *M. persicae* and day 18 on *M. japonica*, as shown on the $l_x m_x$ curves (Figure 2).

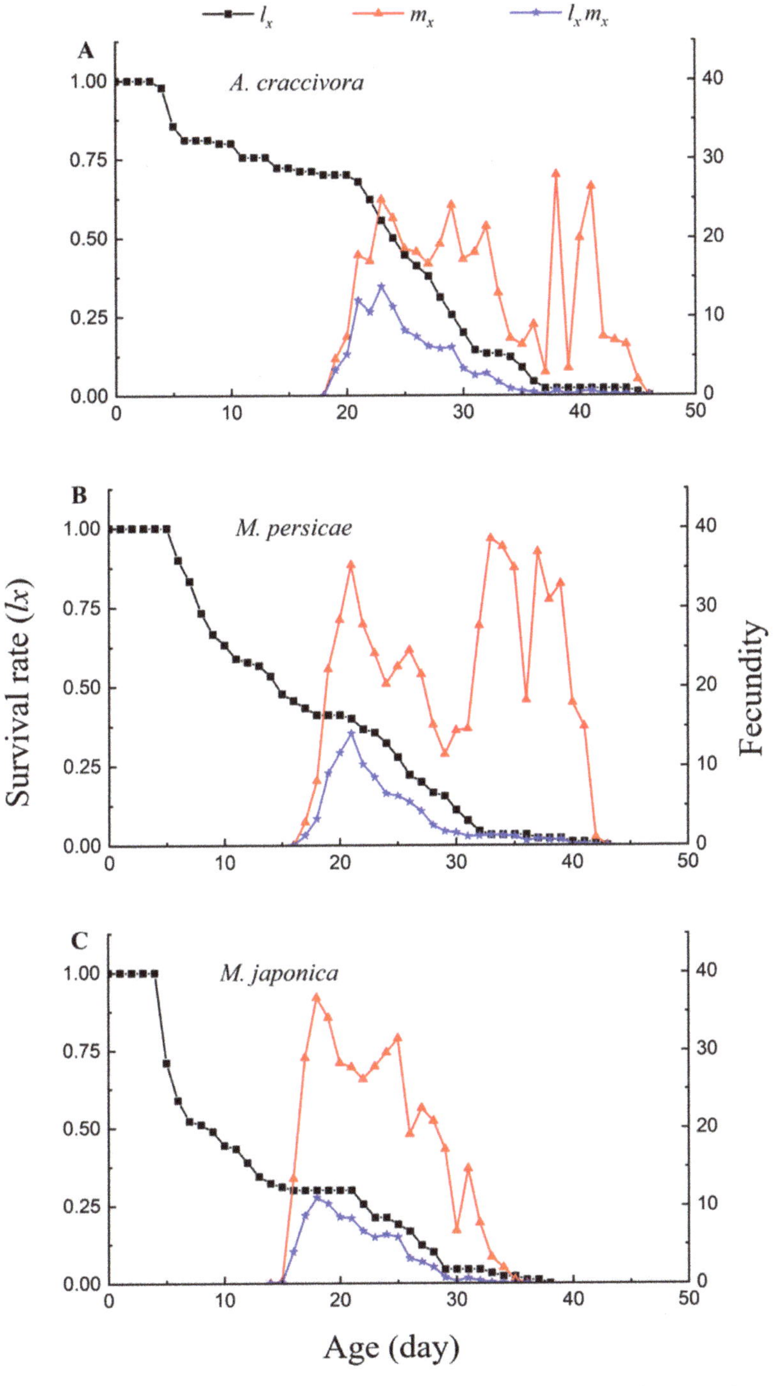

Figure 2. Population age-specific fecundity (m_x), population age-specific survival rate (l_x) and population age-specific maternity (l_xm_x) for *E. corollae* after larvae fed on *A. craccivora* (**A**), *M. persicae* (**B**) and *M. japonica* (**C**) in the laboratory.

3.2. Reproduction of *E. corollae* Fed on Different Aphids

The aphid diets also greatly affected pre-oviposition, total pre-oviposition and number of ovipositions of *E. corollae* (Table 4). On *A. craccivora*, the total pre-oviposition was the longest (20.3 d) and the oviposition number was the least (305.9 eggs per female). On the contrary, the total pre-oviposition was the shortest (16.6 d) and the oviposition number was greatest (596.5 eggs per female) on *M. japonica*.

Table 4. Fecundity variables for *E. corollae* after larvae fed on different aphid species.

Prey	*A. craccivora*	*n*	*M. persicae*	*n*	*M. japonica*	*n*
No. of ovipositions	305.9 ± 36.23 b	31	433.7 ± 73.80 ab	20	596.5 ± 74.85 a	13
Oviposition duration (d)	10.2 ± 1.11 a	30	11.6 ± 1.18 a	19	11.0 ± 0.89 a	13
Total pre-oviposition (d)	20.3 ± 0.18 a	30	18.9 ± 0.29 b	19	16.6 ± 0.37 c	13
Pre-oviposition (d)	3.5 ± 0.11 a	30	3.8 ± 0.26 a	19	2.2 ± 0.17 b	13

Values are mean ± SE. Different lowercase letters on the same row reflect significant differences (paired bootstrap test, $p < 0.05$).

3.3. Flight Performance of *E. corollae* Fed on Different Aphids

There were no significant differences in flight variables (flight duration, flight distance, flight velocity). Adults flew the fastest (0.25 m/s), and the longest distance (624.0 m) and duration (2521.9 s) when they fed on *M. japonica*. The maximum cumulative flight distance (1622.7 m), maximum speed (0.38 m/s), and maximum cumulative duration (6217.2 s) were achieved on *A. craccivora* (Table 5).

Table 5. Flight variables for *E. corollae* after larva fed on different aphids.

Prey		*A. craccivora* (*n* = 32)	*M. persicae* (*n* = 36)	*M. japonica* (*n* = 37)
Duration (s)	Max	6217.2	4305.6	5079.6
	Mean ± SE	2442.3 ± 240.3 a	2037.4 ± 151.3 a	2521.9 ± 186.8 a
Distance (m)	Max	1622.7	950.2	1407.3
	Mean ± SE	582.4 ± 69.2 a	448.2 ± 35.4 a	624.0 ± 52.0 a
Velocity (m/s)	Max	0.38	0.33	0.36
	Mean ± SE	0.23 ± 0.01 a	0.22 ± 0.01 a	0.25 ± 0.01 a

Values are mean ± SE. Different lowercase letters on the same row reflect significant differences (Kruskal–Wallis test, $p < 0.05$).

4. Discussion

Eupeodes corollae depresses aphid populations in ecosystems [34–36]. Our present results show that *E. corollae* completed development and reproduced after consuming the three aphid species. *M. japonica* consumption, however, yielded the highest fecundity, finite rate of increase, intrinsic rate of natural increase and flight performance, and the shortest generation time. Therefore, among the three test species, *M. japonica* is considered to be the most suitable for laboratory propagation of *E. corollae*.

The predatory activity of larvae is closely related to the type, quantity and quality of prey; higher fitness afforded by the prey can be assessed by higher survival rate and faster development of the larval stage of the predator [37–39]. The larval stage of *E. corollae* was shortest when the larvae fed on *M. japonica*, but the larval survival rate was significantly reduced, as also found by Xiong and Dong [21]. Newly hatched *E. corollae* larvae are small and inactive and, thus, feed only on smaller prey. The body of *M. japonica* is larger than that of *A. craccivora* and *M. persicae*; thus, *M. japonica* might have been able to escape predation, contributing to the lower survival rate of *E. corollae* larvae on *M. japonica*. The curves for the age–stage survival rate of various instar larvae after all prey treatments overlapped because individual development was inconsistent, coincident with overlapping generations in the field [40].

Pupal performance and adult fecundity mainly depend on the nutrients consumed by the larvae [41–43]. *Eupeodes corollae* attained the shortest pupal duration and total pre-oviposition and the most ovipositions on *M. japonica*, as well as the lowest pupal emergence rate, as also reported in a previous study [21]. On *A. craccivora*, *E. corollae* attained the longest pupal duration, greatest total pre-oviposition and the fewest ovipositions, as well as the highest pupal emergence rate and pupal mass, which may be associated with aphid nutrient content. High-quality prey improves the performance (e.g., pupal mass, fecundity, longevity) of hoverflies [44], as also found for predatory ladybirds [45,46].

Based on the life table variables (such as R_0, r and λ) for the *E. corollae* population to evaluate the suitability of a prey species [47], r was greater than 0, and λ was greater than 1 for all three aphid diets, revealing that they are suitable prey for *E. corollae*. Further, the highest r and λ (albeit not significantly different) for the *E. corollae* population among the three aphid diets was obtained on *M. japonica*, which suggested that *M. japonica* was the most suitable among the three aphids. The searching and predation activities of *E. corollae* larvae improve with increasing larval age [48]. Because the survival rate of *E. corollae* larvae on *M. japonica* is significantly lower than on *A. craccivora* or *M. persicae*, we recommend that *A. craccivora* or *M. persicae* be used to feed newly hatched larvae and that *M. japonica* be used for second- and third-instar larvae.

For practical applications, the flying ability of predatory insects affects their colonization and reproduction which, in turn, affects their control range and efficacy [49]. Hoverfly activities such as foraging, finding mates and spawning sites are dependent on their flight abilities. As the environment changes, hoverflies—especially the predatory hoverflies, which have poorer flight ability compared with other lepidopteran insects [50,51]—migrate long distances to reproduce [52], requiring good nutrition for the energy needed. Our results showed that the flying performance of *E. corollae* adults was significantly affected by the species of aphid consumed during the larval stage, which helps with *E. corollae* forecasting in the field.

Our systematic evaluation of the development, reproduction and flight ability of *E. corollae* laboratory populations that fed on different species of aphids determined the best aphid species to rear different stages of *E. corollae*. These findings will serve as the basis of large-scale rearing of hoverflies with high fitness and efficacy as biocontrol agents and pollinators. The field environment is much more complex and harsher than the laboratory, and the occurrence of hoverfly populations is highly coincident with that of other natural enemies such as ladybugs and lacewings [53]. Therefore, the fitness of the laboratory *E. corollae* populations reared on aphid diets now need to be evaluated in the field.

5. Conclusions

Herein, our research proved that *E. corollae* larvae can complete development and reproduction on *A. craccivora*, *M. persicae* and *M. japonica*. There is no significant difference in the larval–adult development period of *E. corollae* on the three aphid species. *Eupeodes corollae* attained the highest fecundity, intrinsic rate of natural increase and flight ability and the shortest generation time on *M. japonica*. In conclusion, *M. japonica* is considered to be the most suitable prey for the indoor rearing of *E. corollae* among the three tested aphids; *Aphis craccivora* or *M. persicae* can be used to feed newly hatched larvae of *E. corollae*, and *M. japonica* can be used for second- and third-instar larvae.

Author Contributions: Conceptualization, S.J., H.L. and K.W.; Methodology, S.J., H.L. and K.W.; Software, S.J. and L.H.; Validation, H.L., L.H. and K.W.; Resources, K.W.; Writing—Original Draft Preparation, S.J.; Writing, all authors; Visualization, all authors; Supervision, K.W.; Project Administration, K.W. All authors have read and agreed to the published version of the manuscript.

Funding: This study was funded by The Key R&D Program of Shandong Province (2020CXGC010802) and China Agriculture Research System (CARS-02).

Institutional Review Board Statement: Not applicable.

Informed Consent Statement: Not applicable.

Data Availability Statement: All data analyzed in this study are included in this article.

Conflicts of Interest: The authors declare no conflict of interest.

References

1. Mengual, X.; Ståhls, G.; Rojo, S. Molecular phylogeny of *Allograpta* (Diptera, Syrphidae) reveals diversity of lineages and non-monophyly of phytophagous taxa. *Mol. Phylogenet. Evol.* **2008**, *49*, 715–727. [CrossRef] [PubMed]
2. Mengual, X.; Ståhls, G.; Rojo, S. Phylogenetic relationships and taxonomic ranking of pipizine flower flies (Diptera: Syrphidae) with implications for the evolution of aphidophagy. *Cladistics* **2015**, *31*, 491–508. [CrossRef] [PubMed]
3. Cheng, X.Y. Hoverfly. *Biol. Bull.* **2004**, *2*, 9–12.
4. Tian, J.; Ren, B.Z. Overview of hoverfly research. *J. Jilin Agric. Univ.* **2019**, *41*, 1–10.
5. Ho, K.K.; Zheng, Z.M.; Zhang, H.J. Research progress of Syrphidae in China. *J. Hanzhong Univ.* **2002**, *3*, 70–75.
6. Fischer, O.A.; Mátlová, L.; Dvorská, L.; Švástová, P.; Bartoš, M.; Weston, R.T.; Pavlík, I. Various stages in the life cycle of syrphid flies (*Eristalis tenax;* Diptera: Syrphidae) as potential mechanical vectors of pathogens causing mycobacterial infections in pig herds. *Folia Microbiol.* **2006**, *51*, 147–153. [CrossRef]
7. Ymilie, B.; Marc, F.; Eric, L. Biological control of the foxglove aphid using a banker plant with *Eupeodes americanus* (Diptera: Syrphidae) in experimental and commercial greenhouses. *Biol. Control* **2021**, *155*, 104541.
8. Reemer, M. Review and phylogenetic evaluation of associations between *Microdontinae* (Diptera: Syrphidae) and *Ants* (Hymenoptera: Formicidae). *J. Entomol.* **2013**, *1*, 72–81. [CrossRef]
9. Wotton, K.R.; Gao, B.; Myles, H.M.; Morris, R.K.A.; Ball, S.G.B.; Lim, K.S.; Reynolds, D.R.; Hu, G.; Chapman, J.W. Mass seasonal migrations of hoverflies provide extensive pollination and crop protection services. *Curr. Biol.* **2019**, *29*, 2167–2173. [CrossRef]
10. Li, X.Y.; Luo, Y.Z. Study on the predation effect of Syrphidae on three species of aphids. *J. Agric. Yunnan Univ.* **2001**, *2*, 102–104.
11. Katherine, C.R.B.; Mark, A.G.; Damien, M.H.; Kunin, W.E.; Mitschuna, N.; Morse, H.; Osgathorpe, L.M.; Potts, S.G.; Robertson, K.M.; Scott, A.V.; et al. A systems approach reveals urban pollinator hotspots and conservation opportunities. *Nat. Ecol. Evol.* **2019**, *3*, 363–373.
12. Rader, R.; Cunningham, S.; Howlett, B.; Inouye, D. Non-Bee insects as visitors and pollinators of crops: Biology, ecology, and management. *Annu. Rev. Entomol.* **2020**, *65*, 391–407. [CrossRef] [PubMed]
13. Shi, J.; Cheng, J.; Luo, D.; Luo, Y.B. Using pollination syndrome prediction: Simulated breeding sites of *Paphiopedilum dianthum* to deceive pollination by female hoverflies. *J. Syst. Evol.* **2007**, *4*, 551–560. [CrossRef]
14. He, J.L.; Chu, X.P.; Sun, X.Q.; Ye, W.J. Feeding and utilization evaluation of hoverfly. *J. Agric. Shanghai Univ.* **1994**, *2*, 79–83.
15. Sagar, P. Role of insects in cross pollination of fennel crop at Ludhiana [India]. *J. Res. Punjab Agric. Univ.* **1981**, *18*, 388–392.
16. Burgett, M. Pollination of parsley (*Petroselinum crispum*) grown for seed. *J. Apic. Res.* **1980**, *19*, 79–82. [CrossRef]
17. Pu, D.-Q.; Liu, H.-L.; Gong, Y.Y.; Ji, P.-C.; Li, Y.-J.; Mou, F.-S.; Wei, S.-J. Mitochondrial genomes of the hoverflies *Episyrphus balteatus* and *Eupeodes corollae* (Diptera: Syrphidae), with a phylogenetic analysis of muscomorpha. *Sci. Rep.* **2017**, *7*, 44300. [CrossRef] [PubMed]
18. He, J.L. The biology and ecology of the predatory hoverfly. *J. Agric. Shanghai Univ.* **1989**, *4*, 325–331.
19. Li, H.; Jiang, S.S.; Zhang, H.W.; Geng, T.; Wyckhuys, K.A.G.; Wu, K.M. Two-way predation between immature stages of the hoverfly Eupeodes corollae and the invasive fall armyworm (*Spodoptera frugiperda* J. E. Smith). *J. Integr. Agric.* **2021**, *20*, 829–839. [CrossRef]
20. Yang, F.C.; Li, Z.W.; Gao, H.C.; Hui, F.H.; Lin, G.L. Study on the biology of Syrphidae and its control on wheat aphid. *J. Environ. Entomol.* **1989**, *3*, 116–121.
21. Xiong, H.Z.; Dong, H.F. Experiment on feeding and releasing larvae of Syrphids larvae in greenhouse. *J. Biol. Control Chin.* **1992**, *1*, 6–9.
22. Rob, M.; Sten, B.; Wäckers, F.L.; Pekas, A. Aphid ophagous hoverflies reduce foxglove aphid infestations and improve seed set and fruit yield in sweet pepper. *Pest Manag. Sci.* **2021**, *77*, 2690–2696.
23. Iwai, H.; Niijima, K.; Matsuka, M. An artificial diet for aphidophagous syrphids, *Episyrphus balteatus* (de Geer) and *Eupeodes bucculatus* (Rondani) (Diptera: Syrphidae) using drone honeybee brood powder. *Appl. Entomol. Zool.* **2007**, *42*, 167–172. [CrossRef]
24. Hideki, I.; Keiko, N.; Mitsuo, M. Improvement of artificial diet for aphidophagous syrphids, *Episyrphus balteatus* (de Geer) and *Eupeodes bucculatus* (Rondani) (Diptera: Syrphidae)—additional effects of fatty acids and preservatives. *Appl. Entomol. Zool.* **2009**, *44*, 439–446.
25. Pu, Z.G.; Chi, L. Research progress on biological characteristics and artificial indoor rearing technology of Syrphidae. *Heilongjiang Agric. Sci.* **2006**, *2006*, 120–123.
26. Kan, E. Assessment of aphid colonies by hoverflies. I maple aphids and *Episyrphus balteatus* (de Geer) (Diptera: Syrphidae). *J. Ethol.* **1988**, *6*, 39–48. [CrossRef]
27. Sadeghi, H.; Gilbert, F. Aphid suitability and its relationship to oviposition preference in predatory hoverflies. *J. Anim. Ecol.* **2000**, *69*, 771–784. [CrossRef]
28. Barlow, C.A. On the biology and reproductive capacity of *Syrphus corollae* Fab. (Syrphidae) in the laboratory. *Entomol. Exp. Appl.* **1961**, *4*, 91–100. [CrossRef]

29. Obopile, M.; Ositile, B. Life table and population parameters of cowpea aphid, *Aphis craccivora* Koch (Homoptera: Aphididae) on five cowpea *Vigna unguiculata* (L. Walp.) varieties. *J. Pest Sci.* **2010**, *83*, 9–14. [CrossRef]
30. Chi, H. Life-table analysis incorporating both sexes and variable development rates among individuals. *Environ. Entomol.* **1988**, *17*, 26–34. [CrossRef]
31. Chi, H.; Liu, H. Two new methods for the study of insect population ecology. *Bull. Inst. Zool. Acad. Sin.* **1985**, *24*, 225–240.
32. Ge, S.S.; He, L.M.; He, W.; Xu, R.B.; Sun, X.T.; Wu, K.M. Determination of flying ability of *Spodoptera frugiperda*. *Plant Prot.* **2019**, *45*, 28–33.
33. He, L.M.; Jiang, S.; Chen, Y.C.; Wyckhuy, K.A.G.; Ge, S.S.; He, W.; Gao, X.-W.; Wu, K.-M. Adult nutrition affects reproduction and flight performance of the invasive fall armyworm, *Spodoptera frugiperda* in China. *J. Integr. Agric.* **2021**, *20*, 715–726. [CrossRef]
34. Luo, Q.H. A Preliminary observation on the habits of hoverfly. *Agric. Sci. Guizhou* **1982**, *5*, 21–23.
35. Lan, X.; Luo, J.; Cheng, X.Y. Analysis of predatory hoverfly species and their control effects on soybean aphid in the soybean ecoregion of northeast China. *J. Appl. Entomol.* **2011**, *48*, 1625–1630.
36. He, J.L.; Sun, X.Q.; Gui, L.M.; Ye, W.J. A preliminary study on the biology of hoverfly in Shanghai Area. *J. Agric. Shanghai Univ.* **1990**, *03*, 221–228.
37. Lee, J.H.; Kang, T.J. Functional response of *Harmonia axyridis* (Pallas) (Coleoptera: Coccinellidae) to *Aphis gossypii* Glover (Homoptera: Aphididae) in the Laboratory. *Biol. Control* **2004**, *31*, 306–310. [CrossRef]
38. Moreau, J.; Benrey, B.; Thiéry, D. Grape variety affects larval performance and also female reproductive performance of the European grapevine moth *Lobesia botrana* (Lepidoptera: Tortricidae). *Bull. Entomol. Res.* **2006**, *96*, 205–212. [CrossRef]
39. Yasuda, H. Effect of prey density on behaviour and development of the predatory mosquito, *Toxorhynchites towadensis*. *Entomol. Exp. Appl.* **1995**, *76*, 97–103. [CrossRef]
40. Liu, J.B.; Cui, Y.H.; Gao, J.F.; Wang, W.H.; Cheng, X.H.; Liu, Z.Y. Observation of biological characteristics of *Eupeodes corollae*. *Jilin Agric. Sci.* **2005**, *1*, 38–39.
41. Eduardo, M.S.; Padró, J.; Milla, C.P.; Tureo, D.T.; Carreira, V.P.; Soto, I.M. Pupal emergence pattern in cactophilic *Drosophila* and the effect of host plants. *Insect Sci.* **2017**, *25*, 1108–1118.
42. Li, Y.; Leng, C.M.; Wu, J.X.; Li, Y.P. Effects of different host plants on growth, development and reproduction of *Plutella xylostella*. *Acta Agric. Boreali Occident. Sin.* **2019**, *28*, 475–480.
43. Jabraeil, R.; Bahram, N.; Seyed, A.H. Comparative performance of the cotton bollworm, *Helicoverpa armigera* (Hübner) (Lepidoptera: Noctuidae) on various host plants. *J. Pest Sci.* **2014**, *87*, 29–37.
44. Putra, N.S.; Yasuda, H. Effects of prey species and its density on larval performance of two species of hoverfly larvae, *Episyrphus balteatus* de Geer and *Eupeodes corollae* Fabricius (Diptera: Syrphidae). *Appl. Entomol. Zool.* **2006**, *41*, 389–397. [CrossRef]
45. Sugiura, K.; Takada, H. Suitability of seven aphid species as prey of *Cheilomenes sexmaculata* (Fabricius) (Coleoptera: Coccinellidae). *Appl. Entomol. Zool.* **1998**, *42*, 7–14. [CrossRef]
46. Okamoto, H. Laboratory studies on food ecology of aphidophagous lady beetles (Coleoptera: Coccinellidae). *Mem. Fac. Eng. Kyushu Univ.* **1978**, *32*, 1–94.
47. HE, L.M.; WU, Q.L.; GAO, X.W.; Wu, K.-M. Population life tables for the invasive fall armyworm, *Spodoptera frugiperda* fed on major oil crops planted in China. *J. Integr. Agric.* **2021**, *20*, 745–754. [CrossRef]
48. Dong, K.; Dong, Y.; Li, X.Y.; Luo, Y.Z. Effects of feeding with different aphids on the growth and development of Syrphidae. *J. Environ. Entomol.* **2003**, *4*, 165–168.
49. Yin, Y.Y.; Lin, Q.C.; Zhai, Y.F.; Zheng, L.; Yu, Y.; Chen, H. The flight ability and spreading ability of *Encarsia formosa* Gahan. *J. Environ. Entomol.* **2019**, *41*, 1355–1360.
50. Guo, J.L.; Fu, X.W.; Zhao, X.C.; Wu, K.M. Study on the flying ability of *Agrotis segetum*. *J. Environ. Entomol.* **2016**, *38*, 888–895.
51. Wu, K.M.; Guo, Y.Y. Flying activity of *Helicoverpa armigera*. *J. Ecol. Chin.* **1996**, *6*, 612–617.
52. Nielsen, B.O. On a migration of hoverflies (Dipt., Syrphidae) and sawflies (Hym., Tenthrdinidae) observed in Denmark, August 1967. *Acta Entomol. Sin.* **1968**, *36*, 215–224.
53. Lucas, É.; Coderre, D.; Brodeur, J. Intraguild predation among aphid predators: Characterization and influence of extraguild prey density. *J. Ecol.* **1998**, *79*, 1084–1092. [CrossRef]

insects

MDPI

Article

Efficacy of Biopesticides in the Management of the Cotton Bollworm, *Helicoverpa armigera* (Noctuidae), under Field Conditions

Lawrence N. Malinga [1,2,*] and Mark D. Laing [3]

1 Agricultural Research Council–Industrial Crops, Rustenburg 0300, South Africa
2 South African Sugarcane Research Institute, Mount Edgecombe 4300, South Africa
3 School of Agricultural, Earth and Environmental Sciences, College of Agriculture, Engineering & Science, University of KwaZulu-Natal, Pietermaritzburg 3209, South Africa; laing@ukzn.ac.za
* Correspondence: lawrence.malinga@sugar.org.za; Tel.: +27-84-900-9262

Simple Summary: Cotton remains the most important cash crop in the world. The key insect pests of cotton include the African bollworm *Helicoverpa armigera*. This pest causes damage to crops estimated at greater than USD 2 billion per year worldwide. Excessive use of insecticides to control this pest has a negative effect on the environment, and is expensive for the farmers. The aim of this study is to explore the field efficacy of different biopesticides as an alternative to control *H. armigera*. Four biopesticides—namely, Eco-Bb® (*Beauveria bassiana*), Bb endophyte (*Beauveria bassiana*), Bolldex® (nucleopolyhedrovirus), and Delfin® (*Bacillus thuringiensis*)—were evaluated and compared with the pyrethroid Karate® (lambda-cyhalothrin) and an untreated control. Plots treated with Karate® had significantly lower numbers of *H. armigera* larvae compared to the untreated control, and were comparable to the plots treated with Bolldex®. On average, plots treated with Bolldex® had a high seed cotton yield compared to the other treatments. Biopesticides showed a moderate reduction in the numbers of *H. armigera* larvae, and could thus be used within an integrated pest management programme.

Abstract: Cotton is one of the most valuable materials in the world, popularly used in the clothing industry and other products. However, its production is limited by the high infestation of insect pests. A study was conducted to evaluate the effects of different biopesticides on the control of the African bollworm (*Helicoverpa armigera*) under cotton field conditions. Four biopesticides—namely, Eco-Bb® (*Beauveria bassiana*), Bb endophyte (*Beauveria bassiana*), Bolldex® (Nucleopolyhedrovirus), and Delfin® (*Bacillus thuringiensis*)—were evaluated and compared with the pyrethroid Karate® (lambda-cyhalothrin) and an untreated control against *H. armigera*. Field trials were conducted at the Agriculture Research Council, Rustenburg, in the North West Province of South Africa during the 2017 and 2018 cotton seasons. The results revealed that in plots sprayed with Karate® and Bolldex®, the numbers of *H. armigera* were significantly reduced compared to the untreated controls. Plots treated with Bolldex® had the lowest number of damaged bolls in 2017, while those treated with Karate® had the lowest number of damaged bolls in 2018. All treated plots had significantly fewer damaged bolls when compared to the controls. A seed cotton yield of 5987 kg/ha was recorded in the plots that were treated with Bolldex®—significantly higher than the yields from plots treated with Eco-Bb®, Delfin®, and Bb endophyte—in 2017. However, the yield in treatments with Eco-Bb®, Delfin®, and Bb endophyte was lower than that from the untreated controls during this season. In 2018, plots treated with Bolldex® had the highest yield, at 6818 kg/ha, which was not different from the other treatments. The highest average seed cotton yield of 6400 kg/ha was recorded in the plots treated with Bolldex®, followed by Karate®. In summary, the efficacy of different biopesticides against *H. armigera* varied significantly, while the synthetic pesticide (Karate®) and Bolldex® resulted in more consistent control of this pest. The results suggest that biopesticides may, however, have the potential for use in the sustainable control of cotton bollworms as part of integrated pest management programmes, although further work is required to support this hypothesis.

Keywords: cotton; *Helicoverpa armigera*; biopesticides; Bolldex®; Karate®

1. Introduction

Cotton, *Gossypium hirsutum* L. (Malvaceae), remains the most important cash crop in the world [1]. It is a fibre crop grown in more than 80 tropical and subtropical countries [2,3]. Over a century ago, cotton contributed more than 70% of the world's fibre use—compared to about 30% of current consumption [4]. The decline is due to the development of a high number of synthetic fibres, which increased in the mid-1990s [5]. Furthermore, the decline is also attributed to climate change and pest problems, among other factors [6]. Cotton production is a major economic component in Africa, and is a significant contributor to economic growth [7]. The continent contributes approximately 8% of the cotton produced worldwide from a total of six cotton basins [8], covering over 20 countries in sub-Saharan Africa [9]. In South Africa, cotton is a significant crop produced by 250 commercial and over 2000 small-scale farmers in five provinces, namely, KwaZulu-Natal, Limpopo, Mpumalanga, North West, and Northern Cape [10,11].

The key insect pests of cotton include the African bollworm, *Helicoverpa armigera* Hübner (Noctuidae: Lepidoptera). This bollworm is an indigenous species considered to be a major pest of fibre crops in Africa [12], and ranks as the most important lepidopteran pest in South Africa [13–15]. Four heliothine species are reported as being of economic importance in Africa, but *H. armigera* is the only species of major economic importance [16]. This pest causes damage to crops estimated at greater than USD 2 billion annually in Asia, Europe, Africa, and Australasia [17]. Unlike most other arthropod pests, *H. armigera* is a polyphagous pest that infests more than 200 crop species worldwide, and can adapt to new environments [18,19]. It has a very large range of host plants, including cotton, pepper, corn, tomato, lucerne, soybean, sorghum, and tobacco [20,21]. *H. armigera* is a serious pest because of its polyphagy [22,23], high fecundity [24], high mobility [25], and resistance to chemical insecticides [26–30]. Cotton is mainly attacked by the larval stage of *H. armigera* [31], which causes high yield loss. Because *H. armigera* larvae have a habit of entering the fruit, bolls, or pods, the plant affords them good protection against chemical sprays, making control almost impossible [32]. Low economic damage thresholds in cotton require a high level of control [12], resulting in reliance on synthetic insecticides [33,34].

Worldwide, up to 60% of all commercialised insecticides are used in cotton [35]. Chemical pesticides such as pyrethroids, carbamates, and organophosphates are applied to control *Helicoverpa* pests [36]. Pyrethroids such as Karate® (lambda-cyhalothrin) are non-selective insecticides commonly used against pests—including lepidopterans—in cotton [37,38]. Although chemical pest control is extensively used throughout the world, it has been generally regarded as detrimental to the environment [39]. The use of chemical pesticides to control agricultural pests has resulted in secondary pest outbreaks, as well as reductions in the numbers of beneficial insects and biodiversity [19]. Excessive use of chemicals also causes economic impacts on farmers, and has a negative impact on animals [40]. Before the introduction of Bt cotton with the bacterium *Bacillus thuringiensis* in 1996 [41], cotton production was costly due to the high application of pesticides [42]. However, with the continuous use of Bt cotton on a large scale, *H. armigera* has developed resistance to Bt cotton [43,44]. Today, many agricultural chemical pesticides are under pressure to be banned due to their harmful effects, and some farmers have turned to biopesticides [45,46].

Biopesticides are used to control plant pests, and are made from living micro-organisms or natural products [47]. They are known to be economical, environmentally friendly, and target-specific [48]. Biopesticides have the potential to control yield loss without compromising the quality of the crop [49]. The bacterium *Bacillus thuringiensis* is one of the most used microbial biopesticides. Although *Helicoverpa armigera* nucleopolyhedrovirus (HearNPV) is generally used against cotton bollworms, little is understood about the interaction between the virus and host insects [50]. The use of biopesticides in Africa is minimal, accounting for only 3% of the world's market [51]. South Africa and Kenya are the leading countries in the development and use of biopesticides in Africa [52]. By 2019, South Africa had over 30 registered biopesticides, of which 7 are manufactured locally, and mainly comprise different strains of *Beauveria bassiana* [53].

The idea behind IPM is that combining different practices overcomes the shortcomings of individual practices. The aim is not to eradicate pest populations but, rather, to manage them below levels that cause economic damage [54]. Although biopesticides may not be as rapid as synthetic pesticides, they form part of the crop protection methods used in integrated pest management. Thus, biopesticides are one of the promising alternatives for managing environmental pollution. Integrated control for *H. armigera* that seeks to minimise insecticide use and impacts on non-target organisms needs to be considered. However, in South Africa, there are very few registered biopesticides for use against *H. armigera*. Hence, this study confines itself to evaluating the field efficacy of different biopesticides in the control of *H. armigera*. The importance of this study is to enable scientists to intensify research on the stability of biopesticides under field conditions. This will further provide the bioagents industry with knowledge on biopesticides that have the potential to be registered for use on cotton for the control of *H. armigera*.

2. Materials and Methods

2.1. Trial Site, Layout, and Planting

The trials were conducted in 2017 and repeated in 2018 at the Agricultural Research Council (25°39.0′ S, 27°14.4′ E) in Rustenburg, North West Province, South Africa. Each plot had 6 rows that were 5 m long, with 1 m spacing between rows, 2 m paths between replications, and 20 cm spacing between plants. The trials were completely randomised block designs, and the treatments were replicated four times. A non-GM cotton cultivar, DeltaOPAL, was hand-planted under irrigated conditions, and weeds were manually removed by hand-hoeing. Thinning of seedlings at the fourth true leaf stage was performed to obtain a plant population density of five plants per metre. The trials were planted late in October.

2.2. Treatments and Application

Based on weekly scouting for bollworms, the administration of treatments started 13 weeks after planting, when the first and second instars were observed. Weekly spray applications were performed until 23 weeks after planting, when the cotton bolls were matured. Ten ground applications were administered around 3–4 p.m. due to the UV sensitivity of the biopesticides [55]. The treatments (Table 1) were applied using knapsack sprayers.

Table 1. List of treatments that were used with their active ingredients, formulations, and concentrations.

Trade Name	Active Ingredient	Formulation	Concentration
Eco-Bb®	*Beauveria bassiana* (strain R444)	2×10^9 spores/gram	300 g/ha in 1 g/L water
Bb endophyte	*Beauveria bassiana*	2×10^9 spores/gram	300 g/ha in 1 g/L water
Bolldex®	Nucleopolyhedrovirus (HearNPV)	7.5×10^{12} spores/gram	200 mL/ha in water
Delfin®	*Bacillus thuringiensis* subspecies kurstaki (Btk)	32,000 IU/mg	1 kg/ha in 25 L/ha water
Karate®	Lambda-cyhalothrin	50 g/L	120 mL/ha in 200 L/ha water

2.3. Data Collection

The efficacy of different treatments was assessed based on in situ counts of *H. armigera* larvae. From 12 weeks after planting, 12 plants per plot were scouted weekly. The scouting was conducted using a visual examination of plants in representative locations within a plot. The whole plant was inspected for the presence of all *H. armigera* larval instars, and the total population was counted. The counts were expressed in insect numbers. The seed cotton yields were determined at the end of the season. For accurate yield measurements, and to minimise experimental error, the two middle rows were harvested in each plot. The trials were harvested in May, when over 90% of the bolls had opened. The two middle rows were harvested per plot. Hand-harvesting was performed to ensure that the seed cotton was harvested and weighed accurately.

2.4. Data Analysis

The data were subjected to appropriate analysis of variance (ANOVA). The Shapiro–Wilk test was performed on the standardised residuals to test for deviations from normality [56]. LSDs (least significant differences) were calculated at a 5% significance level to compare the means of significant source effects [57]. The analysis was performed using Genstat Release 18 and SAS version 9.4 statistical software [58]. The seed cotton yields were expressed in percentages (yield in treatment plots–yield in control plots ÷ yield in control plots × 100).

3. Results

In 2017, plots treated with Karate® had significantly lower numbers of *H. armigera* larvae compared to the untreated controls, and were comparable to the plots treated with Bolldex® and Bb endophyte (Figure 1). The controls had the highest numbers of *H. armigera*, and this trend was similar for both the 2017 and 2018 seasons. The *H. armigera* population was significantly lower in all of the treatments compared to the untreated controls during the 2018 season. Plots treated with Karate® had lower numbers of *H. armigera* larvae, which were comparable to the plots treated with Bolldex® and Eco-Bb®. The results shown in Figure 2 reveal that plots treated with Bolldex® had significantly lower numbers of damaged bolls compared to Bb endophyte and the controls in 2017. However, in 2018, none of the active treatments were different from one another, although plots treated with Karate® had the lowest numbers of damaged bolls, followed by Eco-Bb® and Bolldex®. In both seasons, all of the treatments had significantly lower numbers of damaged bolls when compared to the untreated controls. A seed cotton yield of 5987 kg/ha was recorded in the plots that were treated with Bolldex®—significantly higher than the yields from the plots treated with Eco-Bb®, Delfin®, and Bb endophyte—in 2017 (Table 2). However, none of the treatments showed a significantly higher yield than the untreated controls. In 2018, plots treated with Bolldex® had the highest yield, at 6818 kg/ha, which was not different from the other treatments. The yield from the treatments with Eco-Bb®, Delfin®, and Bb endophyte was lower than that from the untreated controls in 2017. The average seed cotton yield was higher for all treatments in 2018 than in 2017. On average, the plots treated with Bolldex® had a significantly higher seed cotton yield (45%) compared to the untreated controls, followed by Karate®. Plots where Karate® was applied had earlier boll opening than the other treatments in 2017.

Table 2. Seed cotton yields from plots with different treatments during the 2017 and 2018 seasons.

Treatment	2017 (kg/ha) *	2018 (kg/ha) *
Eco-Bb®	3055 ± 139.19 b	5961 ± 65.07 ab
Bolldex®	5987 ± 86.56 a	6818 ± 95.59 a
Delfin®	3523 ± 49.24 b	5755 ± 109. 21 ab
Bb endophyte	3100 ± 66.11 b	6409 ± 128.93 a
Karate®	5133 ± 99.34 ab	6405 ± 57.64 a
Untreated control	4168 ± 143.09 ab	4673 ± 124.79 b
LSD (5%)	2373.8	1.6178
CV%	37.94516	17.88032
p-Value	0.1216	0.1436

* Means with the same letter are not significantly different ($p < 0.05$). The comparisons were carried out using the data from the same year.

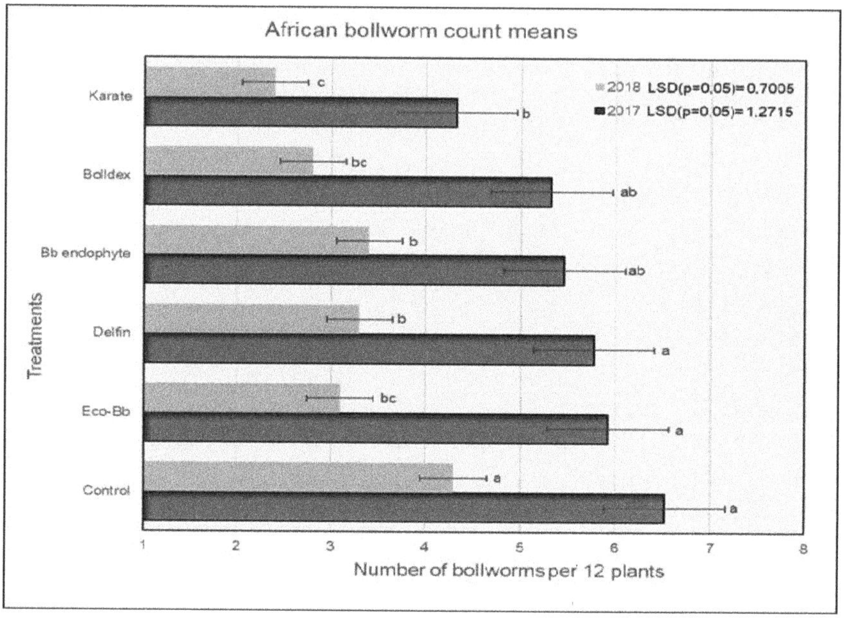

Figure 1. The average number of *H. armigera* larvae found on different treatments during the 2017 and 2018 seasons. Means with the same letter are not significantly different ($p < 0.05$). The comparison was carried out using the data from the same year.

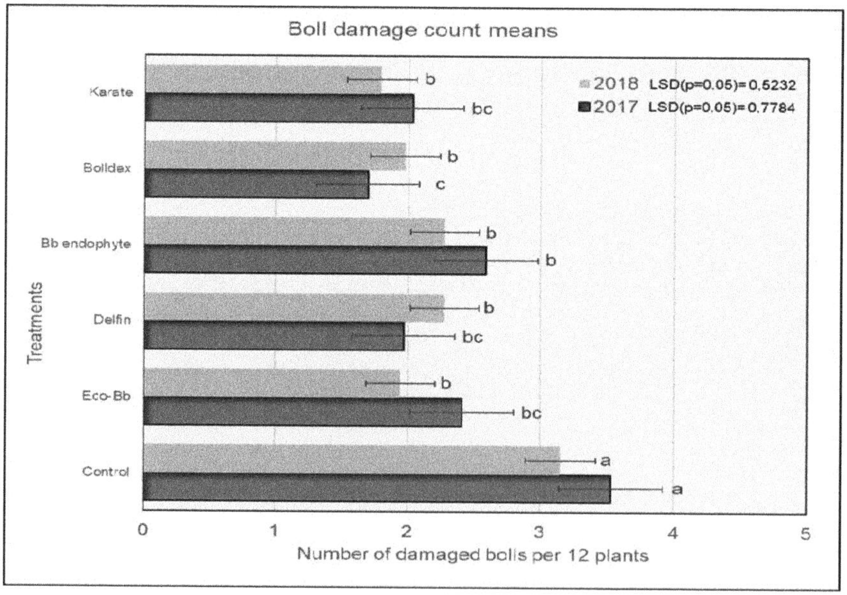

Figure 2. The average number of damaged bolls found on different treatments during the 2017 and 2018 seasons. Means with the same letter are not significantly different ($p < 0.05$). The comparison was carried out using the data from the same year.

4. Discussion

In this study, statistically significant control of *H. armigera* and increases in yield were obtained with some of the biopesticides when compared to untreated controls. However, none of the biopesticides performed significantly better than the chemical control in the reduction in the numbers of *H. armigera*. Unlike greenhouse trials under controlled conditions, field trials are affected by different environmental factors [59], which may influence the effectiveness of the biopesticides. This study clearly shows that the insecticidal action of some of the tested biopesticides was comparable to the commercial synthetic insecticide. Bolldex® warrants more attention, because the reduction it caused in the numbers of *H. armigera* larvae was comparable to that of Karate®. These results demonstrate the potential of biopesticides to reduce *H. armigera* populations, and that they can be introduced as environmentally friendly pesticides in organic and commercial agriculture. This study demonstrates that both Bolldex® and Karate® significantly reduced *H. armigera* populations during the two seasons. Biopesticides have been reported to effectively reduce the larval population of *H. armigera* when combined with parasitoids [60].

Helicoverpa armigera nucleopolyhedrovirus (HaNPV) has been used in several countries and introduced in South Africa for use on several crops [32,61]. Delfin® may have reduced the larval population of *H. armigera* compared to the control in 2018, but there were no significant differences in 2017. Khalique and Ahmed (2001) [62] reported that the mortality response of *H. armigera* larvae to a combination of Karate® and *B. thuringiensis* subsp. kurstaki was synergistic, and that the two products were compatible. While considering the economic injury levels of cotton bollworms per plot [63], the results show that *H. armigera* larvae exceeded the threshold level of five bollworms for all of the treatments except for the Karate® treatment in 2017, and there were less than four bollworms for all of the treatments except for the untreated controls in 2018.

The lowest number of damaged bolls was observed in plots that were treated with Bolldex® and Karate®. The effects of Bolldex® and Karate® on the reduction in boll damage corresponded with the reduction in the numbers of *H. armigera* on the plots where these treatments were applied. These results are also in concordance with the observations of Li et al. (2006) [64], who found that more than 60% parasitism of *H. armigera* decreased boll damage by 80% compared with controls. Joubert (2012) [32] reported that a trial was conducted for the control of *H. armigera* on peaches, and Bolldex® yielded 99% scar-free fruit.

The data on yield revealed that a significantly higher yield of seed cotton was recorded in the treatments with Bolldex®, followed by Karate®, in 2018. In 2017, all of the treatments had yields that were not significantly different from the untreated controls. This may be related to the insect numbers in 2017 being higher than in 2018. In 1997, Cole et al. [65] reported that Karate® increased cotton yield by 12% and provided good pest control whilst maintaining beneficial populations. This is contrary to the findings of Kumar and Stanley (2010) [66], who reported that although lambda-cyhalothrin enhanced seed cotton yields, it caused mortality in both destructive and useful insect species. Sinno et al. (2021) [67] reported that endophytic fungi have the potential to improve plant development and provide protection against pests. However, in this study, the endophytic treatments Eco-Bb® and Bb endophyte only suppressed the *H. armigera* population, but had a lower yield compared to untreated controls in one season. Lotfy and Moustafa (2021) [68] investigated the efficacy of two *B. bassiana* strains against the cotton bollworm, and concluded that *B. bassiana* was efficient against the pest. They further observed that the fungus significantly reduced the numbers of infested bolls. The present study did not evaluate the impact of the pest complex other than the bollworm.

Plots that were treated with Karate® had earlier boll opening than the other treatments in 2017. The additive effects were probably due to multiple mechanisms that affect the pathogens, as opposed to the fewer control mechanisms provided by a single antagonist. Ali (2016) [69] stated that the average number of open bolls/plant is significantly increased by spraying insecticides and salicylic acid.

5. Conclusions

H. armigera remains a major pest of cotton, and its infestation reduces the yield. The main management tool to control this pest is the application of synthetic pesticides. The overuse of such agents subsequently results in pesticide resistance, and this highlights the importance of IPM strategies that include the use of biopesticides. Although biopesticides are safer to use, there are limitations to their full adoption as a pest management tool. The application of biopesticides under field conditions poses some challenges, as high doses are required for good efficacy. This study provides insight that although Eco-Bb® is currently not registered to control *H. armigera*, it has a suppressive potential against the population of *H. armigera* larvae. However, during the first season, the yield in treatments with Eco-Bb®, Delfin®, and Bb endophyte was lower than of untreated controls. Bolldex® provided better control of the pest compared to the untreated controls, and it increased the yield. Although the biopesticides were less effective against the *H. armigera* larvae compared to the chemical pesticide, they have the potential to be used in conjunction with synthetic pesticides to delay the development of pesticide resistance. However, further research that focuses on the efficacy and persistence of different doses and other products is required. More research on the production and application of biopesticides is needed in order to integrate these biological products into cotton production. It is also essential for public and private stakeholders to support and promote the research on biopesticides. Equally important is the training of farmers in the use of biopesticides for the rapid adoption of this technology.

Author Contributions: L.N.M.: Conceptualisation, Investigation, Methodology, Data Collection and Analysis, Validation, Manuscript Preparation, Writing—Corrections, Project Administration, Resources. M.D.L.: Conceptualisation, Supervision, Methodology, Validation, Visualisation. All authors have read and agreed to the published version of the manuscript.

Funding: This research received no external funding.

Institutional Review Board Statement: Not applicable.

Informed Consent Statement: Not applicable.

Data Availability Statement: The data presented in this study are available on request from the corresponding author.

Acknowledgments: The authors would like to acknowledge the Agricultural Research Council for providing the funding and the field to conduct the trials.

Conflicts of Interest: The authors declare no conflict of interest.

References

1. Haider, N.; Ahmed, K.S.; Haidary, A.A.; Afzal, M.; Majeed, M.Z. Field evaluation of different insecticides against spotted bollworm (*Earias* spp.) and comparative yield assessment for BT and non-Bt cotton. *J. Entomol. Zool. Stud.* **2015**, *4*, 33–35.
2. Hussain, M.; Noureen, N.; Fatima, S.; Ghazanfar, M. Cotton mealybug management: A review. *Middle-East J. Sci. Res.* **2016**, *24*, 2424–2430.
3. Ma, H.; Meng, C.; Zhang, K.; Wang, K.; Fan, H.; Li, Y. Study on physiological mechanism of using cottonseed meal to improve salt–alkali tolerance of cotton. *J. Plant Growth Regul.* **2020**, *40*, 126. [CrossRef]
4. Townsend, T. Cotton and Economic Development. Cotton Analytics. 2017. Available online: http://cottonanalytics.com/category/cotton-and-economic-development/ (accessed on 25 June 2021).
5. Dhaliwal, J.S. Natural fibers: Applications. In *Generation, Development and Modifications of Natural Fibers*; Abbas, M., Jeon, H., Eds.; IntechOpen: London, UK, 2019.
6. OECD; FAO. OECD FAO Agricultural Outlook 2019–2028. 2019. Available online: https://doi.org/10.1787/agr_outlook-2019-en (accessed on 25 June 2021). [CrossRef]
7. Vitale, J. Economic Importance of Cotton in Burkina Faso. Food and Agriculture Organization of the United Nations. 2018. Available online: http://www.fao.org/3/i8330en/I8330EN.pdf (accessed on 2 August 2021).
8. Amanet, K.; Chiamaka, E.O.; Quansah, G.W.; Mubeen, M.; Farid, H.U.; Akram, R.; Nasim, W. Cotton production in Africa. In *Cotton Production*; Jabran, K., Chauhan, B.S., Eds.; Wiley: Hoboken, NJ, USA, 2019; pp. 359–369. [CrossRef]
9. IPBO. Cotton in Africa. 2017. Available online: http://www.vib.be/en/about-vib/Documents/vib_fact_CottonAfrican_EN_2017_0901_LR_FINAL.pdf (accessed on 25 June 2021).

10. Van Jaarsveld, M.J. Geographic Susceptibility of *Helicoverpa armigera* (Lepidoptera: Noctuidae) to Insecticidal Proteins in Bt-cotton in South Africa. Master's Thesis, Department of Zoology and Entomology, Rhodes University, Grahamstown, South Africa, 2003.
11. Louw, M. Cotton Farming in South Africa. Field Crops in South Africa. 2020. Available online: http://southafrica.co.za/cotton-farming-in-south-africa.html (accessed on 17 September 2021).
12. Cherry, A.; Cock, M.; Van den Berg, H.; Kfir, R. Biological control of *Helicoverpa armigera* in Africa. In *Biological Control in IPM Systems in Africa*; Neuenschwander, P., Borgemeister, C., Langewald, J., Eds.; CABI Publishing: Wallingford, UK, 2003; pp. 329–346.
13. Moran, V.C. The phytophagous insects and mites of cultivated plants in South Africa: Patterns and pest status. *J. Appl. Ecol.* **1983**, *20*, 439–450. [CrossRef]
14. Bell, M.A.; McGeoch, M.A. An evaluation of the pest status and research conducted on phytophagous Lepidoptera on cultivated plants in South Africa. *Afr. Entomol.* **1996**, *4*, 161–170.
15. Moore, S.; Kirkman, W. HELICOVIR^{TM}: A virus for the biological control of bollworm. *SA Fruit J.* **2010**, *9*, 63–67.
16. Greathead, D.J.; Girling, D.J. Distribution and economic importance of *Heliothis* and of their natural enemies and host plants in southern and eastern Africa. In Proceedings of the Workshop on Biological Control of Heliothis: Increasing the Effectiveness of Natural Enemies, New Delhi, India, 11–15 November 1985; King, E.G., Jackson, R.D., Eds.
17. Tay, W.T.; Soria, M.F.; Walsh, T.; Thomazoni, D.; Silvie, P.; Behere, G.T.; Anderson, C.; Downes, S. A brave new world for an old-world pest: *Helicoverpa armigera* (Lepidoptera: Noctuidae) in Brazil. *PLoS ONE* **2013**, *8*, e80134. [CrossRef]
18. Djihinto, A.C.; Katary, A.; Djaboutou, M.C.; Prudent, P.; Menozzi, P.; Atachi, P. Variation in biological parameters of cypermethrin resistant and susceptible strains of *Helicoverpa armigera* from Benin Republic, West Africa. *Int. J. Biol. Chem. Sci.* **2012**, *6*, 940. [CrossRef]
19. Chen, G.; Liu, H.; Mo, B.C.; Hu, J.; Liu, S.Q.; Bustos-Segura, C.; Xue, J.; Wang, X. Growth and development of *Helicoverpa armigera* (Lepidoptera: Noctuidae) larvae infected by *Heliothis virescens* ascovirus 3i (HvAV-3i). *Front. Physiol.* **2020**, *5*, 93. [CrossRef]
20. Cunningham, J.P.; Zalucki, M.P. Understanding *heliothine* (Lepidoptera: Heliothinae) pests: What is a host plant? *J. Econ. Entomol.* **2014**, *107*, 881–896. [CrossRef]
21. Gu, S.; Han, P.; Ye, Z.; Perkins, L.E.; Li, J.; Wang, H.; Zalucki, M.P.; Lu, Z. Climate change favours a destructive agricultural pest in temperate regions: Late spring cold matters. *J. Pest Sci.* **2018**, *91*, 1191–1198. [CrossRef]
22. Achaleke, J.; Brevault, T.; Blondin, L.; Vassal, J.M. Helicoverpa armigera: Truly Polyphagous. Available online: https://agritrop.cirad.fr/549357/1/document_549357.pdf (accessed on 2 February 2021).
23. Brévault, T.; Nibouche, S.; Achaleke, J.; Carrière, Y. Assessing the role of non-cotton refuges in delaying *Helicoverpa armigera* resistance to Bt cotton in West Africa. *Evol. Appl.* **2011**, *5*, 53–65. [CrossRef] [PubMed]
24. Noor-ul-Ane, M.; Kim, D.; Zalucki, M.P. Fecundity and egg laying in *Helicoverpa armigera* (Lepidoptera: Noctuidae): Model development and field validation. *J. Econ. Entomol.* **2018**, *111*, 2208–2216. [CrossRef]
25. Fitt, G.P. The ecology of *Heliothis* species in relation to agroecosystems. *Annu. Rev. Entomol.* **1989**, *34*, 17–52. [CrossRef]
26. Chaturvedi, I. Status of insecticide resistance in the cotton bollworm, *Helicoverpa armigera* (Hubner). *J. Cent. Eur. Agric.* **2007**, *8*, 171–182.
27. Pretorius, J.D. Status of Resistance of *Helicoverpa armigera* (Lepidoptera: Noctuidae) and *Diparopsis castanea* (Lepidoptera: Noctuidae) to Bt Cotton in South Africa. Master's Thesis, North-West University, Potchefstroom, South Africa, 2011.
28. Yang, Y.; Li, Y.; Wu, Y. Current status of insecticide resistance in *Helicoverpa armigera* after 15 years of Bt cotton planting in China. *J. Econ. Entomol.* **2013**, *106*, 375–381. [CrossRef]
29. Hussain, D.; Saleem, M.; Ghouse, G.; Abbas, M. Insecticide resistance in field populations of *Helicoverpa armigera* (Hübner) (Lepidoptera: Noctuidae). *J. Entomol. Sci.* **2015**, *50*, 119–128. [CrossRef]
30. Tossou, E.; Tepa-Yotto, G.; Kpindou, O.K.; Sandeu, R.; Datinon, B.; Zeukeng, F.; Akoton, R.; Tchigossou, G.M.; Djègbè, I.; Vontas, J.; et al. Susceptibility profiles of *Helicoverpa armigera* (Hübner) (Lepidoptera: Noctuidae) to deltamethrin reveal a contrast between the Northern and the Southern Benin. *Int. J. Environ. Res. Public Health* **2019**, *16*, 1882. [CrossRef]
31. Li, H.; Bouwer, G. Toxicity of *Bacillus thuringiensis* Cry proteins to *Helicoverpa armigera* (Lepidoptera: Noctuidae) in South Africa. *J. Invertebr. Pathol.* **2012**, *109*, 110–116. [CrossRef]
32. Joubert, R. Biological Bollworm Control. Farmers Weekly: 22 May 2012. Available online: https://www.farmersweekly.co.za/agri-technology/farming-for-tomorrow/biological-bollworm-control/ (accessed on 26 January 2021).
33. Mensah, R.K. Development of an integrated pest management programme for cotton. Part 1: Establishing and utilizing natural enemies. *Int. J. Pest Manag.* **2002**, *48*, 87–94. [CrossRef]
34. Safna, M.; Naik, K.V.; Desai, V.S.; Karmarkar, M.S.; Shinde, B.D.; Raut, P.P. Evaluation of the efficacy of some insecticides against fruit borer, *Helicoverpa armigera* (Hubner) infesting tomato. *Int. J. Chem. Stud.* **2018**, *6*, 1158–1163.
35. Yadav, S.; Dutta, S. Evaluation of organophosphorus pesticide residue in cotton of Tijara Tehsil, Alwar, Rajasthan. *Nat. Environ. Pollut. Technol.* **2019**, *18*, 1455–1458.
36. Martin, T.; Ochou, G.O.; Hala-N'klo, F.; Vassal, J.M.; Vaissayre, M. Pyrethroid resistance in the cotton bollworm, *Helicoverpa armigera* (Hubner), in West Africa. *Pest Manag. Sci.* **2000**, *56*, 549–554. [CrossRef]
37. Javaid, I.; Uaine, R.N.; Massua, J. The use of insect growth regulators for the control of insect pests of cotton. *Int. J. Pest Manag.* **1999**, *45*, 245–247. [CrossRef]
38. Machado, A.V.A.; Potin, D.M.; Torres, J.B.; Silva Torres, C.S.A. Selective insecticides secure natural enemies action in cotton pest management. *Ecotoxicol. Environ. Saf.* **2019**, *184*, 109669. [CrossRef]

39. Szewczyk, B.; Rabalski, L.; Krol, E.; Sihler, W.; de Souza, M.L. Baculovirus biopesticides-safe alternative to chemical protection of plants. *J. Biopestic.* **2009**, *2*, 209–216.
40. Patel, S.R.; Patel, K.G.; Ghetiya, L.V. Population dynamics of pod borer (*Helicoverpa armigera* Hubner) infesting chickpea in relation to abiotic factors. *Agres–Int. e-J.* **2015**, *4*, 163–170.
41. James, C. *Global Status of Commercialized Biotech/GM Crops: 2016*; ISAAA Brief No. 52; ISAAA: Ithaca, NY, USA, 2016.
42. Rocha-Munive, M.G.; Soberón, M.; Castañeda, S.; Niaves, E.; Scheinvar, E.; Eguiarte, L.E.; Mota-Sánchez, D.; Rosales-Robles, E.; Nava-Camberos, U.; Martínez-Carrillo, J.L.; et al. Evaluation of the impact of genetically modified cotton after 20 years of cultivation in Mexico. *Front. Bioeng. Biotechnol.* **2018**, *6*, 82. [CrossRef]
43. Ahmad, S.; Cheema, H.M.N.; Khan, A.A.; Khan, R.S.A.; Ahmad, J.N. Resistance status of *Helicoverpa armigera* against Bt cotton in Pakistan. *Transgenic Res.* **2019**, *28*, 199–212. [CrossRef]
44. Zhang, H.N.; Tian, W.; Zhao, J.; Jin, L.; Yang, J.; Liu, C.H.; Yang, Y.; Wu, S.; Wu, K.; Cui, J.; et al. Diverse genetic basis of field-evolved resistance to Bt cotton in cotton bollworm from China. *Proc. Natl. Acad. Sci. USA* **2012**, *109*, 10275–10280. [CrossRef]
45. Maghsoudi, S.; Jalali, E. Noble UV protective agent for *Bacillus thuringiensis* based on a combination of graphene oxide and olive oil. *Sci. Rep.* **2017**, *7*, 11019. [CrossRef] [PubMed]
46. Vilas-Boas, G.T.; Peruca, A.P.S.; Arantes, O.M.N. Biology and taxonomy of *Bacillus cereus*, *Bacillus anthracis*, and *Bacillus thuringiensis*. *Can. J. Microbiol.* **2007**, *53*, 673–687. [CrossRef] [PubMed]
47. Organisation for Economic Co-Operation and Development. Series on Pesticides No. 448. Report of Workshop on the Regulation of Biopesticides: Registration and Communication Issues. 2009. Available online: http://www.oecd.org/dataoecd/3/Collego55/43056580.pdf (accessed on 5 September 2021).
48. Kumar, S.; Singh, A. Biopesticides: Present status and the future prospects. *J. Fertil. Pestic.* **2015**, *6*, 100–129. [CrossRef]
49. Kumar, J.; Ramlal, A.; Mallick, D.; Mishra, V. An overview of some biopesticides and their importance in plant protection for commercial acceptance. *Plants* **2021**, *10*, 1185. [CrossRef]
50. Xing, L.; Yuan, C.; Wang, M.; Lin, Z.; Shen, B.; Hu, Z.; Zou, Z. Dynamics of the interaction between cotton bollworm *Helicoverpa armigera* and nucleopolyhedrovirus as revealed by integrated transcriptomic and proteomic analyses. *Mol. Cell. Proteom.* **2017**, *16*, 1009–1028. [CrossRef] [PubMed]
51. Olson, S. An analysis of the biopesticide market now and where is going. *Outlooks Pest Manag.* **2015**, *26*, 203–206. [CrossRef]
52. Akutse, K.S.; Subramanian, S.; Maniania, N.; Dubois, T.; Ekesi, S. Biopesticide research and product development in Africa for sustainable agriculture and food security–Experiences from the International Centre of Insect Physiology and Ecology (icipe). *Front. Sustain. Food Syst.* **2020**, *4*, 563016. [CrossRef]
53. Hatting, J.L.; Moore, S.D.; Maland, A.P. Microbial control of phytophagous invertebrate pests in South Africa: Current status and future prospects. *J. Invertebr. Pathol.* **2019**, *165*, 54–66. [CrossRef]
54. Chandler, D.; Bailey, A.S.; Tatchell, G.M.; Davidson, G.; Greaves, J.; Grant, W.P. The development, regulation and use of biopesticides for integrated pest management. *Philos. Trans. R. Soc. B Biol. Sci.* **2011**, *366*, 1987–1998. [CrossRef]
55. Zhang, L.; Zhang, X.; Zhang, Y.; Wu, S.; Gelbič, I.; Xu, L.; Guan, X. A new formulation of *Bacillus thuringiensis*: UV protection and sustained release mosquito larvae studies. *Sci. Rep.* **2016**, *6*, 39425. [CrossRef]
56. Shapiro, S.S.; Wilk, M.B. An analysis of variance test for normality (complete samples). *Biometrika* **1965**, *52*, 591–611. [CrossRef]
57. Snedecor, G.W.; Cochran, W.G. *Statistical Methods*, 6th ed.; The Lowa State University: Ames, IA, USA, 1967.
58. SAS. *SAS/STAT User's Guide*; SAS Institute Inc.: Cary, NC, USA, 1999; Volume 8.
59. Tesfagiorgis, H.B. Studies on the Use of Biocontrol Agents and Soluble Silicon against Powdery Mildew of Zucchini and Zinnia. Ph.D. Thesis, University of KwaZulu-Natal, Durban, South Africa, 2008.
60. Sharma, G.D.; Devi, N.; Raj, D.; Oberoi, C.K. Field efficacy of different insecticides and bio-pesticides against fruit borer (*Helicoverpa armigera*) and greenhouse whitefly (*Trialeurodes vaporariorum*). *Himachal J. Agric. Res.* **2008**, *34*, 40–46.
61. Madumbi Sustainable Agriculture. Bolldex®Product Label. Available online: http://www.madumbi.co.za/pdf/Bolldex/Bolldex%20Label.pdf (accessed on 24 January 2021).
62. Khalique, F.; Ahmed, K. Synergistic Interaction between *Bacillus thuringiensis* (Berliner) and lambda-cyhalothrin (Pyrethroid) against, chickpea pod borer, *Helicoverpa armigera* (Hübner). *Pak. J. Biol. Sci.* **2001**, *4*, 1120–1123. [CrossRef]
63. ARC (Agricultural Research Council). *Management Guide for the Cotton Producer*, 2nd ed.; ARC-Institute for Industrial Crops: Rustenburg, South Africa, 2004.
64. Li, J.; Yan, F.; Coudron, T.A.; Pan, W.; Zhang, X.; Liu, X.; Zhang, Q. Field release of the parasitoid *Microplitis mediator* (Hymenoptera: Braconidae) for control of *Helicoverpa armigera* (Lepidoptera: Noctuidae) in cotton fields in Northwestern China's Xinjiang Province. *Environ. Entomol.* **2006**, *35*, 694–699. [CrossRef]
65. Cole, J.F.H.; Pilling, E.D.; Boykin, R.; Ruberson, J.R. Effects of Karate® insecticide on beneficial arthropods in Bollgard® cotton. In Proceedings of the 1997 Proceedings Beltwide Cotton Conferences, New Orleans, LA, USA, 6–10 January 1997; Dugger, C.P., Richter, D.A., Eds.; Volume 2, pp. 1118–1120.
66. Kumar, K.R.; Stanley, S. Bio-efficacy of microbial and chemical insecticides on major lepidopterous pests of cotton and their (insect) natural enemies in cotton ecosystem in Tamil Nadu. *Resist. Pest Manag. Newsl.* **2010**, *20*, 4–7.
67. Sinno, M.; Ranesi, M.; Di Lelio, I.; Iacomino, G.; Becchimanzi, A.; Barra, E.; Molisso, D.; Pennacchio, F.; Digilio, M.C.; Vitale, S.; et al. Selection of endophytic *Beauveria bassiana* as a dual biocontrol agent of tomato pathogens and pests. *Pathogens* **2021**, *10*, 1242. [CrossRef]

68. Lotfy, D.E.; Moustafa, H.Z. Field evaluation of two entomopathogenic fungi; *Beauveria bassiana* and *Metarhizium anisopliae* as a biocontrol agent against the spiny bollworm, *Earias insulana* Boisduval (Lepidoptera: Noctuidae) on cotton plants. *Egypt. J. Biol. Pest Control* **2021**, *31*, 78. [CrossRef]
69. Ali, O.S.M. Effect of Salicylic Acid and Its Mixtures with Three Insecticides on Some Cotton Insect Pests. Ph.D. Thesis, Tanta University, Tanta, Egypt, 2016.

insects

MDPI

Review

Lymantria dispar (L.) (Lepidoptera: Erebidae): Current Status of Biology, Ecology, and Management in Europe with Notes from North America

Maria C. Boukouvala [1,*], Nickolas G. Kavallieratos [1,*], Anna Skourti [1], Xavier Pons [2], Carmen López Alonso [2], Matilde Eizaguirre [2], Enrique Benavent Fernandez [3], Elena Domínguez Solera [3], Sergio Fita [3], Tanja Bohinc [4], Stanislav Trdan [4], Paraskevi Agrafioti [5] and Christos G. Athanassiou [5]

[1] Laboratory of Agricultural Zoology and Entomology, Department of Crop Science, Agricultural University of Athens, 75 Iera Odos Str., 11855 Athens, Greece

[2] Department of Crop and Forest Sciences, Agrotecnio Centre, Universitat de Lleida, Av Rovira Roure 191, 25198 Lleida, Spain

[3] AIMPLAS, Plastics Technology Centre, València Parc Tecnològic, Gustave Eiffel 4, 46980 Paterna, Spain

[4] Department of Agronomy, Biotechnical Faculty, University of Ljubljana, Jamnikarjeva 101, 1000 Ljubljana, Slovenia

[5] Laboratory of Entomology and Agricultural Zoology, Department of Agriculture, Crop Production and Rural Environment, University of Thessaly, Phytokou Str., 38446 Nea Ionia, Greece

* Correspondence: mbouk@aua.gr (M.C.B.); nick_kaval@aua.gr (N.G.K.); Tel.: +30-2105294569 (M.C.B.)

Simple Summary: In the current review, we gathered and summarized the up-to-date information on the life cycle, distribution, outbreaks, control, and health issues to humans and animals of the European Spongy moth. Overall, this noxious species is easily expanded to new areas, causing serious large-scale damage rapidly. The management of this insect is difficult since the chemicals are harmful to human health and the environment, and natural enemies are not able to cause sufficient reduction of the populations of *L. dispar*. Finally, the potential use of biotechnological and physical methods against *L. dispar* is discussed.

Abstract: The European Spongy moth, *Lymantria dispar* (L.) (Lepidoptera: Erebidae), is an abundant species found in oak woods in Central and Southern Europe, the Near East, and North Africa and is an important economic pest. It is a voracious eater and can completely defoliate entire trees; repeated severe defoliation can add to other stresses, such as weather extremes or human activities. *Lymantria dispar* is most destructive in its larval stage (caterpillars), stripping away foliage from a broad variety of trees (>500 species). Caterpillar infestation is an underestimated problem; medical literature reports that established populations of caterpillars may cause health problems to people and animals. Inflammatory reactions may occur in most individuals after exposure to setae, independent of previous exposure. Currently, chemical and mechanical methods, natural predators, and silvicultural practices are included for the control of this species. Various insecticides have been used for its control, often through aerial sprayings, which negatively affect biodiversity, frequently fail, and are inappropriate for urban/recreational areas. However, bioinsecticides based on various microorganisms (e.g., entomopathogenic viruses, bacteria, and fungi) as well as technologies such as mating disruption using sex pheromone traps have replaced insecticides for the management of *L. dispar*.

Keywords: European spongy moth; spread; losses; outbreaks; health issues; control

1. Introduction

Sporadic outbreaks of insect pests are a major challenge to forest health in many countries, leading to serious environmental resource losses and degradations. One of these pests is the spongy moth (formerly known as gypsy moth) *Lymantria dispar* (L.)

(Lepidoptera: Erebidae), a notorious invasive polyphagous pest, causing widespread loss of leaves in forests in Europe, Asia, North America, and parts of Africa (Figure 1a) [1]. This species recently underwent a change to the common name since the word "gypsy" was considered to be an ethnic slur and now refers to the spongy egg masses [2]. It is also considered very harmful to orchards and urban environments [3–5]. In the species name of the *L. dispar*, "dispar" is derived from the Latin word that means "to separate", depicting the conspicuous sexual dimorphism of this species [6]. Females are larger and more brightly coloured than males and do not fly even though they are winged [7]. The International Union for Conservation of Nature (IUCN) has categorized *L. dispar* as one of the 100 worst invaders globally [8]. According to the European and Mediterranean Plant Protection Organization [9], in the United States of America, *L. dispar* has been a quarantine pest since 1989, while in Europe it has been classified in the A1 list in Azerbaijan and Georgia since 2007 and 2018, respectively, and in the A2 list in Russia since 2014.

(a) (b) (c) (d)

Figure 1. *Cont.*

(e)

(f)

Figure 1. (**a**) Severe defoliation of oak by *Lymantria dispar*. (**b**) Egg clusters of *L. dispar*. (**c**) Late instar larva of *L. dispar*. (**d**) Pupa of *L. dispar*. (**e**) Newly emerged adult male of *L. dispar*. (**f**) Female of *L. dispar* laying eggs on a wall.

Lymantria dispar is a univoltine species with four distinct life stages: egg, larva (caterpillar), pupa, and adult (Figure 1b–f) [10,11]. More specifically, the front and vertex of the male's head is pale gray to light brown; the antennae are bipectinates; and the shaft is completely scaled, with a light brown color and some fuscous scales apically. The forewing is 14.5–22 mm long [12] and is color brown. The hindwings are reddish-brown with a dark brown marginal band and light brown fringe [12]. The front and vertex of the female's head is white, the scape is speckled white and brown, the antenna has short pectinations, the shaft and pectinations are fuscous, and the apex is white. The length of forewings in females is 20–30 mm [12], being the color white. The hindwing is white, has a discal spot faint, and has a V-shaped at end of the discal cell [12]. Egg clusters are teardrop shaped, their color is yellowish to brownish, are felt-like in appearance, and turn lighter in color with age. Normally, all individual eggs are hidden by the hairs of the female abdomen [12]. First instar larvae are about 3 mm long [6], are very light in weight, and their color is dark brown to black, while the later instar larvae are more colorful [13,14]. The final larval instar is sometimes up to 90 mm long [6]. Full-grown larvae are hairy and have on the dorsum two rows of blue tubercles on the first five segments and two rows of red tubercles on the following six segments [14]. Pupae are dark brown and teardrop shaped, without a silky cocoon. Female pupae are 15–35 mm long, while males are usually smaller than females with a length of about 15–20 mm [13].

The classification of *L. dispar* has been controversial in the past, but it is now generally accepted that the European spongy moth, *Lymantria dispar dispar* L., differs from the Asian spongy moth [15]. Indeed, apart from the European spongy moth, which is spread throughout Europe and can be found in all environmental zones, with the exception of the Alpine North and part of the Boreal zone [16], two other subspecies have been recognized: (i) *L. dispar asiatica* Vnukovskij, which is found mostly in China, in Korea, and in the east Ural Mountains, and (ii) *L. dispar japonica* Motschulsky, which occurs on the main islands of Japan [12]. However, all three subspecies are polyphagous [17,18], with *L. dispar asiatica* having a wider range of host plants than *L. dispar dispar* [15,19–26]. In addition, females of both Asian subspecies are capable of flight, in contrast to females of the European subspecies that are flightless [10,12,23,27,28], a fact that may impede the spread of the latter subspecies. However, females of the European subspecies have little to no selectivity during oviposition and they can deposit egg masses on available substrates than can be used as vehicles for the spread of this subspecies [29]. Similarly, females of *L. dispar asiatica* that are able to fly are more attracted to lights, especially at well-lit ports, where they lay eggs on ships, which greatly facilitates their dispersal [30,31]. Recently, several molecular tools based on the mitochondrial barcode region have been used for the identification of the three subspecies. It has been found that nucleotide substitutions within regions of mitogenomes of *L. dispar* may be a useful tool to distinguish among the subspecies and/or detect geographical origins [32].

Populations of *L. dispar* usually remain at low densities. However, under some circumstances that are not well understood, densities can increase rapidly and after few generations achieve pest status. It has been reported that the length of the non-outbreak periods may be linked to climate, being shorter in dry forests and longer in wet ones [33]. The projected increase in temperature will strongly affect the ecology and distribution of exothermic organisms and is likely to trigger outbreaks of certain insect pests [34]. Indeed, the increase in mean temperature is expected to have a positive effect on *L. dispar* dispersal in the more northern ecosystems. Moreover, apart from climate change, inadvertent transportation will help this species to colonize new northern European habitats, as has happened in North America [35].

Currently, management of *L. dispar* is mainly based on biopesticides [36,37]. Moreover, several biocontrol agents have been used for the control of *L. dispar* such as entomopathogenic microorganisms alone or in combinations [36,38–48]. Although parasitoids represent appropriate biological control agents [49], they have not been used very often in IPM programs for controlling *L. dispar* in Europe.

Due to a lack of review studies concerning *L. dispar* in Europe, we aimed to gather and present in detail the up-to-date information from the international bibliography, scientific databases, and journals on the following aspects: description of each life stage, lifespan, preferred plant species, distribution, damages and outbreaks, and various control and monitoring methods (i.e., traps, mating disruption, natural enemies, biological and chemical control). In addition to the considerable damage to forests, *L. dispar* larvae are an underestimated problem; established populations of larvae may cause health problems to people and animals. Thus, special emphasis will be given to the health problems and risks arising from the presence of *L. dispar* larvae in forests and urban/recreational areas for humans and the environment.

2. Biology

The lifespan of *L. dispar* adults is about a week, and male emergence usually takes place 1 or 2 days before the females [6]. Specifically, females die about one day after oviposition, whereas males survive for about a week, after mating with different females [11]. *Lymantria dispar* can mate and oviposit soon after adults emerge [10,12,50]. Moreover, females can attract males and mate within two hours of pupal eclosion [51,52]. The courtship behavior is very simple, a male approaches a female, and mating occurs only when the female lifts her wings to allow mating. Copulation lasts for up to an hour, but usually the passage of the spermatophore or sperm packet is accomplished in the first 10 min. After the termination of the sexual interaction, females begin to lay eggs [6,53,54]. Females mate once while males usually mate a few times [55]. Thus, since males are polygamous, almost all females are fertilized, even if males comprise only 17% of the total population [56]. The reproductiive timespan of females is shorter than a week. By the third day after the female's emergence, their attractiveness decreases significantly, perhaps due to a depletion of the pheromone supply. Thus, there is little likelihood a female will mate if a mate has not been selected by then [6].

Mated females lay eggs on the tree trunks usually at less than 1 m from their pupation site because they are functionally flightless [12,23,27,28,57–59]. However, females that have not mated lay eggs that do not hatch and are usually scattered singly or in small, disorganized clusters [39]. Eggs are deposited in clutches directly on the bark or branches of host trees in late summer [12,60–64]. If a female is interrupted during the oviposition, she is able to initiate a new cluster [6]. In addition to host plants, eggs can be found on rocks or other immovable objects (e.g., firewood, recreational vehicles, Christmas trees, boat or cargo containers) and can overwinter there for eight to nine months [10,65]. Generally, females oviposit in the cooler or shady parts of the tree [12]. A female can lay from less than 100 to over 1000 eggs. Under optimal conditions, the mean number of eggs in a cluster is about 750, while at the end of an outbreak when the population begins to decline, it is about 300 [6]. For instance, in Spain, the mean number of eggs per cluster over the entire generation is 250–500 [57–59,66,67]. In addition, the egg clusters are covered with hair-like setae from the female's abdomen to protect them for overwintering [44]. The hairs are long, straight, and generally conical at both ends [68]. Eggs can withstand temperatures below −30 °C (down to −31.7 °C), provided they are deposited in a protected area or insulated with snow [10]. The egg clusters are approximately 3.8 cm long and 1.9 cm wide, whereas the presence of smaller egg clusters indicates that the population of *L. dispar* is declining [69].

The first three weeks of the egg stage consist of embryonation followed by diapause, which is affected by cold and warm conditions [70]. A significant number of studies have been conducted on climatic factors that affect the three stages of diapause (i.e., pre-diapause, diapause, and post-diapause), revealing a complex, non-linear relationship between temperature and progress through the completion of diapause [6,71,72]. The development of the embryo is favored by higher temperatures, completing the prediapause phase in approximately 16 days at 25 °C, but at a lower temperature, this phase takes longer to complete (e.g., 48 d at 15 °C) [73]. This phase is characterized by high respiration rates and abundant morphological development of the embryo [60,73]. The well-developed embryo then enters the next phase, the diapause, as a fully differentiated pharate larva, which lasts several months [60,74], but the role of the temperature is crucial, since at low temperatures, diapause is terminated quickly [60,72,75–77]. This phase is characterized by low respiration and developmental rates favored by low temperatures [72]. After exposure to low temperatures to terminate the diapause, elevated temperatures affect the hatching, where at 25 °C, hatching is completed in 11–18 days, while at 15 °C, more time is needed (i.e., 14–25 days), and the respiration rates are again high during the postdiapause phase [78]. There is no evidence that the photoperiod can control diapause, but 60–150 days of low temperature exposure is required [77].

Egg hatching takes place in early spring to mid-May [10]. However, the temperature strongly affects the hatching and activity of new larvae. Most of them hatch within a week, while hatching of egg clusters deposited in cooler areas or at higher altitudes can extend up to a month [6]. Larvae overwinter as embryos protected in eggshells and emerge in spring [12]. Male larvae usually go through five instars, while females go through six instars [63]. However, it has been reported that larvae may go through as many as 11 instars before pupation [70]. If the newly hatched larvae emerge at temperatures lower than 7 °C, they remain on or near the egg cluster [6]. Heavy rainfall during egg hatch may result in drowning of larvae. Moreover, during rainy weather, the first-instar larvae may delay their migration and accumulate on the underside of leaves [79].

After hatching, the larvae feed on the foliage of the host plants for a period of six to eight weeks [10]. First instar larvae begin feeding by cutting small holes in the surface of leaves, whereas later instars feed on the edge of leaves [6]. A distinct change in the feeding rhythm of *L. dispar* occurs as the larvae develop and mature [80]. The larvae are positively phototropic and negatively geotropic when they leave the egg and spin a silk thread as they move [6]. In addition, phototropic changes shift their behavior, with early instar larvae strongly attracted to light and late-instar larvae indifferent or repelled by light [81]. Early instar larvae (typically first- to third-instar) feed on host leaves during the day and stay on the underside of foliage during the evening [61,82]. Late-instar larvae (fourth- to sixth-instar) feed on the canopy overnight, whereas at daybreak, they move downward in search of cryptic resting places (e.g., bark flaps or crevices, litter under the host tree) [52] as these sites provide protection from predators [83,84]. During the diurnal period of larval movement, it is very common for the late instar larvae to abandon the tree they were feeding on and move to a new one [85]. This is a mechanism that allows late instar larvae to use a wider range of plants than early instar larvae [86]. However, in elevated populations, such as populations of more than 1250 egg masses/ha, all larvae feed in the day and night and do not need to hide in a shelter [87]. Approximately 80% of defoliation is conducted by the fifth and sixth instar larvae [88]. In Central Europe, the larvae begin to feed in late April and continue for up to 10 weeks, until they pupate in late June–early July [89].

The larvae stop feeding just before the prepupal stage, which lasts only for about 2 days, where the larvae void the gut, surround themselves in a sparse silken net, and begin to contract in length. The prepupae remain relatively quiescent inside the silken net [6]. Pupation occurs in hidden places such as the underside of branches, cracks in bark, trunk crevices, and under stones or trunks on the soil [57–59]. The pupal stage lasts 7–14 days depending on climate and sex. Usually, males emerge a few days earlier than females [50,85].

Lymantria dispar adults do not feed, and although moisture is imbibed, their digestive system is not functional [6]. The larvae are the most destructive life stage, as they can feed on a wide range of hosts [17,18,50,90–93]. In addition, voracious larvae can feed on host trees for up to 10 weeks [94]. The consumption of oak foliage by larvae is about 10 mg for each mg gain in larval biomass [95]. Each larva in a dense population is able to consume about 1270 mg of foliage (dry weight) and reach a weight of 114 mg. This is about 170 cm^2 foliage corresponding to three red oak leaves. However, a large (750 mg) female larva typical of a sparse population could consume a total of 1000 cm^2 of foliage [50]. Thus, the sixth instar female larvae are considered the most ravenous feeders and have often twice the size of full-grown male larvae [11].

Forest tree species can be categorized as "susceptible", "resistant", or "immune" to defoliation by *L. dispar* larvae [17,96]. "Susceptible" tree species are those that are consumed by all larval instars; "resistant" are the species that only some larval instars consume, or when susceptible species are not available (Table 1); and "immune" species are rarely, if ever, consumed by any larval instar. Some tree species are considered resistant to *L. dispar* larvae unless in close proximity to susceptible species. In the case that a preferred tree species is not available, the larvae may alternatively feed on the red and sugar maple, *A. rubrum* L. and *A. saccharum* Marshall (Sapindales: Sapindaceae); the American beech, *Fagus grandifolia* Ehrh. (Fagales: Fagaceae); and the American elm, *Ulmus americana* L. (Rosales: Ulmaceae) [10,17].

In Europe, the preferred hosts vary by region, but include species of *Quercus*, *Carpinus*, *Alnus*, *Prunus*, *Populus*, *Gleditsia*, *Tilia*, *Corylus*, and *Robinia* [63,97,98]. For instance, in Lithuania, which is at the northern limits of its range, species of birch (*Betula*) and alder (*Alnus*) are the primary hosts, while in Spain, Portugal, and Sardinia, the cork oak, *Quercus suber* L. (Fagales: Fagaceae), is the dominant host, and stands of this species have been frequently defoliated. The distribution of *L. dispar* in the rest of Europe is associated with the presence of up to seven species of *Quercus*, especially the Austrian oak, *Q. cerris* L.; the pedunculate oak, *Q. robur* L.; and the sessile oak, *Q. petraea* (Matt.) Liebl. (Fagales: Fagaceae), but this latter species is less preferred among the Central European oaks [98]. The hornbeam, *Carpinus betulus* L. (Fagales: Betulaceae), is usually mixed in stands of oaks in Central Europe and is considered an equally preferred host along with species of *Populus*, *Alnus*, and *Salix* [98,99]. In southern France and the Balkans, *Q. suber*, *Q. pubescens*, and *Q. ilex* serve as the main hosts. However, there is an exception to the close relationship between oak species and *L. dispar* populations in the Danube delta of Romania, where 27,000 ha of *Populus* and *Salix* stands serve as preferred hosts for *L. dispar* larvae [98].

In North America, *L. dispar* feed on >300 species of woody plants but prefer the species *Quercus*, *Salix*, *Populus*, and *Betula* [17]. More specifically, 148 host trees have been identified as highly susceptible hosts out of a total of 449 tree species that larvae can feed on [17]. *Quercus* and *Salix* are the most susceptible genera, with the remaining highly susceptible species coming from 28 other tree genera including *Larix*, a deciduous conifer [100]. In both Europe and North America, the variety of hosts plants the larvae utilize expands when they reach the fourth instar, including several conifers, such as some species of pine, spruce, and hemlock [101].

Table 1. Susceptible, preferred, and resistant tree species to *Lymantria dispar* larvae.

Common Name	Scientific Name	Order	Family	Categorization	Reference
European crab apple	*Malus sylvestris* (L.) Mill.	Rosales	Rosaceae	Susceptible	[17]
Bigtooth aspen	*Populus grandidentata* Michaux	Mapighiales	Salicaceae	Susceptible	[17]
Quaking aspen	*P. tremuloides* Michx.	Mapighiales	Salicaceae	Susceptible	[17]
Boxelder	*Acer negundo* L.	Sapindales	Sapindaceae	Susceptible	[17]
American mountain ash	*Sorbus americana* Marshall	Rosales	Rosaceae	Susceptible	[17]
Sweetgum	*Liquidambar styraciflua* L.	Saxifragales	Altingiaceae	Susceptible	[17]
Basswood	*Tilia* spp.	Malvales	Malvaceae	Susceptible	[17]
Birch	*Betula* spp.	Fagales	Betulaceae	Susceptible	[17]
Larch	*Larix* spp.	Pinales	Pinaceae	Susceptible	[17]
Oak	*Quercus* spp.	Fagales	Fagaceae	Susceptible	[17]
Willow	*Salix* spp.	Malpighiales	Salicaceae	Susceptible	[17]
Alder	*Alnus* spp.	Fagales	Betulaceae	Preferred	[10]
Hawthorn	*Crataegus* spp.	Rosales	Rosaceae	Preferred	[10]
Hazelnut	*Corylus* spp.	Fagales	Betulaceae	Preferred	[10]
Hornbeam	*Carpinus* sp.	Fagales	Betulaceae	Preferred	[10]
Serviceberry	*Amelanchier* spp.	Rosales	Rosaceae	Preferred	[10]
Sumac	*Rhus* spp.	Sapindales	Anacardiaceae	Preferred	[10]
Hemlock	*Tsuga canadensis* (L.) Carrière	Pinales	Pinaceae	Resistant	[10,17]
Yellow birch	*Betula alleghaniensis* Britt.	Fagales	Betulaceae	Resistant	[10,17]
Eastern white pine	*Pinus strobus* L.	Pinales	Pinaceae	Resistant	[10,17]

It has been reported that early instar larvae are not able to complete their development on non-deciduous conifers [18,102,103]. However, in outbreak mode, larvae will consume coniferous foliage, and these trees tend to be much less resistant to defoliation and may die after a single attack [104]. On the other hand, hardwood trees that are initially in good condition can produce new leaves after a *L. dispar* attack (which generally takes place in June and early July) and can often withstand several years of defoliation without dying [104]. Consequently, in the case of mixed pine–hardwood stands such as those of eastern North America, defoliation by larvae is largely limited to hardwood hosts, and outbreaks generally do not occur in stands in which the proportion of oaks or other susceptible host plant species is less than 20% of host basal area [105,106].

In Spain, *L. dispar* has been reported on *Quercus* sp., *Castanea* sp., *Corylus* sp., and *Fagus* sp. but also on *Populus* sp., *Ulmus* sp., *Arbutus* sp., *Prunus* sp., *Acer* sp., *Salix* sp., *Betula* sp., *Alnus* sp., or *Pinus* sp. It has been reported to damage the strawberry tree, *Arbutus unedo* L. 1753 (Ericales: Ericaceae), and coniferous species, such as the Aleppo pine; *Pinus halepensis* Mill., in the Balearic Islands and Catalonia; the maritime pine, *Pinus pinaster* Aiton, in central Spain and the Balearic Islands; and the radiata pine, *Pinus radiata* D.Don (Pinales: Pinaceae), in northern Spain [58,66]. This forms an important threat to public health as pine trees constitute a major species in urban/suburban environments and cover large forest areas in Southern Europe, while their presence is common in schools and other public areas, potentially affecting susceptible individuals, such as children.

It has been found in the provinces of Pontevedra (north-western Spain) and Asturias (northern Spain) in the outbreaks during 1952–1953 [107,108]. In northwestern Portugal, 290 ha of a 15-year-old plantation of *P. radiata* were severely defoliated in 1991 [109]. Moreover, Miller and Hanson [110], conducting laboratory tests, found that the first instar larvae originated from egg clusters collected in Oregon, USA, completed their development on different coniferous tree species including the white fir, *Abies concolor* (Gordon) Lindley ex Hildebrand; the blue spruce, *Picea pungens* Engelm.; the ponderosa pine, *Pinus ponderosa*

Douglas ex C.Lawson; and the Douglas fir, *Pseudotsuga menziesii* (Mirbel) Franco (Pinales: Pinaceae). More recently, Castedo-Dorado et al. [111] reported that the *L. dispar* was able to complete development in the field and severely defoliated trees of *P. radiata* in Spain. Radiata pine is native to California, but has spread widely elsewhere, especially in the southern hemisphere, due to its planting for commercial forestry. In 1950, extensive planting of this species began in the Northern Iberian Peninsula [112]. Consequently, the presense of *P. radiata* in this region, along with other southern European countries, are the only areas where both radiata pine and *L. dispar* overlap, since radiata pine is absent from *L. dispar*'s invasion area in North America and the insect pest is absent in the southern hemisphere [111].

3. Ecology of *L. dispar*

3.1. Distribution

Lymantria dispar is a serious insect pest in southern and central Europe [34]. It was intentionally introduced to the east coast of North America from Western Europe in 1869 for silk production [12,50,52,113–116]. Professor L. Trouvelot maintained colonies of this species, attempting to reproduce better silkworm larvae by crossing them with the native silkworm, *Antheraea polyphemus* (Cramer) (Lepidoptera: Saturniidae), but a small number of *L. dispar* larvae escaped from his laboratory [98,115,117]. This species has been established there, reaching nearly the middle of the continent [118]; it has spread throughout New England and the adjoining provinces of Canada. The peak of the infestation has reached Maine, Wisconsin, Illinois, Indiana, Ohio, West Virginia, Virginia, North Carolina, Ontario, Quebec, New Brunswick, and Nova Scotia [27,119]. *Lymantria dispar* has become a much more serious pest in the United States than in its native regions [6,113,115,120–122]. *Lymantria dispar* has recently expanded to several ecosystems globally, where it was previously nonnative, due to high vehicle traffic, international shipping, and global warming, which favors its winter survival and enhances its feeding activity in northern regions, as shown both in research simulations and empirically [35].

Lymantria dispar is widely distributed in Europe, occurring in many environments of this continent, with the exception of Alpine North and partially the Boreal zone [9,16]. It is also found from latitude 60° N in mid-Scandinavia to 35° N on the Mediterranean coast and occurs from the Mediterranean scrub land to the temperate deciduous forests [118]. The northern limit proceeds through southern Sweden and Finland and descends from about 60° to 50° lines of latitude through Europe and Russia. The southern limit begins in the west in northern Morocco, Algeria, and Tunisia and proceeds east to include all of the Mediterranean islands, on a line through Israel into Asia [98]. The latitudinal distribution of *L. dispar* in Canada and the United States lies between 35° N and 48° N [91]. Generally, damage caused by *L. dispar* in Europe increases from west to east and from north to south [123]. Furthermore, due to the projected climate change, *L. dispar* may be a threat to more northerly ecosystems [35].

The spread of *L. dispar* in Europe has been recorded in the database of the European and Mediterranean Plant Protection Organization [9]. According to this database, this species is present in Austria with few occurrences. It is present in Belarus, Belgium, Croatia, Denmark, Italy (including Sardinia and Sicilia), Lithuania, Moldova, North Macedonia, Poland, Romania, Serbia, Slovenia, Turkey, Ukraine, and Azerbaijan, where *L. dispar* has been classified in the A1 list (pests are not present in the EPPO region) since 2007, and in Russia, which has been classified in the A2 (the presence of pests occurs locally in the EPPO region) since 2014. In Spain, there are reports from the XIX century. It is known as "lagarta peluda" in Spain, as "eruga peluda del suro" in Catalonia, and "eruga peluda" in the Balearic Islands. Most of the information available in Spain comes from old pest control books and technical leaflets published by some Spanish Regional Plant Health Services [57,58,66,67].

The dispersal of *L. dispar* is impressive since its females are not able to fly [12,57–59,124,125]. However, the spread of this species in new areas can occur through larvae that can disperse over short distances, as well as through the human-mediated movement, which can help spread some of the life-stages of this pest to over long distances [15,94]. Concretely, dispersal by larvae is accomplished by either their crawling from tree to tree or by larval ballooning, that is, a wind-borne movement, and is therefore limited to short distances (<100 m) [126–129]. Early instar larvae move to tree crowns, where they hang from strands of silk until the wind carries them locally to other trees, particularly in an urban situation [130]. In some cases, larvae may "balloon" several times before they start feeding [131]. However, it has been reported that the newly hatched larvae have been blown to distances of 56 km [61]. In addition, the primary pathway for the long-distance of *L. dispar* is human assisted (i.e., recreational travel; transportation of egg masses on firewood, household goods, and vehicles) [29,125,132,133]. For example, *L. dispar* has spread at a rate of 3 to 29 km per year since its introduction to the USA [134,135], which is based on a dispersal rate of more than 20 km per year for the period between 1966 and 1990 [131]. On the other hand, in the absence of anthropogenic movement, the natural spread that occurs through early instar-larvae ballooning could be as slow as 2.5 km/year [131].

The establishment of *L. dispar* in a new area depends on the temperature. For instance, in warm climates such as southern Florida, this species is not able to complete a full life cycle because there are not suitably low temperatures to complete the diapause [136]. In addition, in areas where the high summer temperatures exceed the optimum temperatures for larval and pupal development, the probabilities of establishing of this species are low [137,138]. Another important parameter for successful establishment in small, isolated populations or those at low density near range edges is the ability of individuals to successfully locate mates [139–141]. More specifically, the failure to establish *L. dispar* infestations ahead of the invasion is because of the unsuccessful detection of females by the males in low-density populations [139].

3.2. Outbreaks

Lymantria dispar populations appear to exist in one of the following four phases: innocuous, release, outbreak, and decline [94]. The innocuous phase is characterized by very low population levels. The release phase usually takes place over the course of one or two years and can result in population density increases of several orders of magnitude. During the outbreak phase, populations are high enough to cause noticeable defoliation to host trees. After this point, the decline phase follows, which is characterized by high levels of *L. dispar* mortality usually due to starvation or disease and the population crashes. Outbreaks in the whole area can last up to 10 years, but generally population densities in localized areas remain high for two or three years [142]. Usually, outbreaks of forest defoliators emerge in a synchronous manner at large spatial scales [143]. In contrast, *L. dispar asiatica*, which is found in Russia [144], exhibits asynchronous population outbreaks. This is because temperature varies considerably in Russia during winter (<-30 °C) vs. other European countries [145]. For several decades, the cyclic population dynamics and their genetic consequences have been an area of interest [146]. *Lymantria dispar* exhibits cyclic population dynamics [98,147,148]. More specifically, in Europe, which is the native territory of this pest, it exhibits cyclical outbreaks every 8–13 years [148]. For instance, in Spain, it is reported that cyclic outbreaks may last for a couple of years, followed by 7–11 years of non-outbreak period [58,66]. On the other hand, in North America, where the *L. dispar* is an invader species, it exhibits varying periodicity, every 4–5 years or 8–10 years, depending on forest type [33]. Severe outbreaks of *L. dispar* larvae can completely defoliate forest canopies, which have many short-term effects. These effects are related to the reduction in productivity or the reduction of seed crops, the increase in light to the forest floor, reduced transpiration that leads to increased water drainage from the forest, and a pulse of nitrogen and labile carbon to the forest floor [6,149,150]. In addition, as the host trees and seedlings can be completely defoliated and killed during outbreaks, this can cause changes in the composition of natural forests [151–153]. After an outbreak, *L. dispar* populations collapse due to viral or fungal disease, parasitism, predation, starvation, or adverse climatic conditions [154]. Apart from the environmental damage caused by the outbreaks of this pest, there are considerable costs in terms of economic losses and the cost of control measures [121,155,156].

Lymantria dispar infestations have positive and negative effects on wildlife. On the one hand, defoliation can lead to increased growth of shrubs, grasses, and herbs, providing an additional habitat for some wildlife species. On the other hand, defoliation may reduce or compromise the habitat for some wildlife species. For instance, bird eggs become vulnerable to predation due to reduced protection following defoliation of the tree leaves. Waterways can also be affected by outbreaks of *L. dispar*. Loss of canopy cover due to defoliation may increase the temperature of streams, which may be detrimental to aquatic organisms [157].

The most numerous and severe outbreaks have occurred in the Balkan peninsula because of the abundance of oak species and the climate (i.e., high temperatures and humidity deficits), which seems to be the optimum for the development and survival of *L. dispar* [98]. Indeed, in an earlier study, Schwenke [158] reported that the climate and the affinity of *L. dispar* for warmth and drought are causing an outbreak in southeastern and southern Europe. However, periodic outbreaks have been recorded since 1600 in Europe [159]. For example, outbreaks were reported in 1600 in Spain, 1750 in Germany, 1840 in Hungary, and 1880 in France [98]. In Serbia, 16 outbreaks were reported between 1862 and 1998, with the largest occurring in 1997 when 500,000 ha were infested [160]. Notably, in the mid-1950s and 1960s, outbreaks of this species were widespread and devastating in Europe, causing 70% defoliation in the Yugoslavian hardwood forests in 1957 [161]. However, the most severe epidemic in Europe took place between 1991 and 1995, affecting seven countries and hundreds of thousands hectares of forests, making control measures imperative [118,162]. During that period, France suffered the most, since this outbreak lasted longer there than in other Central European countries [118], where only in Alsace, an area of 23,500 ha, was defoliated in 1994 [163]. In Austria, 1500 ha were infested by the pest in 1993, causing severe defoliation on 400 ha of oak stands [164], whereas one year earlier in Switzerland, 2000 ha composed mainly of chestnut trees was defoliated. However, after the collapse of the *L. dispar* population in 1993, the trees recovered without any treatment [165]. An outbreak in Germany affected almost 2000 ha in 1992, while one year later, the infestation covered 47,000 ha, and in 1994, before the natural population collapsed, the pest was present in about 80,000 ha [118]. Furthermore, during this period, small outbreaks occurred in England, where the climate is rather unfavorable for *L. dispar* [118,166]. In Hungary, 22 outbreaks have been reported between 1843 and 2007, with the maximum damage (i.e., 212 thousand hectares) recorded in 2005 [98].

Lymantria dispar in Greece is considered one of the most important pests of oak trees. Greece has experienced several population outbreaks [167]. More recently, over the years 2016 and 2017, repeated outbreaks have been reported in different parts of Greece (i.e., Macedonia, Ipirus, and Thessaly) on evergreen pastures, where *Q. coccifera* predominates [130]. In Catalonia (Spain), during the last century, several outbreaks with massive defoliation were recorded, especially on *Q. suber* from the coastal and pre-coastal mountains of the Transversal Mountain System [168]. After a long period without important damage, new outbreaks were recorded in 2018–2020 in several counties, especially in the natural park of Montnegre-Corredor, in Barcelona province. Soldevila [169] and Stefanescu et al. [59] performed a detailed study in some affected forest of the Montnegre mountains of Catalonia (Barcelona province) where they determined the outbreak effects and the role of natural enemies as biocontrol agents. Defoliation of oak forest by *L. dispar* has been especially important in the Montnegre-Corredor park on *Q. rubra* and *Q. ilex*, after more than two decades without records of massive defoliations [65]. In that area in 2019, the damage covered nearly 15,000 ha [170] and it caused social alarm and concern in the forest owners, as more than 90% of the forest area of Catalonia is private. The attack not only affected *Quercus* sp., but also many specimens of *A. unedo*, *P. halepensis*, *Celtis* sp., *Castanea* sp., and other tree species.

During 2006–2008, the territory of the Russian Far East suffered from a *L. dispar* outbreak, which was characterized as the most severe since the 1930s [171]. In the Ugam-Chatkal range of Uzbekistan, 500 ha of various forest and fruit tree species were heavily damaged by *L. dispar* [172]. Furthermore, *L. dispar* outbreaks were reported in Northern Kazakhstan in 2017, damaging 247 ha of forest plants [173]. In North America, the history of outbreaks is related to the dispersal of *L. dispar* in new areas, where the preferred plant species predominate, followed by the rapid spread of populations due to the absence of a natural enemy complex. After a massive outbreak that occurred in 1979–1982, in 1981, at the peak of the outbreak in the USA, *L. dispar* defoliated more than 6 million ha of forests [174]. Later, the next outbreak occurred from 1989 to 1993, affecting forest areas in 12 different states [98].

4. Control of *L. dispar*

Lymantria dispar is considered one of the most destructive invasive pests, ranking third among the most costly invader insect pests worldwide [153]. Defoliation by *L. dispar* larvae can cause tree biomass losses up to 70%, while cumulative effects can result from consecutive defoliations [175,176]. From an economic perspective, *L. dispar* can massively affect timberlands and agricultural products, with high yearly costs on quarantined products. This pest impacts the aesthetics of forests and property values in urban areas, imposing high costs (disinfestation, tree removal, and replacement); it also burdens health systems and may negatively affect tourism. For instance, in the USA, the economic impact of this species was estimated at about USD 250 million per year in 2011 [121], but with the continued spread of *L. dispar* in North America, the economic impact was increased to USD 3.2 billion per year in just a few years [155].

In years with moth outbreaks, cutaneous reactions in humans can reach epidemic proportions in communities located near infested trees, the contact with airborne setae being mainly responsible. It is, therefore, crucial to control and manage *L. dispar* to mitigate the possible ecological, economic, and social impacts. Due to the negative impact of *L. dispar* on the forests, several pest control programs have been conducted against this pest to reduce populations and damage during outbreaks [177,178] or to slow its spread from infested to uninfected areas [134,179–181]. The first tools to control *L. dispar* were crude. Early trapping devices were baited with live females and other control tools were oil-fueled flamethrowers to destroy the life stages and microhabitats of the pest, as well as arsenic-based insecticides (copper acetoarsenite and lead arsenate) [52]. Current control measures include chemical and mechanical methods, natural predators, and silvicultural practices. In addition, the economic damage threshold of 100 egg clusters per 40 trees, which are equivalent to an average number of 2.5 egg clusters/tree, is used to make decisions about the initiation of management tactics [182–184].

The use of conventional insecticides, such as organophosphates, carbamates, and pyrethroids, is common in pest control, especially to prevent the spread of destructive insect species such as *L. dispar*, but this approach can be harmful to human health and the environment [185–187]. Although there is an increasing demand for alternative eco-friendly methods to control *L. dispar* (e.g., viruses, parasites, pheromones, fungi, or bacteria), chemical control is still the most common and effective method [188–191]. Starting from the 1980s, broad-spectrum insecticides used for *L. dispar* control have been replaced with bioinsecticides based on entomopathogenic viruses, bacteria, and fungi, as well as technologies such as mating disruption using sex pheromone traps [192,193].

4.1. Biological Control

4.1.1. Natural Enemies

The infestation levels that *L. dispar* can cause in North America are much higher than those in Europe, due to more suitable biotic and abiotic conditions in the newly invaded areas [98]. Apparently, this phenomenon is partially due to the fact that its natural enemy complex in Europe is much more diverse than that in the Nearctic region, so that the natural regulation (i.e., naturally occurring biological control) of *L. dispar* populations is more efficient in Europe than in North America. The natural enemies include various predators, such as small mammals and birds. However, small mammals and birds do not come in numbers able to control large populations because the pest reproduces much faster than the predators. Pathogenic fungi require specific conditions to germinate and infect *L. dispar* larvae (high humidity, rainfall), making their effectiveness in a given year subject to local weather conditions. The natural enemies of *L. dispar* are presented in Table 2.

Table 2. Predators and parasitoids of *Lymantria dispar* larvae.

Predators			
Species Name	**Order**	**Family**	**Reference**
Peromyscus leucopus (Rafinesque, 1818)	Rodentia	Cricetidae	[194]
Apodemus sylvaticus (Linnaeus, 1758)	Rodentia	Muridae	[59,169]
Crocidura russula (Hermann, 1780)	Eulipotyphla	Soricidae	[59,169]
Parus major Linnaeus, 1758	Passeriformes	Paridae	[59,169]
Garrulus glandarius (Linnaeus, 1758)	Passeriformes	Corvidae	[59,169]
Dendrocopos kizuki (Temminck, 1836)	Passeriformes	Picidae	[59,169]
Sitta eiuropaea Linnaeus, 1758	Passeriformes	Sittidae	[59,169]
Calosoma sycophanta L.	Coleoptera	Carabidae	[56,59,169,195–203]
Parasitoids			
Acropimpla didyma (Gravenhorst)	Hymenoptera	Ichneumonidae	[44]
Aleiodes pallidator Thunberg	Hymenoptera	Braconidae	[44]
Anastatus bifasciatus (Geoffroy)	Hymenoptera	Eupelmidae	[44]
Anastatus catalonicus Bolivar & Pieltain	Hymenoptera	Eupelmidae	[44]
Anastatus japonicus Ashmead	Hymenoptera	Eupelmidae	[44]
Apanteles impurus (Nees)	Hymenoptera	Braconidae	[44]
Apanteles lacteicolor Viereck	Hymenoptera	Braconidae	[44]
Apanteles xanthostigma (Haliday)	Hymenoptera	Braconidae	[44]
Apechthis capulifera (Kriechbaumer)	Hymenoptera	Ichneumonidae	[44]
Apechthis compunctor (L.)	Hymenoptera	Ichneumonidae	[44]
Apechthis quadridentata (Thomson)	Hymenoptera	Ichneumonidae	[44]
Apechthis rufata (Gmelin)	Hymenoptera	Ichneumonidae	[44]
Aphantorhaphopsis samarensis (Villeneuve)	Diptera	Tachinidae	[44]
Banchus falcatorius (Fabricius)	Hymenoptera	Ichneumonida	[44]
Barylypa pallida (Gravenhorst)	Hymenoptera	Ichneumonida	[44]
Baryscapus oophagus (Otten)	Hymenoptera	Eulophidae	[44]
Blepharipa pratensis (Meigen)	Diptera	Tachinidae	[44]
Blepharipa schineri (Mesnil)	Diptera	Tachinidae	[44]
Blondelia nigripes (Fallén)	Diptera	Tachinidae	[44]
Blondelia piniariae (Hartig)	Diptera	Tachinidae	[44]
Bothriothorax altensteinii Ratzeburg	Hymenoptera	Encyrtidae	[44]
Bothriothorax paradoxus Dalman	Hymenoptera	Encyrtidae	[44]
Brachymeria inermis (Fonscolombe)	Hymenoptera	Chalcididae	[44]
Brachymeria minuta (L.)	Hymenoptera	Chalcididae	[44]
Brachymeria secundaria (Ruschka)	Hymenoptera	Chalcididae	[44]
Brachymeria tibialis Walker	Hymenoptera	Chalcididae	[44]
Campoplex difformis (Gmelin)	Hymenoptera	Ichneumonidae	[44]
Carcelia gnava (Meigen)	Diptera	Tachinidae	[44]
Casinaria tenuiventris (Gravenhorst)	Hymenoptera	Ichneumonidae	[44]
Chouioia cunea Yang	Hymenoptera	Eulophidae	[44]
Cirrospilus pictus Nees	Hymenoptera	Eulophidae	[44]
Compsilura concinnata (Meigen)	Diptera	Tachinidae	[44]
Cotesia gastropachae (Bouché)	Hymenoptera	Braconidae	[44]
Cotesia glomerata (L.)	Hymenoptera	Braconidae	[44]

Table 2. *Cont.*

Parasitoids			
Cotesia melanoscela (Ratzeburg)	Hymenoptera	Braconidae	[44]
Cotesia melitaearum (Wilkinson)	Hymenoptera	Braconidae	[44]
Cotesia neustriae (Tobias)	Hymenoptera	Braconidae	[44]
Cotesia ocneriae (Ivanov)	Hymenoptera	Braconidae	[44]
Cotesia praepotens (Haliday)	Hymenoptera	Braconidae	[44]
Cotesia rubripes (Haliday)	Hymenoptera	Braconidae	[44]
Cotesia spuria (Wesmael)	Hymenoptera	Braconidae	[44]
Deuterixys carbonaria (Wesmael)	Hymenoptera	Braconidae	[44]
Doryctes leucogaster (Nees)	Hymenoptera	Braconidae	[44]
Drino gilva (Hartig)	Diptera	Tachinidae	[44]
Drino inconspicua (Meigen)	Diptera	Tachinidae	[44]
Dusona blanda (Förster)	Hymenoptera	Ichneumonidae	[44]
Elachertus charondas Walker	Hymenoptera	Eulophidae	[44]
Elasmus nudus Nees	Hymenoptera	Eulophidae	[44]
Euceros serricornis (Haliday)	Hymenoptera	Ichneumonidae	[44]
Euceros superbus Kriechbaumer	Hymenoptera	Ichneumonidae	[44]
Eulophus cyanescens Bouček	Hymenoptera	Eulophidae	[44]
Eulophus larvarum L.	Hymenoptera	Eulophidae	[44]
Eulophus slovacus Bouček	Hymenoptera	Eulophidae	[44]
Eupelmus annulatus Nees	Hymenoptera	Eulophidae	[44]
Eupelmus urozonus Dalman	Hymenoptera	Eulophidae	[44]
Euplectrus liparidis Ferrière	Hymenoptera	Eulophidae	[44]
Eurytoma appendigaster Swederus	Hymenoptera	Eurytomidae	[44]
Eurytoma goidanichi Bouček	Hymenoptera	Eurytomidae	[44]
Eurytoma verticillata (Fabricius)	Hymenoptera	Eurytomidae	[44]
Exeristes roborator (Fabricius)	Hymenoptera	Ichneumonidae	[44]
Exorista amoena (Mesnil)	Diptera	Tachinidae	[44]
Exorista larvarum (L.)	Diptera	Tachinidae	[44]
Exorista segregata (Rondani)	Diptera	Tachinidae	[44]
Gelis agilis (Fabricius)	Hymenoptera	Ichneumonidae	[44]
Gelis areator (Panzer)	Hymenoptera	Ichneumonidae	[44]
Glyptapanteles porthetriae (Muesebeck)	Hymenoptera	Braconidae	[44]
Glyptapanteles vitripennis (Curtis)	Hymenoptera	Braconidae	[44]
Gregopimpla inquisitor (Scopoli)	Hymenoptera	Ichneumonidae	[44]
Gryon howardi (Mokrzecki and Oglobin)	Hymenoptera	Scelionidae	[44]
Gryon hungaricum (Szabó)	Hymenoptera	Scelionidae	[44]
Gryon lymantriae (Masner)	Hymenoptera	Scelionidae	[44]
Hemiteles pulchellus Gravenhorst	Hymenoptera	Ichneumonidae	[44]
Hyposoter tricoloripes (Viereck)	Hymenoptera	Ichneumonidae	[44]
Ichneumon sarcitorius L.	Hymenoptera	Ichneumonidae	[44]
Iseropus stercorator (Fabricius)	Hymenoptera	Ichneumonidae	[44]
Itoplectis alternans (Gravenhorst)	Hymenoptera	Ichneumonidae	[44]
Itoplectis enslini (Ulbricht)	Hymenoptera	Ichneumonidae	[44]
Itoplectis kolthoffi (Aurivillius)	Hymenoptera	Ichneumonidae	[44]
Itoplectis viduata (Gravenhorst)	Hymenoptera	Ichneumonidae	[44]
Lymantrichneumon disparis (Poda)	Hymenoptera	Ichneumonidae	[44]
Lysibia nana (Gravenhorst)	Hymenoptera	Ichneumonidae	[44]
Melittobia acasta Walker	Hymenoptera	Eulophidae	[44]
Mesochorus confusus Holmgren	Hymenoptera	Ichneumonidae	[44]
Meteorus pendulus (Müller)	Hymenoptera	Braconidae	[44]
Meteorus pulchricornis (Wesmael)	Hymenoptera	Braconidae	[44]
Meteorus versicolor (Wesmael)	Hymenoptera	Braconidae	[44]
**Monodontomerus aereus* Walker	Hymenoptera	Torymidae	[44]
Ooencyrtus kuvanae (Howard)	Hymenoptera	Encyrtidae	[44]
Ooencyrtus masii (Mercet)	Hymenoptera	Encyrtidae	[44]

Table 2. *Cont.*

Parasitoids			
Parasarcophaga uliginosa (Kramer)	Diptera	Sarcophagidae	[44]
Parasetigana silvestris (Robineau-Desvoidy)	Diptera	Tachinidae	[44]
Pediobius cassidae Erdös	Hymenoptera	Eulophidae	[44]
Pediobius crassicornis (Thomson)	Hymenoptera	Eulophidae	[44]
Pediobius foliorum (Geoffroy)	Hymenoptera	Eulophidae	[44]
Pediobius pyrgo (Walker)	Hymenoptera	Eulophidae	[44]
Peribaea tibialis (Robineau-Desvoidy)	Diptera	Tachinidae	[44]
Perilampus neglectus Bouček	Hymenoptera	Perilampidae	[44]
Perilampus ruficornis Fabricius	Hymenoptera	Perilampidae	[44]
Phobocampe lymantriae Gupta	Hymenoptera	Ichneumonidae	[44]
Phobocampe unicincta (Gravenhorst)	Hymenoptera	Ichneumonidae	[44]
Pimpla disparis Viereck	Hymenoptera	Ichneumonidae	[44]
Pimpla rufipes (Miller)	Hymenoptera	Ichneumonidae	[44]
Pimpla turionellae (L.)	Hymenoptera	Ichneumonidae	[44]
Pronotalia carlinarum (Szelényi and Erdös)	Hymenoptera	Eulophidae	[44]
Protapanteles fulvipes (Haliday)	Hymenoptera	Braconidae	[44]
Protapanteles liparidis (Bouché)	Hymenoptera	Braconidae	[44]
Protapanteles nigerrimus (Roman)	Hymenoptera	Braconidae	[44]
Senometopia separata (Rondani)	Diptera	Tachinidae	[44]
Siphona boreata Mesnil	Diptera	Tachinidae	[44]
Tachina magnicornis (Zetterstedt)	Diptera	Tachinidae	[44]
Tachina praeceps Meigen	Diptera	Tachinidae	[44]
Telenomus embolicus Kozlov	Hymenoptera	Scelionidae	[44]
Telenomus laevisculus (Ratzeburg)	Hymenoptera	Scelionidae	[44]
Telenomus longistriatus Kozlov	Hymenoptera	Scelionidae	[44]
Telenomus lymantriae Kozlov	Hymenoptera	Scelionidae	[44]
Telenomus macroceps Szabó	Hymenoptera	Scelionidae	[44]
Telenomus phalaenarum Nees	Hymenoptera	Scelionidae	[44]
Telenomus tetratomus (Thomson)	Hymenoptera	Scelionidae	[44]
Tetrastichomyia clisiocampae Ashmead	Hymenoptera	Eulophidae	[44]
Tetrastichus sp.	Hymenoptera	Eulophidae	[44]
Theronia atalantae (Poda)	Hymenoptera	Ichneumonidae	[44]
Torymus anastativorus Fahringer	Hymenoptera	Torymidae	[44]
Tyndarichus kuriri Fahringer	Hymenoptera	Encyrtidae	[44]
Tyndarichus navae Howard	Hymenoptera	Encyrtidae	[44]
Zenillia libatrix (Panzer)	Diptera	Tachinidae	[44]

When the presence of *L. dispar* in an oak forest is at a high density, it becomes part of the food chain and can be an important food source for birds and small mammals (e.g., rodents) [194,204]. For instance, *P. leucopus*, which feeds on *L. dispar* larvae and pupae, can regulate low-density populations of this pest [194]. Moreover, mice can cause high mortality, for example, they killed 98% of deployed *L. dispar* pupae within 72 h in Ukraine [98] and caused more than 45% mortality in an artificial population of *L. dispar* pupae in Austria [205]. However, the abundance of small mammals and predation rates are affected by forest types and altitudes [206,207].

Recenlty, Soldevila [169] and Stefanescu et al. [59] considered the role of natural enemies in their studies in Catalonia. The authors identified *A. sylvaticus* and *C. russula* as predators with the capacity to maintain populations of *L. dispar* at low densities. As the abundance of small rodents is linked to the abundance of acorns, they suggested that after several years of low abundance of the fruits, the populations of rodents would decrease, allowing an increase in *L. dispar* populations.

In earlier studies, birds had been proven to be one of the most important predators [194,208–211]. For instance, bird predation damaged 77% of the egg masses in Slovakia [208]. The predation rates of egg masses by birds in North America are between 65%

and 89% [98,212]. Soldevila [169] and Stefanescu et al. [59] also reported the potential role of *P. major*, *G. glandarius*, *D. kizuki*, and *S. eiuropaea*.

Calosoma sycophanta is a predatory beetle that is known to occur in high densities during outbreaks of *L. dispar* [56,195–199]. Adults and larvae of this species are the main predator of larvae and pupae of *L. dispar* [98,213,214]. It is distributed in the Western Palearctic, covering the whole of Europe, northwest Africa, and Western and Central Asia [200–203]. It was imported from Europe to the USA as a natural enemy of *L. dispar* in 1906 and is now a well-established predator [214–217]. It has been reported that a pair of *C. sycophanta* adults and their offspring can predate more than 6000 *L. dispar* larvae and pupae within one season [198].

Stefanescu et al. [59] and Soldevila [169] reported the occurrence of *C. sycophanta* in Catalonia, confirming its capacity of this prey upon the larvae and pupae, as also reported by Weseloh [196]. *Calosoma sycophanta* was able to predate a large number of *L. dispar* when the density was high, but its role was much less important when *L. dispar* density was low. The above-named studies also stated that the parasitoids of the genera *Anastasus*, *Cotesia*, *Exorista*, and *Brachymeria* were present in the study area. However, vertebrates are more likely to cause high mortality than invertebrates [211]. In Slovakia, for instance, invertebrates caused 38% of egg mass predation, while vertebrates caused 62% [211]. In general, it has been suggested that predation by small mammals is able to keep *L. dispar* populations at low densities [205,206]. However, outbreaks occur at intervals, presumably due to human transfer of life stages of *L. dispar* from an infested to an uninfested area [132].

In the USA, since the early 20th century, numerous attempts have been carried out to eradicate or suppress the dispersal of *L. dispar*, but none have prevented the continued spread of the pest to the south and west of the country. Among the attempts was the development of a biological control program. The guiding philosophy for biological control in North America was to establish there all the natural enemies that attack *L. dispar* in Europe [218]. Thus, such a program mainly involved the introduction, rearing, and release of *L. dispar* enemies from Europe [198,218]. At least 20 parasitoids and predators have become established, but they have still been unable to control the moth populations [127]. By 1933, nine natural enemies of *L. dispar* had become established in North America. One of these was *C. sycophanta*, as well as some small hymenopterous egg parasitoids and some tachinid flies that attack the large larvae of *L. dispar* [219]. In addition to the other natural enemies in Europe, there are a large number of parasitoids that are able to attack *L. dispar*; of these, 109 species belong to the order Hymenoptera and 56 species belong to the order Diptera [63]. Despite this diversity of parasitoids, efforts have been conducted to establish parasitoids in the invasive range of *L. dispar*, but only a few parasitoid species can be considered as established there [50,83,220,221].

4.1.2. Bioinsecticides or Pathogens

Among the control methods developed against *L. dispar* [222], the application of bioinsecticides has proven to be a successful approach for the control of this pest, with a low environmental impact [36,223]. Aerial applications of formulations based on the entomopathogenic bacterium *Bacillus thuringiensis* Berliner serovar *kurstaki* (*Btk*) exploit the highly specific mode of action of bacterial toxins that selectively target moth larvae [224]. This method is currently the most frequently used and effective to suppress *L. dispar* outbreaks since it has few biological and practical limitations [225]. However, there is some evidence of side effects on non-target Lepidoptera that live in the forest ecosystem [226].

Several factors should be taken into account to increase the effectiveness of *Btk* applications against *L. dispar* [227], including the timing of application relative to the phenology of second instar larvae [228,229], dose [180,230], the size of droplet, and density on foliage of insecticides based on *Btk* [231].

Stefanescu et al. [59] evaluated the defoliation that occurred in plots of a damaged area in 2020 on trees that had been already attacked or not in 2019. The cumulated defoliation that occurred in 2020 (2019 + 2020 damage) was estimated at 20% to 60%. As a measure to reduce damage, the Plant Health Services of the Department of Agriculture of the Catalonia Government planned a program based on the treatment of the most affected forests with *Btk*, and around 2500 ha were sprayed in order to maintain the productivity of the forest, especially areas of *Q. suber* exploited to obtain cork. The cost of spraying was estimated at 50 EUR/ha. The efficacy of such treatments was low and did not have any significant effect on the reduction of defoliation [59], and the use of *Btk* spraying for controlling damage of *L. dispar* was seriously questioned. Several reasons were given: (1) the development of the larval population in one locality is not synchronous and there are several larval instars the at the same time; (2) the negative effect that the treatment may have on the non-target species: although *Btk* is a selective bioinsecticide widely used in integrated pest management, it negatively impacts lepidopteran larvae and some predator species such as the green lacewing *Chrysoperla carnea* (Stephens) (Neuroptera: Chrysopidae) [59]. Indirectly, the reduction of lepidopteran larvae in sprayed *Btk* areas can lead to a reduction in the natural enemies that feed on them. This last aspect of loss of biodiversity is more relevant in protected areas such as the Montnegre of Catalonia. Although the study of Stefanescu et al. [59] did not measure and compare the productivity of sprayed and unsprayed forest, they argue that *Q. suber* forest has a notable resilience to *L. dispar* outbreaks.

During the outbreak of the 1990s in central Europe, control measures against *L. dispar* in France (1995 epidemic) were carried out in an area of 2500 ha using *Btk*. In Slovakia and the Czech Republic in the 1993–1994 outbreaks in oak forests, treatments with *Btk* were necessary. In Germany, in the outbreaks of 1993 and in 1994, *Btk* was applied by helicopter to infested areas. However, mainly due to adverse weather conditions, this treatment failed to provide sufficient protection on the affected forest stands [118]. More recently in Greece, the field treatments with *Btk* against *L. dispar* showed that this biopesticide caused 66% mortality to the second instar larvae after 4–5 days [36].

Lymantria dispar has been treated with other biocontrol agents such as entomopathogenic fungi and entomopathogenic nematodes [36,38–43]. The entomopathogenic fungus *Entomophaga maimaiga* Humber, Shimazu and Soper 1988 (Entomophthoromycota: Entomophthorales) was introduced to the United States from Japan for the biological control of *L. dispar* in the early 1900s; however, its introduction was unsuccessful [232]. Much later, in 1989, there was an unexpectedly high percentage of dead larvae due to this fungus [233,234], probably because it had been accidentally introduced to North America after 1971 [235]. In late 1990s, *E. maimaiga* was introduced to Bulgaria from North America [236]. Since then, it rapidly spread across Europe, indicating that this continent is suitable for the survival and development of the fungus [237]. Field studies in Europe documented that *L. dispar* larvae exhibited 98% mortality when infected by *E. maimaiga* [238–240], suggesting the high potential of this fungus to become an important management tool [237]. The entomopathogenic nematode *Steinernema carpocapsae* (Weiser, 1955) (Rhabditida: Steinernematidae), which was used for the first time in Greece in field trials against this pest, killed 69% of the second instar larvae, indicating that it has a great potential as a control agent [36].

Another group of entomopathogens is the baculoviruses that are very specific microorganisms [241] that cause fatal infections to larvae after the ingestion of viral particles. The bioinsecticidal activity is associated with crystalline occlusion bodies that, after their ingestion by susceptible insects, release occlusion-derived viruses (ODVs) that infect the host midgut epithelial cells. The production of a second type of virions, namely, budded viruses (BVs), responsible for the infection spread in the host body [242]. The *Lymantria dispar* multiple nucleopolyhedrovirus (LdMNPV) is very specific to this pest and is therefore environmentally safe. It was apparently introduced accidentally with *L. dispar* in North America [243]. This virus naturally causes epizootics, resulting in the decline of the pest population after two or three years of outbreaks [244]. For instance, Kurenshchikov et al. [171] reported that LdMNPV suspended the population of *L. dispar* after the outbreak in Khabarovsk Krai (Far East region of Russia) in 2006–2008. However, it can be produced in large quantities only by infecting live *L. dispar* larvae. Thus, it is produced in limited quantities, sufficient only to treat <550 km^2 per year [125]. At first, virus production was costly since efficient host larval breeding techniques were limited [245]. However, research efforts have revealed that virus yields can be increased by using late-instar (i.e., fifth instar) female larvae where, in combination with new larval processing methods (i.e., using freeze-dried, intact larvae), the cost of the production of the virus can be reduced effectively [245]. The USDA Animal and Plant Health Inspection Service and the Forest Service have investigated practical ways to use the virus as biological insecticide. Thus, it was registered with the Environmental Protection Agency in 1978 as Gypchek [243]. Lewis and Yendol [246] reported that Gypchek adequately controlled *L. dispar* in forest trials. In addition, in North America, the virus is considered to be the most significant factor causing the collapse of *L. dispar* populations [98]. More recently, Ruiu et al. [247] suggested a multi-year integrated program that would include the combined use of *Btk* to reduce infestations and nucleopolyhedrovirus in order to regulate population dynamics. The disease is often referred to as "wilt" because of the soft, limp appearance of diseased larvae [11].

Lymantria dispar can host several microsporidian pathogens, some of which have been considered for inoculative releases in North America [248]. *Vairimorpha disparis* (Microsporidia, Burellenidae) and *Nosema lymantriae* (Microsporidia, Nosematidae) infect *L. dispar* larvae when the spores are ingested on food. The developmental cycle begins in the midgut and eventually leads to the formation of primary spores [217]. *Vairimorpha disparis* is a parasite of the fat body. The environmental spores are transmitted between host larvae and can be found in the fat body after 7 days. After 10 days, the fat body is full of spores, and under laboratory conditions, an infected larva dies on average 4 weeks after infection [249]. Horizontal transmission takes place when spores are released from the decomposing cadaver. *Nosema lymantriae* is a systemic parasite that affects the silk glands, the fat body, the gonads, and the Malpighian tubules of a host larva. Most of the infected larvae will die 4 weeks after infection. Horizontal transmission of this species begins after the end of the latent period when the spores are released with feces about two weeks after infection, and also continues after the death of the host larva when its cadaver decomposes [249]. Furthermore, infected females can transmit *Nosema lymantriae* transovarially to their offspring [250].

In Central Europe, it has been reported that pathogens can cause higher mortality than parasitoids [208,251,252]. Since entomopathogens can act against *L. dispar*, Stephanescu et al. [59] reported that it is probable that the baculovirus LdMNPV is present in Catalonia, although there are no records of this. In conclusion, it seems that natural enemies play a role in maintaining *L. dispar* populations under outbreak thresholds or in reducing population levels after outbreaks.

4.2. Traps and Attractants

The development of the sex pheromone *cis*-7,8-epoxy-2-methyloctadecane (disparlure) that attracts *L. dispar* males took place in the 1970s [253,254]. The existence of the pheromone was known in the early 1900s, and the tips of the female abdomen were extracted into organic solvents to provide pheromone that could be used in trapping research as early as 1932 [255]. This pheromone has been used extensively for the detection of this pest [134] and is able to attract all of the different subspecies of *L. dispar* [12]. The commercially available pheromone disparlure provides an extremely sensitive and host-specific research tool that is effective even when moths occur at very low population densities [6,256–258]. Without such a sensitive trapping tool, it would be extremely difficult to engage in management efforts that depend on early detection [125]. The detection of low-density populations of this pest is essential for eradication campaigns. Moreover, pheromone-baited traps are effective at both low and high densities, and this method is considered the most sensitive method for detecting *L. dispar* [15].

The sexual behavior of males [244,259], as well as the physical characteristics of a trap, such as the shape, size, color, and the location of entry ports, must be taken into account when designing a trap [6]. There is a wide variety of pheromone trap designs from disposable cardboard traps with a sticky inner surface to various canister-type or "milk carton" traps made of cardboard or reusable plastic that contain an insecticide to kill the captured moths [11]. Strategies to slow the spread of *L. dispar* are based on pheromone traps for detecting and delimiting small, isolated populations immediately after establishment [193,260]. These infestations can then be targeted for mass trapping, mating disruption, application of microbial insecticides, or other appropriate methods to limit the population growth and spread of the pest [261–263].

Mating disruption (MD) has long been regarded as a viable option for non-chemical control, especially in the case of Lepidoptera [264]. For MD, artificial male pheromone is released in a treated area in order to compete with pheromone produced by calling females, and this can drastically reduce mate finding by males and thus mating [265]. Currently, this method is used to control a large number of insect pest species, mainly lepidopteran species, but not exclusively so [264–272]. MD has been successfully marketed for a very wide range of moth species of economic importance [273–275].

With this method, mating may be delayed, which contributes to reduced fecundity [276,277] and egg fertilization [277]. Moreover, MD can reduce the risk of insecticide resistance because mating among resistant individuals is decreased [278–280]. Since the target organism is not killed by this method, non-targeted effects are believed to be rare [179]. It is also compatible with other pest management strategies, such as biological control [261]. The discovery of disparlure also helped to develop and optimize host-specific MD tactics that are also considered a very useful control tool [193,254,269].

MD has been proven to be effective against low-density populations of *L. dispar* in the USA [179,281–284]. On the other hand, it may not work in high-density populations where female moths are abundant and easy for males to locate, making it ineffective without the combined use of other control measures. As a univoltine species, *L. dispar* is ideal for MD, as a single annual application may lead to rapid population suppression, without the need to reapply the method; however, the timing of the application is critical, i.e., before males mate with females. Although *L. dispar* is a serious threat for both the environment and human health, the MD method has never been evaluated on a large scale in Europe as a reliable management tool.

An alternative control method is the use of trunk barriers, which can take advantage of the dispersal behavior of the larvae [80]. The use of burlap bands wrapped around the trunk is an old technique that helped to collect and destroy the larvae [52]. In the early 1900s, the protection of trees from *L. dispar* in Massachusetts was based on the extensive use of sticky barrier bands [285]. Later, burlap bands and similar devices were used to detect and monitor *L. dispar* populations [84,286,287]. For instance, burlap bands that were placed around the base of the trunk have been used to trap and monitor late-instar larvae; the larvae use the burlap bands as a preferred resting site during the day [288]. Moreover, the first-instar larvae that move upward on the tree stem after hatching can be trapped in the sticky barrier bands [289,290].

4.3. Chemical and Biorational Control

There are only a few insecticides registered for *L. dispar* control in Europe, such as insect growth regulators (IGRs), which are biorational products, and Spinosad, a naturally derived product [36,37]. Other insecticides may be used to effectively control *L. dispar*, but since their potential harm to other species is considered too high, they are not recommended for widespread use. In the past, the application of pesticides to prevent *L. dispar* attack has been shown to have a greater effect on the bird populations than on *L. dispar* themselves [291]. Moreover, since this species may occur in urban/recreational areas, chemical control is even more challenging as many of the registered pesticides cannot be used due to risks of exposure to mammals. Prior to 1966, area-wide eradication programs by spraying with dichloro diphenyl trichloro ethane (DDT) were considered effective. However, DDT and similar insecticides were eventually banned due to environmental concerns about their toxic effects becoming recognized, leading to research in finding alternatives to broad-spectrum synthetic pesticides [292].

In North America, massive aerial applications of pesticides have been made against *L. dispar* populations, aiming largely at suppressing their outbreaks [192]. For instance, the application of chemical pesticides (e.g., DDT) to millions of acres of forest in the United States took place annually during the 1940s and 1950s [293]. In the United States, the chemical insecticide diflubenzuron, which is a molting disruptor, was widely used during the 1980s [182] and in some European forests until recently [294]. For example, in Slovakia and the Czech Republic in the 1993–1994 outbreaks, the application of diflubenzuron in oak forests was used to control this pest. In Germany, the same treatment was applied to three-quarters of the infested area, which was proven to be more effective than treatment with *Btk* [118]. Although this insecticide is not toxic to vertebrates, concerns have been raised about its adverse effects on invertebrates [295] and the potential effects of the 4-chloroalinine metabolite on human health [296]. Later, tebufenozide, which is a molting hormone agonist, was approved as an alternative to diflubenzuron due to its specific action on Lepidoptera [178]. This compound now plays an important role in *L. dispar* management in the United States [89]. It is also the preferred option for some European countries (e.g., Germany) because of its reliable suppression of *L. dispar* populations [297].

In a recent study conducted in northern Greece, Papadopoulou et al. [36] reported that a product containing 24% metaflumizone had a rapid action (1–2 days) against second-instar larvae of *L. dispar*, causing 88% mortality, while an IGR containing 15% teflubenzuron was less effective, killing 76% of second instar larvae. In Spain, the authorized insecticides comprise azadiractin, cypermetrin, and indoxacarb [66]. Spraying with aircraft or some other flight system can only be performed by government Plant Health Services. The semi-synthetic bioinsecticide emamectin benzoate (EMB), which is derived from naturally occurring avermectin, is a widely used control agent for agricultural and forestry pests [298,299]. The determination of the sublethal concentrations of EMB and their effects on *L. dispar* was conducted by Xu et al. [300]. These authors found that sublethal concentrations of EMB can inhibit the growth of larvae, which is related to midgut damage, digestive dysfunction, and nutritional metabolism disorders. They also provided a theoretical basis for understanding the sublethal effect of EMB and its application to the prevention and control of *L. dispar*. According to Cannon et al. [166], an eradication program was applied in the UK, including three axes: (1) investigation for detecting egg clusters or larvae, and captures of adults using pheromone traps; (2) chemical treatments of larvae; and (3) extensive information of the public about the importance of the pest. This procedure resulted in a decrease in the captures of adult males in comparison with the captures of the first two years of the program.

In the past decade, increasing attention has been paid to the development of insecticides based on unmodified nucleic acid fragments, with an emphasis on antisense DNA fragments [301,302] and double-stranded RNA fragments [303,304]. These insecticides are considered to be the next-generation control agents that have numerous advantages over broad-spectrum products. They can combine the affordability and rapid action of chemical insecticides combined with the selectivity of biological preparations. The nucleic acid synthesis technologies are becoming less expensive, making DNA insecticides and RNA preparations increasingly economical, and their affordability can be compared to that of chemical insecticides. So far, antisense DNA-based insecticides are the only nucleic acid preparations being developed to control *L. dispar* [305].

Recently, Oberemok et al. [306] suggested a novel biotechnology for pest control, using a DNA insecticide that has improved insecticidal action based on a new antisense oligoRIBO-11 sequence from the 5.8S ribosomal RNA gene. This novel insecticide caused high mortality among *L. dispar* larvae reared in the laboratory and those collected from the forest. Furthermore, the insecticidal potential of three different 10–12 nucleotide long antisense sequences from the 5.8S ribosomal RNA gene of *L. dispar* against its larvae was compared. The results showed that antisense fragments of 10 and 11 nucleotides (oligoRIBO-10 and oligoRIBO-11) caused higher larval mortality than the 12 nucleotide long fragment (oligoRIBO-12) [307]. In addition, this oligoRIBO-11 insecticide was more affordable and acted faster than the previous preparations developed by Oberemok et al. [307], according to longer antisense fragments of anti-apoptosis genes of the baculovirus–host system.

4.4. Essential Oils

Many insecticides have been removed from the market due to the development of resistance and the severe environmental disturbances (i.e., persistence of residues and adverse effects on non-target organisms) [308–313]. Intensive work is being done to find alternative environmentally friendly pest control methods that will be effective [314–319]. In this context, plant-based products have been considered as potential control agents for pests, including *L. dispar* [320,321]. Recently, plant essential oils (EOs), which are complex mixtures of compounds that act as a defense agents against pests and pathogens and provide protection to the plant from heat and cold have attracted attention [322,323]. EOs are also generally recognized as safe biopesticidal agents [324–328].

Several EOs can be promising "green" alternatives to chemical insecticides for the control of *L. dispar,* as they have been examined under laboratory conditions. For instance, Moretti et al. [329] found a high digestive toxicity of rosemary, *Rosmarinus officinalis* L., and thyme, *Thymus herba-barona* Loisel. (Lamiales: Lamiaceae), EO emulsions to second and third instar larvae of *L. dispar* after 3 days of exposure. In addition, the EOs from basil, *Ocimum basilicum* L. (Lamiales: Lamiaceae); *Athamanta haynaldii* L. (Apiales: Apiaceae); and nutmeg, *Myristica fragrans* Houtt. (Magnoliales: Myristicaceae) had low to moderate residual contact and digestive toxicity, but they had good antifeedant activity against second instar larvae [3,330].

More recently, Devrnja et al. [331] examined the effect of the EO of the tansy, *Tanacetum vulgare* L. (syn. *Chrysanthemum vulgare* L.) (Asterales: Asteraceae), at three concentrations (i.e., 0.1, 0.5, and 1% v/v), on the survival and molting of second-instar larvae, as well as on the nutritional indices of the fourth-instar larvae of *L. dispar*. Exposure of the second instar larvae to tansy EO (residual contact toxicity) caused low mortality (<10%), but larval development was significantly extended, i.e., the proportion of larvae that molted into the third instar was decreased after 120 h of exposure in comparison with the control larvae (92% molted into the third instar). Consequently, delayed ecdysis, which is associated with prolonged growth, could extend the period for which larvae are exposed to their natural enemies. However, in a digestive toxicity assay in which tansy EO was incorporated into the diet, the highest concentration of the EO (i.e., 1% v/v) caused high mortality and a lack of molting after 120 h of consumption. Oxygenated monoterpenes were the predominant group of compounds with 93.5% in tansy EO [331,332]. Terpenes can act as deterrents or toxicants to *L. dispar* larvae [333], indicating that compounds from EOs could be isolated and used for the control of this species. However, the sensitivity of exposed larvae depends on the terpene structures and larval age [331].

Kostić et al. [321] evaluated the impact of different concentrations of essential oils (EOs) from the seeds of three Apiaceae plants, namely, anise, *Pimpinella anisum* L.; dill, *Anethum graveolens* L.; and fennel, *Foeniculum vulgare* Mill. (Apiales: Apiaceae), on the behavior, mortality, molting, and nutritional physiology of *L. dispar* larvae. The authors also compared EO efficacy with the commercial insecticide NeemAzal®-T/S (neem). Their results showed that the tested EOs may be a promising strategy for *L. dispar* control since they provided strong negative effects on survival and consumption in second-instar larvae and impairment of nutritional physiology in fourth-instar larvae. Anise EO was proven to be the best antifeedant, while dill EO caused the highest mortality. In addition, at the concentration of 0.5%, the three EOs performed better than the commercial insecticide neem in reducing relative growth rate, efficiency of conversion of ingested food, approximate digestibility, and efficiency of conversion of digested food of the fourth instar larvae of *L. dispar*. Although essential oils represent an eco-friendly management tool, their cost is high.

5. Public Health Concerns

Envenomation (i.e., infusion of an insect secretion into a human's body) caused by moth or butterfly larvae in humans is known as "erucism" or "caterpillar dermatitis", coming from the Latin "eruca", which means caterpillar [334]. This kind of dermatitis has been noted since ancient Greek times [88,113,335]. Caterpillar envenomation constitutes a serious public health issue of international importance [336].

The setae of the caterpillars, which are filled with toxins (such as proteolytic enzymes, histamine, and other pro-inflammatory substances), are responsible for the allergic reactions they cause to humans or to animals [337–340]. The bristles are able to penetrate the subcutaneous tissue and release toxins [336,341], causing cutaneous reactions, such as immediate severe pain, erythema, and edema [336]. Of the three types of urticating hairs that exist in Lepidoptera, *L. dispar* larvae have modified setae [342]. The presence of histamine has been reported in the setae of *L. dispar* larvae [343–347]. In addition, histamine or histamine analogues have been isolated in other lepidopteran species of medical importance found in Europe, such as the pine-tree lappet moth, *Dendrolimus pini* (L.) (Lepidoptera, Lasiocampidae), and the browntail moth, *Euproctis chrysorrhoea* L. (Lepidoptera: Lymantriidae) [348]. Moreover, first instar larvae of *L. dispar* carry setae that are impregnated with nicotine, while this substance occurs in a lower concentration in last instar larvae [342,349].

It has long been recognized that *L. dispar* can cause dermatitis in humans [350], especially in laboratory staff [351]. In addition, the early instar larvae are considered more allergenic than mature larvae [130,352]. Several studies have shown that a large proportion of workers in breeding facilities of this species for experimental purposes developed pruritic urticaria on exposed skin after contact with its larvae [353,354]. Other symptoms associated with allergic reactions to larvae are eye irritation, rhinitis, and shortness of breath [354,355].

Very little information is available on the prevalence of cutaneous reactions focusing on children and in the general adult population. The first community-wide outbreak of *L. dispar* dermatitis was reported in the United States in 1981 [343]. That year, a massive outbreak of larvae was recorded in northeastern United States during spring [356], where people outdoors came in to contact with first instar larvae [352]. Thousands of people presented skin irritation, which was described as unusual pruritic dermatitis with stinging, while some people had respiratory difficulties [343,345]. Moreover, the skin lesions of the patients' occurred within 12 h of contact with *L. dispar* larvae [343,345,352,357]. After the outbreak of 1981, there were no other reports of an allergic reaction to this pest [357] until the spring of 1990, when six new cases with clinical and histopathological features were described [88]. In addition to skin irritation, respiratory problems (e.g., rhinitis or shortness of breath and eye irritation) were reported [88,358].

Although larvae cause dermatitis with a pruritic eruption that lasts from 4 to 7 days, very few clinical and epidemiological studies have been conducted in this species [348,352,359]. Moreover, the etiology of *L. dispar* erucism and lepidopterism has not been fully clarified [115, 357]. An epidemiologic study was carried out during the outbreak of *L. dispar* in 1981 that compared a severely infested and a lesser infested area [353]. The authors found that the highest risk factors for developing *L. dispar* dermatitis were previous history of hay fever, a history of a similar rash a year ago and direct physical contact or indirect exposure (e.g., hanging laundry outdoors) [352,353,355]. Furthermore, Beaucher and Farnham [345] conducted patch testing using the hairs of *L. dispar* larvae in 8 patients with a history of *L. dispar* dermatitis as well as in 11 persons without history that were used as controls. Generally, patch testing with moth or larvae setae revealed the presence of an immediate hypersensitivity, delayed-type hypersensitivity, or both [360]. In all patients with a history of dermatitis, patch testing caused delayed papulovesicular reactions, while only 1 out of 11 controls reacted, indicating a delayed-type hypersensitivity response [345]. Consequently, the mechanism of *L. dispar* erucism and lepidopterism probably involves local and pulmonary histamine release and delayed-type hypersensitivity reactions in susceptible persons [115,358]. In cases of *L.*

dispar dermatitis, patients receive topical and parenteral antihistamines, oral or parenteral corticosteroids, and bronchodilators (as indicated for bronchospasm) [115].

6. Conclusions and Future Perspectives

Lymantria dispar is a pest of economic importance that can seriously disrupt forest ecosystems worldwide [361]. Considering the elevated environmental and economic impact of this species, a holistic approach is required rather than ad hoc interventions. In support of this, the European Directive 2009/128/CE of the European Parliament and of the Council of the European Union have established a framework for action so as to achieve sustainable use of pesticides, i.e., reducing their risks and impacts on human health and the environment. This framework promotes the use of integrated pest management (IPM) and the use of alternative non-chemical approaches or techniques [362]. Risks from exposure to chemicals in public parks, gardens, sports and recreational areas, school grounds, children's playgrounds, or areas close to healthcare facilities are high and should therefore be minimized or prohibited. New approaches regarding monitoring and predicting outbreaks and controlling high-density populations should be introduced in forest and urban habitats. For example, novel trap types should be designed and tested extensively for the capture of larvae and adults. These trap devices can be incorporated in trap monitoring systems that will optimize trap captures at the very early beginning of the presence of the insect. Thus, effective management approaches should be applied accurately, such as the mating disruption method. For this purpose, the European Commission has initiated financial support to universities, research institutes, and private companies to work simultaneously in large-scale field experiments towards the development effective management methods against this serious pest.

Author Contributions: Conceptualization, M.C.B., N.G.K. and C.G.A.; validation, M.C.B., N.G.K. and C.G.A.; investigation, M.C.B., N.G.K., A.S., X.P., C.L.A., M.E., T.B., S.T., E.B.F., E.D.S., S.F., P.A. and C.G.A.; data curation, M.C.B., N.G.K. and A.S.; writing—original draft preparation, M.C.B. and N.G.K.; writing—review and editing, M.C.B., N.G.K., A.S., X.P., C.L.A., M.E., T.B., S.T., E.B.F., E.D.S., S.F., P.A. and C.G.A.; visualization, M.C.B., N.G.K. and C.G.A.; supervision, N.G.K.; project administration, M.C.B. and N.G.K.; funding acquisition, M.C.B., N.G.K. and A.S. All authors have read and agreed to the published version of the manuscript.

Funding: This study was supported by the EU grant "Using smart traps and pheromones to control the gypsy moth: ecofriendly control in practice" (eGYMER: LIFE20 ENV/GR/000801).

Data Availability Statement: Data are contained within the article.

Conflicts of Interest: The authors declare no conflict of interest.

References

1. De Beurs, K.M.; Townsend, P.A. Estimating the effect of gypsy moth defoliation using MODIS. *Remote Sens. Environ.* **2008**, *112*, 3983–3990. [CrossRef]
2. Mull, A.; Spears, L.R. *Spongy Moth (Lymantria dispar dispar Linnaeus)*; Utah State University Extension and Utah Plant Pest Diagnostic Laboratory: Logan, UT, USA, 2022.
3. Kostić, M.; Popović, Z.; Brkić, D.; Milanović, S. Larvicidal and antifeedant activity of some plant-derived compounds to *Lymantria dispar* L. (Lepidoptera, Lymantriidae). *Bioresour. Technol.* **2008**, *99*, 7897–7901. [CrossRef] [PubMed]
4. Milanović, S.; Lazarević, J.; Popović, Z.; Miletić, Z.; Kostić, M.; Radulović, Z.; Karadžić, D.; Vuleta, A. Preference and performance of the larvae of *Lymantria dispar* (Lepidoptera: Lymantriidae) on three species of European oaks. *Eur. J. Entomol.* **2014**, *111*, 371–378. [CrossRef]
5. Cao, C.; Sun, L.; Wen, R.; Shang, Q.; Ma, L.; Wang, Z. Characterization of the transcriptome of the Asian gypsy moth *Lymantria dispar* identifies numerous transcripts associated with insecticide resistance. *Pestic. Biochem. Phys.* **2015**, *119*, 54–61. [CrossRef]
6. Doane, C.C.; McManus, M.L. *The Gypsy Moth: Research toward Integrated Pest Management (No. 1584)*; US Department of Agriculture: Washington, DC, USA, 1981.
7. Mihajlović, L.J.; Grbić, C.; Vandić, D. The latest outbreaks of gypsy moth, *Lymantria dispar* L., in the region of Serbia in the period 1995–1998. The gypsy moth outbreaks in Serbia. *Acta Entomol. Serb.* **1998**, 80–88.
8. Lowe, S.; Brone, M.; Boudjelas, S.; De Poorter, M. *100 of the World's Worst Invasive Alien Species. A Selection from the Global Invasive Species Database*; Hollands Printing Ltd.: Auckland, New Zealand, 2000.

9. EPPO (European and Mediterranean Plant Protection Organization). EPPO Global Data Base. *Lymantria dispar*. Available online: https://gd.eppo.int/taxon/LYMADI/categorization (accessed on 6 September 2022).
10. Fabel, S. Effects of *Lymantria dispar*, the Gypsy moth, on broadleaved forests in eastern North America. *Restor. Reclam. Rev.* **2000**, *6*, 1–15.
11. Nealis, V.G.; Erb, S. *A Sourcebook for Management of the Gypsy Moth*; Canadian Forestry Service, Great Lakes Forestry Centre: Sault Ste. Marie, ON, Canada, 1993; p. 47.
12. Pogue, M.; Schaefer, P.W. *A Review of Selected Species of Lymantria Hübner (1819) (Lepidoptera: Noctuidae: Lymantriinae) from Subtropical and Temperate Regions of Asia, Including the Descriptions of Three New Species, Some Potentially Invasive to North America*; Forest Health Technology Enterprise Team: Washington, DC, USA, 2007.
13. Humble, L.; Stewart, A.J. Gypsy Moth. Natural Resources. Available online: http://cfs.nrcan.gc.ca/pubwarehouse/pdfs/3456.pdf (accessed on 6 September 2022).
14. Keena, M.A. Identification of Gypsy Moth Larval Color Forms. NE/NA-INF-123-94. Available online: http://www.forestpests.org/gypsymth (accessed on 6 September 2022).
15. Hajek, A.E.; Tobin, P.C. North American eradications of Asian and European gypsy moth. In *Use of Microbes for Control and Eradication of Invasive Arthropods*; Hajek, A.E., Glare, T.R., O'Callaghan, M., Eds.; Springer: New York, NY, USA, 2009; pp. 71–89.
16. Metzger, M.J.; Bunce, R.G.H.; Jongman, R.H.G.; Mucher, C.A.; Watkins, J.W. A climatic stratification of the environment of Europe. *Glob. Ecol. Biogeogr.* **2005**, *14*, 549–563. [CrossRef]
17. Liebhold, A.M.; Gottschalk, K.W.; Muzika, R.M.; Montgomery, M.E.; Young, R.; O'Day, K.; Kelley, B. Suitability of North American tree species to the gypsy moth: A summary of field and laboratory tests. In *U.S. Department of Agriculture Forest Service NE Forest Experimental Station General Technical Bulletin, NE-211*; U.S. Department of Agriculture: Washington, DC, USA, 1995.
18. Tobin, P.C.; Liebhold, A.M. "Gypsy moth". In *Encyclopedia of Biological Invasions*; Simberloff, D., Rejmanek, M., Eds.; University of California Press: Berkeley, CA, USA, 2011; pp. 298–304.
19. Kozhanchikov, I.V. Gypsy moth. In *Fauna Sssr. Nasekomye Cheshuekrylyye. Volnyanki Orgyidae*; Izd.AN SSSR: Moscow/Leningrad, Russia, 1950; Volume 12, p. 582.
20. Kim, C.H.; Nam, S.H.; Lee, S.M. *Insecta. (vlll)*; Ministry of Education: Seoul, Korea, 1982.
21. Schaefer, P.W.; Weseloh, R.M.; Sun, X.L.; Wallner, W.E.; Yan, J.J. Gypsy-moth, *Lymantria* (=Ocneria) *dispar* (L.) (Lepidoptera: Lymantriidae), in the People's Republic of China. *Environ. Entomol.* **1984**, *13*, 1535–1541. [CrossRef]
22. Schaefer, P.W.; Ikebe, K.; Higashiura, Y. Gypsy moth, *Lymantria dispar* (L.) and its natural enemies in the Far East (especially Japan). In *Annotated Bibliography and Guide to the Literature through 1986 and Host Plant List for Japan*; University of Delaware, Agricultural Experiment Station: Newark, DE, USA, 1988; p. 160.
23. Baranchikov, Y.N. Ecological basis of the evolution of host relationships in Eurasian gypsy moth populations. In *Proceedings, Lymantriidae: A Comparison of Features of New and Old World Tussock Moths. GTR-NE-123*; U.S. Department of Agriculture, Forest Service, Northeastern Forest Experiment Station: Broomall, PA, USA, 1989; pp. 319–338.
24. Baranchikov, Y.N.; Montgomery, M.E. *Tree Suitability for Asian, European and American Populations of Gypsy Moth*; General Technical Reports NE-188; USDA Forest Service: Washington, DC, USA, 1994; p. 4.
25. Gninenko, Y.I.; Orlinskii, A.D. Outbreaks of *Lymantria dispar* in Russian forests during the 1990s. *EPPO Bull.* **2003**, *33*, 325–329. [CrossRef]
26. Johns, R.C.; Tobita, H.; Hara, H.; Ozaki, K. Adaptive advantages of dietary mixing different-aged foliage within conifers for a generalist defoliator. *Ecol. Res.* **2015**, *30*, 793–802. [CrossRef]
27. Keena, M.A.; Côté, M.J.; Grinberg, P.S.; Wallner, W.E. World distribution of female flight and genetic variation in *Lymantria dispar* (Lepidoptera: Lymantriidae). *Environ. Entomol.* **2008**, *37*, 636–649. [CrossRef]
28. Yang, F.; Luo, Y.; Huang, D.; Cui, X.; Yang, H.; Liu, X.; Shi, J. A preliminary study on flight ability among Chinese populations of Asian gypsy moth, *Lymantria dispar*. *Chin. Agr. Sci. Bull.* **2012**, *28*, 53–57.
29. Bigsby, K.M.; Tobin, P.C.; Sills, E.O. Anthropogenic drivers of gypsy moth spread. *Biol. Invasions* **2011**, *13*, 2077. [CrossRef]
30. Walsh, P.J. Asian gypsy moth: The risk to New Zealand. *N. Z. For.* **1993**, *38*, 41–43.
31. Schaefer, P.W.; Strothkamp, K.G. Mass flights of *Lymantria dispar* japonica and *Lymantria mathura* (Erebidae: Lymantriinae) to commercial lighting, with notes on female viability and fecundity. *J. Lepid. Soc.* **2014**, *68*, 124–129.
32. Djoumad, A.; Nisole, A.; Zahiri, R.; Freschi, L.; Picq, S.; Gundersen-Rindal, D.E.; Sparks, M.E.; Dewar, K.; Stewart, D.; Maaroufi, H.; et al. Comparative analysis of mitochondrial genomes of geographic variants of the gypsy moth, *Lymantria dispar*, reveals a previously undescribed genotypic entity. *Sci. Rep.* **2017**, *7*, 14245. [CrossRef]
33. Johnson, D.M.; Liebhold, A.M.; Bjorsnstad, O.N. Geographical variation in periodicity of gypsy moth outbreaks. *Ecography* **2006**, *29*, 367–374. [CrossRef]
34. Lindner, M.; Maroschek, M.; Netherer, S.; Kremer, A.; Barbati, A.; Garcia-Gonzalo, J.; Seidl, R.; Delzon, S.; Corona, P.; Kolström, M.; et al. Climate change impacts, adaptive capacity, and vulnerability of European forest ecosystems. *For. Ecol. Manag.* **2010**, *259*, 698–709. [CrossRef]
35. Vanhanen, H.; Veteli, T.O.; Päivinen, S.; Kellomäki, S.; Niemela, P. Climate change and range shifts in two insect defoliators: Gypsy moth and nun moth—A model study. *Silva Fenn.* **2007**, *41*, 621–638. [CrossRef]

36. Papadopoulou, S.; Chryssochoides, C.; Avtzis, D. Genetic Diversity of *Lymantria dispar* Linnaeus (Lepidoptera: Lymantriidae) in Northern Greece and Evaluation of the Effectiveness of Novel Insecticides. *Biotechnol. Biotechnol. Equip.* **2012**, *26*, 2976–2980. [CrossRef]
37. Wanner, K.W.; Helson, B.V.; Harris, B.J. Laboratory evaluation of two novel strategies to control first-instar gypsy moth larvae with spinosad applied to tree trunks. *Pest Manag. Sci.* **2002**, *58*, 817–824. [CrossRef] [PubMed]
38. Dubois, N.R.; Dean, D.H. Synergism between CryIA insecticidal crystal proteins and spores of *Bacillus thuringiensis*, other bacterial spores, and vegetative cells against *Lymantria dispar* (Lepidoptera: Lymantriidae) larvae. *Environ. Entomol.* **1995**, *24*, 1741–1747. [CrossRef]
39. Georgiev, G.; Hubenov, Z.; Georgieva, M.; Mirchev, P.; Matova, M.; Solter, L.F.; Pilarska, D.; Pilarski, P. Interactions between the introduced fungal pathogen *Entomophaga maimaiga* and indigenous tachinid parasitoids of gypsy moth *Lymantria dispar* in Bulgaria. *Phytoparasitica* **2013**, *41*, 125–131. [CrossRef]
40. Georgieva, M.; Georgiev, G.; Pilarska, D.; Pilarski, P.; Mirchev, P.; Papazova-Anakieva, I.; Matova, M. First record of *Entomophaga maimaiga* (Entomophthorales: Entomophthoraceae) in *Lymantria dispar* populations in Greece and the former yugoslavian republic of Macedonia. *Ŝumarski List* **2013**, *137*, 307–311.
41. Papadopoulou, S.; Chryssochoides, C.; Katanos, J. Control of *Lymantria dispar* L. for eliminating the risk of forage production loss for small ruminants. In *Nutritional and Foraging Ecology of Sheep and Goats*; Papachristou, T.G., Parissi, Z.M., Ben Salem, H., Morand-Fehr, P., Eds.; CIHEAM: Zaragoza, Spain; FAO: Rome, Italy; NAGREF: Thermi, Greece, 2009; pp. 197–199.
42. Pilarska, D.; McManus, M.; Pilarski, P.; Georgiev, G.; Mirchev, P.; Linde, A. Monitoring the establishment and prevalence of the fungal entomopathogen *Entomophaga maimaiga* in two *Lymantria dispar* L. populations in Bulgaria. *J. Pest Sci.* **2006**, *79*, 63–67. [CrossRef]
43. Shapiro, M.; McLane, W.; Belli, R. Laboratory evaluation of selected chemicals as antidesiccants for the protection of the entomogenous nematode, *Steinernema feltiae* (Rhabditidae: Steinernematidae), against *Lymantria dispar* (Lepidoptera: Lymantriidae). *J. Econ. Entomol.* **1985**, *78*, 1437–1441. [CrossRef]
44. Žikić, V.; Stanković, S.S.; Kavallieratos, N.G.; Athanassiou, C.; Grorgiou, P.; Tschorsnig, H.P.; Achterberg, C.V. Parasitoids associated with *Lymantria dispar* (Lepidoptera: Erebidae) and *Malacosoma neustria* (Lepidoptera: Lasiocampidae) in Greece and comparative analysis of their parasitoid spectrum in Europe. *Zool. Anz.* **2017**, *270*, 166–175. [CrossRef]
45. Zhang, J.; Lapointe, R.; Thumbi, D.; Morin; Lucarotti, C.J. Molecular comparisons of alphabaculovirus-based products: Gypchek with Disparvirus (*Lymantria dispar*) and TM BioControl-1 with Virtuss (*Orgyia pseudotsugata*). *Can. Entomol.* **2010**, *142*, 546–556. [CrossRef]
46. Harrison, R.L.; Rowley, D.L. Complete genome sequence of the strain of *Lymantria dispar* multiple nucleopolyhedrovirus found in the gypsy moth biopesticide Virin-ENSh. *Genome Announc.* **2015**, *3*, e01407-14. [CrossRef]
47. Hajek, A.E.; van Frankenhuyzen, K. Use of entomopathogens against forest pests. In *Microbial Control of Insect and Mite Pests: From Theory to Practice*; Lacey, L.A., Ed.; Elsevier: Cambridge, MA, USA, 2017; pp. 313–330.
48. Akhanaev, Y.; Pavlushin, S.; Polenogova, O.; Klementeva, T.; Lebedeva, D.; Okhlopkova, O.; Kolosov, A.; Martemyanov, V. The effect of mixtures of Bacillus thuringiensis-based insecticide and multiple nucleopolyhedrovirus of *Lymantria dispar* L. in combination with an optical brightener on *L. dispar* larvae. *BioControl* **2022**, *67*, 331–343. [CrossRef]
49. Greathead, D. Parasitoids in classical biological control. In *Insect Parasitoids, Proceedings of the 13th Symposium of the Royal Entomological Society of London, Department of Physics Lecture Theatre, Imperial College, London, UK, 18–19 September 1985*; Waage, J., Greathead, D., Eds.; London Academic Press: London, UK, 1986; pp. 289–318.
50. Montgomery, M.E.; Wallner, W.E. The Gypsy Moth. In *Dynamics of Forest Insect Populations. Population Ecology*; Berryman, A.A., Ed.; Springer: Boston, MA, USA, 1988; pp. 353–375.
51. Doane, C.C. Aspects of mating behavior of the gypsy moth. *Ann. Entomol. Soc. Am.* **1968**, *61*, 768–773. [CrossRef]
52. Forbush, E.H.; Fernald, C.H. *The Gypsy Moth*; Wrightand Potter Press: Boston, MA, USA, 1896.
53. Cardé, R.T. Precopulatory sexual behavior of the adult gypsy moth. In *The Gypsy Moth: Research toward Integrated Pest Management*; Doane, C.C., McManus, M.L., Eds.; USDA Technical Bulletin 1584; USDA: Washington, DC, USA, 1981; pp. 572–587.
54. Charlton, R.E.; Cardé, R.T. Behavioral interactions in the courtship of *Lymantria dispar* (Lepidoptera: Lymantriidae). *Ann. Entomol. Soc. Am.* **1990**, *83*, 89–96. [CrossRef]
55. Timms, L.L.; Smith, S.M. Effects of gypsy moth establishment and dominance in native caterpillar communities of northern oak forests. *Can. Entomol.* **2011**, *143*, 479–503. [CrossRef]
56. Campbell, R.W. The analysis of numerical change in gypsy moth populations. *For. Sci.* **1967**, *13*, a0001–z0001.
57. Closa, S.; Núñez, L.; Parga, E. *Eruga Peluda (Lymantria dispar), L'Insecte Defoliador de les Alzines. Quadern de Natura*; Conselleria de Medi Ambien CAIB, Govern de les Illes Balears: Palma, Spain, 2008; p. 26.
58. Junta de Andalucía. *Plan de Lucha Integrada Contra la Lagarta Peluda Lymantria dispar (Linnaeus, 1978) en la Comunidad Autónoma de Andalucía*; Consejería de Medio Ambiente y Ordenación del Territorio: Seville, Spain, 2013; p. 45.
59. Stefanescu, C.; Soldevila, A.; Gutiérrez, C.; Torre, I.; Ubach, A.; Miralles, M. Explosions demogràfiques de l'eruga peluda del suro, *Lymantria dispar* (Linnaeus, 1758), als boscos del Montnegre el 2019 i 2020: Possibles causes, impactes i idoneitat dels tractaments per combatre la plaga. *Butlletí Inst. Catalana D'història Nat.* **2020**, *84*, 267–279.
60. Leonard, D.E. Diapause in the gypsy moth. *J. Econ. Entomol.* **1968**, *61*, 596–598. [CrossRef]
61. Leonard, D.E. Recent developments in ecology and control of the gypsy moth. *Ann. Rev. Entomol.* **1974**, *19*, 197–229. [CrossRef]

62. Leonard, D.E. Bioecology of the gypsy moth. *Bull. U.S. Dep. Agric.* **1981**, *1584*, 9–29.
63. Grijpma, P. Overview of research on lymantrids in eastern and western Europe. In *Proceedings, Lymantriidae: A Comparison of Features of New and Old World Tussock Moth*; GTR-NE-123; U.S. Department of Agriculture, Forest Service, Northeastern Forest Experiment Station: Broomall, PA, USA, 1989; pp. 21–49.
64. Sawyer, A.J.; Tauber, M.J.; Tauber, C.A.; Ruberson, J.R. Gypsy moth (Lepidoptera: Lymantriidae) egg development: A simulation analysis of laboratory and field data. *Ecol. Model.* **1993**, *66*, 121–155. [CrossRef]
65. Zharkov, D.G.; Tvaradze, M.S. Gypsy moth and its entomophages in the forests of Georgia. In *Neparnyy Shelkopryad: Itogi I Perspektivy Issledovaniy*; Institut Lesa I Drevesiny SO AN SSSR: Krasnoyarsk, Russia, 1988; p. 22.
66. Ministerio de Agricultura, Pesca y Alimentación. *Plagas de Insectos en las Masas Forestales Españolas*; Publicaciones del Ministerio de Agricultura, Pesca y Alimentación; Secretaría General Técnica, Servicio de Publicaciones Agrarias: Madrid, Spain, 1981; p. 254.
67. Hernandez-Alonso, R.; Martín-Bernal, E.; Cañada-Martín, J.F.; Pérez-Fortea, V.; Ibarra-Ibáñez, N.; Soriano-Giménez, M. *Oruga Defoliadora de las Frondosas, Lymantria dispar L. (Lepidóptero, Fam. Lymantriidae)*; Informaciones Técnicas 3/2001; Dirección General del Medio Natural, Departamento de Medio Ambiente, Gobierno de Aragón: Zaragoza, Spain, 2001; p. 4.
68. Roonwal, M.L. Structure of the egg-masses and their hairs in some species of *Lymantria* of importance to forestry (Insecta: Lepidoptera: Lymnantriidae). *India For.* **1954**, *8*, 265–276.
69. Katovich, S.; Haack, R. Gypsy moth in the northern hardwood forest. In *Northern Hardwood Notes*; Hutchinson, J.G., Ed.; U.S. Department of Agriculture, Forest Service, North Central Forest Experiment Station: St. Paul, MN, USA, 1991; Section 7.10.
70. Keena, M.A.; O'Dell, T.M.; Tanner, J.A. Effects of diet ingredient source and preparation method on larval development of laboratory-reared gypsy moth (Lepidoptera, Lymantriidae). *Ann. Entomol. Soc. Am.* **1995**, *88*, 672–679. [CrossRef]
71. Gray, D.R. Age-dependent postdiapause development in the gypsy moth (Lepidoptera: Lymantriidae) life stage model. *Environ. Entomol.* **2009**, *38*, 18–25. [CrossRef] [PubMed]
72. Gray, D.R.; Ravlin, F.W.; Braine, J.A. Diapause in the gypsy moth: A model of inhibition and development. *J. Insect Physiol.* **2001**, *47*, 173–184. [CrossRef]
73. Gray, D.R.; Logan, J.A.; Ravlin, F.W.; Carlson, J.A. Toward a model of gypsy moth egg phenology: Using respiration rates of individual eggs to determine temperature–time requirements of prediapause development. *Environ. Entomol.* **1991**, *20*, 1645–1652. [CrossRef]
74. Bell, R.A. Manipulation of diapause in the gypsy moth, *Lymantria dispar* L., by application of KK-42 and precocious chilling of eggs. *J. Insect Physiol.* **1996**, *42*, 557–563. [CrossRef]
75. Giese, R.L.; Cittadino, M.L. *Relationship of the Gypsy Moth to the Physical Environment*; Staff Paper #6; Department of Forestry, University of Wisconsin-Madison: Madison, WI, USA, 1977; p. 13.
76. Pantyukhov, G.A. The effect of low temperatures on different populations of the brown-tail moth *Euproctis chrysorrhoea* (L.) and the gypsy moth *Lymantria dispar* (L.) (Lepidoptera: Orgyidae). *Entomol. Rev.* **1964**, *43*, 47–55.
77. Masaki, S. The effect of temperature on the termination of diapause in the egg of *Lymantria dispar* Linné (Lepidoptera: Lymantriidae). *Jpn. J. Appl. Zool.* **1956**, *21*, 148–157.
78. Gray, D.R.; Ravlin, F.W.; Régnière, J.; Logan, J.A. Further advances toward a model of gypsy moth (*Lymantria dispar* (L.)) egg phenology: Respiration rates and thermal responsiveness during diapause, and age-dependent developmental rates in postdiapause. *J. Insect Physiol.* **1995**, *41*, 247–256. [CrossRef]
79. Marr, J.; Director, P.W. *Gypsy Moth Monitoring Program*; BioForest: Toronto, ON, Canada, 2020.
80. Wanner, K.W.; Helson, B.V.; Harris, B.J. Laboratory and field evaluation of spinosad against the gypsy moth, *Lymantria dispar*. *Pest Manag. Sci.* **2000**, *56*, 855–860. [CrossRef]
81. Weseloh, R.M. Behavioural responses of gypsy moth (Lepidoptera:Lymantridae) larvae to abiotic environmental factors. *Environ. Entomol.* **1989**, *18*, 361–367. [CrossRef]
82. Leonard, D.E. Feeding rhythm in larvae of the gypsy moth. *J. Econ. Entomol.* **1970**, *63*, 1454–1457. [CrossRef]
83. Campbell, R.W.; Sloan, R.J. Influence of behavioral evolution on gypsy moth pupal survival in sparse populations. *Environ. Entomol.* **1976**, *5*, 1211–1217. [CrossRef]
84. Wallner, W.E. Gypsy moth host interactions: A concept of room and board. In *Proceedings, Forest Defoliator-Host Interactions: Comparison between Gypsy Moth and Spruce Budworms, New Haven, CT, USA, 5–7 April 1983*; General Technical Report, NE-85; Talerico, R.L., Montgomery, M., Eds.; U.S. Department of Agriculture, Forest Service, Northeastern Forest Experiment Station: Broomall, PA, USA, 1983; pp. 5–8.
85. USDA. *Gypsy Moth Management in the United States: A Cooperative Approach*; Draft Environmental Impact Statement; USDA Forest Service: Washington, DC, USA, 1995.
86. Barbosa, P. Distribution of an endemic level gypsy moth population among various tree species. *Environ. Entomol.* **1978**, *7*, 526–527. [CrossRef]
87. Brooks, C.; Hall, D. *Gypsy Moth Silvicultural Guidelines for Wisconsin*; Wisconsin DNR PUB-FR-123 97; Wisconsin Department of Natural Resources: Madison, WI, USA, 2005; p. 14.
88. Allen, V.T.; Miller, O.F.; Tyler, W.B. Gypsy moth caterpillar dermatitis-revisited. *J. Am. Acad. Dermatol.* **1991**, *24*, 979–981. [CrossRef]

89. Leroy, B.M.L.; Lemme, H.; Braumiller, P.; Hilmers, T.; Jacobs, M.; Hochrein, S.; Kienlein, S.; Müller, J.; Pretzsch, H.; Stimm, K.; et al. Relative impacts of gypsy moth outbreaks and insecticide treatments on forest resources and ecosystems: An experimental approach. *Ecol. Solut. Evid.* **2021**, *2*, e12045. [CrossRef]
90. Lance, D.R. Host-seeking behavior of the gypsy moth: The influence of polyphagy and highly apparent host plants. In *Host-Seeking Behavior and Mechanisms*; Herbivorous Insects; Ahmad, S., Ed.; Academic Press: New York, NY, USA, 1983; pp. 210–224.
91. APHIS (Animal and Plant Health Inspection Service). Asian Gypsy Moth. Available online: https://www.aphis.usda.gov/aphis/ourfocus/planthealth/plant-pest-and-disease-programs/pests-and-diseases/gypsy-moth/CT_Gypsy_Moth (accessed on 6 September 2022).
92. Mosher, F.H. Food plants of the gypsy moth in America. *USDA Bull.* **1915**, *250*, 1–39.
93. Shields, V.D.C.; Broomell, B.P.; Salako, J.O.B. Host selection and acceptability of selected tree species by gypsy moth larvae, *Lymantria dispar* (L.). *Ann. Entomol. Soc. Am.* **2003**, *96*, 920–926. [CrossRef]
94. Elkinton, J.S.; Liebhold, A.M. Population dynamics of Gypsy moth in North America. *Annu. Rev. Entomol.* **1990**, *35*, 571–596. [CrossRef]
95. Valentine, H.T.; Talerico, R.L. Gypsy moth larval growth and consumption on red oak. *For. Sci.* **1980**, *26*, 599–605.
96. Montgomery, M.E. Variation in the susceptibility of tree species for gypsy moth. In Proceedings of the U.S. Department of Agriculture Interagency Gypsy Moth Review 1990, GTR-NE-146, East Windsor, CT, USA, 22–25 January 1990; United States Department of Agriculture, Forest Service, Northeastern Forest Experiment Station: Radnor, PA, USA, 1991; pp. 1–13.
97. Fuester, R.W.; Drea, J.J.; Gruber, F.; Herard, F. Explorations in Europe and Iran by the ARS European Parasite Laboratory: 1972–77. The gypsy moth: Research toward integrated pest management. *USDA Tech. Bull.* **1981**, *1584*, 324–340.
98. McManus, M.; Csóka, G. History and impact of gypsy moth in north America and comparison to the recent outbreaks in Europe. *Acta Silv. Lignaria Hung.* **2007**, *3*, 47–64.
99. Hirka, A. *A 2004. Évi Biotikus És Abiotikus Erdőgazdasági Károk, Valamint a 2005-Ben Várható Károsítások*; Hungarian Forest Research Institute: Budapest, Hungary, 2005.
100. Keena, M.A.; Richards, J.Y. Comparison of survival and development of gypsy moth *Lymantria dispar* L. (Lepidoptera: Erebidae) populations from different geographic areas on North American conifers. *Insects* **2020**, *11*, 260. [CrossRef] [PubMed]
101. Lechowicz, M.J.; Mauffette, Y. Host preference of the gypsy moth in eastern North America versus European forests. *Rev. Entomol. Que.* **1986**, *31*, 43–51.
102. Rossiter, M. Use of a secondary host by non-outbreak populations of the gypsy moth. *Ecology* **1987**, *68*, 857–868. [CrossRef]
103. Strom, B.L.; Hain, F.P.; Ayres, M.P. Field performance of F1-sterile gypsy moth larvae (Lepidoptera: Lymantriidae) on loblolly pine and sweetgum. *Environ. Entomol.* **1996**, *25*, 749–756. [CrossRef]
104. Lovett, G.M.; Camja, C.D.; Arthur, M.A.; Weathers, K.C.; Fitzhugh, R.D. Forest ecosystem responses to exotic pests and pathogens in Eastern North America. *BioScience* **2006**, *56*, 395–405. [CrossRef]
105. Davidson, C.B.; Johnson, J.E.; Gottschalk, K.W.; Amateis, R.L. Prediction of stand susceptibility and gypsy moth defoliation in Coastal Plain mixed pine hardwoods. *Can. J. For. Res.* **2001**, *31*, 1914–1921.
106. Campbell, R.W.; Garlo, A.S. Gypsy moth in New Jersey pine-oak. *J. For.* **1982**, *80*, 89–90.
107. Romanyk, N. Les gradations de *Lymantria dispar* L. en Espagne. *Zast. Bilja.* **1973**, *24*, 285–288.
108. Romanyk, N.; Rupérez, A. Principales parásitos observados en los defoliadores de España con atención particular de la *Lymantria dispar* L. *Entomophaga* **1960**, *5*, 229–239. [CrossRef]
109. Leite, R.M.M.R. *Ocorrência de Lymantria dispar L. em Pinus radiata D. Don: Estudo do Ciclo de Vida e Comportamento da Praga Neste Hospedeiro: Medidas de Proteção e Combate*; Instituto Politécnico de Castelo Branco: Castelo Branco, Portugal, 1993.
110. Miller, J.C.; Hanson, P.E. Laboratory studies on development of gypsy moth, *Lymantria dispar* (L.) (Lepidoptera: Lymantriidae), larvae on foliage of gymnosperms. *Can. Entomol.* **1989**, *121*, 425–429. [CrossRef]
111. Castedo-Dorado, F.; Lago-Parra, G.; Lombardero, M.J.; Liebhold, A.M.; Álvarez-Taboada, M.F. European gypsy moth (*Lymantria dispar dispar* L.) completes development and defoliates exotic radiata pine plantations in spain. *N. Z. J. For. Sci.* **2016**, *46*, 18. [CrossRef]
112. Mead, D.J. *Sustainable Management of Pinus Radiata Plantations*; FAO: Rome, Italy, 2013.
113. Aber, R.; De Melfi, T.; Gill, T.; Healey, B.; McCarthy, M.A. Rash illnesses associated with gypsy moth caterpillars, Pennsylvania. *Morb. Mortal. Wkly. Rep.* **1982**, *31*, 169–170.
114. Liebhold, A.M.; Mastro, V.; Schaefer, P.W. Learning from the legacy of Léopold Trouvelot. *Bull. Entomol. Soc. Am.* **1989**, *35*, 20–22. [CrossRef]
115. Diaz, J.H. The Evolving Global Epidemiology, Syndromic Classification, Management, and Prevention of Caterpillar Envenoming. *Am. J. Trop. Med. Hyg.* **2005**, *72*, 347–357. [CrossRef]
116. Wu, Y.; Molongoski, J.J.; Winograd, D.F.; Bogdanowicz, S.M.; Louyakis, A.S.; Lance, D.R.; Mastro, V.C.; Harrison, R.G. Genetic structure, admixture and invasion success in a Holarctic defoliator, the gypsy moth (*Lymantria dispar*, Lepidoptera: Erebidae). *Mol. Ecol.* **2015**, *24*, 1275–1291. [CrossRef]
117. Glare, T.R.; Walsh, P.J.; Kay, M.; Barlow, N.D. *Strategies for the Eradication or Control of Gypsy Moth in New Zealand*; Lincoln, Agresearch: Lincoln, New Zealand, 2003; pp. 30–31.
118. Wulf, A.; Graser, E. Gypsy moth outbreaks in Germany and neighboring countries. *Nachr. Dtsch. Pflanzenschutzd.* **1996**, *48*, 265–269.

119. Grayson, K.L.; Johnson, D.M. Novel insights on population and range edge dynamics using an unparalleled spatiotemporal record of species invasion. *J. Anim. Ecol.* **2018**, *87*, 581–593. [CrossRef]
120. Tobin, P.C.; Robinet, C.; Johnson, D.M.; Whitmire, S.L.; Bjornstad, O.N.; Liebhold, A.M. The role of Allee effects in gypsy moth, *Lymantria dispar* (L.), invasions. *Popul. Ecol.* **2009**, *51*, 373–384. [CrossRef]
121. Aukema, J.E.; Leung, B.; Kovacs, K.; Chivers, C.; Britton, K.O.; Englin, J.; Frankel, S.J.; Haight, R.G.; Holmes, T.P.; Liebhold, A.M.; et al. Economic impacts of non-native forest insects in the continental United States. *PLoS ONE* **2011**, *6*, e24587. [CrossRef]
122. Morin, R.S.; Liebhold, A.M. Invasive forest defoliator contributes to the impending downward trend of oak dominance in eastern North America. *Forestry* **2016**, *89*, 284–289. [CrossRef]
123. McNamara, D.G. EPPO's perspective on the gypsy moth in Europe. In Proceedings of the 1995 Annual Gypsy Moth Review, Traverse City, MI, USA, 5–8 November 1995; Michigan Department of Agriculture: Lansing, MI, USA, 1995; pp. 60–65.
124. Keena, M.A.; Grinberg, P.S.; Wallner, W.E. Inheritance of female flight in *Lymantria dispar* (Lepidoptera: Lymantriidae). *Environ. Entomol.* **2007**, *36*, 484. [CrossRef] [PubMed]
125. Tobin, P.C.; Bai, B.B.; Eggen, D.A.; Leonard, D.S. The ecology, geopolitics, and economics of managing *Lymantria dispar* in the United States. *Int. J. Pest Manag.* **2012**, *58*, 195–210. [CrossRef]
126. Weseloh, R.M. Evidence for limited dispersal of larval gypsy moth, *Lymantria dispar* L. (Lepidoptera: Lymantriidae). *Can. Entomol.* **1997**, *129*, 355–361. [CrossRef]
127. Campbell, R.W. Population dynamics. In *The Gypsy Moth: Research toward Integrated Pest Management*; USDA Forest Service Technical Bulletin 1584; Government Printing Office: Washington, DC, USA, 1981; pp. 65–214.
128. Lance, D.R.; Barbosa, P. Host tree influences on the dispersal of late instar gypsy moths, *Lymantria dispar*. *Oikos* **1982**, *38*, 1–7. [CrossRef]
129. Deml, R.; Dettner, J.K. "Balloon hairs" of gypsy moth larvae (Lepidoptera: Lymantriidae): Morphology and comparative chemistry. *Comp. Biochem. Physical.* **1995**, *112B*, 673–681. [CrossRef]
130. McManus, M.L. *The Role of Behavior in the Disperal of Newly Hatch Gypsy Moth Larvae*; Research Paper NE-267; USDA Forest Service, Northeastern Forest Experiment Station: Upper Darby, PA, USA, 1973.
131. Liebhold, A.M.; Halverson, J.A.; Elmes, G.A. Gypsy moth invasion in North America: A quantitative analysis. *J. Biogeogr.* **1992**, *19*, 513–520. [CrossRef]
132. McFadden, M.W.; McManus, M.E. An insect out of control? The potential for spread and establishment of the gypsy moth in new forest areas in the United States. In *Insect Guilds: Patterns of Interaction with Host Trees*; Baranchikov, Y.N., Mattson, W.J., Hain, F.P., Payne, T.L., Eds.; USDA Forest Service, Northeastern Forest Experiment Station: Radnor, PA, USA, 1991; pp. 172–186.
133. Kearns, D.N.; Tobin, P.C. Oregon vs. the gypsy moth: Forty years of battling an invasive species. *Am. Entomol.* **2020**, *66*, 50–58. [CrossRef]
134. Tobin, P.; Blackburn, L.M. *Slow the Spread: A National Program to Manage the Gyspy Moth*; General Technical Reports NRS-6; US Department of Agriculture, Forest Service, Northern Research Station: Newton Square, PA, USA, 2007.
135. Picq, S.; Keena, M.; Havill, N.; Stewart, D.; Pouliot, E.; Boyle, B.; Levesque, R.C.; Hamelin, R.C.; Cusson, M. Assessing the potential of genotyping-by-sequencing-derived single nucleotide polymorphisms to identify the geographic origins of intercepted gypsy moth (*Lymantria dispar*) specimens: A proof-of-concept study. *Evol. Appl.* **2018**, *11*, 325–339. [CrossRef]
136. Gray, D.R. The gypsy moth life stage model: Landscapewide estimates of gypsy moth establishment using a multigenerational phenology model. *Ecol. Model.* **2004**, *176*, 155–171. [CrossRef]
137. Tobin, P.C.; Gray, D.R.; Liebhold, A.M. Supraoptimal temperatures influence the range dynamics of a non-native insect. *Divers. Distrib.* **2014**, *20*, 813–823. [CrossRef]
138. Tobin, P.C.; Kean, J.M.; Suckling, D.M.; McCullough, D.G.; Herms, D.A.; Stringer, L.D. Determinants of successful arthropod eradication programs. *Biol. Invasions* **2014**, *16*, 401–414. [CrossRef]
139. Contarini, M.; Onufrieva, K.S.; Thorpe, K.W.; Raffa, K.F.; Tobin, P.C. Mate-finding failure as an important cause of Allee effects along the leading edge of an invading insect population. *Entomol. Exp. Appl.* **2009**, *133*, 307–314. [CrossRef]
140. Yamanaka, T.; Liebhold, A.M. Mate-location failure, the Allee effect, and the establishment of invading populations. *Popul. Ecol.* **2009**, *51*, 337–340. [CrossRef]
141. Rhainds, M. Size-dependent realized fecundity in two lepidopteran capital breeders. *Environ. Entomol.* **2015**, *44*, 1193–1200. [CrossRef]
142. Cloyd, R.A.; Nixon, P.L.; Gypsy Moth. University of Illinois Extension. Available online: https://extension.illinois.edu/gypsymoth/biology.cfm (accessed on 6 September 2022).
143. Myers, J.H.; Cory, J.S. Population Cycles in Forest Lepidoptera Revisited. *Annu. Rev. Ecol. Evol. Syst.* **2013**, *44*, 565–592. [CrossRef]
144. Roy, A.S.; McNamara, D.G.; Smith, I.M. Situation of *Lymantria dispar* in Europe. *EPPO Bull.* **1995**, *25*, 611–616. [CrossRef]
145. Ananko, G.G.; Kolosov, A.V. Asian gypsy moth (*Lymantria dispar* L.) populations: Tolerance of eggs to extreme winter temperatures. *J. Therm. Biol.* **2021**, *102*, 103123. [CrossRef]
146. Elton, C.S. Periodic fluctuations in the numbers of animals: Their causes and effects. *Br. J. Exp. Biol.* **1924**, *2*, 119–163. [CrossRef]
147. Johnson, D.M.; Liebhold, A.M.; Bjornstad, O.N.; Mcmanus, M.L. Circumpolar variation in periodicity and synchrony among gypsy moth populations. *J. Anim. Ecol.* **2005**, *74*, 882–892. [CrossRef]

148. Hlasny, T.; Trombik, J.; Holusa, J.; Lukasova, K.; Grendar, M.; Turcani, M.; Zubrik, M.; Tabakovic-Tosic, M.; Hirka, A.; Buksha, I.; et al. Multi-decade patterns of gypsy moth fluctuations in the Carpathian Mountains and options for outbreak forecasting. *J. Pest Sci.* **2016**, *89*, 413–425. [CrossRef]
149. Lovett, G.M.; Christenson, L.M.; Groffman, P.M.; Jones, C.G.; Hart, J.E.; Mitchell, M.J. Insect defoliation and nitrogen cycling in forests. *BioScience* **2002**, *52*, 335–341. [CrossRef]
150. Kosola, K.R.; Durall, D.M.; Robertson, G.P.; Dickmann, D.I.; Parry, D.; Russell, C.A.; Paul, E.A. Resilience of mycorrhizal fungi on defoliated and fertilized hybrid poplars. *Can. J. Bot.* **2004**, *82*, 671–680. [CrossRef]
151. Campbell, R.W.; Sloan, R.J. Forest stand responsesto defoliation by the gypsy moth. *For. Sci.* **1977**, *19*, 1–34.
152. Gottschalk, K.W. Gypsy moth effects on mast production. In Proceedings of the Workshop: Southern Appalachian Mast Management, Knoxville, TN, USA, 14–16 August 1989; McGee, C.E., Ed.; University of Tennessee: Knoxville, TN, USA, 1990; pp. 42–50.
153. Wu, Y.; Bogdanowicz, S.M.; Andres, J.A.; Vieira, K.A.; Wang, B.; Cossé, A.; Pfister, S.E. Tracking invasions of a destructive defoliator, the gypsy moth (Erebidae: *Lymantria dispar*): Population structure, origin of intercepted specimens, and Asian introgression into North America. *Evol. Appl.* **2020**, *13*, 2056–2070. [CrossRef]
154. McCullough, D.M.; Bauer, L.S. *Bt: One Option for Gypsy Moth Management*; Ext. Bull. E-2724; Michigan State University Extension: East Lansing, MI, USA, 2000; p. 4.
155. Bradshaw, C.J.A.; Leroy, B.; Bellard, C.; Roiz, D.; Albert, C.; Fournier, A.; Barbet-Massin, M.; Salles, J.M.; Simard, F.; Courchamp, F. Massive yet grossly underestimated global costs of invasive insects. *Nat. Commun.* **2016**, *7*, 12986. [CrossRef]
156. Harrison, R.L.; Rowley, D.L.; Keena, M.A. Pathology and genome sequence of a Lymantria dispar multiple nucleopolyhedrovirus (LdMNPV) isolate from Heilongjiang, China. *J. Invertebr. Pathol.* **2020**, *177*, 107495. [CrossRef]
157. Gottschalk, K.W. *Silvicultural Guidelines for Forest Stands Threatened by the Gypsy Moth*; GTR NE-171; Department of Agriculture Forest Service Northeastern Forest Experiment Station: Radnor, PA, USA, 1993.
158. Schwenke, W. *Die Forstschädlinge Europas, Ein Handbuch in Fünf Bäden. Autorenkollektiv: Band 3. Schmetterlinge*, 1st ed.; Paul Parey: Hamburg/Berlin, Germany, 1978; p. 467.
159. Keremidchiev, M.T. Dynamics of outbreaks of the gypsy moth (*Lymantria dispar* L.) in the People's Republic of Bulgaria. In Proceedings of the 13th International Congres of Entomology, Moscow, Russia, 2–9 August 1968; Bei-Bienko, G.Y., Ed.; Volume 3, pp. 51–54.
160. Milenković, M.; Ducić, V. The solar activity cycles and the outbreaks of the Gypsy Moth—*Lymantria dispar* L. (Lepidoptera: Lymantriidae) in Serbia. *Ecol. Montenegrina* **2016**, *7*, 538–545. [CrossRef]
161. Živojinović, S. Introduction. Gypsy-moth-Results of the activity on its study and control in our country in the course of 1957. *Zast. Bilja* **1957**, *52/53*, 3–6.
162. Lipa, J.J.; Kolk, A. The recent situation of the gypsy moth (*Lymantria dispar*) and other Lymantariids in Poland. *Bull. OEPP* **1995**, *25*, 623–629. [CrossRef]
163. Landmann, G.; Barthod, C. La santé des forêts françaises en 1994. *Rev. For. Française* **1996**, *48*, 101. [CrossRef]
164. Krehan, H. Schwammspinner-Bekämpfung: Ist sie in Österreich möglich, ist sie notwendig? *Forstsch. Aktuell* **1994**, *15*, 13.
165. Anon. *PBMD Bulletin Forstschutz-Überblick 1994*; Eidgenössische Forschungsanstalt für Wald, Schnee und Landschaft: Birmensdorf, Switzerland, 1995.
166. Cannon, R.J.C.; Koerper, D.; Ashby, S.; Baker, R.; Bartlett, P.W.; Brookes, G.; Burgess, R.; Cheek, S.; Evans, H.F.; Hammon, R.; et al. Gypsy moth, *Lymantria dispar*, outbreak in Northeast London, 1995–2003. *Int. J. Pest. Sci.* **2004**, *50*, 259–273. [CrossRef]
167. Kailidis, D.S. *Forest Entomology and Zoology*, 4th ed.; Christodoulidis Press: Thessaloniki, Greece, 1991; p. 536.
168. Blanch Rissech, F.; Massana Ribas, R.M. Les industries sureres a Llagostera. *Crònica* **1992**, *6*, 1–12.
169. Soldevila, A. Causes de les Explosions Demogràfiques i Mètodes de Seguiment de la Papallona *Lymantria dispar* Als Boscos del Montnegre. Bachelor Thesis, University of Girona, Girona, Spain, 2020.
170. Tardà, A.; Corbera, J.; Riera, R. *Estudi de L'àrea D'afectació de L'eruga Peluda del Suro al Massís del Montnegre a Partir D'imatges Sentinel-2*; Institut Cartogràfic i Geològic de Catalunya i Diputació de Barcelona: Generalitat de Catalunya, Spain, 2021; p. 18.
171. Kurenshchikov, D.K.; Martemyanov, V.V.; Imranova, E.L. Features of the far eastern gypsy moth (*Lymantria dispar* L.) population outbreak. *Contemp. Probl. Ecol.* **2020**, *13*, 172–179. [CrossRef]
172. Shamuratova, N.G. Gypsy moth (*Lymantria dispar* L.) population in Uzbekistan and pathogenicity of the nuclear polyhedrosis virus. *Eur. Sci. Rev.* **2018**, *11*, 204–206.
173. Kozhevnikova, L.N.; Levykh, A.Y.; Panchenko, V.Y. Distribution of *Lymantria dispar* in forests of CSI «Kyzylzharskoe forestry» of the North Kazakhstan region. *Karaganda Univ. Bull. Ser. Biomed. Geogr.* **2018**, *91*, 33–39.
174. Krcmar-Nozic, E.; Wilson, B.; Arthur, L. *The Potential Impacts of Exotic Forest Pests in North America: A Synthesis of Research*; Pacific Forestry Centre: Victoria, BC, Canada, 2000; pp. 12–13.
175. Pernek. M.; Pilaš. I.; Vrbek. B.; Benko. M.; Hrašovec. B.; Milković. J. Forecasting the impact of the gypsy moth on lowland hardwood forests by analyzing the cyclical pattern of population and climate data series. *For. Ecol. Manag.* **2008**, *255*, 1740–1748. [CrossRef]
176. Zhang, J.; Cong, Q.; Rex, E.A.; Hallwachs, W.; Janzen, D.H.; Grishin, N.V.; Gammon, D.B. Gypsy moth genome provides insights into flight capability and virus–host interactions. *Proc. Natl. Acad. Sci. USA* **2019**, *116*, 1669–1678. [CrossRef]

177. Smitley, D.R.; Davis, T.W. Aerial application of *Bacillus thuringiensis* for suppression of gypsy moth (Lepidoptera: Lymantriidae) in Populus–Quercus forests. *J. Econ. Entomol.* **1993**, *86*, 1178–1184. [CrossRef]
178. Luciano, P.; Lentini, A. Ten years of microbiological control program against lepidopterous defoliators in Sardinian cork oak forests. *IOBC/WPRS Bull.* **2012**, *76*, 175–178.
179. Tobin, P.C.; Whitmire, S.L. Spread of gypsy moth (Lepidoptera: Lymantriidae) and its relationship to defoliation. *Environ. Entomol.* **2005**, *34*, 1448–1455. [CrossRef]
180. Tobin, P.C.; Liebhold, A.M.; Roberts, E.A.; Blackburn, L.M. Estimating Spread Rates of Non-native Species: The Gypsy Moth as a Case Study. In *Invasive Alien Species: Pest Risk Modelling and Mapping*; Venette, R.C., Ed.; CABI: Wallingford, UK, 2015; pp. 131–144.
181. Onufrieva, K.S.; Hickman, A.D.; Leonard, D.S.; Tobin, P.C. Relationship between efficacy of mating disruption and gypsy moth density. *Int. J. Pest Manag.* **2019**, *65*, 44–52. [CrossRef]
182. Luciano, P.; Prota, R. La dinamica di popolazione di *Lymantria dispar* L. in Sardegna. Indicatori della gradazione ricavati dale ovideposizioni. *Studi Sassar. III* **1981**, *27*, 137–160.
183. Luciano, P.; Prota, R. Osservazioni sulla densità di popolazione di *Lymantria dispar* L. nelle principali aree subericole della Sardegna. *Studi Sassar. III* **1982**, *28*, 168–179.
184. Mannu, R.; Olivieri, M.; Cocco, A.; Lentini, A. Development of Enumerative and Binomial Sequential Sampling Plans for Monitoring *Lymantria dispar* (L.) (Lepidoptera Erebidae) in Mediterranean Oak Forests. *Agronomy* **2022**, *12*, 1501. [CrossRef]
185. Herms, D.A. Assessing management options for gypsy moth. *Pestic. Outlook* **2003**, *14*, 14–18. [CrossRef]
186. Kumar, S. Biopesticides: A need for food and environmental safety. *J. Biofertil. Biopestic.* **2012**, *3*, e107. [CrossRef]
187. Nicolopoulou-Stamati, P.; Maipas, S.; Kotampasi, C.; Stamatis, P.; Hens, L. Chemical pesticides and human health: The urgent need for a new concept in agriculture. *Front. Public Health* **2016**, *4*, 148. [CrossRef]
188. Rausell, C.; Martinez-Ramirez, A.C.; Garcia Roblez, I.; Real, M.D. The toxicity and physiological effects of *Bacillus thuringiensis* toxins and formulations on *Thaumetopoea pityocampa*, the pine processionary caterpillar. *Pestic. Biochem. Physiol.* **1999**, *65*, 44–54. [CrossRef]
189. Kalender, Y.; Uzunhisarcikli, M.; Ogutcu, A.; Suludere, Z.; Kalender, S. Effects of endosulfan on *Thaumetopoea pityocampa* (Lepidoptera: Thaumetopoeidae) larvae. *Folia Biol.* **2005**, *53*, 3–4. [CrossRef]
190. Ogutcu, A.; Suludere, Z.; Uzunhisarcikli, M.; Kalender, Y. Effects of Bacillus thuringiensis kurstaki on Malpighian tubule cells of Thaumetopoea pityocampa (Lepidoptera: Thaumetopoeidae) larvae. *Folia Biol.* **2005**, *53*, 7–11. [CrossRef]
191. Aslanturk, A.; Kalender, S.; Uzunhisarcikli, M.; Kalender, Y. Effects of methidathion on antioxidant enzyme activities and malondialdehyde level in midgut tissues of *Lymantria dispar* (lepidoptera) larvae. *J. Entomol. Res. Soc.* **2011**, *13*, 27–38.
192. Liebhold, A.; McManus, M. The evolving use of insecticides in gypsy moth management. *J. For.* **1999**, *97*, 20–23.
193. Sharov, A.A.; Leonard, D.; Liebhold, A.M.; Clemens, N.S. Evaluation of preventive treatments in low-density gypsy moth populations. *J. Econ. Entomol.* **2002**, *95*, 1205–1215. [CrossRef]
194. Jones, C.G.; Ostfeld, R.S.; Richard, M.P.; Schauber, E.M.; Wolff, J.O. Chain reactions linking acorns to gypsy moth outbreaks and Lyme disease. *Science* **1998**, *279*, 1023–1026. [CrossRef]
195. Bess, H.A. Population ecology of the gypsy moth *Porthetria dispar* L. (Lepidoptera: Lymantriidae). *Conn. Agric. Exp. Stn. Bull.* **1961**, *646*, 1–43.
196. Weseloh, R.M. Changes in population size, dispersal behaviour, and reproduction of *Calosoma sycophanta* (Coleoptera: Carabidae), associated with changes in gypsy moth, *Lymantria dispar* (Lepidoptera: Lymantriidae), abundance. *Environ. Entomol.* **1985**, *14*, 370–377. [CrossRef]
197. Weseloh, R.M. Predation by *Calosoma sycophanta* L (Coleoptera, Carabidae) evidence for a large impact on gypsy moth *Lymantria dispar* pupae. *Can. Entomol.* **1985**, *117*, 1117–1126. [CrossRef]
198. Burgess, A.F.; Crossman, S.S. *Imported Insect Enemies of the Gypsy Moth and the Browntail Moth*; USDA Technical Bulletin 86; USDA: Washington, DC, USA, 1929.
199. Hoch, G.; Kahlbacher, G.; Schopf, A. Gypsy moth revisited: Studies on the natural enemy complex of *Lymantria dispar* L. (Lep., Lymantriidae) during an outbreak in a well known gypsy moth area. *Mitt. Dtsch. Ges. Allg. Angew. Entomol.* **2006**, *15*, 201–204.
200. Kryzhanovsky, O.L. Calosoma sycophanta. In *Areas of Insects in the European Part of the USSR*; Gorodkov, K.B., Ed.; Atlas: Leningrad, Russia, 1981; p. 20.
201. Kryzhanovsky, O.L.; Belousov, I.A.; Kabak, I.I.; Kataev, B.M.; Makarov, K.V.; Shilenkov, V.G. *A Checklist of the Ground-Beetles of Russia and Adjacent Lands (Insecta, Coleoptera, Carabidae)*; Pensoft Publishers: Moscow, Russia, 1995; p. 271.
202. Kryzhanovsky, O.L.; Obydov, D.V. *Krasotel Pakhuchiy Calosoma sycophanta*; Red Data Book of the Russian Federation: Moscow, Russia, 2001; p. 122.
203. Bespalov, A.N.; Dudko, R.Y.; Lyubechanskii, I.I. Additions to the ground beetle fauna (Coleoptera, Carabidae) of the Novosibirsk Oblast: Do the southern species spread to the north? *Evraziatskii Entomol. Zhurnal* **2010**, *9*, 625–628.
204. Ostfeld, R.S.; Jones, C.G.; Wolff, J.O. Of mice and mast: Ecological connections in eastern deciduous forests. *BioScience* **1996**, *46*, 323–330. [CrossRef]
205. Gschwantner, T.; Hoch, G.; Schopf, A. Impact of predators on artificially augmented populations of *Lymantria dispar* L. Pupae (Lep., Lymantriidae). *J. Appl. Entomol.* **2002**, *126*, 66–73. [CrossRef]

206. Liebhold, A.M.; Higashiura, Y.; Unno, A. Forest type affects predation on gypsy moth (Lepidoptera: Lymantriidae) pupae in Japan. *Environ. Entomol.* **1998**, *27*, 858–862. [CrossRef]
207. Liebhold, A.M.; Raffa, K.F.; Diss, A.L. Forest type affects predation on gypsy moth pupae. *Agric. For. Entomol.* **2005**, *7*, 179–185. [CrossRef]
208. Turcani, M.; Novotny, J.; Zubrik, M.; McManus, M.L.; Pilarksa, D.; Maddox, J. The role of biotic factors in gypsy moth population dynamics in Slovakia: Present knowledge. In Proceedings of the Integrated Management and Dynamics of Forest Defoliating Insects, Victoria, BC, Canada, 15–19 August 1999; Liebhold, A.M., McManus, M.L., Otvos, I.S., Fosbroke, S.L.C., Eds.; General Technical Reports NE-277. USDA Forest Service, Northeastern Research Station: Newtown Square, PA, USA, 2001; pp. 152–167.
209. Reichart, G. Birds destroying eggs of *Lymantria dispar* L. *Aquila* **1959**, *77*, 315–317.
210. Higashiura, Y. Survival of eggs in the gypsy-moth *Lymantria dispar* L. Predation by birds. *J. Anim. Ecol.* **1989**, *58*, 403–412. [CrossRef]
211. Turcani, M.; Liebhold, A.; McManus, M.; Novotny, J. Preliminary results on predation of gypsy moth egg masses in Slovakia. In Proceedings of the Ecology, Survey and Management of Forest Insects, Kraków, Poland, 1–5 September 2002; McManus, M.L., Liebhold, A.M., Eds.; General Technical Reports NE-311. USDA Forest Service, Northeastern Research Station: Newtown Square, PA, USA, 2002; pp. 115–120.
212. Cooper, R.J.; Smith, H.R. Predation on gypsy-moth (Lepidoptera, Lymantriidae) egg masses by birds. *Environ. Entomol.* **1995**, *24*, 571–575. [CrossRef]
213. Weseloh, R.M. Behaviour of the gypsy-moth predator, *Calosoma sycophanta* L (Carabidae, coleoptera), as influenced by time of day and reproductive status. *Can. Entomol.* **1993**, *125*, 887–894. [CrossRef]
214. Weseloh, R.; Bernon, G.; Butler, L.; Fuester, R.; McCullough, D.; Stehr, F. Releases of *Calosoma sycophanta* (Coleoptera: Carabidae) near the edge of gypsy moth (Lepidoptera: Lymantriidae) distribution. *Environ. Entomol.* **1995**, *24*, 1713–1717. [CrossRef]
215. Burgess, A.F. *Calosoma sycophanta*: Its life history, behavior, and successful colonization in New England. *Bull. U.S. Dep. Agric.* **1911**, *101*, 1–94.
216. Evans, A.V. The forest caterpillar hunter, *Calosoma sycophanta*, an Old World species confirmed as part of the Virginia beetle fauna (Coleoptera: Carabidae). *Banisteria* **2009**, *34*, 33–37.
217. Goertz, D.; Hoch, G. Influence of the forest caterpillar hunter *Calosoma sycophanta* on the transmission of microsporidia in larvae of the gypsy moth *Lymantria dispar*. *Agric. For. Entomol.* **2013**, *15*, 178–186. [CrossRef]
218. Howard, L.O.; Fiske, W.F. The importation into the United States of the parasites of the gypsy moth and brown-tailed moth. *Bull. U.S. Dep. Agric.* **1911**, *91*, 312.
219. Hoy, M.A. Establishment of gypsy moth parasitoids in North America: An evaluation of possible reasons for establishment or non-establishment. In *Perspectives in Forest Entomology*; Anderson, J.F., Kaya, H.K., Eds.; Academic Press: New York, NY, USA, 1976; pp. 215–232.
220. Reardon, R.C. Parasite incidence and ecological relationships in field populations of gypsy moth Lepidoptera-Lymantriidae larvae and pupae. *Environ. Entomol.* **1976**, *5*, 981–987. [CrossRef]
221. Glare, T.R.; Barlow, N.D.; Walsh, P.J. Potential agents for eradication or control of gypsy moth in New Zealand. In Proceedings of the New Zealand Plant Protection Conference, Hamilton, New Zealand, 12 August 1998; pp. 224–229.
222. Webb, R.E.; Peiffer, R.; Fuester, R.W.; Thorpe, K.W.; Calabrese, L.; McLaughlin, J.M. An evaluation of the residual activity of traditional, safe, and biological insecticides against the gypsy moth. *J. Arboric.* **1998**, *24*, 286–293. [CrossRef]
223. Martin, J.C.; Bonneau, X. *Bacillus thuringiensis* 30 ans de lutte contre les chenilles defoliatrices en foret. *Phytoma Défense Végétaux* **2006**, *590*, 4–7.
224. Crickmore, N. Beyond the spore—Past and future developments of *Bacillus thuringiensis* as a biopesticide. *J. Appl. Microbiol.* **2006**, *101*, 616–619. [CrossRef]
225. Mannu, R.; Cocco, A.; Luciano, P.; Lentini, A. Influence of *Bacillus thuringiensis* application timing on population dynamics of gypsy moth in Mediterranean cork oak forests. *Pest Manag. Sci.* **2020**, *76*, 1103–1111. [CrossRef]
226. Scriber, J.M. Non-target impacts of forest defoliator management options: Decision for no spraying may have worse impacts on non-target Lepidoptera than *Bacillus thuringiensis* insecticides. *J. Insect Conserv.* **2004**, *8*, 243–263. [CrossRef]
227. Bateman, R. Application of Biopesticides. In *Pesticide Application Methods*, 4th ed.; Matthews, G.A., Bateman, R., Miller, P., Eds.; John Wiley & Sons: Oxford, UK, 2014; pp. 411–427.
228. Reardon, R.C.; Podgwaite, J.D. Summary of efficacy evaluations using aerially applied Gypchek® against gypsy moth in the U.S.A. *Environ. Sci. Health* **1994**, *B29*, 739–756. [CrossRef]
229. Lentini, A.; Mannu, R.; Cocco, A.; Ruiu Pino, A.; Cerboneschi, A.; Luciano, P. Long-term monitoring and microbiological control programs against lepidopteran defoliators in the cork oak forests of Sardinia (Italy). *Ann. Silvic. Res.* **2019**, *45*, 21–30.
230. Glare, T.R.; O'Callaghan, M. *Bacillus thuringiensis: Biology, Ecology and Safety*; John Wiley & Sons: New York, NY, USA, 2000.
231. Maczuga, S.A.; Mierzejewski, K.J. Droplet size and density effects of *Bacillus thuringiensis* kurstaki on gypsy moth (Lepidoptera: Lymantriidae) larvae. *J. Econ. Entomol.* **1995**, *88*, 1376–1379. [CrossRef]
232. Speare, A.T.; Colley, R.H. *The Artificial Use of the Brown-Tail Fungus in Massachusetts, with Practical Suggestions for Private Experiment, and a Brief Note on a Fungous Disease of the Gypsy Caterpillar*; Wright and Potter Printing Co.: Boston, MA, USA, 1912.
233. Andreadis, T.G.; Weseloh, R.M. Discovery of *Entomophaga maimaiga* in North American gypsy moth, *Lymantria dispar*. *Proc. Natl. Acad. Sci. USA* **1990**, *87*, 2461–2465. [CrossRef]

234. Hajek, A.E.; Humber, R.A.; Griggs, M.H. Decline in virulence of *Entomophaga maimaiga* (Zygomycetes: Entomophthorales) with repeated in vitro subculture. *J. Invertebr. Pathol.* **1990**, *56*, 91–97. [CrossRef]
235. Nielsen, C.; Milgroom, M.G.; Hajek, A.E. Genetic diversity in the gypsy moth fungal pathogen *Entomophaga maimaiga* from founder populations in North America and source populations in Asia. *Mycol. Res.* **2005**, *109*, 941–950. [CrossRef]
236. Pilarska, D.; McManus, M.; Hajek, A.; Herard, F.; Vega, F.; Pilarski, P.; Markova, G. Introduction of the entomopathogenic fungus *Entomophaga maimaiga* Hum., Shim. & Sop. (Zygomycetes: Entomophtorales) to a *Lymantria dispar* (L.) (Lepidoptera: Lymantriidae) population in Bulgaria. *J. Pest Sci.* **2000**, *73*, 125–126.
237. Zúbrik, M.; Hajek, A.; Pilarska, D.; Spilda, I.; Georgiev, G.; Hrasovec, B.; Hirka, A.; Goertz, D.; Hoch, G.; Barta, M.; et al. The potential for *Entomophaga maimaiga* to regulate gypsy moth *Lymantria dispar* (L.) (Lepidoptera: Erebidae) in Europe. *J. Appl. Entomol.* **2016**, *140*, 565–579. [CrossRef]
238. Zúbrik, M.; Barta, M.; Pilarska, D.; Goertz, D.; Úradník, M.; Galko, J.; Vakula, J.; Gubka, A.; Rell, S.; Kunca, A. First record of *Entomophaga maimaiga* (Entomophthorales: Entomophthoraceae) in Slovakia. *Biocontrol Sci. Technol.* **2014**, *24*, 710–714. [CrossRef]
239. Tabaković-Tošić, M.; Georgiev, G.; Mirchev, P.; Tošić, D.; Golubović-Ćurguz, V. *Entomophaga maimaiga*—New entomopathogenic fungus in the Republic of Serbia. *Afr. J. Biotechnol.* **2012**, *34*, 8571–8577. [CrossRef]
240. Tabaković-Tošić, M.; Georgieva, M.; Hubenov, Z.; Georgiev, G. Impact of tachinid parasitoids of gypsy moth (*Lymantria dispar*) after the natural spreading and introduction of fungal pathogen *Entomophaga maimaiga* in Serbia. *J. Entomol. Zool. Stud.* **2014**, *2*, 262–266.
241. Cory, J.S.; Myers, J.H. The ecology and evolution of insect baculoviruses. *Annu. Rev. Ecol. Evol. Syst.* **2003**, *34*, 239–272. [CrossRef]
242. Clem, R.J.; Passarelli, A.L. Baculoviruses: Sophisticated pathogens of insects. *PLoS Pathog.* **2013**, *9*, e1003729. [CrossRef]
243. Weseloh, R.M. People and the gypsy moth: A story of human interactions with an invasive species. *Am. Entomol.* **2003**, *49*, 180–190. [CrossRef]
244. Doane, C.C. Ecology of pathogens of the gypsy moth. In *Perspectives in Forest Entomology*; Anderson, J.F., Kaya, H.K., Eds.; Academic Press: New York, NY, USA, 1976; pp. 285–293.
245. Shapiro, M.; Bell, R.A.; Owens, C.D. In vivo mass production of gypsy moth nucleopolyhedrosis virus. In *The Gypsy Moth; Research toward Integrated Pest Management*; Doane, C.C., McManus, M.L., Eds.; USDA Technical Bulletin 1584; USDA: Washington, DC, USA, 1981; pp. 633–655.
246. Lewis, F.B.; Yendol, W.G. Efficacy [of virus]. *U.S. Dep. Agric. For. Serv. Tech. Bull.* **1981**, *1584*, 503–512.
247. Ruiu, L.; Mannu, R.; Olivieri, M.; Lentini, A. Gypsy Moth Management with LdMNPV Baculovirus in Cork Oak Forest. *Forests* **2021**, *12*, 495. [CrossRef]
248. McManus, M.L.; Solter, L.F. Microsporidian pathogens in European gypsy moth populations. In *Ecology, Survey and Management of Forest Insects*; General Technical Report NE-311; McManus, M.L., Liebhold, A.M., Eds.; USDA Forest Service: Newton Square, PA, USA, 2003; pp. 44–51.
249. Goertz, D.; Hoch, G. Horizontal transmission pathways of terrestrial microsporidia: A quantitative comparison of three pathogens infecting different organs in *Lymantria dispar* L. (Lep.: Lymantriidae) larvae. *Biol. Control* **2008**, *44*, 196–206. [CrossRef]
250. Goertz, D.; Hoch, G. Vertical transmission and overwintering of microsporidia in the gypsy moth, Lymantria dispar. *J. Invertebr. Pathol.* **2008**, *99*, 43–48. [CrossRef]
251. Hoch, G.; Zubrik, M.; Novotny, J.; Schopf, A. The natural enemy complex of the gypsy moth, *Lymantria dispar* (Lep., Lymantriidae) in different phases of its population dynamics in eastern Austria and Slovakia—A comparative study. *J. Appl. Entomol.* **2001**, *125*, 217–227. [CrossRef]
252. Bathon, H. Biologische Bek€ Bekämpfung des Schwammspinners: Räuber und Parasitoide. In *Schwammspinner-Kalamität im Forst*; Wulf, A.B., Berendes, K.H., Eds.; Mitteilungen aus der Biologischen Bundesanstalt für Land- und Forstwirtschaft: Berlin, Germany, 1993; pp. 117–124.
253. Bierl, B.A.; Beroza, M.; Collier, C.W. Potent sex attractant of the gypsy moth: Its isolation, identification and synthesis. *Science* **1970**, *170*, 87–89. [CrossRef]
254. Beroza, M.; Knipling, E.F. Gypsy moth control with the sex attractant pheromone. *Science* **1972**, *177*, 19–27. [CrossRef] [PubMed]
255. Brown, R.C.; Sheals, R.A. The present outlook on the gypsy moth problem. *J. For.* **1944**, *42*, 393–407.
256. Mastro, V.C.; Richerson, J.V.; Cameron, E.A. An evaluation of gypsy moth pheromone-baited traps using behavioral observations as a measure of trap efficiency. *Environ. Entomol.* **1977**, *6*, 128–132. [CrossRef]
257. Elkinton, J.S.; Childs, R.D. Efficiency of two gypsy moth (Lepidoptera: Lymantriidae) pheromone-baited traps. *Environ. Entomol.* **1983**, *12*, 1519–1525. [CrossRef]
258. Thorpe, K.W.; Ridgway, R.L.; Leonhardt, B.A. Relationship between gypsy moth (Lepidoptera: Lymantriidae) pheromone trap catch and population density: Comparison of traps baited with 1 and 500 mg (þ)-disparlure lures. *J. Econ. Entomol.* **1993**, *86*, 86–92. [CrossRef]
259. Richerson, J.V. Pheromone-mediated behavior of the gypsy moth. *J. Chem. Ecol.* **1977**, *3*, 291–301. [CrossRef]
260. Liebhold, A.M.; Tobin, P.C. Population ecology of insect invasions and their management. *Annu. Rev. Entomol.* **2008**, *53*, 387–408. [CrossRef]
261. Suckling, D.M.; Tobin, P.C.; McCullough, D.G.; Herms, D.A. Combining tactics to exploit Allee effects for eradication of alien insect populations. *J. Econ. Entomol.* **2012**, *105*, 1–13. [CrossRef]

262. Epanchin-Niell, R.S.; Haight, R.G.; Berec, L.; Kean, J.M.; Liebhold, A.M. Optimal surveillance and eradication of invasive species in heterogeneous landscapes. *Ecol. Lett.* **2012**, *15*, 803–812. [CrossRef]
263. Tobin, P.C.; Onufrieva, K.S.; Thorpe, K.W. The relationship between male moth density and female mating success in invading populations of *Lymantria dispar*. *Entomol. Exp. Appl.* **2013**, *146*, 103–111. [CrossRef]
264. Miller, J.R.; Gut, L.J. Mating disruption for the 21st century: Matching technology with mechanism. *Environ. Entomol.* **2015**, *44*, 427–453. [CrossRef] [PubMed]
265. Cardé, R.; Minks, A.K. Control of moths by mating disruption: Successes and constraints. *Annu. Rev. Entomol.* **1995**, *40*, 559–585. [CrossRef]
266. Rice, R.E.; Kirsch, P. *Mating Disruption of Oriental Fruit Moth in the United States. Behavior-Modifying Chemicals for Insect Management*; Marcel Dekker: New York, NY, USA, 1990; pp. 193–211.
267. Bengtsson, M.; Karg, G.; Kirsch, P.A.; Lofqvist, J.; Sauer, A.; Witzgall, P. Mating Disruption of Pea Moth *CydiaNigricana* F (Lepidoptera, Tortricidae) by a Repellent Blend of Sex-Pheromone and Attraction Inhibitors. *J. Chem. Ecol.* **1994**, *20*, 871–887. [CrossRef] [PubMed]
268. Witzgall, P.; Backman, A.C.; Svensson, M.; Koch, U.; Rama, F.; ElSayed, A.; Brauchli, J.; Arn, H.; Bengtsson, M.; Lofqvist, J. Behavioral observations of codling moth, *Cydia pomonella*, in orchards permeated with synthetic pheromone. *Biocontrol* **1999**, *44*, 211–237. [CrossRef]
269. Thorpe, K.; Reardon, R.; Tcheslavskaia, K.; Leonard, D.; Mastro, V. A Review of the Use of Mating Disruption to Manage Gypsy Moth, *Lymantria dispar* (L.). Available online: http://www.fs.fed.us/foresthealth/technology/pdfs/GMComplete.pdf (accessed on 29 June 2022).
270. Walton, V.M.; Daane, K.M.; Bentley, W.J.; Millar, J.G.; Larsen, T.E.; Malakar-Kuenen, R. Pheromone-based mating disruption of *Planococcus ficus* (Hemiptera: Pseudococcidae) in California vineyards. *J. Econ. Entomol.* **2006**, *99*, 1280–1290. [CrossRef]
271. Stelinski, L.L.; Gut, L.J.; Miller, J.R. An attempt to increase efficacy of moth mating Disruption by Co-releasing pheromones with kairomones and to understand possible underlying mechanisms of this technique. *Environ. Entomol.* **2013**, *42*, 158–166. [CrossRef]
272. Trematerra, P.; Colacci, M.; Athanassiou, C.G.; Kavallieratos, N.G.; Rumbos, C.; Boukouvala, M.C.; Nikolaidou, A.J.; Kontodimas, D.C.; Benavent-Fernández, E.; Gálvez-Settier, S. Evaluation of Mating disruption for the control of *Thaumetopoea pityocampa* (Lepidoptera: Thaumetopoeidae) in suburban recreational areas in Italy and Greece. *J. Econ. Entomol.* **2019**, *112*, 2229–2235. [CrossRef]
273. Gordon, D.; Zahavi, T.; Anshelevich, L.; Harel, M.; Ovadia, S.; Dunkelblum, E.; Harari, A.R. Mating disruption of *Lobesia botrana* (Lepidoptera: Tortricidae): Effect of pheromone formulations and concentrations. *J. Econ. Entomol.* **2005**, *98*, 135–142. [CrossRef]
274. Stelinski, L.L.; Miller, J.R.; Ledebuhr, R.; Siegert, P.; Gut, L.J. Season-long mating disruption of *Grapholita molesta* (Lepidoptera: Tortricidae) by one machine application of pheromone in wax drops (SPLAT-OFM). *J. Pest Sci.* **2007**, *80*, 109–117. [CrossRef]
275. Hoshi, H.; Takabe, M.; Nakamuta, K. Mating disruption of a carpenter moth, *Cossus insularis* (Lepidoptera: Cossidae) in apple orchards with synthetic sex pheromone, and registration of the pheromone as an agrochemical. *J. Chem. Ecol.* **2016**, *42*, 606–611. [CrossRef]
276. Mori, B.A.; Evenden, M.L. When mating disruption does not disrupt mating: Fitness consequences of delayed mating in moths. *Entomol. Exp. Appl.* **2013**, *146*, 50–65. [CrossRef]
277. Tobin, P.C.; Bolyard, J.L.; Onufrieva, K.S.; Hickman, A.D. The effect of male and female age on *Lymantria dispar* (Lepidoptera: Lymantriidae) fecundity. *J. Econ. Entomol.* **2014**, *107*, 1076–1083. [CrossRef] [PubMed]
278. Hide, R.; Suckling, D. Decision analysis of insecticide resistance in light-brown apple moth. *N. Z. J. Exp. Agric.* **1988**, *16*, 219–224. [CrossRef]
279. Suckling, D.; Shaw, P.; Khoo, J.; Cruickshank, V. Resistance management of lightbrown apple moth, *Epiphyas postvittana* (Lepidoptera: Tortricidae) by mating disruption. *N. Z. J. Crop Hortic. Sci.* **1990**, *18*, 89–98. [CrossRef]
280. Caprio, M.A.; Suckling, D.M. *Mating Disruption Reduces the Risk of Resistance Development to Transgenic Apple Orchards: Simulations of the Lightbrown Apple Moth: New Zealand*; Plant Protection Society: Rotorua, New Zeland, 1995; pp. 52–58.
281. Schwalbe, C.P.; Mastro, V.C. Gypsy moth mating disruption: Dosage effects. *J. Chem. Ecol.* **1988**, *14*, 581–588. [CrossRef]
282. Leonhardt, B.A.; Mastro, V.C.; Leonard, D.S.; McLane, W.; Reardon, R.C.; Thorpe, K.W. Control of low-density gypsy moth (Lepidoptera: Lymantriidae) populations by mating disruption with pheromone. *J. Chem. Ecol.* **1996**, *22*, 1255–1272. [CrossRef]
283. Thorpe, K.W.; Mastro, V.C.; Leonard, D.S.; Leonhardt, B.A.; McLane, W.; Reardon, R.C.; Talley, S.E. Comparative efficacy of two controlled-release gypsy moth mating disruption formulations. *Entomol. Exp. Appl.* **1999**, *90*, 267–277. [CrossRef]
284. Onufrieva, K.S.; Thorpe, K.W.; Hickman, A.D.; Leonard, D.S.; Mastro, V.C.; Roberts, E.A. Gypsy moth mating disruption in open landscapes. *Agric. For. Entomol.* **2008**, *10*, 175–179. [CrossRef]
285. Collins, C.W.; Hood, C.E. *Gypsy Moth Tree Banding Material: How to Make, Use, and Apply It*; Bulletin 899; United States Department of Agriculture: Washington, DC, USA, 1920.
286. McManus, M.L.; Houston, D.R.; Wallner, W.E. *The Homeowner and the Gypsy Moths: Guidelines for Control*; Home and Garden Bulletin No. 227; USDA: Washington, DC, USA, 1980; p. 34.
287. McManus, M.L.; Smith, H.R. *Effectiveness of Artificial Bark Flaps in Mediating Migration of Late-Instar Gypsy Moth Larvae*; Forest Service Res. NE-316; U.S. Department of Agriculture: Washington, DC, USA, 1984.
288. Liebhold, A.M.; Elkinton, J.S.; Wallner, W.E. Effect of burlap bands on between-tree movement of late-instar gypsy moth, *Lymantria dispar* (Lepidoptera:Lymantriidae). *Environ. Entomol.* **1986**, *15*, 373–379. [CrossRef]

289. Blumenthal, E.M.; Hoover, C.R. Gypsy moth (Lepidoptera: Lymantriidae) population control using mechanical barriers and contact insecticides applied to tree stems. *J. Econ. Entomol.* **1986**, *79*, 1394–1396. [CrossRef]
290. Thorpe, K.W.; Webb, R.E.; Ridgway, R.L.; Venables, L.; Tatman, K.M. Sticky barrier bands affect density of Gypsy moth (Lepidoptera:Lymantriidae) and damage in oak canopies. *J. Econ. Entomol.* **1993**, *86*, 1479–1501. [CrossRef]
291. Bell, J.L.; Whitmore, R.C. Bird populations and habitat in *Bacillus thuringiensis* and Dimilin-treated and untreated areas of hardwood forest. *Am. Midl. Nat.* **1997**, *137*, 239–250. [CrossRef]
292. Conis, E. Polio, DDT, and disease risk in the United States after World War II. *Environ. Hist.* **2017**, *22*, 696–721. [CrossRef]
293. Allstadt, A.J.; Haynes, K.J.; Liebhold, A.M.; Johnson, D.M. Long-term shifts in the cyclicity of outbreaks of a forest-defoliating insect. *Oecologia* **2013**, *172*, 141–151. [CrossRef]
294. Schönfeld, F. Dimilin im Eichenwald—Insektizideinsatz mit Nebenwirkungen. *LWF Aktuell* **2009**, *70*, 58–60.
295. Durkin, P.R. *Control/Eradication Agents for the Gypsy Moth—Human Health and Ecological Risk Assessment for Bacillus thuringiensis var. Kurstaki (B.t.k.)*; SERA Technical Reports, 03-43-05-02d, (4-1–4-25); U.S. Department of Agriculture, Forest Service, Forest Health: Arlington, WA, USA, 2004.
296. Protection European Food Safety Authority. Conclusion on the peer review of the pesticide risk assessment of confirmatory data submitted for the active substance diflubenzuron. *EFSA J.* **2012**, *10*, 2870. [CrossRef]
297. Lemme, H.; Lobinger, G.; Müller-Kroehling, S. Schwammspinner—Massenvermehrung in Franken. *LWF Aktuell* **2019**, *121*, 37–43.
298. Sial, A.A.; Brunner, J.F. Toxicity and residual efficacy of Chlorantraniliprole, Spinetoram, and Emamectin benzoate to *Obliquebanded Leafroller* (Lepidoptera: Tortricidae). *J. Econ. Entomol.* **2010**, *103*, 1277–1285. [CrossRef]
299. Ishtiaq, M.; Razaq, M.; Saleem, M.A.; Anjumm, F.; Noor ul An, M.; Raza, A.M.; Wright, D.J. Stability, cross-resistance and fitness costs of resistance to emamectin benzoate in a re-selected field population of the beet armyworm, *Spodoptera exigua* (Lepidoptera: Noctuidae). *Crop Prot.* **2014**, *65*, 227–231. [CrossRef]
300. Xu, Z.; Bai, J.; Li, L.u.; Liang, L.; Ma, X.; Ma, L. Sublethal concentration of emamectin benzoate inhibits the growth of gypsy moth by inducing digestive dysfunction and nutrient metabolism disorder. *Pest Manag. Sci.* **2021**, *77*, 4073–4083. [CrossRef]
301. Oberemok, V.V.; Laikova, K.V.; Zaitsev, A.S.; Nyadar, P.M.; Gninenko, Y.I.; Gushchin, V.A.; Makarov, V.V.; Agranovsky, A.A. Topical treatment of LdMNPV-infected gypsy moth caterpillars with 18 nucleotides long antisense fragment from LdMNPV IAP-3 gene triggers higher levels of apoptosis in the infected cells and mortality of the pest. *J. Plant Prot. Res.* **2017**, *57*, 18–24. [CrossRef]
302. Oberemok, V.V.; Laikova, K.V.; Repetskaya, A.I.; Kenyo, I.M.; Gorlov, M.V.; Kasich, I.N.; Krasnodubets, A.M.; Gal'chinsky, N.V.; Fomochkina, I.I.; Zaitsev, A.S.; et al. A half-century history of applications of antisense oligonucleotides in medicine, agriculture and forestry: We should continue the journey. *Molecules* **2018**, *23*, 1302. [CrossRef] [PubMed]
303. Wang, Y.; Zhang, H.; Li, H.; Miao, X. Second-generation sequencing supplies an effective way to screen RNAi targets in large scale for potential application in pest insect control. *PLoS ONE* **2011**, *6*, e18644. [CrossRef] [PubMed]
304. Gu, L.; Knipple, D.C. Recent advances in RNA interference research in insects: Implications for future insect pest management strategies. *Crop Prot.* **2013**, *45*, 36–40. [CrossRef]
305. Oberemok, V.V.; Laikova, K.V.; Useinov, R.Z.; Gal'chinsky, N.V.; Novikov, I.A.; Yurchenko, K.A.; Volkov, M.E.; Gorlov, M.V.; Brailko, V.A.; Plugatar, Y.V. Insecticidal activity of three 10–12 nucleotides long antisense sequences from 5.8S ribosomal RNA gene of gypsy moth *Lymantria dispar* L. against its larvae. *J. Plant Prot. Res.* **2019**, *59*, 561–564.
306. Oberemok, V.V.; Laikova, K.V.; Gal'chinsky, N.V.; Useinov, R.Z.; Novikov, I.A.; Temirova, Z.Z.; Shumskykh, M.N.; Krasnodubets, A.M.; Repetskaya, A.I.; Dyadichev, V.V.; et al. DNA insecticide developed from the *Lymantria dispar* 5.8S ribosomal RNA gene provides a novel biotechnology for plant protection. *Sci. Rep.* **2019**, *9*, 6197. [CrossRef]
307. Oberemok, V.V.; Laikova, K.V.; Zaitsev, A.S.; Shumskykh, M.N.; Kasich, I.N.; Gal'chinsky, N.V.; Bekirova, V.V.; Makarov, V.V.; Agranovsky, A.A.; Gushchin, V.A.; et al. Molecular alliance of *Lymantria dispar* multiple nucleopolyhedrovirus and a short unmodified antisense oligonucleotide of its anti-apoptotic IAP-3 gene: A novel approach for gypsy moth control. *Int. J. Mol. Sci.* **2017**, *18*, 2446. [CrossRef]
308. Stenersen, J. *Chemical Pesticides Mode of Action and Toxicology*, 1st ed.; CRC Press: Boca Raton, FL, USA, 2004.
309. Devine, G.J.; Furlong, M.J. Insecticide use: Contexts and ecological consequences. *Agric. Hum. Values* **2007**, *24*, 281–306. [CrossRef]
310. Guedes, R.N.C.; Walse, S.S.; Throne, J.E. Sublethal exposure, insecticide resistance, and community stress. *Curr. Opin. Insect Sci.* **2017**, *21*, 47–53. [CrossRef]
311. Brevik, K.; Schoville, S.D.; Mota-Sanchez, D.; Chen, Y.H. Pesticide durability and the evolution of resistance: A novel application of survival analysis. *Pest Manag. Sci.* **2018**, *74*, 1953–1963. [CrossRef]
312. Umina, P.A.; McDonald, G.; Maino, J.; Edwards, O.; Hoffmann, A.A. Escalating insecticide resistance in Australian grain pests: Contributing factors, industry trends and management opportunities. *Pest Manag. Sci.* **2019**, *75*, 1494–1506. [CrossRef] [PubMed]
313. Dar, M.A.; Kaushik, G.; Chiu, J.F.V. Pollution status and biodegradation of organophosphate pesticides in the environment. In *Abatement of Environmental Pollutants*; Singh, P., Kumar, A., Borthakur, A., Eds.; Elsevier: Amsterdam, The Netherlands, 2020; pp. 25–66.
314. Senthil-Nathan, S. A review of bio pesticides and their mode of action against insect pests. In *Environmental Sustainability—Role of Green Technologies*; Thangavel, P., Sridevi, G., Eds.; Springer: New Delhi, India, 2015; pp. 49–63.
315. Kumar, V. A review on efficacy of biopesticides to control the agricultural insect's pest. *Int. J. Agric. Sci. Res.* **2015**, *4*, 168–179.
316. Anwer, M.A. *Biopesticides and Bioagents: Novel Tools for Pest Management*, 1st ed.; CRC Press: Boca Raton, FL, USA, 2017.

317. Isman, M.B. Commercial development of plant essential oils and their constituents as active ingredients in bioinsecticides. *Phytochem. Rev.* **2020**, *19*, 235–241. [CrossRef]
318. Stanković, S.; Kostić, M.; Kostić, I.; Krnjajić, S. Practical approaches to pest control: The use of natural compounds. In *Pests, Weeds and Diseases in Agricultural Crop and Animal Husbandry Production*; Kontogiannatos, D., Ed.; IntechOpen: London, UK, 2020; pp. 43–60.
319. Shahzad, K.; Manzoor, F. Nanoformulations and their mode of action in insects: A review of biological interactions. *Drug Chem. Toxicol.* **2021**, *44*, 1–11. [CrossRef]
320. Helson, B. Naturally derived insecticides: Prospects for forestry use. *For. Chron.* **1992**, *68*, 349–354. [CrossRef]
321. Kostić, I.; Lazarević, J.; Šešlija Jovanović, D.; Kostić, M.; Marković, T.; Milanović, S. Potential of Essential Oils from Anise, Dill and Fennel Seeds for the Gypsy Moth Control. *Plants* **2021**, *10*, 2194. [CrossRef]
322. Carson, F.; Hammer, K.A. Chemistry and bioactivity of essential oils. In *Lipids and Essential Oils as Antimicrobial Agents*; Thormar, H., Ed.; John Wiley & Sons: London, UK, 2011; Volume 25, pp. 203–238.
323. Koul, O. Phytochemicals and insect control: An antifeedant approach. *Crit. Rev. Plant Sci.* **2008**, *27*, 1–24. [CrossRef]
324. Trumble, J.T. Caveat emptor: Safety considerations for natural products used in arthropod control. *Am. Entomol.* **2002**, *48*, 7–13. [CrossRef]
325. Trongtokit, Y.; Rongsriyam, Y.; Komalamisra, N.; Apiwathnasorn, C. Comparative repellency of 38 essential oils against mosquito bites. *Phytother. Res.* **2005**, *19*, 303–309. [CrossRef]
326. Katz, M.; Miller, H.; Hebert, A. Insect repellents: Historical perspectives and new developments. *J. Am. Acad. Dermatol.* **2008**, *58*, 865–871. [CrossRef]
327. Regnault-Roger, C.; Vincent, C.; Arnason, J.T. Essential oils in insect control: Low-risk products in a high-stakes world. *Annu. Rev. Entomol.* **2012**, *57*, 405–424. [CrossRef]
328. Pavela, R.; Benelli, G. Essential oils as ecofriendly biopesticides? Challenges and constraints. *Trends. Plant Sci.* **2016**, *21*, 1000–1007. [CrossRef]
329. Moretti, L.; Sanna-Passino, G.; Demontis, S.; Bazzoni, E. Essential oil formulations useful as a new tool for insect pest control. *AAPS Pharm. Sci. Tech.* **2002**, *3*, e13. [CrossRef] [PubMed]
330. Kostić, I.; Petrović, O.; Milanović, S.; Popović, Z.; Stanković, S.; Todorović, G.; Kostić, M. Biological activity of essential oils of *Athamanta haynaldii* and *Myristica fragrans* to gypsy moth larvae. *Ind. Crop Prod.* **2013**, *41*, 17–20. [CrossRef]
331. Devrnja, N.; Kostić, I.; Lazarević, J.; Savić, J.; Ćalić, D. Evaluation of tansy essential oil as potential 'green' alternative for gypsy moth control. *Environ. Sci. Pollut. Res.* **2020**, *27*, 11958–11967. [CrossRef] [PubMed]
332. Devrnja, N.; Anđelković, B.; Aranđelović, S.; Radulović, S.; Soković, M.; Krstić-Milošević, D.; Ristić, M.; Ćalić, D. Comparative studies on the antimicrobial and cytotoxic activities of *Tanacetum vulgare* L. essential oil and methanol extracts. *S. Afr. J. Bot.* **2017**, *111*, 212–221. [CrossRef]
333. Berry, E.; Moldenke, F.; Miller, C.; Wernz, G. Toxicity of diflubenzuron in larvae of gypsy moth (Lepidoptera: Lymantriidae): Effects of host plant. *J. Econ. Entomol.* **1993**, *86*, 809–814. [CrossRef]
334. Traupman, J.C. *The Bantam New College Latin and English Dictionary*, 2nd ed.; Bantam Books: New York, NY, USA, 1995.
335. Hellier, F.F.; Warin, R.P. Caterpillar dermatitis. *Br. Med. J.* **1967**, *2*, 346–348. [CrossRef]
336. Haddad, V., Jr.; Lastória, J.C. Envenomation by caterpillars (erucism): Proposal for simple pain relief treatment. *J. Venom Anim. Toxins Incl. Trop. Dis.* **2014**, *20*, 21. [CrossRef]
337. Goldman, L.; Sawyer, F.; Levine, A.; Goldman, J.; Goldman, S.; Spinanger, J. Investigative studies of skin irritations from caterpillars. *J. Investig. Dermatol.* **1960**, *34*, 67–79. [CrossRef]
338. Haddad, V., Jr.; Cardoso, J.L.C. Erucismo e lepidopterismo. In *Animais Peçonhentos No Brasil: Biologia, Clínica e Terapêutica Dos Acidentes*; Cardoso, J.L.C., Franca, F.O.S., Wen, F.H., Malaque, C.M.S., Haddad, V., Jr., Eds.; Sarvier: São Paulo, Brazil, 2003; pp. 236–239.
339. Cardoso, A.E.C.; Haddad, V., Jr. Acidentes por lepidópteros (larvas e adultos de mariposas): Estudo dos aspectos epidemiológicos, clínicos e terapêuticos. *An. Bras. Dermatol.* **2005**, *80*, 571–578. [CrossRef]
340. Haddad, V., Jr.; Cardoso, J.L.; Lupi, O.; Tyring, S.K. Tropical dermatology: Venomous arthropods and human skin: Part I: Insecta. *J. Am. Acad. Dermatol.* **2012**, *67*, 331. [CrossRef] [PubMed]
341. Stipetic, M.E.; Rosen, P.B.; Borys, D.J. A retrospective analysis of 96 "Asp" (*Megalopyge opercularis*) envenomations in central Texas during 1996. *J. Toxicol. Clin. Toxicol.* **1999**, *37*, 457–462. [CrossRef] [PubMed]
342. Battisti, A.; Holm, G.; Fagrell, B.; Larsson, S. Urticating Hairs in Arthropods: Their Nature and Medical Significance. *Annu. Rev. Entomol.* **2011**, *56*, 203–210. [CrossRef]
343. Shama, S.K.; Etkind, P.H.; Odell, T.M.; Canada, A.T.; Finn, A.M.; Soter, N.A. Gypsy-moth-caterpillar-dermatitis. *N. Engl. J. Med.* **1982**, *306*, 1300–1301. [CrossRef]
344. Beaucher, W.N.; Farnham, J.E. Gypsymoth-caterpillar dermatitis. *N. Engl. J. Med.* **1982**, *21*, 1301–1302. [CrossRef]
345. De Jong, M.C.; Bleumink, E. Investigative studies of the dermatitis caused by the larva of the brown-tail moth, *Euproctis chrysorrhoea* L. (Lepidoptera, Lymantriidae). *Arch. Dermatol. Res.* **1977**, *259*, 247–262. [CrossRef]
346. Croitoru, D.O.; Brooks, S.G.; Pon, K. Dermatitis after exposure to *Lymantria dispar dispar*. *Can. Med. Assoc. J.* **2022**, *194*, E500. [CrossRef]
347. Deml, R. Pyrrolidinyl and pyridyl alkaloids in *Lymantria dispar*. *Z. Naturforsch. C Biosci.* **2003**, *58*, 860–866. [CrossRef]

348. Villas-Boas, I.M.; Bonfá, G.; Tambourgi, D.V. Venomous caterpillars: From inoculation apparatus to venom composition and envenomation. *Toxicon* **2018**, *153*, 39–52. [CrossRef]
349. Aldrich, J.R.; Schaefer, P.W.; Oliver, J.E.; Puapoomchareon, P.; Lee, C.J.; Vander Meer, R.K. Biochemistry of the exocrine secretion from gypsy moth caterpillars (Lepidoptera: Lymantriidae). *Ann. Entomol. Soc. Am.* **1997**, *90*, 75–82. [CrossRef]
350. Teutschlaender, O. Ueber die durch Rauperhaare Verursachten Erkrankungen. *Arch. Augenheilkd.* **1908**, *61*, 117–118.
351. Perlman, F.; Press, E.; Googins, J.A.; Malley, A.; Poarea, H. Tussockosis: Reactions to Douglas fir tussock moth. *Ann. Allergy* **1976**, *36*, 302–307. [PubMed]
352. Anderson, J.F.; Furniss, W.E. Epidemic of urticaria associated with first-instar larvae of the gypsy moth (Lepidoptera: Lymantriidae). *J. Med. Entomol.* **1983**, *20*, 146–150. [CrossRef]
353. Tuthill, R.W.; Canada, A.T.; Wilcock, K.; Etkind, P.H.; O'Dell, T.M.; Sharma, S.K. An epidemiologic study of gypsy moth rash. *Am. J. Public Health* **1984**, *74*, 799–803. [CrossRef] [PubMed]
354. Wirtz, R.A. Occupational allergies to arthropods—Documentation and prevention. *Bull. Entomol. Soc. Am.* **1980**, *26*, 356–360. [CrossRef]
355. Etkind, P.H.; Odell, T.M.; Canada, A.T.; Shama, S.K.; Finn, A.M.; Tuthill, R.W. The gypsy moth caterpillar: A significant new occupational and public health problem. *J. Occup. Med.* **1982**, *24*, 659–662. [CrossRef] [PubMed]
356. Steen, C.J.; Carbonaro, P.A.; Schwartz, R.A. Arthropods in dermatology. *J. Am. Acad. Dermatol.* **2004**, *50*, 819–842. [CrossRef]
357. Haq, M.; O'Toole, A.; Beecker, J.; Gooderham, M.J. Return of *Lymantria dispar dispar* (gypsy moth): A case report. *SAGE Open Med. Case Rep.* **2021**, *9*, 2050313X211057926. [CrossRef]
358. Gooderham, M.; Haq, M.; Beecker, J.; O'Toole, A. *Lymantria dispar dispar* (Gypsy) moth dermatitis. *J. Cutan. Med. Surg.* **2021**, *25*, 555–556. [CrossRef]
359. Kikuchi, T.; Kobayashi, K.; Sakata, K.; Akasaka, T. Gypsy moth-induced dermatitis: A hospital review and community survey. *Eur. J. Dermatol.* **2012**, *22*, 384–390. [CrossRef] [PubMed]
360. Hossler, E.W. Caterpillars and moths. Part I: Dermatologic manifestations of encounters with Lepidoptera. *J. Am. Acad. Dermatol.* **2010**, *62*, 1–10. [CrossRef] [PubMed]
361. Lin, W.; Yu, Y.; Zhou, P.; Zhang, J.; Dou, L.; Hao, Q.; Chen, H.; Zhu, S. Identification and Knockdown of the Olfactory Receptor (OrCo) in Gypsy Moth, *Lymantria dispar*. *Int. J. Biol. Sci.* **2015**, *11*, 772–780. [CrossRef] [PubMed]
362. European Parliament. Directive 2009/128/CE of the European Parliament and of the Council of 21 October 2009 Establishing a Framework for Community Action to Achieve the Sustainable Use of Pesticides. Available online: https://eur-lex.europa.eu/LexUriServ/LexUriServ.do?uri=OJ:L:2009:309:0071:0086:en:PDF (accessed on 6 September 2022).

insects

MDPI

Article

Effect of Entomopathogenic Fungi, *Beauveria bassiana* (Cordycipitaceae), on the Bark Beetle, *Ips typographus* (L.), under Field Conditions

Ciprian George Fora [1,*], Nicușor Boja [1], Mihaela Moatăr [1], Ferenc Tóth [2] and Adalbert Balog [3,*]

[1] Department of Forestry, Faculty of Horticulture and Forestry, Banat's University of Agricultural Sciences and Veterinary Medicine "King Michael I of Romania" from Timișoara, Calea Aradului 119, 300645 Timișoara, Romania
[2] Research Institute of Organic Agriculture, Miklós tér 1., 1033 Budapest, Hungary
[3] Department of Horticulture, Faculty of Technical and Human Sciences, Sapientia Hungarian University of Transylvania, Aleea Sighișoarei 1C, Corunca, 540485 Târgu Mureș, Romania
* Correspondence: ciprian.fora@usab-tm.ro (C.G.F.); adalbert.balog@ms.sapientia.ro (A.B.)

Simple Summary: The effect of *B. bassiana* fungal treatment on spruce bark beetle, *Ips typographus* (L.), was tested under field conditions. Altogether, it was detected that the effect of *B. bassiana* preparations on bark beetle infestations in terms of possible environmental effects is minimal compared to the severe damage and hence yield loss caused by this insect species worldwide. Even if the relatively low effects on I. typographus stages, including adults, larvae, pupae and even next-generation young adults, were reported, it remains clear that the present novel approach using a natural biological control agent rather than toxic synthetic insecticides is timely and may be potentially beneficial to the forestry industry as a rational control approach to combat this pest.

Abstract: The spruce bark beetle, *Ips typographus* (L.), attack progression (adult and larval galleries) and parental and offspring mortality rate were assessed in managed forests of the Poiana Ruscă and Bihor Mountains, along with Western Romanian Carpathians using fungal (*Beauveria bassiana*) treatments. The results show that the effect of *B. bassiana* on adult (maternal) gallery length was similar to the untreated variant and was less effective than the synthetic insecticide lambda-cyhalothrin applied at a dose of 50 g/L. Additionally, its effect on the mean larval gallery number per maternal gallery was low. *B. bassiana* did not have a significant influence on the attack progression. Significant correlations between log diameter and *I. typographus* attack progression were detected; such differences were higher than the effect of any treatment. Altogether, abiotic (low humidity, high temperature) and biotic factors (log diameter) influenced the bark beetles' attack progression and reduced the entomopathogenic fungal effects. Considering the efficacy of the *B. bassiana* treatment on logs infested with *I. typographus*, the results showed that parents and offspring were infected, but even if the fungal treatment was applied in high concentrations, the mortality rate remained relatively low. Further research is necessary to test if different *B. bassiana* strains and their commercially recommended concentrations might be more effective under dry and warm climate conditions, respectively.

Keywords: attack progression; biological control; fungal treatment; synthetic insecticide; climate change; mortality rate

1. Introduction

Forests are one of the planet's most valuable resources, being constantly under abiotic, biotic and anthropogenic pressure. One of the factors significantly influencing forest dynamics is drought, which results in reduced tree growth and accelerated drying. In Europe, the area of forest affected by drought and drying exceeded 500,000 ha between 1987 and 2016 [1]. Together with drought, the increase in multiannual mean temperature has also had a significant an impact on forests. The vegetation and the photosynthesis period have duly increased, as well as the intensity of respiration. As a consequence, the formation of buds and leaves and their maturation have been disturbed [2].

A result of such obvious climate change is the periodic occurrence of storms that affect deciduous stands and, especially, coniferous forests, spruce trees being highly susceptible to climate-induced destruction and disease in Europe [3]. The frequency of major storms is continuing to intensify year by year and is likely to increase in the coming decades. Thus, while at the beginning of the century, storms were less intense and rare, as we approach the present day, storms are becoming more powerful and more frequent, and the damage caused amounts to millions of cubic meters of forest on hundreds of thousands of hectares. In Europe, an average of some 128,000 ha of forest are affected by storms each year. In addition to storms, fires have a major impact on forests. Thus, in Europe as a whole, around 142,000 ha of forest are affected by fires every year [3].

Prolonged droughts, rising temperatures, severe storms and fires are factors that influence forest dynamics differentially from region to region and create favorable conditions for plant diseases and pests. One of the most damaging insect pest species under the above-mentioned abiotic conditions remains the eight-toothed spruce bark beetle, *Ips typographus* (L.) (Coleoptera: Scolytidae). The species is a secondary pest that can produce mass outbreaks under certain conditions and can even attack healthy standing spruce trees (e.g., Norway spruce, *Picea abies* L.) during such outbreaks. Several methods have been investigated to control outbreaks, ranging from felling and extracting attacked trees or peeling dry trees to using synthetic pheromones to capture adults or spraying logs with synthetic pyrethroids. Each of these methods has its drawbacks, which is why new, innovative, highly effective and environmentally friendly solutions are urgently required and need to be tested. The use of microbiological preparations, such as entomopathogenic fungi, *B. bassiana*, may prove to be a solution in forest management to limit bark beetle attacks.

B. bassiana is a naturally occurring pathogen of *I. typographus* [4]. Its pathogenicity to bark beetles has been demonstrated in a few studies [5–7]; however, its efficacy is known to decrease over time [8]. The effect of this fungus on bark beetles is still limited, and further research is needed to investigate the effects, especially including the ideal concentration and comparisons with commercial synthetic insecticides under the aforementioned abiotic conditions.

Therefore, in this study, we have attempted to answer some of the questions related to the effective use of *B. bassiana*-based products, the followings questions being posed: (1) How does the commercial product Metab (*B. bassiana* product) applied at different concentrations influence the length of maternal galleries and the number of larval/maternal galleries? (2) Does *B. bassiana* cause mortality in adults and offspring? (3) Does *B. bassiana* cause mortality in young adults in the next generation? (4) Is Metab more effective than a commercially used synthetic insecticide? In this light, the bark beetle attack development was assessed under fungal and synthetic insecticide treatments in managed forest plots and in buffer zones of protected areas.

2. Material and Methods

2.1. Data Collection

In order to answer the questions, three experiments were set up between 2019 and 2021 and conducted in three sites (OSAL 1, 2 and 3, where OSAL is the official name of the Forest Management Agency in Romania, Timiș, Romania) parallel in the vicinity of stands in which spruce represented 90% of trees and larch represented 10% of trees. The distance between the experimental sites was about 200 m. The sites were located outside the natural area, in the Poiana Ruscă Mountains of the Western Romanian Carpathians, at an altitude of 1100 m above sea level (a.s.l.) (45°40′45″ N; 22°17′16″ E). The forest stands had an average age of 110 years, a consistency of 0.9 and a production class of II (Figure 1A,B).

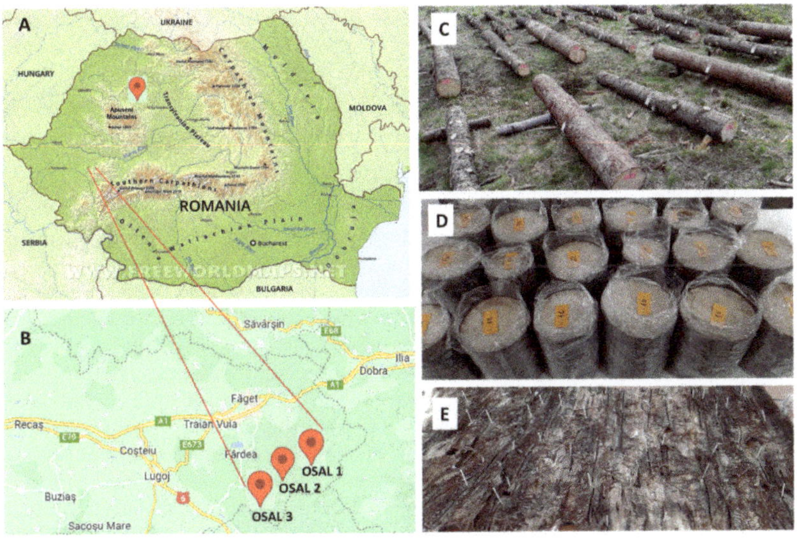

Figure 1. Experimental sites (**A**,**B**) (Google Map); plots with stems of branches of 1 m-long sections with diameters between 16 and 22 cm (**C**,**D**); galleries of *I. typographus* adults and larvae (**E**).

Trees were selected as wind felled trees during the winter period, which still had green crowns and no obvious spruce bark beetle infestations. They were cleaned of branches and sectioned every two meters, and the resulting logs were placed horizontally (so that they did not touch the ground) in Latin rectangle test platforms with six variations and four repetitions for each experiment. The logs were placed randomly and labelled. The diameter of the logs was then measured at the middle, and bark area was calculated for each one. Each sample was primed at the middle with the commercial pheromone Atratyp. The samples were sprayed as follows: variant 1 (V1) with water; variant 2 (V2) with Metab 75 mL/10 L water; variant 3 (V3) with Metab 105 mL/10 L water; variant 4 (V4) with Metab 135 mL/10 L water; variant 5 (V5) with Metab 165 mL/10 L water; variant 6 (V6) with lambda-cyhalothrin 50 g/L (Karate Zeon) 50 mL/10 L water. The spray solutions were prepared one day prior to the use of the microbiological product, according to the manufacturer's instructions. To activate the microorganisms, the Metab control product and the Nutriaction product were always added in double the amount of the test product. Thus, for example, for (V2), 150 mL Nutriaction was added; for (V3), 210 mL Nutriaction was added; for (V4), 270 mL Nutriaction was added; for (V5), 330 mL Nutriaction was added. The Zeon Karate solution was prepared shortly before spraying. The amount of solution applied was 300 mL/m^2 bark.

The first treatment application was made on 19 April 2019; the first set of samples for laboratory analysis was collected on 7 June 2019. Also on this latter date, the second treatment was applied, and, again, on 21 June 2019, the second set of samples was collected. These collections were made to avoid the flight of live insects from the experimental logs. The logs were transported to the laboratory two weeks after the last spraying period. There, the samples were peeled, and the following assessments were made: (1) The number of entries per bark sample subsequently expressed per 1 m^2 of bark; (2) 10 attack systems were chosen from the middle of each bark sample, and the lengths of 10 maternal galleries per sample were measured; (3) existing larval galleries were counted in the measured maternal galleries (Figure 1C–E).

The choice of the site for the experiments in the year 2020 was based on reports of outbreaks of *I. typographus* in the Apuseni Mountains (46°63′19″ N; 22°74′25″ E) (Figure 1A). In this area, after the massive felling caused by a great storm which occurred in September 2017 and which blew down about 150,000 spruce trees, bark beetle attacks were high. In the conservation area of the park, no intervention was made to clear the debris, and in the buffer zone, the debris was gradually exploited. The stand was selected in the buffer zone of the Apuseni Nature Park (managed forest), where spruce accounts for around 90% of the trees and beech accounts for 10% of the trees. The average age of trees in the stand was ~120 years; they had a consistency of 0.6 and were located at an altitude of 1170 m a.s.l. (Figure 1).

Of the trees felled during the storm and following the winter period, those that were not yet dry and had no insect infestation were chosen for the experiment. After clearing the stems of branches, they were cut into one-meter-long sections with diameters between 16 and 22 cm. The area of existing bark was calculated for each section. The sampled logs were randomly placed in the forest by hanging them upright on a dry wooden support. This was followed by labelling the four variants with six repetitions set for testing. The samples were investigated on 17 May 2020 with the onset of spring flight and sprayed on 17 May, 2 June and 16 June 2020 with the following experimental variants: variant 1 (V1) was sprayed with water; variant 2 (V2) was sprayed with *B. bassiana* 50 g/L; variant 3 (V3) was sprayed with *B. bassiana* 100 g/L; variant 4 (V4) was sprayed with *B. bassiana* 200 g/L. The amount of solution applied was 200 mL/m^2 of bark. The concentration of *B. bassiana* powder was 2 × 109 CFU/g. Throughout the experimental period, the air temperature and humidity were recorded using the Hexo device (Figure S1 Supplementary Materials). From the middle of each log, a 30 cm-long sample of stem was collected on 30 June 2020. The resulting samples were peeled and analyzed in the laboratory. Live and dead (infected) beetles from each sample were counted following the developmental stages of larva, pupa and adults. At the time of hatching, there were no young adults (considered the next generation) in the samples.

A new experiment was performed in the same experimental area as the previous one. Here, the aim was to observe the effects of the *B. bassiana* preparation (2 × 109 CFU (colony forming unit)/g)) on the adults of the new generations of *I. typographus*. Among the trees with broken, calcined tops, those with no insect entry were chosen for the experiment. After cleaning the stems from branches, they were sectioned into one-meter-long sections with diameters between 16 and 21 cm. The area of existing bark was calculated for each section. Logs were randomly placed in the forest by hanging vertically from a dry wood support. This was followed by labelling the four variants with six repetitions set for testing. The samples were installed on 30 June 2021, before the onset of the summer flight, and sprayed on 30 June and 14 July 2021 with the following experimental variants: variant 1 (V1) was sprayed with water; variant 2 (V2) was sprayed with *B. bassiana* 50 g/L; variant 3 (V3) was sprayed with *B. bassiana* 100 g/L; variant 4 (V4) was sprayed with *B. bassiana* 200 g/L. The amount of solution applied was 200 mL/m^2 of bark. Other experimental conditions and procedures were as previously detailed above.

2.2. Data Analyses

The data of the first experiment, relating to the average length of the maternal gallery, and the data related to the mean larval/maternal galley were normally distributed; therefore, ANOVA followed by Tukey's test were used for each experimental site (OSAL 1, 2, 3) and for the two sets of assessments separately to compare treatments.

Principal Component Analyses (PCoA were performed for each site in order to test the influence of log diameter on treatment effects (determined by the maternal gallery length), where the logs' diameter was considered as the main component, and the adult gallery length was considered as the variable. Linear correlations were also computed between variables (log diameter and maternal gallery length). The analyses were conducted in PAST, version 4.2. Data from the second and third experiments, following parent adults, larvae, pupae and young adults' survival rate, did not meet the assumption of normality; therefore, a Kruskal–Wallis test followed by a Wilcoxon test were used to compare survival rates. The data were also tested using PAST, version 4.2 (Oslo, Norway), with statistical differences assessed at a significance level of $p < 0.01$.

3. Results

In total, 148 logs of 50 cm in length were assessed in the first experiment, and 1480 maternal galleries and corresponding larval galleries were assessed and compared for each treatment. The effect of lambda-cyhalothrin 50 g/L was significant; smaller maternal galleries were detected after the first treatment in OSAL 1, 2. No effect between treatments at OSAL 3 was observed (Figure 2A–C, Table S1 Supplementary Materials).

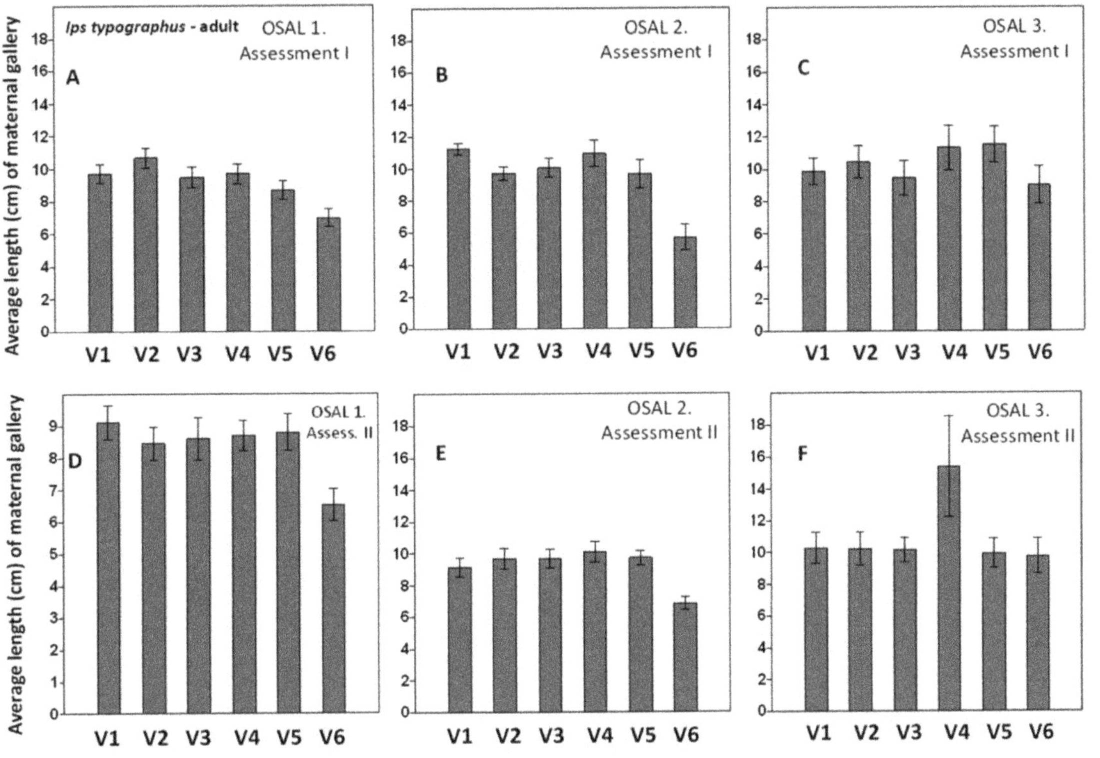

Figure 2. Average length (cm) of the adult *I. typographus* gallery in the first (**A–C**) and second (**D–F**) assessments: variants V1–V6 (V1, control, sprayed with water; V2, Metab 75 mL/10 L water; V3, Metab 105 mL/10 L water; V4, Metab 135 mL/10 L water; V5, Metab 165 mL/10 L water; V6, lambda-cyhalothrin 50 g/L (Karate Zeon) 50 mL/10 L water). Lines above bars represent standard errors (SE). OSAL is the official name of the Forest Management Agency in Romania.

The same results and a significant effect of the synthetic insecticide were observed at assessment 2 in sites (OSAL) 1 and 2, while the effect of treatment V4 (Metab 135 mL/10 L water) was less obvious than the others in site 3. No differences between the control treatment and the other treatments were detected (Figure 2D–F, Table S1 Supplementary Materials).

Considering the treatment effects of the (mean larval/maternal galleries), at the first assessment, a significant influence of the lambda-cyhalothrin was observed at sites 2 and 3, while in site 1, the effects of Metab at 75 and 105 mL (V2 and V3) were significant ((Figure 3A–C, Table S2 Supplementary Materials). For the second assessment, the effect of lambda-cyhalothrin was obvious in sites 1 and 2, whereas no differences in site 3 were detected (Figure 3D–F, Table S2 Supplementary Materials).

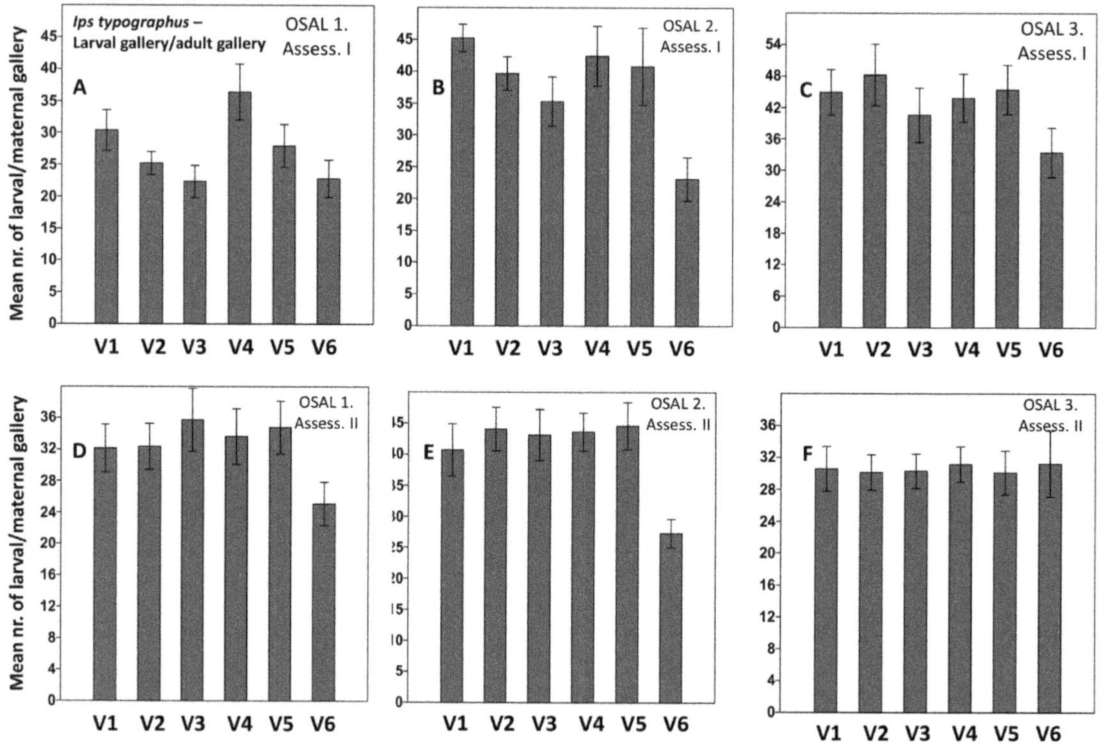

Figure 3. Average number of larval/adult gallery at first (**A–C**) and second (**D–F**) assessment. (V1) control, sprayed with water; variant 2 (V2) with Metab 75 mL/10 L water; variant 3 (V3) with Metab 105 mL/10 L water; variant 4 (V4) with Metab 135 mL/10 L water; variant 5 (V5) with Metab 165 mL/10 L water; variant 6 (V6) with lambda-cyhalothrin 50 g/L (Karate Zeon) 50 mL/10 L water. Lines above bars represent SE. OSAL is the official name of the Forest Management Agency in Romania.

Log diameter had positive effect on adult gallery length at each treatment, log diameter determined more than 90% the presence of bark beetles, and was highly correlated with adult gallery length (Figure 4 and Figure S2B,C Supplementary Materials).

Altogether 4775 individual beetles were assessed to test the mortality rate. No mortality were detected at all stages when treatments were made with water (control). Adult mortality rate in treated plots changed between ~7% at the lowest, and ~77% at the highest *B. bassiana* concentration. The mortality rate increased in accordance with *B. bassiana* increasing concentrations for all developmental stages and was found to be between 28.5% and 46.8% for larvae, 7.5% and 34.3% for pupae, and 16.1% to 29.4% for young adults. By assessing the adults, larvae, pupae and the young adult (next genera-tions) survival rate, it was observed that *B. bassiana* treatment was less effective. By comparing the rate of survived and dead individuals, the survived rate was higher for larvae, pupae and for the young adults (Figure 5 and Table S3 Supplementary Materials).

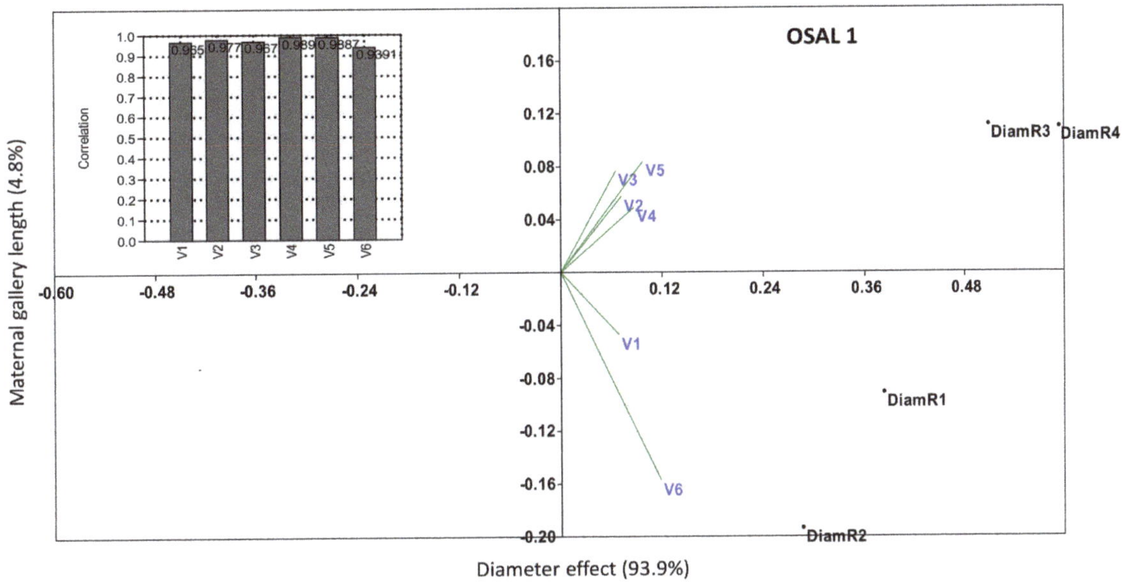

Figure 4. PCoA and linear correlation (bars) to test the influence of log diameter on treatments effects (determined by maternal gallery length), where the log length was considered as the main component, and the maternal gallery length was considered as the variable.

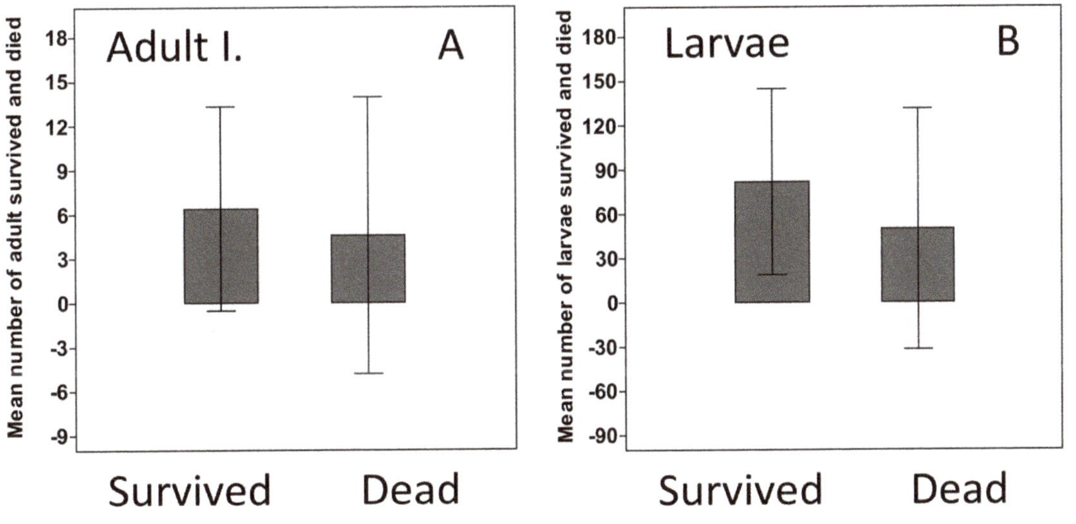

Figure 5. *Cont.*

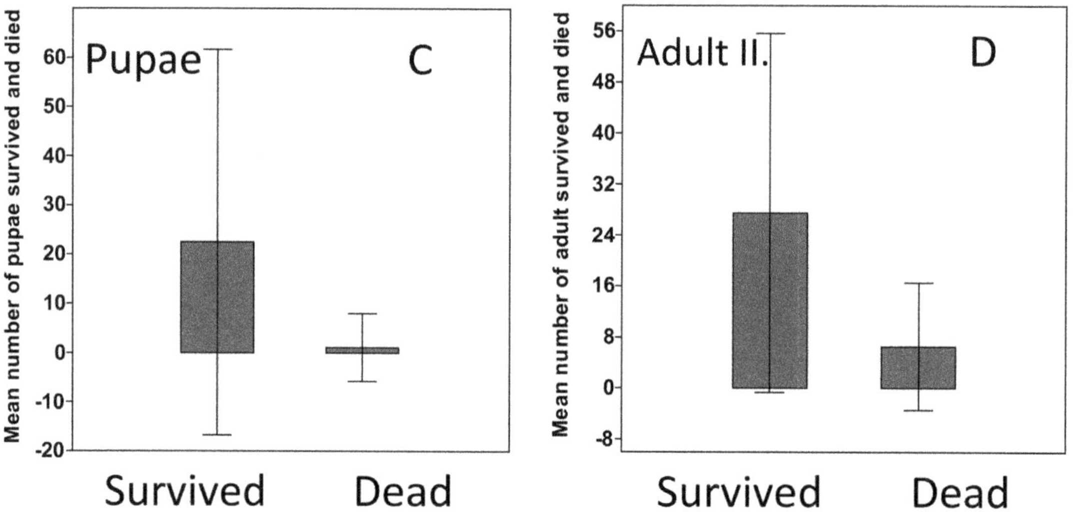

Figure 5. Adult, larval, pupae and next-generation adult survival rate under *B. bassiana* treatment. A Kruskal–Wallis test followed by a Wilcoxon test were used to compare survival rates. (**A**)–Adults from the start of the experiment, (**B**)–larval survival, (**C**)–Pupae survival, (**D**)–Adults from the next generations.

4. Discussion

According to our results, the effect of *B. bassiana* on the maternal gallery length and the average larvae gallery number per maternal gallery was less obvious than the synthetic insecticide lambda-cyhalothrin 50 g/L. The fungal treatment effect was rather similar to the untreated variant/s. The rates of infection detected in this experiment were seemingly highly, influenced by the favorable environmental conditions for bark beetles, which was also mentioned by other authors [9]. In our experiment, the relatively low humidity during the whole experimental period probably had a positive effect on bark beetles infestation and might have reduced the effect on fungal treatment. In addition, it can be seen that log diameter was the most influential factor, influencing more than 90% of the presence of bark beetle in all research areas. Other similar studies have also reported that log diameter has the most important and positive effect on *I. typographus* infestation since, as the diameter increases, the infestation rate also increases [10,11]. In our case, highly significant correlations were detected for all sites and treatments between log diameter and *I. typographus* infection, this being higher that the effect of the fungal treatments. Beside log diameters, high temperatures also positively influenced *I. typographus* infestation and reduced the effect of *B. bassiana*, as similarly demonstrated by other authors [12–14]. Research in the USA has shown that bark beetle outbreaks significantly influence forest ecosystem dynamics, warming summer and winter temperatures being major drivers of beetle population outbreaks across the US [15].

Overall, these factors (low humidity, high log diameter and high temperature) have highly positive effects on *I. typographus* infestation capacity and can be used as broad indicators when forecasting probable tree loss. These conditions were present during the whole experimental period (Figure S1; Supplementary Materials).

The application of *B. bassiana* under field conditions had a relatively low effect on *I. typographus* stages, including adults, larvae, pupae and even next-generation young adults. Even when high product concentrations were used, the mortality rate remained relatively low. This result is in accordance with the results obtained by other studies [16]. In contrast, other similar studies demonstrated that the application of *B. bassiana* preparations in the field can cause 93% [5] to 100% [6] mortality among bark beetles in the field and 99% mortality in the laboratory [16]. Further research is necessary to test whether *B. bassiana* preparations can be applied terrestrially, aerially or in combination with pheromones in traps, as previously suggested [17]. It remains clear that the effect of *B. bassiana* preparations on bark beetle infestations in terms of possible environmental effects is minimal compared to the severe damage and hence yield loss caused by this insect species worldwide, especially in Europe and North America [18]. The economic damage caused by this insect pest is huge, so the present novel approach of using a natural biological control agent rather than toxic synthetic insecticides is hopefully timely, as it is potentially beneficial to the forestry industry as a rational control approach to combatting this major pest.

Supplementary Materials: The following supporting information can be downloaded at: www.mdpi.com/xxx/s1. **Figure S1**. Air temperature and humidity recorded during the experiment using the Hexo device. **Table S1**. Statistical values related to the average length of maternal gallery. ANOVA followed by Tukey test were used for each experimental site (OSAL 1, 2, 3) separately to compare treatments. Significant *p*-values are green marked. **Table S2**. Statistical values related to the average number of maternal/larval gallery. ANOVA followed by Tukey test were used for each experimental site (OSAL 1, 2, 3) separately to compare treatments. Significant *p*-values are green marked. **Figure S2**. Principal Component Analyses (PCoA) and linear correlation to test the influence of trunk diameter on treatments effects (determined by adult gallery length) where the trunk length was considered as main component, and adult gallery length as variables at OSAL 2 and 3. **Table S3**. The adult, larvae, pupae and next generation adult survival rate were tested with Kruskal-Wallis followed by Wilcoxon test. Significant *p*-values are green marked.

Author Contributions: Conceptualization, C.G.F. and M.M; methodology, C.G.F.; software, A.B.; validation, C.G.F., F.T., M.M. and A.B.; formal analysis, C.G.F., F.T. and A.B.; investigation, C.G.F., N.B. and M.M.; resources, C.G.F. and F.T.; data curation, C.G.F. and A.B.; writing—original draft preparation, C.G.F. and A.B.; writing—review and editing, C.G.F. and A.B.; visualization, C.G.F., N.B. and A.B.; supervision, C.G.F. and A.B.; project administration, C.G.F.; funding acquisition, C.G.F. and A.B. All authors have read and agreed to the published version of the manuscript.

Funding: This research was partially funded by the project PN-III-P4-PCE-2021-0543.

Data Availability Statement: Fora, Ciprian George; Boja, Nicușor; Moatăr, Mihaela; Tóth, Ferenc; Balog, Adalbert (2022): Entomopathogenic fungi *Beauveria bassiana* (Cordycipitaceae) effect on the bark beetle *Ips typographus* (L.) under field conditions. figshare. Dataset. https://doi.org/10.6084/m9.figshare.20072207.v1, accessed on 12 September 2020.

Acknowledgments: We are grateful for *B. bassiana* products to BIOVÉD 2005 Kft. and for his director, Gyula Bohár support. We thank to the ROMSILVA administration for technical support and general permission to sample insects. We thank Hugh D. Loxdale for his helpful editorial comments on the manuscript of this paper.

Conflicts of Interest: The authors of this article have no financial or other conflict of interest to declare.

References

1. Senf, C.; Buras, A.; Zang, C.S.; Rammig, A.; Seidl, R. Excess Forest Mortality Is Consistently Linked to Drought across Europe. *Nat. Commun.* **2020**, *11*, 6200. [CrossRef] [PubMed]
2. Sendall, K.M.; Reich, P.B.; Zhao, C.; Jihua, H.; Wei, X.; Stefanski, A.; Rice, K.; Rich, R.L.; Montgomery, R.A. Acclimation of Photosynthetic Temperature Optima of Temperate and Boreal Tree Species in Response to Experimental Forest Warming. *Glob. Change Biol.* **2015**, *21*, 1342–1357. [CrossRef] [PubMed]
3. Senf, C.; Seidl, R. Storm and Fire Disturbances in Europe: Distribution and Trends. *Glob. Change Biol.* **2021**, *27*, 3605–3619. [CrossRef] [PubMed]
4. Wegensteiner, R.; Stradner, A.; Händel, U. Occurrence of Pathogens in Ips Typographus (Coleoptera: Curculionidae) and in Other Spruce Bark Beetles from the Wilderness Reserve Dürrenstein (Lower Austria). *Biologia* **2014**, *69*, 92–100. [CrossRef]
5. Kreutz, J.; Zimmermann, G.; Vaupel, O. Horizontal Transmission of the Entomopathogenic Fungus Beauveria Bassiana among the Spruce Bark Beetle, Ips Typographus (Col., Scolytidae) in the Laboratory and under Field Conditions. *Biocontrol. Sci. Technol.* **2004**, *14*, 837–848. [CrossRef]
6. Batta, Y.A. Biocontrol of Almond Bark Beetle (Scolytus Amygdali Geurin-Meneville, Coleoptera: Scolytidae) Using Beauveria Bassiana (Bals.) Vuill. (Deuteromycotina: Hyphomycetes). *J. Appl. Microbiol.* **2007**, *103*, 1406–1414. [CrossRef] [PubMed]

7. Reay, S.D.; Brownbridge, M.; Cummings, N.J.; Nelson, T.L.; Souffre, B.; Lignon, C.; Glare, T.R. Isolation and Characterization of Beauveria Spp. Associated with Exotic Bark Beetles in New Zealand Pinus Radiata Plantation Forests. *Biol. Control.* **2008**, *46*, 484–494. [CrossRef]
8. Yan, Z.; Sun, J.; Don, O.; Zhang, Z. The Red Turpentine Beetle, Dendroctonus Valens LeConte (Scolytidae): An Exotic Invasive Pest of Pine in China. *Biodivers. Conserv.* **2005**, *14*, 1735–1760. [CrossRef]
9. Krischik, V.A.; Davidson, J. IPM (Integrated Pest Management) of Midwest Landscapes. In *Minnesota Agricultural Experiment Station (MAES)*; Minnesota Agricultural Experiment Station (MAES): St. Paul, MN, USA, 2004.
10. Simionescu, A. *Protecţia Pădurilor*; Muşatinii: Suceava, Romania, 2000.
11. Simionescu, A.; Mihalache, G.; Mihalciuc, V.; Ciornei, C.; Chira, D.; Olenici, N.; Lupu, D.; Nețoiu, C.; Vlăduleasa, A.; Iliescu, M.; et al. *Protecția Pădurilor*; Mușatinii: Suceava, Romania, 2003.
12. Yeo, H.; Pell, J.K.; Alderson, P.G.; Clark, S.J.; Pye, B.J. Laboratory Evaluation of Temperature Effects on the Germination and Growth of Entomopathogenic Fungi and on Their Pathogenicity to Two Aphid Species. *Pest. Manag. Sci.* **2003**, *59*, 156–165. [CrossRef] [PubMed]
13. Bugeme, D.M.; Maniania, N.K.; Knapp, M.; Boga, H.I. Effect of Temperature on Virulence of Beauveria Bassiana and Metarhizium Anisopliae Isolates to Tetranychus Evansi. *Exp. Appl. Acarol.* **2008**, *46*, 275–285. [CrossRef] [PubMed]
14. Mann, A.J.; Davis, T.S. Plant Secondary Metabolites and Low Temperature Are the Major Limiting Factors for Beauveria Bassiana (Bals.-Criv.) Vuill. (Ascomycota: Hypocreales) Growth and Virulence in a Bark Beetle System. *Biol. Control.* **2020**, *141*, 104130. [CrossRef]
15. Bentz, B.J.; Régnière, J.; Fettig, C.J.; Hansen, E.M.; Hayes, J.L.; Hicke, J.A.; Kelsey, R.G.; Negrón, J.F.; Seybold, S.J. Climate Change and Bark Beetles of the Western United States and Canada: Direct and Indirect Effects. *BioScience* **2010**, *60*, 602–613. [CrossRef]
16. Nemčovič, M.; Šalamon, M.; Hyblerova, S.; Mazáň, M.; Salamon, I. Entomopathogenic Fungus Species Beauveria Bassiana (Bals.) and Metarhizium Anisopliae (Metsch.) Used as Mycoinsecticide Effective in Biological Control of Ips Typographus (L.). *J. Microbiol. Biotechnol. Food Sci.* **2013**, *2*, 2469.
17. Zúbrik, M.; Raši, R.; Vakula, J.; Varínsky, J.; Nikolov, C.; Novotný, J. Bark Beetle (Ips Typographus L., Pityogenes Chalcographus L., Col.: Scolytidae) Pheromone Traps Spatial Distribution Optimisation in Central Slovakian Mountains. *Lesn. Časopis-For. J.* **2008**, *2008*, 235–248.
18. Steinwender, B.; Krenn, H.; Wegensteiner, R. Different Effects of the Insectpathogenic Fungus Beauveria Bassiana (Deuteromycota) on the Bark Beetle Ips Sexdentatus (Coleoptera: Curculionidae) and on Its Predator Thanasimus Formicarius (Coleoptera: Cleridae). *J. Plant. Dis. Prot.* **2010**, *117*, 33–38. [CrossRef]

insects

MDPI

Communication

Survival Rate of the Neotropical Stingless Bees *Nannotrigona perilampoides* and *Frieseomelitta nigra* after Exposure to Five Selected Insecticides, under Controlled Conditions

Cristian Góngora-Gamboa, Esaú Ruiz-Sánchez *, Horacio S. Ballina-Gómez, Alejandra González-Moreno and Roberto Zamora-Bustillos *

División de Estudios de Posgrado e Investigación, TecNM/Instituto Tecnológico de Conkal, Av. Tecnológico s/n, Conkal 97345, Mexico

* Correspondence: esau.rs@conkal.tecnm.mx (E.R.-S.); roberto.zb@conkal.tecnm.mx (R.Z.-B.)

Simple Summary: In the neotropics, stingless bees co-exist with plant pests in agroecosystems. The use of chemical insecticides used to control sap-sucking insects may pose a risk to the communities of stingless bees. To gain insight into the potential risk of insecticides commonly used for farmers in horticultural crops, we evaluated under laboratory conditions the effects of oral exposure to five insecticides on the survival of two species of stingless bees, *Nannotrigona perilampoides* and *Frieseomelitta nigra*. The results showed that some insecticides have a significant negative impact on the survival of stingless bees under laboratory conditions. These results suggest that stingless bees may be negatively affected when foraging flowers of crops that have been treated with insecticides.

Abstract: Insecticides used in agricultural pest management pose survival risks to the stingless bees that forage on crops in tropical and subtropical regions. In the present study, we evaluated, under laboratory conditions, the acute oral toxicity of five selected insecticides (dinotefuran, imidacloprid, flupyradifurone, spirotetramat, and cyantraniliprole) to two species of neotropical stingless bees: *Nannotrigona perilampoides* and *Frieseomelitta nigra*. At field recommended doses, dinotefuran, imidacloprid, and flupyradifurone caused the highest mortality in both bee species. These insecticides also caused the largest decrease in the survival rate when exposed to a 10-fold dilution of the field recommended doses. Notably, dinotefuran exerted a high effect even at 100-fold dilution (100% mortality). In contrast, cyantraniliprole had a low effect and spirotetramat was virtually nontoxic. These results suggest that some insecticides used to control sap-sucking insects may have a significant negative impact on the communities of stingless bees.

Keywords: stingless bees; neonicotinoid insecticides; diamide insecticide; butenolide insecticide; cyclic keto-enol insecticide; nAChR-acting insecticides

1. Introduction

Stingless bees (Hymenoptera: Meliponini) are important pollinators to wild and cultivated plants [1]. In some regions of southeastern Mexico, stingless bees are particularly associated with economically important local species crops. For example, in the Yucatán Peninsula, Mexico, a region with a high diversity and abundance of stingless bees, *Nannotrigona perilampoides* and *Frieseomelitta nigra* are efficient pollinators of peppers, tomatoes, and avocado [2–4], and of important native plants such as annatto, also known as "*achiote*" (*Bixa orellana*) and *Jatropha curcas*, used for the production of biofuel [5,6]. Furthermore, the stingless bees *Melipona beecheii* pollinate *achiote* [6] and *Scaptotrigona pectoralis*, and *Trigona fulviventris* pollinate chili peppers (*Capsicum chinense*) [7], which shows that stingless bees play an important role in maintaining the diversity of the flora.

In the last couple of decades, a reduction in the richness and abundance of native stingless bees has been experienced, which has been associated with various factors, such as the anthropogenic disturbance and high dominance of the Africanized honeybee [8], the loss of forests [9,10], and the presence of insecticides in the environment [11].

In intensively managed agroecosystems, stingless bees are exposed to a variety of chemical insecticides when foraging flowers [12,13]. In this sense, to control highly damaging sap-sucking insects, farmers of the region have made intense use of four groups of insecticides: neonicotinoids (imidacloprid, thiamethoxam, and dinotefuran), butenolides (flupyradifurone), cyclic keto-enols (spirotetramat), and diamides (cyantraniliprole) [13–15]. The use of these novel groups of insecticides has been questioned due to their negative impact upon and threat to the bee community [16].

To gain insight into the potential risks of insecticide use in horticultural crops, we evaluated under laboratory conditions the effects of oral exposure to dinotefuran, imidacloprid, flupyradifurone, spirotetramat, and cyantraniliprole to the survival of two species of stingless bees, *Nannotrigona perilampoides* and *Frieseomelitta nigra*.

2. Materials and Methods

2.1. Insects and Insecticides

The research was carried out at the Tecnológico Nacional de México, Campus Conkal, at Conkal, Yucatán, Mexico. The colonies of *N. perilampoides* and *F. nigra* were maintained in wooden hives in the field (five hives per bee species) and allowed to forage naturally. The vegetation surrounding the hives included *Piscidia piscipula*, *Leucaena leucocephala*, *Tithonia diversifolia*, and *Parthenium hysterophorus*. In the feeding zone of the stingless bees, the absence of agricultural crops under the application of agrochemicals was verified. Additionally, twice a week, 10 mL of a 50% sucrose solution (1 g of sucrose: 1 mL of distilled water) was provided to each hive as a diet source.

The commercial formulations of the insecticides were purchased from local agrochemical suppliers. All the insecticides evaluated (Table 1) are recommended to control sap-sucking insects (Aleyrodidae and Aphididae) in horticultural crops. The concentrations of the insecticides used for the evaluation were selected based on the application rate recommended by the manufacturer for the whitefly (*Bemisia tabaci*). From these field recommended doses, 10-fold and 100-fold dilutions were prepared for the experiment of survival analysis.

Table 1. Details of the insecticides evaluated in the present experiment.

Insecticide (Chemical Group)	Concentration (Formulation) [a]	Evaluated Concentrations (mg L^{-1}) *	Trade Name (Manufacturer)	Mode of Action
Dinotefuran (Neonicotinoid)	200 g/kg (SG)	600, 60, 6	Venom 20 SG (Valent)	Agonist of Nicotinic Acetylcholine receptor (nAChR)
Imidacloprid (Neonicotinoid)	350 g a.i. L^{-1} (SC)	700, 70, 7	Confial (Quimica Sagal)	Agonist of Nicotinic Acetylcholine receptor (nAChR)
Flupyradifurone (Butenolide)	200 g a.i. L^{-1} (SL)	680, 68, 6.8	Sivanto Prime (Bayer Crop Science)	Agonist of Nicotinic Acetylcholine receptor (nAChR)
Spirotetramat (Cyclic keto-enol)	150 g a.i. L^{-1} (OD)	300, 30, 3	Movento 150 OD (Bayer Crop Science)	Inhibitor of acetyl CoA carboxilase
Cyantraniliprole (Anthranilic diamide)	200 g L^{-1} (SC)	200, 20, 2	Benevia (FMC)	Activator of Ryanodine Receptor (RyR)

* The highest concentrations of the insecticides used for the evaluation were based on the field recommended doses for the whitefly (*Bemisia tabaci*) and their 10-fold and 100-fold dilution. [a] Concentration and (formulation). SG: soluble granules; SC: suspension concentrate; SL: soluble concentrate; OD: oil dispersion.

2.2. Oral Toxicity Bioassay

Groups of 10 foragers were collected from the entrance of the hives of the stingless bees. Each group, obtained from a different hive, was placed in a plastic bottle (500 mL) with filter paper on the bottom and a 20 cm^2 section of organza mesh was fixed to the side of the bottle to facilitate the air exchange. In the middle part of the bottle (the organza mesh section) we placed a microtube (1.5 mL) for food supply, which consisted of sucrose solution (sucrose/water 1:1) ad libitum. To minimize the stress caused by confinement, prior to the bioassays the bees remained in adaptation for approximately 12 h (one night and part of the next morning) at room conditions, 25 °C and 70% R.H.

The oral acute toxicity test was conducted in adult foragers as described by Botina et al. [17]. Prior to insecticide exposure, bees were starved for 2 h. Following this, 1 mL of insecticide-contaminated diet was placed in the microtube placed at the middle part of the bottle. Insecticide-free diet was used as control. The mortality was recorded after 2, 4, and 24 h of offering the diet. Individuals were considered dead if they did not react when stimulated with a fine hairbrush [18,19]. The experiment was set in a completely randomized design with five replicates per insecticide, each experimental unit consisting of 10 foragers confined in a 500 mL plastic bottle.

2.3. Statistical Analysis

Analysis of variance (ANOVA) was used to compare the mortality caused by the field recommended doses of the insecticides over 24 h. Comparison of means was carried out with Tukey post hoc test at $p < 0.05$. The normality and homoscedasticity of the data were confirmed prior to their analysis. Survival curves were obtained with the Kaplan–Meier estimators, which were generated from the percentage of mortality of bees over 2, 4, and 24 h evaluation periods with 10-fold and 100-fold dilution of the insecticides. The estimated survival functions were compared with a nonparametric Log-Rank test (Holm–Sidak method). All the statistical analyses were performed using the SigmaPlot 11.0 software (Systat Software, San Jose, CA, USA).

3. Results

3.1. Mortality

Oral exposure for 24 h to the field recommended doses of insecticides caused significant differences in mortality in *N. perilampoides* ($F = 14.03$, df = 5, $p < 0.0001$) and *F. nigra* ($F = 76.06$, df = 5, $p < 0.0001$). In both bee species, dinotefuran, imidacloprid, and flupyradifurone were the most toxic. Bee mortality ranged from 90–100% when exposed to dinotefuran and imidacloprid, and from 75–90% when exposed to flupyradifurone (Figure 1). Mortality caused by cyantraniliprole was lower (30–45%). Spirotetramat was not toxic to either bee species (Figure 1).

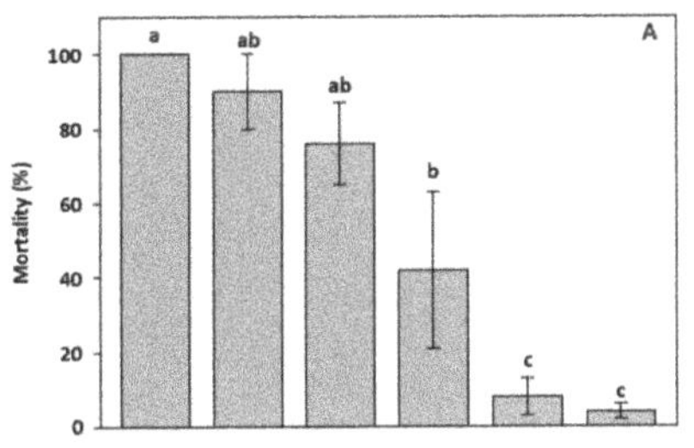

Figure 1. *Cont.*

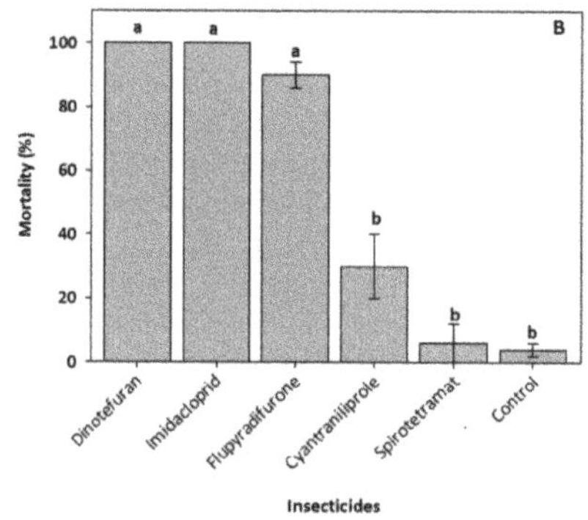

Figure 1. Mortality percentage (± standard error) of the stingless bees *Nannotrigona perilampoides* (**A**) and *Frieseomelitta nigra* (**B**), orally exposed for 24 h to the insecticides at field recommended doses, dinotefuran (600 mg a.i. L^{-1}), imidacloprid (700 mg a.i. L^{-1}), flupyradifurone (680 mg a.i. L^{-1}), spirotetramat (300 mg a.i. L^{-1}), and cyantraniliprole (200 mg a.i. L^{-1}). Bars with different letters are statically different (Tukey test, n = 5, $p < 0.05$).

3.2. Survival Analysis

Exposure to the insecticides, in a 10-fold dilution and 100-fold dilution, resulted in significantly distinctive effects in both bee species (Tables S1–S4): *N. perilampoides* ($\chi^2 = 300.8$, df = 5, $p < 0.001$; Figure 2A) and *F. nigra* ($\chi^2 = 432.9$, df = 5, $p < 0.001$; Figure 2B), depending on the agrochemical used. Overall, in both species, dinotefuran, and imidacloprid produced the largest decrease in the survival rate, followed by flupyradifurone. Cyantraniliprole had a slight, but significant effect, and spirotetramat showed no toxicity (Figure 2).

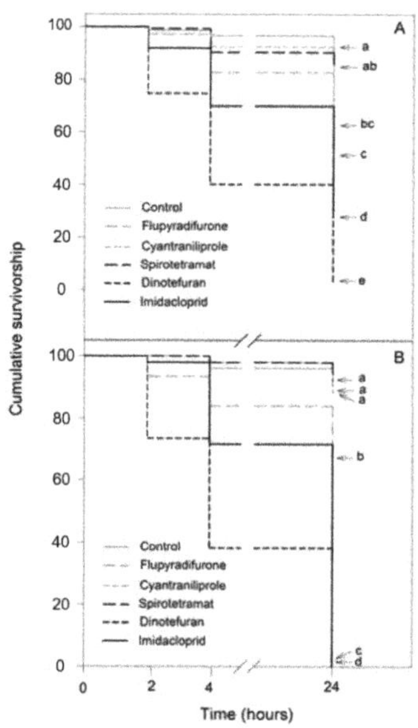

Figure 2. Kaplan–Meier survivorship curves of stingless bees orally exposed to 10-fold diluted commercial formulation of the selected insecticides: dinotefuran, 60 mg L^{-1}; imidacloprid, 70 mg L^{-1}; flupyradifurone, 68 mg L^{-1}; spirotetramat, 30 mg L^{-1}; and cyantraniliprole, 20 mg L^{-1}. The figure displays *Nannotrigona perilampoides* (**A**) and *Frieseomelitta nigra* (**B**). The estimated survival functions were compared with a nonparametric Log-Rank test (Holm–Sidak method). Significant differences ($p < 0.05$) among insecticide treatments are indicated by different letters.

For the 100-fold dilution of the recommended application rate, significant differences in the effects of the insecticides were observed depending on the bee species (Figure 3). In *N. perilampoides* ($\chi^2 = 322.1$, df = 5, $p < 0.001$), flupyradifurone and dinotefuran caused significant decreases in the survival rates: the highest effect was produced by dinotefuran (90% mortality), while cyantraniliprole, spirotetramat, and imidacloprid had no effect (Figure 3A). On the other hand, in *F. nigra* ($\chi^2 = 466.6$, df = 5, $p < 0.001$) dinotefuran produced a complete depletion of the survival rate (100% mortality). The other insecticides (flupyradifurone, cyantraniliprole, spirotetramat, and imidacloprid) had no effect (Figure 3B).

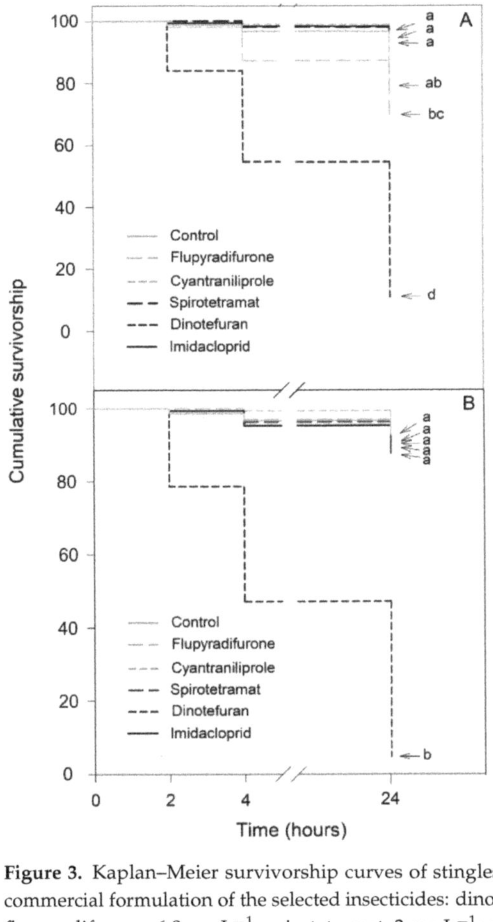

Figure 3. Kaplan–Meier survivorship curves of stingless bees orally exposed to 100-fold diluted commercial formulation of the selected insecticides: dinotefuran, 6 mg L^{-1}; imidacloprid, 7 mg L^{-1}; flupyradifurone, 6.8 mg L^{-1}; spirotetramat, 3 mg L^{-1}; and cyantraniliprole, 2 mg L^{-1}. The figure displays *Nannotrigona perilampoides* (**A**) and *Frieseomelitta nigra* (**B**). The estimated survival functions were compared with a nonparametric Log-Rank test (Holm–Sidak method). Significant differences ($p < 0.05$) among insecticide treatments are indicated by different letters.

4. Discussion

In the present study, we evaluated the effects of oral exposure to selected insecticides on the survival of two species of neotropical stingless bees, *N. perilampoides* and *F. nigra*, under laboratory conditions. It is worth considering that oral exposure may negatively impact not only the survival of the directly exposed bees, but also the bees in the colonies, given that foragers, who collect contaminated pollen or nectar, carry them to the hives [20–22].

Here, we observed that at field recommended doses, dinotefuran, imidacloprid, and flupyradifurone (nAChR-acting insecticides) exerted the strongest lethal effects in both bee species. For dinotefuran and imidacloprid, a dramatic decrease in the survival rate (approximately 50–60% mortality) was observed as quickly as 4 h after exposure to the 10-fold dilution of the field recommended doses. Moreover, we observed that even at 100-fold dilution, dinotefuran was able to dramatically deplete the bee survival rate (100% mortality). In this sense, previous studies have shown that nAChR-acting insecticides are highly toxic to various species of bees [23–25], and their lethal effects at realistic concentrations are observed as quickly as 1 h after insect exposure [26]. In the present work, we also found that flupyradifurone caused a significant decrease in the survival rate of *N. perilampoides* and *F. nigra* when used at the 10-fold dilution. Flupyradifurone is considered a highly toxic insecticide to the European bee (*Apis mellifera*), which may indicate that this insecticide would be even more toxic to stingless bees, given their high susceptibility to insecticides [27,28].

The anthranilic diamide, cyantraniliprole, showed low to moderate toxicity to both species of bees at the 10-fold dilution, but had no effect at all at the 100-fold dilution. To our knowledge, this insecticide has not been previously evaluated on stingless bees. However, Tomé et al. [29] found that chlorantraniliprole (3 mg L^{-1}), another anthranilic diamide compound, affects the flight activity of *Partamona helleri* and *Scaptotrigona xanthotrica*. In *Apis mellifera*, compounds from this group of insecticides have moderate effects when applied topically [30]. In agriculture, cyantraniliprole is recommended to control phytophagous species of the families Thripidae, Liviidae, Aphididae, Aleyrodidae, Noctuidae, Plutellidae, and Gelechiidae [14,31–33]. In contrast to nAChR-acting insecticides, anthranilic diamide insecticides act on ryanodine receptors (RyR), which are responsible for intracellular calcium regulation within the sarcoplasmic reticulum of insect muscle cells [34], but with differential effects among insect orders and even among species [35,36]. This difference in susceptibility has been attributed to the binding affinity of the insecticide molecule to the target site [37]. It is reasonable to infer that the low to moderate toxicity of cyantraniliprole to the stingless bees, shown in the present work, may be in part due to this pharmacological characteristic of the anthranilic diamide insecticides.

The cyclic keto-enol, spirotetramat, had no acute oral effects on both species of stingless bees. Spirotetramat has been recommended to control both immature and adult stages of phytophagous mites [38,39], and immature Hemiptera of the families Aleyrodidae, Aphididae, and Triozidae [40,41]. Spirotetramat is particularly effective against the juvenile stages of pest insects; the compound also reduces fecundity and fertility in adults [40]. Owing to its mode of action (the disruption of lipid biosynthesis by inhibiting the Acetyl-CoA-carboxylase), spirotetramat has low toxicity to adult insects [14,42], which would explain why this insecticide was not toxic to either species of stingless bee. This outcome is in agreement with other studies that showed the nontoxic effects of spirotetramat on adults of *A. mellifera* [43,44], and adults of *Apis cerana*, *Apis florea*, and *Trigona iridipennis* [44]. In this work, we did not evaluate the effects of spirotetramat on the immature stages of the stingless bees, thus we could not rule out the potential effects of this compound when ingested by larvae in contaminated nectar/pollen carried by foragers to the hives.

5. Conclusions

The acute oral exposure of *N. perilampoides* and *F. nigra* to the nAChR-acting insecticides dinotefuran, imidacloprid, and flupyradifurone, negatively impacted their survival rate. Even at a low concentration (100-fold of the recommended application rate), dinotefuran produced a dramatic impact on bee survival, whereas insecticides that act on ryanodine receptors (cyantraniliprole) or inhibit lipid biosynthesis (spirotetramat) caused moderate and nontoxic acute effects, respectively. The results show that the intensive use of the nAChR-acting insecticides (dinotefuran, imidacloprid and flupyradifurone) to control sap-sucking insects in horticultural crops poses a threat to the communities of neotropical stingless bees that forage in the area. It is important to evaluate the impact of these insecticides on the expression of genes involved in metabolic pathways as well as to evaluate the epigenetic changes after long term exposure, in order to gain insight on the effects of these insecticides at physiological and molecular levels in stingless bees.

Supplementary Materials: The following are available online at https://www.mdpi.com/article/10.3390/insects13100961/s1, Tables S1–S4. Evaluation of 10–100-fold dilution of the recommended application rate.

Author Contributions: Conceptualization, E.R.-S.; methodology, C.G.-G., and E.R.-S.; validation, C.G.-G., E.R.-S., H.S.B.-G., A.G.-M., and R.Z.-B.; formal analysis, H.S.B.-G.; investigation, C.G.-G.; resources, E.R.-S. and R.Z.-B.; data curation, C.G.-G.; writing—original draft preparation, C.G.-G., and E.R.-S.; writing—review and editing, E.R.-S., H.S.B.-G., A.G.-M., and R.Z.-B.; visualization, E.R.-S., and R.Z-B.; supervision, E.R.-S. All authors have read and agreed to the published version of the manuscript.

Funding: This research was funded by Tecnológico Nacional de México (TecNM) grant number 13817.22-P.

Data Availability Statement: The data are available upon request from the corresponding authors esau.rs@conkal.tecnm.mx and roberto.zb@conkal.tecnm.mx.

Acknowledgments: To the National Council of Science and Technology (CONACyT, Mexico) for the postgraduate scholarship granted to the first author. Likewise, thanks to Marcos Cua-Basulto for his technical assistance.

Conflicts of Interest: The authors declare no conflict of interest.

References

1. Meléndez-Ramírez, V.; Ayala, R.; Delfín-González, H. Crop pollination by stingless bees. In *Pot-Pollen in Stingless Bee Melittology*, 1st ed.; Vit, P., Pedro, S.R.M., Roubik, D.W., Eds.; Springer Cham: Cham, Switzerland, 2018; pp. 139–153. [CrossRef]
2. Cauich, O.; Quezada-Euán, J.J.G.; Macias-Macias, J.O.; Reyes-Oregel, V.; Medina-Peralta, S.; Parra-Tabla, V. Behavior and pollination efficiency of *Nannotrigona perilampoides* (Hymenoptera: Meliponini) on greenhouse tomatoes (*Lycopersicon esculentum*) in subtropical Mexico. *J. Econ. Entomol.* **2004**, *97*, 475–481. [CrossRef] [PubMed]
3. Can-Alonzo, C.; Quezada-Euán, J.J.G.; Ancona-Xiu, P.; Moo-Valle, H.; Valdovinos-Nuñez, G.R.; Medina-Peralta, S. Pollination of 'criollo' avocados (*Persea americana*) and the behaviour of associated bees in subtropical Mexico. *J. Apic. Res.* **2005**, *44*, 3–8. [CrossRef]
4. Cauich, O.; Quezada-Euán, J.J.G.; Meléndez-Ramírez, V.; Valdovinos-Nuñez, G.R.; Moo-Valle, H. Pollination of habanero pepper (*Capsicum chinense*) and production in enclosures using the stingless bee *Nannotrigona perilampoides*. *J. Apic. Res.* **2006**, *45*, 125–130. [CrossRef]
5. Romero, M.J.; Quezada-Euán, J.J.G. Pollinators in biofuel agricultural systems: The diversity and performance of bees (Hymenoptera: Apoidea) on *Jatropha curcas* in Mexico. *Apidologie* **2013**, *44*, 419–429. [CrossRef]
6. Caro, A.; Moo-Valle, H.; Alfaro, R.; Quezada-Euán, J.J.G. Pollination services of Africanized honey bees and native *Melipona beecheii* to buzz-pollinated annatto (*Bixa orellana* L.) in the neotropics. *Agric. For. Entomol.* **2016**, *19*, 274–280. [CrossRef]
7. Landaverde-González, P.; Quezada-Euán, J.J.G.; Theodorou, P.; Murray, T.E.; Husemann, M.; Ayala, R.; Moo-Valle, H.; Vandame, R.; Paxton, J.R. Sweat bees on hot chillies: Provision of pollination services by native bees in traditional slash-and-burn agriculture in the Yucatán Peninsula of tropical Mexico. *J. Appl. Ecol.* **2017**, *54*, 1814–1824. [CrossRef]
8. Cairns, C.E.; Villanueva-Gutiérrez, R.; Koptur, S.; Bray, B.D. Bee populations, forest disturbance, and africanization in Mexico. *Biotropica* **2005**, *37*, 686–692. [CrossRef]
9. Brosi, B.J. The complex responses of social stingless bees (Apidae: Meliponini) to tropical deforestation. *For. Ecol. Manag.* **2009**, *258*, 830–1837. [CrossRef]
10. Mayes, D.M.; Bhatta, C.P.; Shi, D.; Brown, J.C.; Smith, D.R. Body size influences stingless bee (Hymenoptera: Apidae) communities across a range of deforestation levels in Rondônia, Brazil. *J. Insect Sci.* **2019**, *19*, 23. [CrossRef]
11. Main, A.R.; Webb, E.B.; Goyne, K.W.; Mengel, D. Reduced species richness of native bees in field margins associated with neonicotinoid concentrations in non-target soils. *Agric. Ecosyst. Environ.* **2020**, *287*, 106693. [CrossRef]
12. Macias-Macias, O.; Chuc, P.; Ancona-Xiu, P.; Cauich, O.; Quezada-Euán, J.J.G. Contribution of native bees and Africanized honeybees (Hymenoptera: Apoidea) to Solanaceae crop pollination in tropical México. *J. Appl. Entomol.* **2009**, *133*, 456–465. [CrossRef]
13. Montejo-Canul, E.; Aguiñaga-Bravo, A.; Ruiz-Sánchez, E.; Ballina-Gómez, H.; González-Moreno, A.; Latournerie-Moreno, L.; Martín-Mex, R.; Garruña-Hernández, R. Effects of the inclusion of biorational insecticides for pest management on phytophagous insects, fruit yield, and bee abundance in tomato and tomatillo. *Arch. Phytopathol. Pflanzenschutz* **2019**, *52*, 17–18. [CrossRef]

14. Chen, J.-C.; Wang, Z.-H.; Cao, L.-J.; Gong, Y.-J.; Hoffmann, A.A.; Wei, S.-J. Toxicity of seven insecticides to different developmental stages of the whitefly *Bemisia tabaci* MED (Hemiptera: Aleyrodidae) in multiple field populations of China. *Ecotoxicology* **2018**, *27*, 742–751. [CrossRef] [PubMed]
15. Monroy-Reyes, B.; Carrillo-Gutiérrez, T.; Beas-Zarate, C.; Posos-Ponce, P.; Castro-Rodriguez, M.; Enciso-Cabral, J.G.; Flores-Galano, G. Evaluación de efectividad biológica del insecticida Benevia 10 OD (ciantraniliprol) para el control de *Frankliniella occidentalis* pergande, (Thysanoptera: Thripidae) en aguacate. *Entomología Mexicana* **2019**, *6*, 188–193.
16. Siviter, H.; Muth, F. Do novel insecticides pose a threat to beneficial insects? *Proc. R. Soc. B* **2020**, *287*, 20201265. [CrossRef]
17. Botina, L.L.; Bernardes, R.C.; Barbosa, W.F.; Lima, M.A.P.; Guedes, R.N.C.; Martins, G.F. Toxicological assessments of agrochemical effects on stingless bees (Apidae, Meliponini). *MethodsX* **2020**, *7*, 100906. [CrossRef]
18. Lourenço, C.T.; Carvalho, S.M.; Malaspina, O.; Nocelli, R.C.F. Oral toxicity of fipronil insecticide against the stingless bee *Melipona scutellaris* (Latreille, 1811). *Bull. Environ. Contam. Toxicol.* **2012**, *89*, 921–924. [CrossRef]
19. Jacob, C.R.O.; Malaquias, J.B.; Zanardi, O.Z.; Silva, C.A.S.; Jacob, J.F.O.; Yamamoto, P.T. Oral acute toxicity and impact of neonicotinoids on *Apis mellifera* L. and *Scaptotrigona postica* Latreille (Hymenoptera: Apidae). *Ecotoxicology* **2019**, *28*, 744–753. [CrossRef]
20. Sanchez-Bayo, F.; Goka, K. Pesticide residues and bees–a risk assessment. *PLoS ONE* **2014**, *9*, e94482. [CrossRef]
21. Lima, M.A.P.; Martins, G.F.; Oliveira, E.E.; Guedes, R.N.C. Agrochemical-induced stress in stingless bees: Peculiarities, underlying basis, and challenges. *J. Comp. Physiol. A* **2016**, *202*, 733–747. [CrossRef]
22. Cham, K.O.; Nocelli, R.C.F.; Borges, L.O.; Viana-Silva, F.E.C.; Tonelli, C.A.M.; Menezes, O.M.C.; Rosa-Fontana, A.S.; Blochtein, B.; Freitas, B.M.; Pires, C.S.S.; et al. Pesticide exposure assessment paradigm for stingless bees. *Environ. Entomol.* **2018**, *20*, 36–48. [CrossRef] [PubMed]
23. Cresswell, J.E. A meta-analysis of experiments testing the effects of a neonicotinoid insecticide (imidacloprid) on honey bees. *Ecotoxicology* **2011**, *20*, 149–157. [CrossRef] [PubMed]
24. Goulson, D. An overview of the environmental risks posed by neonicotinoid insecticides. *J. Appl. Ecol.* **2013**, *50*, 977–998. [CrossRef]
25. Arena, M.; Sgolastra, F. A meta-analysis comparing the sensitivity of bees to pesticides. *Ecotoxicology* **2014**, *23*, 324–334. [CrossRef]
26. Tosi, S.; Nieh, J.C. Lethal and sublethal synergistic effects of a new systemic pesticide, flupyradifurone (Sivanto®), on honeybees. *Proc. R. Soc B* **2019**, *286*, 20190433. [CrossRef] [PubMed]
27. Nauen, R.; Jeschke, P.; Velten, R.; Beck, M.E.; Ebbinghaus-Kintscher, U.; Thielert, W.; Wölfel, K.; Hass, M.; Kunz, K.; Raupach, G. Flupyradifurone: A brief profile of a new butenolide insecticide. *Pest Manag. Sci.* **2015**, *71*, 850–862. [CrossRef] [PubMed]
28. Hesselbach, H.; Scheiner, R. Effects of the novel pesticide flupyradifurone (Sivanto) on honeybee taste and cognition. *Sci. Rep.* **2018**, *8*, 4954. [CrossRef] [PubMed]
29. Tomé, H.V.V.; Barbosa, W.F.; Corrêa, A.S.; Gontijo, L.M.; Martins, G.F.; Guedes, R.N.C. Reduced-risk insecticides in neotropical stingless bee species: Impact on survival and activity. *Ann. Appl. Biol.* **2015**, *167*, 186–196. [CrossRef]
30. Kadala, A.; Charreton, M.; Charnet, P.; Collet, C. Honey bees long-lasting locomotor deficits after exposure to the diamide chlorantraniliprole are accompanied by brain and muscular calcium channels alterations. *Sci. Rep.* **2019**, *9*, 2153. [CrossRef]
31. Zhang, Z.; Xu, C.; Ding, J.; Zhao, Y.; Lin, J.; Liu, F.; Mu, W. Cyantraniliprole seed treatment efficiency against *Agrotis ipsilon* (Lepidoptera: Noctuidae) and residue concentrations in corn plants and soil. *Pest Manag. Sci.* **2019**, *75*, 1464–1472. [CrossRef]
32. Li, Y.; Dou, Y.; An, J.; Tu, X.; Lv, H.; Pan, W.; Dang, Z.; Gao, Z. Temperature-dependent variations in toxicity of diamide insecticides against three lepidopteran insects. *Ecotoxicology* **2020**, *29*, 607–612. [CrossRef] [PubMed]
33. Renkema, J.M.; Krey, K.; Devkota, S.; Liburd, O.E.; Funderburk, J. Efficacy of insecticides for season-long control of thrips (Thysanoptera: Thripidae) in winter strawberries in Florida. *Crop Prot.* **2020**, *127*, 104945. [CrossRef]
34. Selby, T.P.; Lahm, G.P.; Stevenson, T.M.; Hughes, K.A.; Cordova, D.; Annan, I.B.; Barry, J.D.; Benner, E.A.; Currie, M.J.; Pahutski, T.F. Discovery of cyantraniliprole, a potent and selective anthranilic diamide ryanodine receptor activator with cross-spectrum insecticidal activity. *Bioorg. Med. Chem. Lett.* **2013**, *23*, 6341–6345. [CrossRef] [PubMed]
35. Isaacs, A.K.; Qi, S.; Sarpong, R.; Casida, J.E. Insect ryanodine receptor: Distinct but coupled insecticide binding sites for N-C(3)H(3) chlorantraniliprole, flubendiamide, and (3)H ryanodine. *Chem. Res. Toxicol.* **2012**, *25*, 1571–1573. [CrossRef] [PubMed]
36. Dinter, A.; Samel, A. Cyantraniliprole: Pollinator profile of the novel insecticides under laboratory, semi-field and field conditions. In Proceedings of the Hazards of Pesticides to Bees–12th International Symposium of the ICP-Bee Protection Group, Ghent, Belgium, 15–17 September 2014.
37. Qi, S.; Casida, J.E. Species differences in chlorantraniliprole and flubendiamide insecticide binding sites in the ryanodine receptor. *Pestic. Biochem. Phys.* **2013**, *107*, 321–326. [CrossRef]
38. Lümmen, P.; Khajehali, J.; Luther, K.; Van Leeuwen, T. The cyclic keto-enol insecticide spirotetramat inhibits insect and spider mite acetyl-CoA carboxylases by interfering with the carboxyltransferase partial reaction. *Insect Biochem. Mol. Biol.* **2014**, *55*, 1–8. [CrossRef]
39. Popov, S.Y.; Alyokhin, A. Gender-Specific acaricidal properties and sexual transmission of spirotetramat in two-spotted spider mite (Tetranychidae: Acariformes). *J. Econ. Entomol.* **2019**, *112*, 2186–2192. [CrossRef]
40. Nauen, R.; Reckmann, U.; Thomzik, J.; Thielert, W. Biological profile of spirotetramat (Movento®)–a new two-way systemic (ambimobile) insecticide against sucking pest species. *Bayer Cropsci. J.* **2008**, *61*, 245–278.

41. Brück, E.; Elbert, A.; Fischer, R.; Krueger, S.; Kühnhold, J.; Klueken, A.M.; Nauen, R.; Niebes, J.F.; Reckmann, U.; Schnorbach, H.J.; et al. Movento®, an innovative ambimobile insecticide for sucking insect pest control in agriculture: Biological profile and field performance. *Crop Prot.* **2009**, *28*, 838–844. [CrossRef]
42. Marčić, D.; Perić, P.; Petronijević, S.; Prijović, M.; Drobnjaković, T. Cyclic ketoenols: Acaricides and insecticides with a novel mode of action. *Pestic. I Fitomed.* **2011**, *26*, 185–195. [CrossRef]
43. Maus, C. Ecotoxicological profile of the insecticide spirotetramat. *Bayer Cropsci. J.* **2008**, *61*, 159–180.
44. Vinothkumar, B.; Kumaran, N.; Boomathi, N.; Saravanan, P.A.; Kuttalam, S. Toxicity of spirotetramat 150 OD to honeybees. *Madras Agric. J.* **2010**, *97*, 86–87.

insects

MDPI

Article

Trunk Injection with Insecticides Manages *Xylotrechus chinensis* (Chevrolat) (Coleoptera: Cerambycidae)

Nickolas G. Kavallieratos *, Maria C. Boukouvala *, Anna Skourti, Erifili P. Nika and Georgios Th. Papadoulis

Laboratory of Agricultural Zoology and Entomology, Department of Crop Science, Agricultural University of Athens, 75 Iera Odos Str., 11855 Athens, Greece

* Correspondence: nick_kaval@aua.gr (N.G.K.); mbouk@aua.gr (M.C.B.); Tel.: +30-210-529-4569 (N.G.K.)

Simple Summary: Over a two-year period, the trunk injection method was adopted for the management of *Xylotrechus chinensis* (Chevrolat) (Coleoptera: Cerambycidae) infesting mulberries. Three conventional insecticides, i.e., fipronil, imidacloprid, and spirotetramat, were injected into their trunks and the number of exit holes was recorded after the application. Abamectin was used as a positive control and untreated trees were selected as controls. In the first year (late April–late June 2021), after the insecticidal applications, the emerging holes of *X. chinensis* were significantly decreased compared to the holes recorded before the treatment (late April–late June 2020). The same trend was observed with the second round of applications of the method (mid June–mid August 2021), where the number of exit holes (late April–late June 2022) was further reduced, reaching 71.8, 76.1, and 85.6% reductions in the cases of fipronil, imidacloprid, and abamectin, respectively. In contrast, spirotetramat did not cause a significant reduction in the emerging holes of *X. chinensis* throughout the experimental period.

Abstract: *Xylotrechus chinensis* (Chevrolat) (Coleoptera: Cerambycidae) is a serious wood-boring insect of mulberry trees (*Morus* spp.). Larvae of this species enter the trunk of the tree and feed on woody tissues. *Xylotrechus chinensis* is endemic in several Asian countries, while, in the last decade, it invaded Europe. In the present work, we evaluated trunk injection against this pest. The systemic insecticides fipronil, imidacloprid, and spirotetramat were used in the trials. Abamectin was used as a positive control since it has been found to be effective for the management of *X. chinensis*. Imidacloprid and fipronil proved to be highly effective against this species in both years (9.5 and 12.1 exit holes/tree in 2021, 5.0 and 8.8 exit holes/tree in 2022, respectively), while spirotetramat was the least effective. The lowest mean number of exit holes was recorded when abamectin was applied in both years (4.7 exit holes/tree in 2021 and 3.3 exit holes/tree in 2022). The percentage of exit holes was reduced by 76.1, 71.8, and 85.6% in trees treated with imidacloprid, fipronil, and abamectin, respectively, after two years of application, while spirotetramat caused a 37.4% reduction. Trunk injection with imidacloprid, fipronil, and abamectin could be used against *X. chinensis* for long-term control of mulberry trees.

Keywords: tiger longicorn beetle; mulberry trees; insecticides; pest management

1. Introduction

The tiger longicorn beetle, *Xylotrechus chinensis* (Chevrolat) (Coleoptera: Cerambycidae) is an invasive wood-boring insect pest of economic importance, which mainly infests the mulberry trees (*Morus* spp.) [1,2]. It is endemic to China, Japan, South Korea, and the Republic of China [3–5], while it has spread exclusively on Europe's mulberry trees [1,2,6,7]. However, the Asian populations of this species have been reported to infest a wider range of host plants compared to the European populations, including the common grape vine, *Vitis vinifera* L. (Vitales: Vitaceae), as well as species of the genera *Malus* and *Pyrus* (Rosaceae) [3,8–12].

Xylotrechus chinensis overwinters as a larva inside wood of infested trees, where it subsequently becomes a pupa. The adult individuals emerge between May and August through cyclic exit holes with a diameter of 5–6 mm [7]. However, adults emerge earlier in Greece (May–June) than in Spain (July–August) [1,2]. After emergence, adults move towards the top of the tree to feed with the stems and leaves, and finally become sexually mature [1,2]. Mating and oviposition occurs on the bark of the tree, where each female can lay approximately 80 eggs in her lifetime [7,13]. After hatching, larvae create galleries under the bark and enter the woody tissues, gradually emphasizing the lower parts of the trunk and roots [1,2]. The feeding activity of larvae results in a gradual collapse of the tree through the disruption of water and nutrients, leading finally to its death [13]. Infestation symptoms include many exit holes, mainly located in the southeastern parts of the tree, as well as insect droppings and frass on the trunk surface [1,2]. Apart from the severe damage to the trees, there are concerns about the safety of people in urban parks and streets, since heavily infested branches may fall down [2], an issue that forces the municipal authorities to take measures regarding the drastic management of *X. chinensis*.

The first record of *X. chinensis* in Europe was in 2007 in Bavaria (Germany), where two beetles were detected in packing box made of wood, originating from China, and were intercepted [14]. In July 2017, another interception of this pest was reported in Rhineland-Palatinate (Germany), in a container from China carrying woody decorative items [5]. In 2013, this species was reported in Catalonia, (Spain), but it is believed to have been established in the country in 2012 or earlier [2,7]. Since then, *X. chinensis* spread rapidly in Catalonia (Spain) covering an infestation area of 44.1 km^2 in four cities in 2018, which further increased two years later to 378.1 km^2 in 12 cities [15]. In 2018, new outbreaks were recorded in Spain, this time in Valencia [16]. *Xylotrechus chinensis* was detected for the first time in Greece, in the island of Crete, during the spring of 2017, infesting the trunk of a single mulberry tree [1]. After two years, a serious outbreak of this pest occurred in Athens (Greece), where 1,300 out of 20,000 mulberry trees, were heavily infested [6,17]. The first official report of *X. chinensis* in France took place in October 2018 [18], where a specimen of the pest was found on a balcony in a residence in Gironde [19,20], while the next year, infested mulberry trees were observed in Occitanie [7,20].

Although *X. chinensis* has not been yet listed as a quarantine organism in Europe [19] it fulfills the criteria for further consideration as a potential Union quarantine insect species [7]. In 2018, *X. chinensis* was incorporated into the Alert List of European and Mediterranean Plant Protection Organization (EPPO), to accentuate the risks incurred to EPPO member countries in the case of its introduction into their territories [13].

The detection of wood-boring insects faces difficulties due to the fact that they pupate inside the trunk; consequently, the tree itself becomes the host, providing a natural refuge [1,2]. These insects cause detrimental effects on tree health and, in many cases, as in the case of *X. chinensis*, attack and kill trees [2]. Preventive measures such as the removal and destruction of heavily infected mulberry trees are recommended for the management of *X. chinensis* [2,19]. Another suggested method is to spray the bark of the trees with contact insecticides during summer, targeting females, oviposition sites, and first instar larvae [2]. However, this practice is not recommended in urban areas due to the risk for public health [21].

A way to control *X. chinensis* is related to trunk injection with insecticides. An insecticide is injected into the xylem of the tree, which is transferred with the xylem sap upwards through vessels to phloem, where larvae can feed [1]. Thus, the insecticide is confined to the treated tree, making it useful and safe in urban areas [22]. This method has been applied in Barcelona, Spain, in 2018 by injecting abamectin in 107 mulberry trees [15]. However, there are no data on the effectiveness of imidacloprid, spirotetramat, and fipronil against *X. chinensis*. Thus, the objective of the current study was to test the three active ingredients (a.i.) against *X. chinensis* through the trunk injection method over a period of two years.

2. Materials and Methods

2.1. Formulations

The following three insecticidal formulations were used for the experiment: i. Movento OD, which is an oil-dispersion containing 15.78% w/v spirotetramat as a.i. (provided by Bayer Hellas, Greece), ii. Termidor SC, which is a suspension concentrate that contains 9.1% w/w fipronil as a.i. (provided by BASF Hellas, Greece), and iii. Confidor SL with 20% w/v imidacloprid as a.i. (provided by Bayer Hellas, Greece). Vertimec EC, an emulsifiable concentrate that contains 1.8% w/v abamectin as a.i. (provided by Syngenta, Greece), was used as positive control, since it was documented as an effective insecticide against *X. chinensis* using the endotherapy method [15].

2.2. Application of Insecticidal Formulations

The insecticidal formulations were applied to 40 infested *Morus* spp. trees (mainly *Morus alba* L./few *Morus nigra* L.) in the campus of the Agricultural University of Athens (Greece) as follows: 10 trees were treated with each insecticide. Ten more infested trees were chosen as untreated controls. Preliminary observations revealed a severe infestation of mulberry trees by *X. chinensis* in this area. The average height of the selected trees was 2.5 m. Before the application of the insecticides to the selected trees, the exit holes of each tree were recorded and marked with a red color (Figure 1a). The perimeter of each tree was measured to be 1.30 m above the ground to estimate the diameter of the trunk at breast height (Diameter Breast Height, DBH). The perimeter at 1.30 m above the ground was divided by 3.14 to give the DBH [23]. Then, the number of insecticide injection points/tree was calculated, dividing the DBH by 2 [23]. Since the selected formulations are not registered for trunk injection; therefore, there are no recommended label doses, so we relied on the study of VanWoerkom [24] to determine the concentration of the a.i. that was to be injected into the trunk. The author examined 2 rates, 0.2 and 0.4 g, of a.i. per trunk DBH of each tree. In the present study, the higher ratio (0.4 g) was chosen for the experimentation. Based on the concentration of each a.i. in the corresponding formulation, the quantity of each insecticidal formulation that contained 0.4 g of a.i. was calculated. Consequently, the corresponding DBH of the tree trunk was multiplied by 0.4, and the total quantity of insecticide per tree in mL was estimated. Then, the volume of each insecticide per injection point was determined, dividing the total quantity of insecticide per tree by the number of injection points/tree [23].

The trunk injection was carried out from 8.00 a.m. to 2.00 p.m. daily between 15 July and 10 August 2020, because the newly emerged larvae start feeding with the phloem of the tree. Thus, it is considered the best period to apply this method, achieving the maximum insecticide effectiveness [15]. For each a.i., 10 infested mulberry trees were randomly selected. An impact drill (Metabo Cordless Hammer Drill, SB 18 LTX BL Q I, Garafas and Co. SA, Lykovryssis, Greece) with a 15 cm long drill and a 7 mm diameter drill bit, was used to create holes 20–25 cm above the ground [15]. The holes were angled at 45° so they could hold the quantity of each insecticide until absorbed by the tree. To inject an appropriate volume of each insecticide into the holes, 10 mL syringes were used for the a.i. imidacloprid, fipronil, and spirotetramat, while for the a.i. abamectin, 60 mL feeding syringes were used. After injecting the insecticides, the holes were covered with inoculation paste to prevent contamination of the trees with pathogens (Figure 1b).

In late April to late July 2021, when *X. chinensis* adults began to leave the trees [1,2,13], the new exit holes were recorded as follows: i. total number of exit holes/tree, ii. number of exit holes/tree < 1.5 m above the ground and iii. number of exit holes/tree > 1.5 m above the ground. The new exit holes were marked with yellow color to distinguish them from the old ones marked with red color (Figure 1c). Afterward, from 15 July to 10 August 2021, new quantities of each insecticide were injected into the respective tree. Trunk injection was carried out at the same injection points of each tree, removing the inoculation paste and adding the corresponding volume of insecticide, as described above. Then, the injection points were shielded again with the inoculation paste. The counts of the new exit holes of *X. chinensis* adults were carried out at the end of April 2022 and colored blue (Figure 1d).

Figure 1. (**a**) Marked exit holes of *X. chinensis* before the application of trunk injection, in 2020. (**b**) Injection holes covered with inoculation paste; (**c**) Exit holes of *X. chinensis* before (marked in red, 2020) and after trunk injection (marked in yellow, 2021); (**d**) Exit holes of *X. chinensis* during the entire experimental period, i.e., with red exit holes before the injection of insecticides (2020), with yellow exit holes after the first application of trunk injection (2021); with blue exit holes after the second application of the method (2022).

The percentage of reduction in exit holes was based on the following formula: exit hole reduction (%) = (number of exit holes before the application of trunk injection–number of exit holes the first or second year after application)/number of exit holes before the application of trunk injection × 100% [25].

2.3. Statistical Analysis

Before analysis, data were transformed to log (x + 1) to normalize variances and standardize means [26,27]. Concerning the effectiveness of the trunk injection method and the effectiveness of insecticides, analysis of data was performed using one-way ANOVA [28]. Response variable was the number of exit holes, while the main effects were year of application or insecticides. As far as the efficacy of insecticide related to the height of the trees and infestation, the repeated-measures model was followed [28]. Year of application was the repeated factor, response variable was number of exit holes, and main effects were insecticide and height. Means were discriminated by the Tukey–Kramer HSD test at a significance level of 0.05 [29]. The two-tailed *t*-test discriminated means between tree heights at 0.05 level of significance. All analyses were conducted with JMP 16.2 software [30].

3. Results

3.1. The Overall Effectiveness of the Trunk Injection Method

The total performance of the four a.i. is presented in Figure 2. Before the application of trunk injection to the selected trees, the mean number of exit holes was 25.9 exit holes/tree. In the first year, after trunk injection, a significant reduction in the exit holes was recorded, reaching 12.8 exit holes/tree. This trend continued the following year of the experiment, where the mean number of emerging holes did not surpass 8.8 exit holes/tree. In contrast, no significant differences were reported in the mean number of exit holes in the control trees between the three years of observations (Figure 3). However, a slight decrease in infestation was noticed (i.e., 29.9 exit holes/tree in 2020, 27.4 exit holes/tree in 2021 and 21.5 exit holes/tree in 2022).

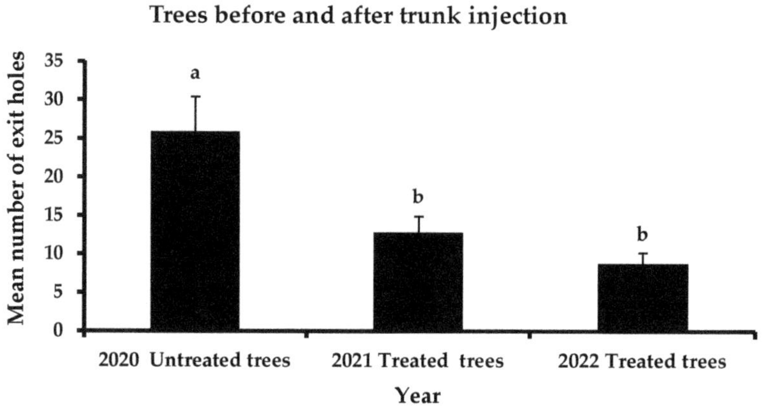

Figure 2. Mean number (+ Standard Error) of *Xylotrechus chinensis* exit holes from the tested trees before the trunk injection method is applied, in 2020, and after the application of trunk injection method, in 2021 and 2022. Different letters above columns indicate significant differences ($F = 8.0$, DF = 2, 119, $p = 0.01$).

3.2. Effectiveness of Insecticides in a Two-Year Period and Performance of Each Insecticide during the Trials

Regarding the efficacy of each insecticide separately, imidacloprid and fipronil proved to be more effective than spirotetramat against *X. chinensis* during the two-year application of trunk injection (Figure 4). Concretely, in 2021, the mean number of exit holes was 9.5 and 12.1 exit holes/tree for imidacloprid and fipronil, respectively (Figure 4a). The mean number of the exit holes of the treated trees with spirotetramat was lower than the aforementioned insecticides while no significant differences were recorded between spirotetramat-treated and control trees (24.7 and 27.4 exit holes/tree for spirotetramat and control, respectively). The positive control, abamectin, significantly reduced the mean number of exit holes (4.7 exit holes/tree) compared to imidacloprid, fipronil and spirotetramat.

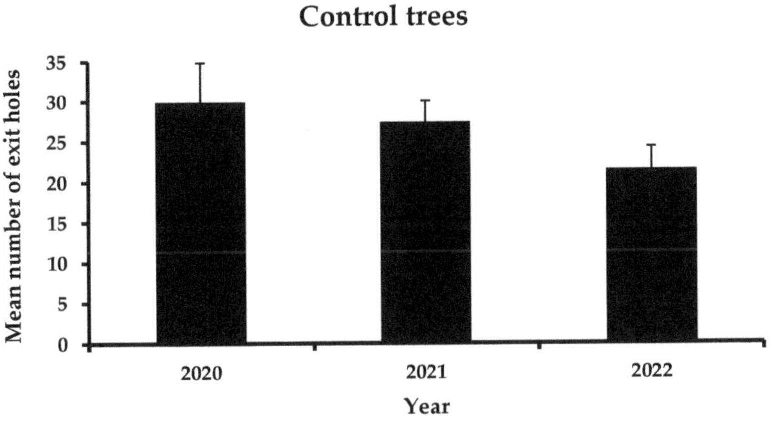

Figure 3. Mean number (+ Standard Error) of *Xylotrechus chinensis* exit holes from the control trees during 2020–2022. (F = 1.7, DF = 2, 29, p = 0.21). No letters above columns indicate no significant differences.

In 2022, the mean number of exit holes was further reduced in treated trees as well as in control trees (Figure 4b). The exit holes of spirotetramat (17.9 exit holes/tree) were significantly higher than those of imidacloprid and fipronil (5.0 and 8.8 exit holes/tree, respectively). Furthermore, no significant differences were recorded between imidacloprid and control (21.5 exit holes/tree). Abamectin resulted in the lowest mean number of exit holes (3.3 exit holes/tree).

The total number of emerging holes per treatment or control, before and during the evaluation of the trunk injection method, as well as the percentage of the reduction in infestation, is shown in Table 1. During 2021, the proportion of exit holes decreased by 79.6, 61.2, and 54.5% in the trees treated with abamectin, fipronil and imidacloprid, respectively. In the case of spirotetramat, a 13.6% reduction in the emerging holes was recorded. The next year, the percentage of the reduction in exit holes ranged between 37.4 and 85.6%, where abamectin was the most effective a.i., followed by imidacloprid, fipronil, and spirotetramat.

Table 1. Total number of *Xylotrechus chinensis* exit holes before (2020) and after the application of trunk injection (2021, 2022). Percentage of reduction of exit holes after the two-year application of trunk injection in comparison with the year before treatments.

Treatment	Trees before Trunk Injection 2020	Trees after Trunk Injection 2021	Trees after Trunk Injection 2022	% Reduction of Exit Holes in 2021 (in Comparison with 2020)	% Reduction of Exit Holes in 2022 (in Comparison with 2020)
Imidacloprid	209	95	50	54.5%	76.1%
Spirotetramat	286	247	179	13.6%	37.4%
Fipronil	312	121	88	61.2%	71.8%
Abamectin	230	47	33	79.6%	85.6%
Control	299	274	215	8.4%	28.1%

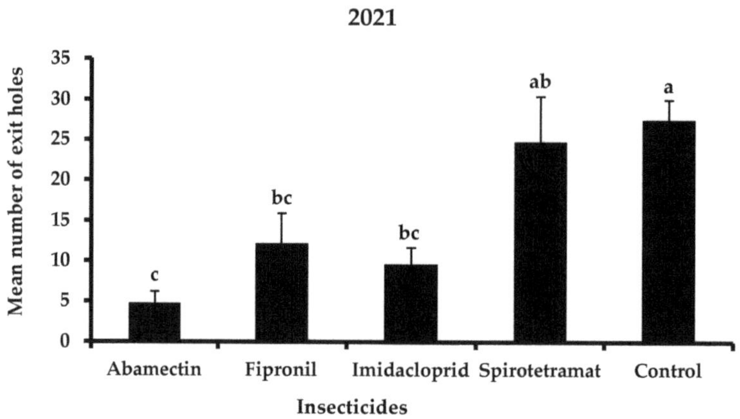

(a)

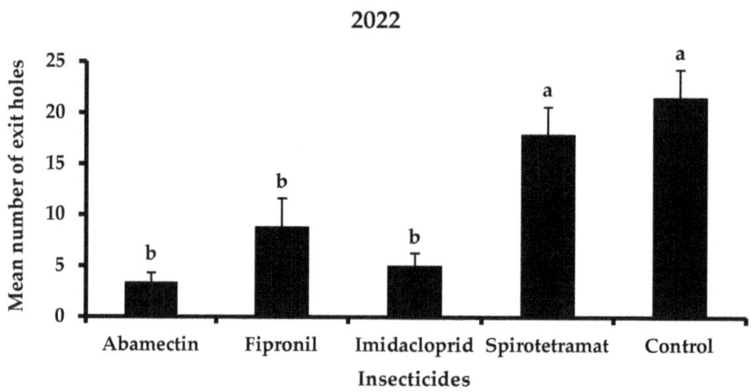

(b)

Figure 4. Mean number (+ Standard Error) of *Xylotrechus chinensis* exit holes from the trees treated with abamectin, fipronil, imidacloprid and spirotetramat or untreated trees (controls) during 2021 (**a**) and 2022 (**b**). Different letters above columns indicate significant differences (2021: $F = 8.0$, DF = 4, 49, $p < 0.01$; 2022: $F = 9.9$, DF = 4, 49, $p < 0.01$).

3.3. *Insecticidal Activity in Relation to Tree Height and Infestation*

Between years of application, the main effect of insecticide was significant ($p < 0.05$) (Table 2). Within years of application, the main effect year and the interaction year × insecticide × height were significant. No significant differences were recorded in the mean number of exit holes of *X. chinensis* between the two heights for fipropil, spirotetramat, and abamectin in both years of application (Table 3). For imidacloprid, the mean number of exit holes was significantly higher for a height > 1.5 m (6.3 exit holes/tree) from the ground compared to a height < 1.5 m (2.2 exit holes/tree) from the ground in 2021, while one year later, no significant differences were observed between the heights. More exit holes were found > 1.5 m from the ground in trees treated with imidacloprid in 2021, spirotetramat in 2022, and abamectin in both years, in comparison with the height < 1.5 m.

Table 2. MANOVA parameters for main effects and interactions for exit holes of *Xylotrechus chinensis* adults between and within years of application (error DF = 72).

Effect	Exit Holes		
Between years of application			
Source	DF	*F*	*p*
Intercept	1	217.0	0.01
Insecticide	3	10.2	<0.01
Height	1	0.5	0.49
Insecticide × height	3	1.1	0.34
Within years of application			
Source	DF	*F*	*p*
Year	1	10.5	0.03
Year × insecticide	3	1.0	0.42
Year × height	1	0.2	0.69
Year × insecticide × height	3	4.8	0.01

Table 3. Mean number (± Standard Errors) of *Xylotrechus chinensis* exit holes from the trees treated with abamectin, fipronil, imidacloprid, and spirotetramat or untreated trees (controls), below or above the 1.5 m from the ground, during 2021 and 2022. Within each row, significant differences are indicated by asterisks (two-tailed *t*-test at *p* = 0.05). Within column, means followed by the same lowercase letter do not differ significantly (Tukey–Kramer HSD test at *p* = 0.05). No asterisks indicate no significant differences.

Height	<1.5 m	>1.5 m			
Year	2021				
Treatment			DF	*t*	*p*
Fipronil	7.6 ± 2.2 bc	4.5 ± 1.8 b	18	−1.1	0.28
Imidacloprid	2.2 ± 0.7 c	6.3 ± 1.3 ab *	18	2.5	0.03
Spirotetramat	12.4 ± 4.0 ab	12.3 ± 2.8 a	18	0.5	0.66
Abamectin	2.0 ± 0.7 c	2.7 ± 1.2 b	18	0.4	0.72
Control	18.7 ± 2.0 a	14.8 ± 2.5 a	18	−1.5	0.17
F	10.3	7.5			
p	<0.01	0.01			
	2022				
Fipronil	5.1 ± 1.6 bc	3.7 ± 1.4 b	18	−0.7	0.47
Imidacloprid	2.5 ± 0.9 c	2.5 ± 0.8 b	18	0.1	0.97
Spirotetramat	8.1 ± 1.7 b	9.8 ± 1.4 a	18	1.1	0.27
Abamectin	1.0 ± 0.4 c	2.3 ± 0.9 b	18	1.3	0.20
Control	18.6 ± 1.8 a	16.0 ± 2.0 a	18	−1.1	0.28
F	15.7	13.5			
p	<0.01	<0.01			

4. Discussion

The four insecticides that were evaluated by the trunk injection method against *X. chinensis* belong to different chemical groups. Fipronil is a wide spectrum systemic insecticide with neurotoxic action, belonging to the chemical family of phenylpyrazoles [31]. Fipronil disrupts the activity of the central nervous system of insects, targeting the GABA and the glutamate-activated chloride channel, ultimately causing their death [32,33]. The neonicotinoid systemic insecticide imidacloprid affects the central nervous system of insects, through agonism of the post-synaptic nicotinic acetylcholine (nAChRs) receptors [34,35]. Consequently, imidacloprid causes a collapse of the nervous system and paralysis of the insects that eventually die [36,37]. Sirotetramat, the spirocyclic tetronic/tetramic acid derivative, is a systemic insecticide that targets a wide range of sucking pests [38–40]. It inhibits the acetyl-CoA carboxylase (ACC) significantly reducing the biosynthesis of lipids [41–44]. Abamectin, also known as Avermectin B_1 [45], belongs to the family of avermectins, which are derived from the soil bacterium *Streptomyces avermitilis* (ex. Burg et al. 1979) Kim and Godfellow 2002 (Streptomycetales: Streptomycetaceae) [46,47]. This is used to control a wide range of pests such as insects and mites [48,49]. Abamectin, through its action on GABA (γ-aminobutyric acid) receptors and glutamate-gated chloride channels, causes an increase in the chloride permeability of the chloride channels, leading the target organisms to paralysis and, eventually, to death [45,50,51].

Trunk injection is the method by which plant protection compounds (e.g., biopesticides or chemical pesticides) are directly delivered to the vasal system of the trees through bark drilling and systemically transported with the xylem [52–54]. It is considered a promising, highly efficient application pesticide method for numerous tree species, as it directly targets pests. At the same time, this method reduces the negative impact that insecticides have on the environment [53,55,56]. Furthermore, trunk injection reduces the exposure of workers and non-target organisms to the toxicants and can be applied to orchards, forests, recreational, urban or suburban areas where chemical spraying is prohibited [56–62]. This method has been used for the management of numerous pests of trees in forest and agricultural landscape. Pine, oak, spruce, elm, palms and ash are examples of trees found in urban areas or woodland that have been treated with trunk-injected pesticides [22,63–69]. In addition, trunk injection has been evaluated for the protection of economically important tree crops, such as apple, pear, vines, avocado, walnut, and date palm [61,70–81]. Trunk injection method was successfully evaluated against several noxious insect species by using different insecticidal formulations. For example, when the imidacloprid was injected into the trunk of apple trees, it showed >80.0% and >90% suppression of the green citrus aphid, *Aphis citricola* van der Goot (Hemiptera: Aphididae) 14 and 25 days post-injection [79]. Endotherapy with thiamethoxam is used to control the larvae of the red-necked longhorn beetle, *Aromia bungii* Faldermann (Coleoptera: Cerambycidae) infesting cherry trees in Japan [82,83]. It should be noted that thiamethoxam and imidacloprid provided a considerable reduction (85.71 and 76.78%, respectively) in the penetration holes formed by larvae of the leopard moth, *Zeuzera pyrina* L. (Lepidoptera: Cossidae) to Persian walnut [77].

In the present study, imidacloprid exhibited elevated efficacy against *X. chinensis* after a two-year application. This is because it reduced the number of the exit holes of adults by 76.1%, indicating that this compound could be a candidate for the control of this species. Similar findings were obtained by Archer et al. [84] against the Asian citrus psyllid, *Diaphorini citri* Kuwayama (Hemiptera: Sternorrhyncha). The authors recorded 63% mortality of adults after one week of application and 80% reduction in offspring survival, but the effectiveness of imidacloprid decreased over a period of two months. In our study, the time of application [15,85] played a crucial role in the performance of imidacropid, since the young larvae of *X. chinensis* [15] were targeted. Recently, spirotetramat and imidacloprid were injected into the new flower buds of banana trees to manage the banana flower thrips, *Thrips hawaiiensis* (Morgan) (Thysanoptera: Thripidae) [86]. Both compounds provided more than 90% protection against this species, while no residues were found in mature fruits. Here, spirotetramat was the least effective insecticide against *X. chinensis*, with a performance almost equal to untreated trees. This could be attributed to the fact that spirotetramat is suggested for the management of immature sucking insects, such as whiteflies, scales, and aphids [87]. Previous reports have documented that it exhibits toxicity and various side effects to coleopterans, such as low fecundity to the ladybeetle, *Menochilus sexmaculatus* Fabricius (Coleoptera: Coccinellidae), low reproduction percentages, and extensions in the development of the immatures of the Adonis ladybird, *Hippodamia variegata* Goeze (Coleoptera: Coccinellidae) [88,89]. However, in an earlier study on the ladybird beetle, *Cryptolaemus montrouzieri* Mulsant (Coleoptera: Coccinellidae), Planes et al. [90] reported that spirotetramat did not affect the survival of exposed adults and larvae, and their progeny, as well as the longevity, fertility, and egg hatchability. On the basis of our results regarding the effects on spirotetramat to coleopterans, further investigation is needed to clarify its actual efficacy to xylophagous beetles.

Fipronil also demonstrated increased efficacy against *X. chinensis*, reducing the exit holes by 61.2% in the first year of application, and by 71.8% in the second year of field trials. It is widely used for the management of pests of both agricultural and public health pests, such as grasshoppers, termites, mosquitoes, ticks, cockroaches and stored-product insects [91–93]. Due to the mode of action of fipronil, it is suitable for insects resistant to other common insecticides (e.g., organophosphates and pyrethroids) [92,94]. Recent research efforts against the red palm weevil, *Rhynchophorus ferrugineus* (Olivier) (Coleoptera: Curculionidae) showed that the injected fipronil in date palms at label dose was not harmful to human health compared to the spray method [95]. Thus, taking into account the performance of fipronil in mulberry trees, as well as the low risk to humans when injected into the trunk could be an effective option for *X. chinensis* control.

Avermectins have been widely used against many pests by the trunk injection method. For instance, when emamectin benzoate applied to willows, it reduced populations of larvae of the Asian longhorned beetle, *Anoplophora glabripennis* (Coleoptera: Cerambycidae) by 99%, while no exit holes were observed in the second year after treatment [96]. In addition, a single application of emamectin benzoate in the trunk of 99 ash trees controlled almost 100% of the larvae of the emerald ash borer, *Agrilus planipennis* Obenberger (Coleoptera: Buprestidae), offering a 2–3 year protection [97]. Interestingly, abamectin almost suppressed nymphs (0.3 nymphs per leaf) of the pear psylla, *Cacopsylla pyricola* (Forster) (Hemiptera: Psyllidae) when it was applied as injection in the trunks. Additionally, the injection of abamectin in trunks of walnut trees resulted in the low infestation of husks (<11%) by the walnut husk fly, *Rhagoletis completa* (Diptera: Tephritidae) [76,78]. Recently, in Spain, abamectin endotherapy was carried out to protect mulberries from the invasion of *X. chinensis* [15]. The authors reported a significant reduction in the exit holes of *X. chinensis*. We agree with Sarto i Monteys et al. [15], since we recorded a 79.6% reduction in exit holes one year after application, while the rate of exit holes further decreased to 85.6% in the following year compared to controls.

No significant differences were noted in the number of exit holes of *X. chinensis* regarding the two different heights for fipropil, spirotetramat and abamectin in both years of application and imidacloprid in the second year. In contrast, significantly more holes were noted above 1.5 m for imidacloprid in 2021. Therefore, due to these controversial findings, there is no clear evidence as to whether height plays a role in the effectiveness of insecticides against *X. chinensis*. The moderate height of 2.5 m of the tested trees may allow the insecticides to perform equally in the entire length of the trees. Whether the efficacy of insecticides used with trunk injection method is differentiated correspondingly to the height of the trees merits further investigation. Last, but not least, several insecticides (e.g., imidacloprid) are negatively associated with performance of pollinators [98]. However, potential risk to pollinators is low here, since the tested mulberries are wind-pollinated [99,100]. Furthermore, pesticide residues, including imidacloprid, in nectar and pollen, minimally introduce acute risk or they do not affect honey bees, *Apis mellifera* L. (Hymenoptera: Apidae) [101].

5. Conclusions

In conclusion, imidacloprid and fipronil can be used against *X. chinensis*, since they reduce the emergence of adults from mulberry trees, while spirotetramat is not recommended against this species, because its effectiveness was almost equal to that of untreated trees. The fact that abamectin has already been proposed for the management of *X. chinensis* [2,15] could be a temporal solution. This is because the spread of this alien species will trigger extensive management tactics that include abamectin, an issue which may lead to the development of resistance issues [102,103]. We propose the consideration of the three aforementioned insecticides to be used as a barrier to the development of potential resistance of this pest to insecticidal treatments. More insecticides with different modes of action should be investigated against *X. chinensis*, including the promising trunk injection method, for the long-term protection of mulberry trees.

Author Contributions: Conceptualization, M.C.B., N.G.K. and G.T.P.; Methodology, N.G.K., M.C.B. and G.T.P.; Validation, N.G.K., M.C.B., A.S., E.P.N. and G.T.P.; Formal Analysis, N.G.K. and M.C.B.; Investigation, N.G.K., M.C.B., A.S., E.P.N. and G.T.P.; Data Curation, N.G.K., M.C.B., A.S. and E.P.N.; Writing—Original Draft Preparation, N.G.K. and M.C.B.; Writing—Review and Editing, N.G.K., M.C.B., A.S., E.P.N. and G.T.P.; Visualization, N.G.K., M.C.B. and G.T.P.; Supervision, N.G.K. and G.T.P.; Project Administration; N.G.K., M.C.B. and G.T.P.; Resources, N.G.K. and G.T.P.; Funding Acquisition, N.G.K. and G.T.P. All authors have read and agreed to the published version of the manuscript.

Funding: This study was supported by the project "Management of *Xylotrechus chinensis* (Coleoptera: Cerambycidae) with pheromones and insecticides in Municipality of Athens", 33.0256 (Municipality of Athens).

Data Availability Statement: Data are contained within the article.

Conflicts of Interest: The authors declare no conflict of interest.

References

1. Leivadara, E.; Leivadaras, I.; Vontas, I.; Trichas, A.; Simoglou, K.; Roditakis, E.; Avtzis, D.N. First record of *Xylotrechus chinensis* (Coleoptera, Cerambycidae) in Greece and in the EPPO region. *EPPO Bull.* **2018**, *48*, 277–280. [CrossRef]
2. Sarto i Monteys, V.; Torras i Tutusaus, G. A new alien invasive longhorn beetle, *Xylotrechus chinensis* (Cerambycidae), is infesting mulberries in Catalonia (Spain). *Insects* **2018**, *9*, 52. [CrossRef] [PubMed]
3. Han, Y.; Lyu, D. Taxonomic review of the genus *Xylotrechus* (Coleoptera: Cerambycidae: Cerambycinae) in Korea with a newly recorded species. *Korean J. Appl. Entomol.* **2010**, *49*, 69–82. [CrossRef]
4. Sama, G.; Löbl, I. Cerambycinae. In *Catalogue of Palaearctic Coleoptera. Chrysomeloidea*; Löbl, I., Smetana, A., Eds.; Apollo Books: Stenstrup, Denmark, 2010; pp. 143–207.
5. Schrader, G. Express—PRA zu *Xylotrechus chinensis*. Available online: https://pflanzengesundheit.julius-kuehn.de/dokumente/upload/b75e3_xylotrechus-chinensis_express-pra.pdf (accessed on 29 October 2022).
6. Demetriou, J.; Kalaentzis, K.; Kazilas, C.; Koutsoukos, E.; Avtzis, D.; Georgiadis, C. Revisiting the non-native insect fauna of Greece: Current trends and an updated checklist. *NeoBiota* **2021**, *65*, 93–108. [CrossRef]
7. Bragard, C.; Baptista, P.; Chatzivassiliou, E.; Di Serio, F.; Gonthier, P.; Jaques Miret, J.A.; Justesen, A.F.; Magnusson, C.S.; Milonas, P.; Navas-Cortes, J.A. Pest categorisation of *Xylotrechus chinensis*. *EFSA J.* **2021**, *19*, e07022. [PubMed]
8. Cho, P.S. A study on the damaged plants of longicorn beetles in Korea (Cerambycidae). *Collect. Pap. Shinheung Coll.* **1959**, *2*, 355–386.
9. Cherepanov, A.I. *Cerambycidae of Northern Asia. Lamiinae. Part 1*; Kopthekar, V.S., Ed.; Amerind Publishing Co. Pvt. Ltd.: New Delhi, India, 1990.
10. Hua, L.Z. *List of Chinese Insects*; University Press: Guangzhou, China, 2002.
11. Kang, E.Y.; Oh, H.Y.; Oh, H.Y. A larval host plant list of the Cerambycidae (Coleoptera) in South Korea. *Lucanus* **2002**, *3*, 1–5.
12. Lim, J.; Jung, S.Y.; Lim, J.S.; Jang, J.; Kim, K.M.; Lee, Y.M.; Lee, B.W. A review of host plants of Cerambycidae (Coleoptera: Chrysomeloidea) with new host records for fourteen Cerambycids, including the Asian longhorn beetle (*Anoplophora glabripennis* Motschulsky), in Korea. *Korean J. Appl. Entomol.* **2014**, *53*, 111–133. [CrossRef]
13. EPPO (European and Mediterranean Plant Protection Organization). Alert list—*Xylotrechus chinensis* (Coleoptera: Cerambycidae). Available online: https://www.eppo.int/ACTIVITIES/plant_quarantine/alert_list_insects/xylotrechus_chinensis (accessed on 29 October 2022).
14. Benker, U. Stowaways in wood packaging material. Current situation in Bavaria. *Forstsch. Aktuell* **2008**, *44*, 30–31.
15. Sarto i Monteys, V.; Ribes, A.C.; Savin, I. The invasive longhorn beetle *Xylotrechus chinensis*, pest of mulberries, in Europe: Study on its local spread and efficacy of abamectin control. *PLoS ONE* **2021**, *16*, e0245527. [CrossRef]
16. Sarto i Monteys, V. El Escarabajo-Avispa, Nueva Especie Invasora en Europa. Adelantos Digital. Available online: https://www.adelantosdigital.com/web/escarabajo-avispa-nueva-especie-invasora-europa/ (accessed on 29 October 2022).
17. EPPO (European and Mediterranean Plant Protection Organization). New data on quarantine pests and pests of the EPPO alert list. *EPPO Rep. Serv.* **2020**, *91*, 5.
18. EPPO (European and Mediterranean Plant Protection Organization). First report of *Xylotrechus chinensis* in France. *EPPO Rep. Serv.* **2018**, *220*, 11.
19. Cocquempot, C.; Desbles, F.; Mouttet, R.; Valladares, L. *Xylotrechus chinensis* (Chevrolat, 1852), nouvelle espèce invasive pour la France métropolitaine (Coleoptera: Cerambycidae: Clytini). *Bull. Soc. Entomol. Fr.* **2019**, *124*, 27–32. [CrossRef]
20. EPPO (European and Mediterranean Plant Protection Organization). Invasive alien wood borers: Trapping studies in France. *EPPO Rep. Serv.* **2021**, *157*, 7.
21. Trematerra, P.; Athanassiou, C.; Stejskal, V.; Sciarretta, A.; Kavallieratos, N.; Palyvos, N. Large-scale mating disruption of *Ephestia* spp. and *Plodia interpunctella* in Czech Republic, Greece and Italy. *J. Appl. Entomol.* **2011**, *135*, 749–762. [CrossRef]

22. Fettig, C.J.; Grosman, D.M.; Munson, A.S. Efficacy of abamectin and tebuconazole injections to protect lodgepole pine from mortality attributed to mountain pine beetle (Coleoptera: Curculionidae) attack and progression of blue stain fungi. *J. Entomol. Sci.* **2013**, *48*, 270–278. [CrossRef]
23. Fishel, F.M. Pesticide Injection and Drenching. Available online: https://edis.ifas.ufl.edu/publication/PI274/ (accessed on 29 October 2022).
24. VanWoerkom, A.H. Trunk Injection: A New and Innovative Technique for Pesticide Delivery in Tree Fruits. Master's Thesis, Michigan State University, East Lansing, MI, USA, 2012.
25. Kou, H.; Sun, Y.; Dong, Z.; Zhang, Z. Comparison between sustained effects of spray and injection thiamethoxam on apple aphids and non-target insects in apple orchard. *Ecotoxicol. Environ. Saf.* **2021**, *207*, 111307. [CrossRef]
26. Sall, J.; Lehman, A.; Creighton, L. JMP start statistics. In *A Guide to Statistics and Data Analysis Using JMP and JMP in Software*; Duxbury Press: Belmont, ON, Canada, 2001.
27. Zar, J.H. *Biostatistical Analysis*; Pearson Education Limited: Essex, UK, 2014.
28. Scheff, D.S.; Arthur, F.H. Fecundity of *Tribolium castaneum* and *Tribolium confusum* adults after exposure to deltamethrin packaging. *J. Pest Sci.* **2018**, *91*, 717–725. [CrossRef]
29. Sokal, R.R.; Rohlf, F.J. *Biometry*; Freeman & Company: New York, NY, USA, 1995.
30. SAS Institute Inc. *Using JMP 16.2*; SAS Institute Inc.: Cary, NC, USA, 2021.
31. Simon-Delso, N.; Amaralrogers, V.; Belzunces, L.P.; Bonmatin, J.M.; Chagnon, M.; Downs, C.A.; Furlan, L.; Gibbons, D.W.; Giorio, C.; Girolami, V.; et al. Systemic insecticides (neonicotinoids and fipronil): Trends, uses, mode of action and metabolites. *Environ. Sci. Pollut. Res.* **2015**, *22*, 5–34. [CrossRef]
32. Gant, D.B.; Chalmers, A.E.; Wolff, M.A.; Hoffman, H.B.; Bushey, D.F. Fipronil: Action at the GABA receptor. *Rev. Toxicol.* **1998**, *2*, 147–156.
33. Caboni, P.; Sammelson, R.E.; Casida, J.E. Phenylpyrazole insecticide photochemistry, metabolism, and GABAergic action: Ethiprole compared with fipronil. *J. Agric. Food Chem.* **2003**, *51*, 7055–7061. [CrossRef] [PubMed]
34. Jeschke, P.; Nauen, R.; Schindler, M.; Elbert, A. Overview of the status and global strategy for neonicotinoids. *J. Agric. Food Chem.* **2010**, *59*, 2897–2908. [CrossRef] [PubMed]
35. Ijaz, M.; Shad, S.A. Risk of resistance and cross-resistance development to selection with imidacloprid and level of heritability in *Oxycarenus hyalinipennis* Costa (Hemiptera: Lygaeidae): A potential pest of cotton. *Phytoparasitica* **2021**, *49*, 287–297. [CrossRef]
36. Kumar, A.; Verma, A.; Kumar, A. Accidental human poisoning with a neonicotinoid insecticide, imidacloprid: A rare case report from rural India with a brief review of literature. *Egypt. J. Food Sci.* **2013**, *3*, 123–126. [CrossRef]
37. Ijaz, M.; Shad, S.A. Inheritance mode and realized heritability of resistance to imidacloprid in *Oxycarenus hyalinipennis* Costa (Hemiptera: Lygaeidae). *Crop Prot.* **2018**, *112*, 90–95. [CrossRef]
38. Brück, E.; Elbert, A.; Fischer, R.; Krueger, S.; Kühnhold, J.; Klueken, A.M.; Nauen, R.; Niebes, J.F.; Reckmann, U.; Schnorbach, H.J.; et al. Movento®®, an innovative ambimobile insecticide for sucking insect pest control in agriculture: Biological profile and field performance. *Crop Prot.* **2009**, *28*, 838–844. [CrossRef]
39. Nauen, R.; Slater, R.; Sparks, T.C.; Elbert, A.; McCaffery, A. IRAC: Insecticide resistance and mode-of-action classification of insecticides. In *Modern Crop Protection Compounds*; Wiley: Weinheim, Germany, 2019; pp. 995–1012.
40. Pan, Y.; Wen, S.; Chen, X.; Gao, X.; Zeng, X.; Liu, X.; Tian, F.; Shang, Q. UDP-glycosyltransferases contribute to spirotetramat resistance in *Aphis gossypii* Glover. *Pestic. Biochem. Physiol.* **2020**, *166*, 104565. [CrossRef] [PubMed]
41. Van Pottelberge, S.; Van Leeuwen, T.; Khajehali, J.; Tirry, L. Genetic and biochemical analysis of a laboratory-selected spirodiclofen-resistant strain of *Tetranychus urticae* Koch (Acari: Tetranychidae). *Pest Manag. Sci.* **2009**, *65*, 358–366. [CrossRef]
42. Khajehali, J.; Van Nieuwenhuyse, P.; Demaeght, P.; Tirry, L.; Van Leeuwen, T. Acaricide resistance and resistance mechanisms in *Tetranychus urticae* populations from rose greenhouses in the Netherlands. *Pest Manag. Sci.* **2011**, *67*, 1424–1433. [CrossRef]
43. Lümmen, P.; Khajehali, J.; Luther, K.; Leeuwen, T.V. The cyclic keto-enol insecticide spirotetramat inhibits insect and spider mite acetyl-CoA carboxylases by interfering with the carboxyltransferase partial reaction. *Insect Biochem. Mol. Biol.* **2014**, *55*, 1–8. [CrossRef] [PubMed]
44. Assouguem, A.; Kara, M.; Mansouri, I.; Imtara, H.; AlZain, M.N.; Mechchate, H.; Conte, R.; Squalli, W.; Farah, A.; Lazraq, A. Evaluation of the Effectiveness of spirotetramat on the diaspine Scale *Parlatoria pergandii* in citrus orchards. *Agronomy* **2021**, *11*, 1562. [CrossRef]
45. Bai, S.H.; Ogbourne, S. Eco-toxicological effects of the avermectin family with a focus on abamectin and ivermectin. *Chemosphere* **2016**, *154*, 204–214. [CrossRef] [PubMed]
46. Burg, R.W.; Miller, B.M.; Baker, J.; Birnbaum, E.E.; Currie, S.A.; Hartman, R.R.; Kong, Y.L.; Monaghan, R.L.; Olsen, G.; Putter, I.; et al. Avermectins, new family of potent anthelmintic agents: Producing organism and fermentation. *Antimicrob. Agents Chemother.* **1979**, *15*, 361–367. [CrossRef] [PubMed]
47. Siddique, S.; Syed, Q.; Adnan, A.; Nadeem, M.; Irfan, M.; Qureshi, F.A. Production of avermectin B1b from *Streptomyces avermitilis* 41445 by batch submerged fermentation. *Jundishapur J. Microbiol.* **2013**, *6*, e7198. [CrossRef]
48. Putter, I.; Mac Connell, J.G.; Preiser, F.A.; Haidri, A.A.; Ristich, S.S.; Dybas, R.A. Avermectins: Novel insecticides, acaricides and nematicides from a soil microorganism. *Cell. Mol. Life Sci.* **1981**, *37*, 963–964. [CrossRef]
49. Wang, M.; Liu, X.; Shi, L.; Liu, J.; Shen, G.; Zhang, P.; Lu, W.; He, L. Functional analysis of UGT201D3 associated with abamectin resistance in *Tetranychus cinnabarinus* (Boisduval). *Insect Sci.* **2018**, *27*, 276–291. [CrossRef]

50. Lasota, J.A.; Dybas, R.A. Avermectins, a novel class of compounds: Implications for use in arthropod pest control. *Annu. Rev. Entomol.* **1991**, *36*, 91–117. [CrossRef]
51. Jansson, R.K.; Brown, R.; Cartwright, B.; Cox, D.; Dunbar, D.M.; Dybas, R.A.; Eckel, C.; Lasota, J.A.; Mookerjee, P.K.; Norton, J.A.; et al. Emamectin benzoate: A novel avermectin derivative for control of lepidopterous pests. In Proceedings of the 3rd International Workshop Management of Diamondback Moth Other Crucifer Pests, Kuala Lumpur, Malaysia, 29 October–1 November 1997; pp. 172–177.
52. Hu, J.; Jiang, J.; Wang, N. Control of citrus Huanglongbing via trunk injection of plant defense activators and antibiotics. *Phytopathology* **2018**, *108*, 186–195. [CrossRef]
53. Werrie, P.Y.; Burgeon, C.; Le Goff, G.J.; Hance, T.; Fauconnier, M.L. Biopesticide trunk injection into apple trees: A proof of concept for the systemic movement of mint and cinnamon essential oils. *Front. Plant Sci.* **2021**, *12*, 650132. [CrossRef]
54. Archer, L.; Crane, J.H.; Albrecht, U. Trunk injection as a tool to deliver plant protection materials—An overview of basic principles and practical considerations. *Horticulturae* **2022**, *8*, 552. [CrossRef]
55. Doccola, J.J.; Wild, P.M. Tree Injection as an Alternative Method of Insecticide Application. In *Insecticides—Basic and Other Applications*; Soloneski, S., Ed.; InTech: Rijeka, Croatia, 2012; pp. 61–78.
56. Wise, J.C.; VanWoerkom, A.H.; Acimovic, S.G.; Sundin, G.W.; Cregg, B.M.; Vandervoort, C.V. Trunk injection: An alternative technique for pesticide delivery in apples. *Crop Prot.* **2014**, *65*, 173–185.
57. Sánchez Zamora, M.A.; Fernández Escobar, R. Injector-size and the time of application affects uptake of tree trunk-injected solutions. *Sci. Hortic.* **2000**, *84*, 163–177. [CrossRef]
58. Kobza, M.; Juhásová, G.; Adamčíková, K.; Onrušková, E. Tree injection in the management of horse-chestnut leaf miner, *Cameraria Ohridella* (Lepidoptera: Gracillariidae). *Gesunde Pflanz.* **2011**, *62*, 139–143. [CrossRef]
59. Ferracini, C.; Alma, A. How to preserve horse chestnut trees from *Cameraria ohridella* in the urban environment. *Crop Prot.* **2008**, *27*, 1251–1255. [CrossRef]
60. Doccola, J.J.; Hascher, W.; Aiken, J.J.; Wild, P.M. Treatment strategies using imidacloprid in hemlock woolly adelgid (*Adelges tsugae* Annand) infested eastern hemlock (*Tsuga Canadensis* Carriere) trees. *Arboric. Urban For.* **2012**, *38*, 41–49. [CrossRef]
61. Coslor, C.C.; Vandervoort, C.; Wise, J.C. Control of insect pests using trunk injection in a newly established apple orchard. *Int. J. Fruit Sci.* **2019**, *19*, 151–164. [CrossRef]
62. Berger, C.; Laurent, F. Trunk injection of plant protection products to protect trees from pests and diseases. *Crop Prot.* **2019**, *124*, 104831. [CrossRef]
63. Pajares, J.A.; Lanier, G.N. Pyrethroid insecticides for control of European elm bark beetle (Coleoptera: Scolytidae). *J. Econ. Entomol.* **1989**, *82*, 873–878. [CrossRef]
64. Svihra, P.; Crosby, D.F.; Duckles, B. Emergence suppression of bark and ambrosia beetles in infested oaks. *J. Arboric.* **2004**, *30*, 62–66.
65. Herms, D.A.; Mccullough, D.G.; Smitley, D.R.; Sadof, C.S.; Cranshaw, W. *Insecticide Options for Protecting Ash Trees from Emerald Ash Borer*, 2nd ed.; North Central IPM Center Bulletin: Champaign, IL, USA, 2014.
66. Grosman, D.M.; Fettig, C.J.; Jorgensen, C.L.; Munson, A.S. Effectiveness of two systemic insecticides for protecting western conifers from mortality due to bark beetle attack. *West J. Appl. For.* **2010**, *25*, 181–185. [CrossRef]
67. Mashal, M.M.; Obeidat, B.F. The efficacy assessment of emamectin benzoate using micro injection system to control red palm weevil. *Heliyon* **2019**, *5*, e01833. [CrossRef] [PubMed]
68. McCullough, D.G.; Poland, T.M.; Tluczek, A.R.; Anulewicz, A.; Wieferich, J.; Siegert, N.W. Emerald ash borer (Coleoptera: Buprestidae) densities over a 6-yr period on untreated trees and trees treated with systemic insecticides at 1-, 2-, and 3-yr intervals in a central Michigan forest. *J. Econ. Entomol.* **2019**, *112*, 201–212. [CrossRef] [PubMed]
69. Chihaoui-Meridja, S.; Harbi, A.; Abbes, K.; Chaabane, H.; la Pergola, A.; Chermiti, B.; Suma, P. Systematicity, Persistence and efficacy of selected insecticides used in endotherapy to control the red palm weevil Rhynchophorus ferrugineus (Olivier, 1790) on Phoenix canariensis. *Phytoparasitica* **2020**, *48*, 75–85. [CrossRef]
70. Düker, A.; Kubiak, R. Stem application of metalaxyl for the protection of *Vitis vinifera* L. ('Riesling') leaves and grapes against downy mildew (*Plasmopara viticola*). *Vitis J. Grapevine Res.* **2009**, *48*, 43–48.
71. Byrne, F.J.; Urena, A.A.; Robinson, L.J.; Krieger, R.I.; Doccola, J.; Morse, J.G. Evaluation of neonicotinoid, organophosphate and avermectin trunk injections for the management of avocado thrips in California avocado groves. *Pest Manag. Sci.* **2012**, *68*, 811–817. [CrossRef] [PubMed]
72. Aćimović, S.G.; Vanwoerkom, A.H.; Reeb, P.D.; Vandervoort, C.; Garavaglia, T.; Cregg, B.M.; Wise, J.C. Spatial and temporal distribution of trunk-injected imidacloprid in apple tree canopies. *Pest Manag. Sci.* **2014**, *70*, 1751–1760. [CrossRef]
73. Khalaf, M.Z.; Alrubeai, H.F. Chemical control of date palm tree borers, *Oryctes* species (Coleoptera: Scarabaeidae: Dynastinae). *Pak. Entomol.* **2016**, *38*, 1–5.
74. Coslor, C.C.; Sundin, G.W.; Wise, J.C. The efficacy of trunk injections of emamectin benzoate and phosphorous acid for control of obliquebanded leafroller and apple scab on semi-dwarf apple. *Crop Prot.* **2019**, *118*, 44–49. [CrossRef]
75. Byrne, F.J.; Almanzor, J.; Tellez, I.; Eskalen, A.; Grosman, D.M.; Morse, J.G. Evaluation of trunk-injected emamectin benzoate as a potential management strategy for kuroshio shot hole borer in avocado trees. *Crop Prot.* **2020**, *132*, 105136. [CrossRef]
76. Wheeler, C.E.; Vandervoort, C.; Wise, J.C. Organic control of pear psylla in pear with trunk injection. *Insects* **2020**, *11*, 650. [CrossRef]

77. Mokhtaryan, A.; Sheikhigarjan, A.; Arbab, A.; Mohammadipour, A.; Ardestanirostami, H. The efficiency of systemic insecticides and complete fertilizer by trunk injection method against leopard moth in infested walnut trees. *J. Basic Appl. Zool.* **2021**, *82*, 55. [CrossRef]
78. Kiss, M.; Hachoumi, I.; Nagy, V.; Ladányi, M.; Gutermuth, Á.; Szabó, Á.; Sörös, C. Preliminary results about the efficacy of abamectin trunk injection against the walnut husk fly (Rhagoletis completa). *J. Plant Dis. Prot.* **2021**, *128*, 333–338. [CrossRef]
79. Zhang, F.M.; Jiang, Z.W.; Yu, Q. Control of *Aphis citricola* by trunk injection. *Hebei Fruits* **2000**, *2*, 41.
80. Laxmi, A. Management of cerambycid wood borer, *Celosterna scabrator* Fab. In grape vines. *J. Entomol. Zool. Stud.* **2021**, *9*, 446–450.
81. Sunitha, N.D. Studies on management of grape stem borer *Celosterna scabrator* Fab. (Cerambycidae: Coleoptera). *J. Entomol. Zool. Stud.* **2021**, *9*, 399–404.
82. Funaki, Y. The emergency approach to pesticide registration for the red–necked longhorn beetle, Aromia bungii. *Plant Prot.* **2019**, *73*, 7–13.
83. Sunamura, E.; Tamura, S.; Shoda-Kagaya, E. Efficacy of insecticide trunk injection against larvae of invasive red–necked longhorn beetle Aromia bungii in cherry blossom trees. *Jpn. J. Environ. Entomol. Zool.* **2020**, *31*, 13–19.
84. Archer, L.; Qureshi, J.; Albrecht, U. Efficacy of trunk injected imidacloprid and oxytetracycline in managing huanglongbing and Asian citrus psyllid in infected sweet orange (*Citrus sinensis*) trees. *Agriculture* **2022**, *12*, 1592. [CrossRef]
85. Herms, D.A.; McCullough, D.G. Emerald ash borer invasion of North America: History, biology, ecology, impacts, and management. *Annu. Rev. Entomol.* **2014**, *59*, 13–30. [CrossRef]
86. Fu, B.; Qiu, H.; Li, Q.; Tang, L.; Zeng, D.; Liu, K.; Gao, Y. Flower injection of imidacloprid and spirotetramat: A novel tool for the management of banana thrips Thrips hawaiiensis. *J. Pest Sci.* **2020**, *93*, 1073–1084. [CrossRef]
87. Salazar-López, N.J.; Aldana-Madrid, M.L.; Silveira-Gramont, M.I.; Aguiar, H.L. Spirotetramat—An Alternative for the Control of Parasitic Sucking Insects and its Fate in the Environment. In *Insecticide Resistance*; Tardan, S., Ed.; IntechOpen: London, UK, 2016; pp. 41–55.
88. Azod, F.; Shahidi-Noghabi, S.; Mahdian, K.; Smagghe, G. Lethal and sublethal effects of spirotetramat and abamectin on predatory beetles (*Menochilus sexmaculatus*) via prey (*Agonoscena pistaciae*) exposure, important for integrated pest management in pistachio orchards. *Belg. J. Zool.* **2016**, *146*, 113–122. [CrossRef]
89. Zeinadini, M.; Sahebzadeh, N.; Ravan, S.; Basirat, M. Side effects of spirotetramat and imidacloprid on *Hippodamia variegata* Goezee feeding on *Agonoscena pistaciae* Burckhardt & Lauterer. *J. Nuts* **2019**, *10*, 35–45.
90. Planes, L.; Catalán, J.; Tena, A.; Porcuna, J.L.; Jacas, J.A.; Izquierdo, J.; Urbaneja, A. Lethal and sublethal effects of spirotetramat on the mealybug destroyer, *Cryptolaemus montrouzieri*. *J. Pest Sci.* **2013**, *86*, 321–327. [CrossRef]
91. Tingle, C.C.D.; Rother, J.A.; Dewhurst, C.F.; Lauer, S.; King, W.J. Fipronil: Environmental fate, ecotoxicology, and human health concerns. *Rev. Environ. Contam. Toxicol.* **2003**, *176*, 1–66.
92. Singh, N.S.; Sharma, R.; Singh, S.K.; Singh, D.K. A comprehensive review of environmental fate and degradation of fipronil and its toxic metabolites. *Environ. Res.* **2021**, *199*, 111316. [CrossRef]
93. Wakil, W.; Kavallieratos, N.G.; Ghazanfar, M.U.; Usman, M. Laboratory and field studies on the combined application of *Beauveria bassiana* and fipronil against four major stored-product coleopteran insect pests. *Environ. Sci. Pollut. Res.* **2022**, *29*, 34912–34929. [CrossRef]
94. Aajoud, A.; Ravanel, P.; Tissut, M. Fipronil metabolism and dissipation in a simplified aquatic ecosystem. *J. Agric. Food. Chem.* **2003**, *51*, 1347–1352. [CrossRef]
95. Abbassy, M.A.; Salim, Y.M.M.; Shawir, M.S.; Nassar, A.M.K. Disappearance and hazard quotient of chlorpyrifos-methyl, fipronil, and imidacloprid insecticides from dates. *J. Consum. Prot. Food Saf.* **2017**, *12*, 223–230. [CrossRef]
96. Wang, J.H.; Che, S.C.; Qiu, L.F.; Li, G.; Shao, J.L.; Zhong, L.; Zhang, G.F.; Xu, H. Efficacy of emamectin benzoate trunk injection against the Asian long-horned beetle [*Anoplophora glabripennis* (Coleoptera: Cerambycidae)]. *J. Econ. Entomol.* **2020**, *113*, 340–347. [CrossRef]
97. Smitley, D.R.; Doccola, J.J.; Cox, D.L. Multiple-year protection of ash trees from emerald ash borer with a single trunk injection of emamectin benzoate, and single-year protection with an imidacloprid basal drench. *Arboric. Urban For.* **2010**, *36*, 206–211. [CrossRef]
98. Zhang, Y.; Zeng, D.; Li, L.; Hong, X.; Li-Byarlay, H.; Luo, S. Assessing the toxicological interaction effects of imidacloprid, thiamethoxam, and chlorpyrifos on *Bombus terrestris* based on the combination index. *Sci. Rep.* **2022**, *12*, 6301. [CrossRef] [PubMed]
99. Bajpai, P.K.; Warghat, A.R.; Sharma, R.K.; Yadav, A. Structure and genetic diversity of natural populations of *Morus alba* in the trans-himalayan Ladakh region. *Biochem. Genet.* **2014**, *52*, 137–152. [CrossRef] [PubMed]
100. Leit, V.G.; Kjellberg, F.; Pereira, R.A.S.; Teixera, S.P. What makes a fig: Insights from a comparative analysis of inflorescence morphogenesis in Moraceae. *Ann. Bot.* **2021**, *127*, 621–631. [CrossRef]
101. Démares, F.J.; Schmehl, D.; Bloomquist, J.R.; Cabrera, A.R.; Huang, Z.Y.; Lau, P.; Rangel, J.; Sullivan, J.; Xie, X.; Ellis, J.D. Honey bee (*Apis mellifera*) exposure to pesticide residues in nectar and pollen in urban and suburban environments from four regions of the United States. *Environ. Toxicol. Chem.* **2022**, *41*, 991–1003. [CrossRef] [PubMed]

102. Clark, J.; Scott, J.; Campos, F.; Bloomquist, J. Resistance to avermectins: Extent, mechanisms, and management implications. *Annu. Rev. Entomol.* **1995**, *40*, 1–30. [CrossRef] [PubMed]
103. Wang, R.; Fang, Y.; Che, W.; Zhang, Q.; Wang, J.; Luo, C. Metabolic resistance in abamectin-resistant *Bemisia tabaci* Mediterranean from northern China. *Toxins* **2022**, *14*, 424. [CrossRef]

insects

MDPI

Article

Characterization of *Leptoglossus occidentalis* Eggs and Egg Glue

Eva Sánchez-Hernández [1,*], Pablo Martín-Ramos [1,*], Jonatan Niño-Sánchez [2,3], Sergio Diez-Hermano [2,3], Flor Álvarez-Taboada [4], Rodrigo Pérez-García [2,3], Alberto Santiago-Aliste [1], Jesús Martín-Gil [1] and Julio Javier Diez-Casero [2,3]

1 Department of Agricultural and Forestry Engineering, ETSIIAA, University of Valladolid, Avenida de Madrid 44, 34004 Palencia, Spain; alberto.santiago@estudiantes.uva.es (A.S.-A.); jesus.martin.gil@uva.es (J.M.-G.)

2 Instituto Universitario de Investigación en Gestión Forestal Sostenible (iuFOR), Universidad de Valladolid, Avenida de Madrid 57, 34071 Palencia, Spain; jonatan.nino@uva.es (J.N.-S.); sergio.diez.hermano@uva.es (S.D.-H.); rodri.pg1993@gmail.com (R.P.-G.); juliojavier.diez@uva.es (J.J.D.-C.)

3 Departamento de Producción Vegetal y Recursos Forestales, ETSIIAA, Universidad de Valladolid, Avenida de Madrid 57, 34071 Palencia, Spain

4 School of Agrarian and Forest Engineering, DRACONES, Universidad de León, Avenida de Portugal 41, 24401 Ponferrada, Spain; flor.alvarez@unileon.es

* Correspondence: eva.sanchez.hernandez@uva.es (E.S.-H.); pmr@uva.es (P.M.-R.)

Simple Summary: This study explored the chemical components of the egg glue used by the Western Conifer Seed Bug (*Leptoglossus occidentalis* Heidemann, 1910) to agglutinate eggs and adhere to pine needles. Results showed that the adhesive secretion includes plasticizers and thermoplastic elastomer resins with semiochemical properties in an oily matrix containing proteins. This knowledge of the egg glue composition can be used to develop new control strategies for *L. occidentalis*, potentially limiting the economic impact caused by this pest insect that reduces the production of pine nuts by up to 25%.

Abstract: The western conifer seed bug (*Leptoglossus occidentalis* Heidemann, 1910, Heteroptera: Coreidae) has a significant economic impact due to the reduction in the quality and viability of conifer seed crops; it can feed on up to 40 different species of conifers, showing a clear predilection for *Pinus pinea* L. in Europe. Its incidence is especially relevant for the pine nut-producing industry, given that the action of this pest insect can reduce the production of pine nuts by up to 25%. As part of ongoing efforts aimed at the design of control strategies for this insect, this work focuses on the characterization (by scanning electron microscopy–energy-dispersive X-ray spectroscopy, Fourier-transform infrared spectroscopy, and gas chromatography–mass spectroscopy, GC–MS) of the compounds released by these insects during oviposition, with emphasis on the adhesive secretion that holds *L. occidentalis* eggs together. Elemental analysis pointed to the presence of significant amounts of compounds with high nitrogen content. Functional groups identified by infrared spectroscopy were compatible with the presence of chitin, scleroproteins, LNSP-like and gelatin proteins, shellac wax analogs, and policosanol. Regarding the chemical species identified by GC–MS, eggs and glue hydromethanolic extracts shared constituents such as butyl citrate, dibutyl itaconate, tributyl aconitate, oleic acid, oleamide, erucamide, and palmitic acid, while eggs also showed stearic and linoleic acid-related compounds. Knowledge of this composition may allow advances in new strategies to address the problem caused by *L. occidentalis*.

Keywords: eggs; GC–MS; FTIR; glue; semiochemicals; SEM-EDS; oviposition; western conifer seed bug; WCSB; pine nuts; edible pine

1. Introduction

1.1. On Compounds Released by Insects

The secretions of insects and the effects that they trigger in host plants have been the subject of research since the early 2000s [1]. Some of these secretions fulfill a defensive function (e.g., osmeterial secretions), others convey specific chemical messages that modify behavior or physiology (e.g., semiochemicals), but most are involved in egg attachment.

Osmeterial-like secretions (which are odoriferous liquids secreted by epidermal glands present in the prothoracic segment of papilionid larvae when the larva feels threatened) contain monoterpene hydrocarbons, sesquiterpenic compounds, or a mixture of aliphatic acids and esters: isobutyric acid, 2-methylbutyric acid, and their methyl esters, methyl 3-hydroxybutanoate, 3-hydroxybutanoic acid, α-pinene, myrcene, limonene, 2-methylpropanoic acid, 2-methylbutanoic acid, and their methyl and ethyl esters.

Semiochemicals are organic compounds used by insects to locate the potential host community (kairomones) or for attractive sexual purposes (pheromones). For example, *Phthia picta* (Drury, 1773) male-produced sex pheromone contains 5,9,17-trimethylhenicosane [2], but, in some cases, the male-released *trans*-2-hexenyl *trans*-2-hexenoate, *trans*-2-hexenyl *cis*-3-hexenoate, and myristyl isobutyrate play the double role of attractant pheromone for adults as well as aggregation pheromone.

Concerning extracts of egg clusters (such as those from *Telenomus podisi* Ashmead, 1893), previous studies showed that the major components were saturated and unsaturated fatty acids, including hexadecanoic acid, linoleic acid, and *cis*-9-octadecenoic acid [3]. In regards to egg glues, hydrogel glues proved to be rich in glycine, serine, and/or proline, and some contained substantial levels of 4-hydroxyproline [4]. Egg glue proteins (EGPs), produced by female insects, can make the eggs firmly attached to the oviposition sites. For instance, the full-length sequence and secondary structure of silkworm EGP are that of a repetitive pentapeptide motif, G-G-N/K/D-Q/E/K-Q/P, repeated 346 times, forming a hydrophilic and elastic β-spiral structure [5].

The most complex fluid composition an insect can produce is one that comprises four key components: an aqueous portion rich in amino acids and carbohydrates; oily nanodroplets containing hydrocarbons; an emulsifier (e.g., cholesterol, monoglycerides, etc.) to stabilize the mixture; and volatile organic compounds like kairomones or pheromones.

1.2. Description and Importance of *Leptoglossus occidentalis*

The western conifer seed bug, *Leptoglossus occidentalis* Heidemann (1910) (Hemiptera: Coreidae), is a pest that causes significant damage to *Pinus pinea* L. nut production (one of the most profitable non-wood forest products in several Mediterranean countries) [6,7] and to other host conifers such as *Pinus contorta* Douglas ex Loudon, *Pinus glauca* Moench, and *Pseudotsuga menziesii* (Mirb.) Franco. Both nymphs and adults feed on seeds by piercing their stylets into cones, causing conelet abortion and seed loss. Based on previously determined damage estimates for each life stage of feeding during three periods of cone development, a hypothetical density of one seed bug per tree early in the season will result in an expected seed loss of approximately 310 seeds [8]. Due to globalization, this pest has colonized almost all European countries and beyond, including marine islands, from Portugal to the south of Russia and Kazakhstan, and from Norway to the Maghreb, Turkey, Lebanon, and the Syrian Golan Heights, as well as other areas of the world such as Japan, Korea, Chile, or Uruguay [9]. The ruin of businesses by this pest has led many companies and self-employed to abandon the activity.

Phenology is dependent on weather conditions, which may vary between years. The life cycle of the western conifer seed bug in the Mediterranean area has two annual generations, with the first laying in mid-May and the second in August [10]. According to Barta [11], a single female in the field can lay up to 30 eggs during her life cycle; in laboratory observations, clutches range from 9 to 16 eggs in a row. Eggs are laid on pine needles, from which the first-instar nymphs emerge. After emergence, the newly hatched nymphs remain together on a shoot close to where the clutch is located [12]. Young nymphs are orange and brown, turning reddish-brown as they develop [13]. Approximately 9% of them survive until adulthood [8]. *L. occidentalis* nymphs go through five instar stages, which last between one and three weeks, before transforming into the first generation of adults (mid-July). These adults lay further clutches, which give rise to the second generation of nymphs during the second half of the summer and to new adults from late August and September onward. These adults do not reproduce and overwinter in groups until the following spring [10].

Environmentally friendly management of this species is the main objective of the different research groups working with *L. occidentalis*. Recently, a new Tachinid parasitizing western conifer seed bugs was found in Portugal [14]. However, the biological control of this species is still far away. Under this scenario, the use of semiochemicals for the management of this species may be a promising strategy for the control of this pest. To the best of our knowledge, there is no effective trapping technique for *L. occidentalis*. Traps with different colors and conformations were tested, but all performed poorly to attract this bug [15], and although *L. occidentalis* males produce an aggregation pheromone to which both males and females respond [16–18], it has not yet been identified. However, important advances have been recently made by Millar et al. [19]: leptotriene (i.e., 1*S*,8*R*,*E*)-9,9-dimethyl-2-methylene-5-vinylbicyclo[6.2.0]dec-4-ene, a unique sesquiterpene hydrocarbon produced by sexually mature males of the leaffooted bugs *Leptoglossus zonatus* (Dallas, 1852) and *L. occidentalis*, was recently described and identified by electroantennogram response.

Applications of dimethoate or carbaryl, used to control other insects, have been suggested as possible treatments for the control of conifer seed bugs. Another control option, using traps that emit infrared light wavelengths, was explored by researchers at Simon Fraser University and the British Columbia Ministry of Forests and Range [20]. The traps that achieved the best results for *L. zonatus*, a relative species of *L. occidentalis*, were green and blue, according to Franco-Archundia et al. [21], and yellow ones with a polytetrafluoroethylene (Fluon®) lubricant coating [22]. In a recent article by Inoue et al. [23], the identification of the alarm and sex pheromones from nymphs and adults of the leaf-footed bug, *L. zonatus*, was elucidated as 1-hexanol, hexanoic acid, hexyl acetate, octyl acetate, 5-ethyl-2(5H)-furanone, *cis*-β-ocimene, *cis*-allo-ocimene, decanal, α-*trans*-bergamotene, and *trans*-β-farnesene.

1.3. Aim of the Study

As part of ongoing efforts aimed at the design of control strategies for the western conifer seed bug, the work presented herein attempts to: (1) characterize the egg attachment glue of *L. occidentalis*; (2) identify volatile compounds in *L. occidentalis* eggs analogous to those of *L. zonatus*; and (3) determine if any odorous volatile component of pine bark or pine nuts can be related to the unknown secretion released by *L. occidentalis*. This information may provide leads that can be developed for the control of this insect.

2. Material and Methods

2.1. Egg and Egg Glue Samples

Fresh eggs of *L. occidentalis* were used for this study. The eggs and glue samples were collected from the pine needles of 2-3-year-old potted *Pinus halepensis* Mill. saplings (provided as ovipositional substrate) at the Forest Pathology Lab at the University of Valladolid (Palencia, Spain), where the western conifer seed bugs are reproduced and maintained in cages under controlled laboratory conditions at a temperature of 21 ± 2 °C, a relative humidity of 40 ± 10%, and under daylight photoperiod. The individuals were housed in rearing boxes measuring 47.5 cm × 47.5 cm × 93 cm, with a transparent plastic front panel with two openings for manual handling and side walls made of 1 mm × 0.3 mm light polyester mesh. *Ad libitum* shelled *P. pinea* kernels were provided as food [19].

The aforementioned colony was reared from adults collected from a heavily infected *P. pinea* forest in the province of Valladolid ('El Molinillo', Tordesillas, Valladolid, Spain).

Needle removal from the potted pine saplings was carried out during the months of May and June, selecting needles with fully hatched clutches. For this purpose, the presence of an operculum due to the nymph's exit (or a color change indicating the abortion of the embryo) was checked for each egg. The selected needles were cut using scissors, placed in Petri dishes, and sealed with Parafilm.

2.2. Characterization

The morphology and multi-elemental composition of the egg and glue samples from the abovementioned colony reared in captivity were analyzed by scanning electron microscopy with energy-dispersive X-ray spectroscopy (SEM-EDX), using an EVO HD 25 (Carl Zeiss, Oberkochen, Germany) apparatus. Measurement conditions are indicated in each micrograph.

The infrared spectra were recorded using a Nicolet iS50 (ThermoFisher Scientific; Waltham, MA, USA) FTIR spectrometer with a diamond attenuated total reflection (ATR) module. Operative conditions were: room temperature; a 400–4000 cm^{-1} range; a 0.5 cm^{-1} spectral resolution; and the co-addition of 64 scans.

Egg/egg glue samples were mixed in a 1:20 (w/v) ratio with a methanol/water solution (1:1 v/v) and heated in a water bath at 50 °C for 30 min. The mixture was then sonicated in pulse mode for 5 min with a 1-min stop every 2.5 min, using a model UIP1000hdT probe-type ultrasonicator (Hielscher Ultrasonics; Teltow, Germany). The solution was centrifuged at 9000 rpm for 15 min, and the supernatant was filtered through Whatman No. 1 paper before being freeze-dried to obtain the solid residue. For subsequent gas chromatography-mass spectrometry (GC–MS) analysis, 25 mg of the freeze-dried extract was dissolved in 5 mL of HPLC-grade methanol (gradient grade, suitable as an ACS-grade LC reagent, ≥99.9; Sigma-Aldrich) to obtain a 5 $mg \cdot mL^{-1}$ solution, which was further filtered.

The GC–MS characterization was outsourced to the University of Alicante's Research Support Services (STI) and was conducted using a model 7890A gas chromatograph coupled to a model 5975C quadrupole mass spectrometer (both from Agilent Technologies; Santa Clara, CA, USA). The chromatographic conditions were as follows: 1 µL injection volume; 280 °C injector temperature; splitless mode; 60 °C initial oven temperature, kept for 2 min, followed by a 10 $°C \cdot min^{-1}$ ramp up to a final temperature of 300 °C, maintained for 15 min. For the separation of the compounds, an Agilent Technologies HP-5MS UI chromatographic column (30 m in length, 0.250 mm in diameter, and 0.25 µm film) was used. In regards to the mass spectrometer conditions, the temperature of the electron impact source of the mass spectrometer and the quadrupole were 230 and 150 °C, respectively, with a 70 eV ionization energy. For the identification of components, their mass spectra and retention times were compared with those of the authentic compounds and with the database of the National Institute of Standards and Technology. Siloxane-derivatives, which may arise from contamination from a variety of sources such as inlet septa, columns, valves, solvents, and vial cap seals, represented less than 0.9% of the peak area in the egg and glue sample chromatograms and were removed from the identified phytochemical lists.

3. Results and Discussion

3.1. SEM–EDS Characterization

Scanning electron micrographs of both the hemicylindrical eggs, which are laid in long chains, end to end, along the needles of the host tree [24], and the egg glue are presented in Figure 1. Figure 1a shows an intact egg, while Figure 1b shows a hatched egg. Figure 1c shows the egg glue.

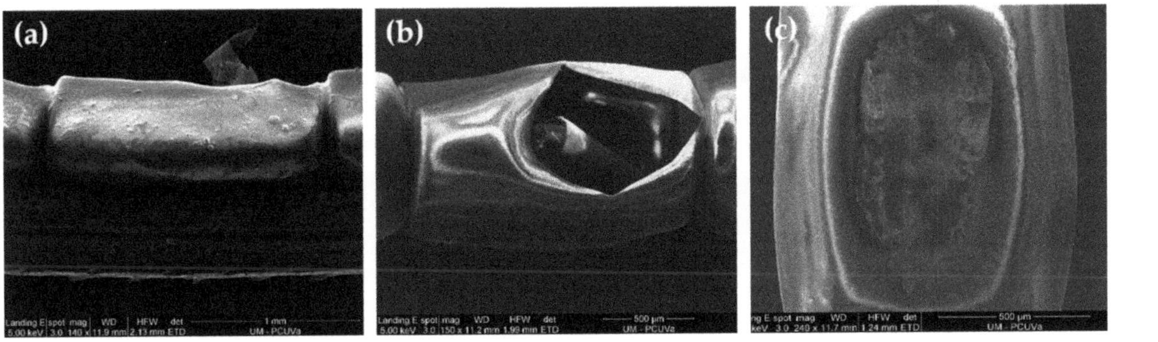

Figure 1. Scanning electron micrographs of *L. occidentalis* eggs and egg glue: (**a**) an egg on a pine needle; (**b**) a hatched egg; (**c**) egg glue.

The EDS analysis results are summarized in Table 1, along with the elemental composition of several potentially related organic chemicals.

Table 1. EDS elemental analyses and C/N, C/O, and N/O ratios of *L. occidentalis* eggs and egg glue components, together with those of potentially associated organic chemicals.

	C	N	O	S	Si	C/N	C/O	N/O	$C_xN_yO_z$
Eggs (inner surface)	56.8	25.1	18.0	-	-	2.27	3.15	1.40	C_5N_2O
Eggs (outer surface)	72.0	9.7	18.3	-	-	7.42	5.57	0.53	$C_{10}NO_2$
Eggs glue	65.5	15.0	19.5	-	-	4.36	3.36	0.77	$C_{10}N_2O_2$ or C_xNO
Gelatin	50.3	17.8	25.6	-	-	2.82	1.97	0.69	$C_{102}[H_{151}]N_{31}O_{39}$
Lignin	46.7	9.1	27.6	1.4	-	5.13	1.69	0.33	$C_{18}[H_{13}]N_3O_8S_2$
Arginine	41.4	32.2	18.4	-	-	1.28	2.25	1.75	$C_6[H_{14}]N_4O_2$
Chitin	47.2	6.9	39.4	-	-	6.84	1.20	0.18	$C_6[H_{14}]NO_5$
4-Hydroxyproline	45.8	10.7	35.1	-	-	4.28	1.31	0.30	$C_5[H_9]NO_3$
Shellac	61.4	-	30.0	-	-	-	2.04	-	$C_{30}[H_{50}]O_{11}$
Hexadecanoic acid	75.0	-	12.5	-	-	-	6.00	-	$C_{16}[H_{32}]O_2$
Oleic acid	76.6	-	14.8	-	-	-	5.17	-	$C_{18}[H_{34}]O_2$
Pine needle	67.0	-	28.5	-	4.5	-	2.35	-	$C_{10}[H_{11}]O_3$

3.2. Vibrational Characterization of the Eggs and Their Glue

The absorption bands of *L. occidentalis* eggs and glue (as shown in Table 2) correspond to functional groups compatible with the components found in the cuticle (chitin and scleroproteins), Louse Nit Sheath Protein (LNSP)-like and gelatin proteins, shellac wax analogs, and related products (such as policosanol). These components are all essential for maintaining the viability of the egg, besides the adhesive function.

The prediction of protein secondary structure from FTIR spectra typically relies on the amide I–amide II region, specifically the band at approximately 1630 cm^{-1}. The characterization of the nit sheath of the human head louse (*Pediculus humanus* subsp. *capitis* de Geer, 1778) reveals that the glue consists mainly of proteins, such as LNSP 1 and 2 [25], which predominantly contain Gly (25.4%), Glx (Glu or Gln, 24.4%), Ala (20.2%), and Val (10.0%) residues. In the amide spectral range, arginine has a strong absorption near 1673 cm^{-1} that is assigned to $\nu_{as}(CN_3H_5^+)$, a less strong one near 1633 cm^{-1} that is assigned to $\nu_s(CN_3H_5^+)$, and a weaker one at 1522 cm^{-1} that is assigned to $\delta_s(CN_3H_5^+)$. Furthermore, the deconvoluted amide I bands are centered around 1610, 1626, 1651, and 1691 cm^{-1}.

The band of gelatin is generally located at 1630 cm^{-1}, and its shift to 1634 cm^{-1} may indicate complexation, such as through the grafting of a polypeptide chain (gelatin) onto chitin. Other bands, such as those at 1445 and 1156 cm^{-1}, appear to correspond to PAla.

Bands at 2922, 2858, 1729, 1455, 1226, 1173, 1015, and 729 cm^{-1} have been previously associated with policosanol-rich insect waxes [26,27], such as shellac, and with saffron [28]. Saffron's composition includes carotenoids, safranal, and primarily lanierone (2-hydroxy-4,4,6-trimethylcyclohexa-2,5-dien-1-one), which serves as both an odorous volatile component of saffron and a pheromone emitted by the pine engraver (*Ips* spp.).

Table 2. Main infrared absorption bands of western conifer seed bug eggs and egg glue and possible assignments, together with bands reported in the literature (for comparison purposes).

Eggs	Egg Glue	Gum Moth Eggs Glue [4]	Sorghum Bugs Gelatin [29]	Shellac Wax [27]	Saffron [28]	Assignment
	3422 3386				3365	O-H stretching in alcohols
3270	3269		3274	3269		N-H stretching/typical of geraniol
3068	3106	3081				hydrocarbon chains in polymeric films/ =C-H stretching in aromatics
2965	2960	2953		2955		C-H methoxyl, cuticles (chitins)
2922	2918		2927	2916	2924	CH_2– asymmetrical stretching, in butyl itaconate/ $COCH_3$, in chitins/policosanol
2858	2849			2850	2854	CH_2– symmetrical stretching, in policosanol and cuticles (chitins)
1729	1736			1738	1741	C=O stretching, in policosanol, fatty acids, and lipids/C=C, phorone
1634	1626	1653	1633		1653	amide I (C=O of N-acetyl group), in gelatin and chitins
	1547	1544	1542		1543	C–N stretching/amide II (N-H bend), in chitins and chitosan
1515	1472					aromatic squeletal vibrations
1455	1463 1445			1462	1462	C–H bending (scissor)/aromatic –C=C stretching, in PAla and policosanol
1385	1377	1378	1399	1374	1400 1376	–OH bending/-C-O-H in-plane bending/–CH_3 out-of-plane bending/CH_2– wagging and twisting/in chitins
	1308				1317	CH_3 wagging, in amide III, proteins
1226	1245	1265		1246	1227	C-O-C chitosan/C-O polyols
	1173			1173		C(O)O policosanol
	1156	1150		1148	1157	C-O stretch in C-O-C linkages)/-C-H in chitosan and PAla
	1115			1115	1119	asymmetric in-phase ring stretching
	1073	1065	1067	1081	1075	C-H bending/C–O–C asym. stretching, in saccharide rings
1015	1028		1042	1029	1020	C-OH stretching/aromatics, phenols
	953	945		943	945	CH_3 wagging along the chain
		837	862			*trans* =C–H out-of-plane bending
	759 729	759		720	740	*cis* =C-H out-of-plane bending/C=C in aromatic structures

3.3. GC–MS Characterization of Chemical Species Present in Eggs and Their Glue

In the hydroalcoholic extracts of eggs (Figure S1), the organic compounds listed in Table S1 were identified by GC–MS. The major components were tributyl acetylcitrate (33.3%) and butyl citrate (7.3%); 9,12-octadecadienoic acid (linoleic acid) and its methyl ester (25.6%); n-hexadecanoic acid (palmitic acid) and its methyl ester (6%); 1-propene-1,2,3-tricarboxylic acid tributyl ester (tributyl aconitate) (6%); octadecanoic acid (stearic acid) (4.9%); dibutyl itaconate (3.8%); and 9-octadecenoic acid (oleic acid) methyl ester (2.7%). Additionally, secondary constituents such as 9-octadecenamide (oleamide), *cis*-13-docosenamide (erucamide), squalene, and heptaethylene glycol were identified. Sex-specific elicitors such as nonanal, decanal, ocimene, caryophyllene, farnesene, and leptotriene [19] were not detected.

The egg glue hydroalcoholic extract was analyzed by GC-MS (Figure S2), and the organic compounds identified are listed in Table S2. Butyl citrate (61.4%) was found to be the major component, followed by heptaethylene glycol (6.5%), dibutyl itaconate (4.7%), 1-propene-tributyl aconitate (3.8%), and oleic acid (3.2%). The minor components identified were oleamide, erucamide, dianhydromanitol, squalene, adipic acid, 2-ethylhexyl isobutyl ester, and palmitic acid.

Based on the information provided, it appears that eggs and egg glue share many chemical species in common, as shown in Figure 2. These include butyl citrate, dibutyl itaconate, tributyl itaconate, oleic acid (or its methyl ester), oleamide, erucamide, squalene, palmitic acid, and heptaethylene glycol. However, eggs also contain stearic and linoleic acid-related compounds that are not present in egg glue.

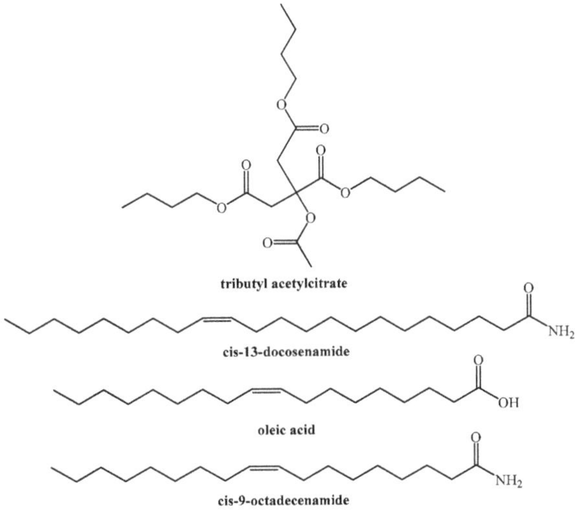

Figure 2. Chemical species detected by GC−MS both in *L. occidentalis* eggs and egg glue.

4. Discussion

4.1. On Elemental Analysis Results

In a first approximation, if a simplified binary mixture approach is chosen and some constituents identified by other techniques are taken as a reference, the C and O data for eggs (Table 1) could be justified by, for instance, a combination of oleic acid (ca. 84%) and chitin or 4-hydroxyproline (ca. 16%). Additionally, the data from egg glue would be compatible with a tentative combination of ca. 60% gelatin and ca. 40% oleic acid or ca. 60% arginine and ca. 40% oleic acid. However, it should be noted that the actual combinations would not be binary but would involve multiple nitrogenated and non-nitrogenated constituents. In any case, apart from speculative considerations, elemental analyses demonstrated that the egg shells and egg glue contained a significant proportion of nitrogen, indicating the presence of peptides, proteins, or other compounds with high nitrogen content.

4.2. On Compounds Identified by GC−MS

Concerning the accuracy of the identification of the compounds, it is worth noting that methyl esters, such as hexadecanoic acid methyl ester, 9,12-octadecadienoic acid methyl ester, and 9-octadecenoic acid methyl ester, could be artifacts associated with the use of methanol as a solvent for GC−MS analyses [30].

Regarding the known functions of the chemical species identified by GC−MS (Tables S1 and S2), several of them are plasticizers. Tributyl citrate and tributyl acetylcitrate are hydrophobic plasticizers and lacquer-forming agents [31]. 1-Propene-1,2,3-tricarboxylic acid tributyl ester (tributyl aconitate) is a hydrolysis product of tributyl acetylcitrate [32], while dibutyl itaconate is used for making plasticizers, adhesives, lubricants, and gelation accelerators [33]. They all exhibit high plasticization efficiency, excellent intermiscibility, outstanding light, water, and cold resistance, low volatility, great ductility, strong weatherability, and good biodegradability. On the other hand, other organic compounds identified in *L. occidentalis* eggs and egg glue have been previously identified as semiochemicals. High levels of *cis*-13-docosenamide (an elastomer resin) were found on the surface of females of the lone star tick, *Amblyomma americanum* L. [34]. Hexadecanoic acid methyl ester is known to be attractive to female moths of *Ostrinia furnacalis* Guenée, 1854 [35], while oleic acid, the second most abundant fatty acid in pine nut oil, is a necromone across arthropod taxa [36]. Additionally, *cis*-9-octadecenamide has been reported to be an oviposition stimulant and attractive to female moths of *O. furnacalis* [35]. Concerning palmitic, oleic, and linoleic acids, they have been previously reported in extracts of egg clusters of *T. podisi* [3].

With regards to the detection of the aforementioned 'plasticizers', initially, reasonable doubts arose as they were generally considered non-natural products resulting from contamination. However, they were not present in black controls (ruling out contamination from the extraction and injection methods used), and in additional GC-MS analyses conducted for new samples, their presence was confirmed. Unlike other plasticizers, such as phthalate esters that may result from solvent storage in plastic containers, their presence as contaminants in extracts has not been referred to in the review paper by Venditti [30]. It should be noted that butyl-related compounds, such as isobutyric and 2-methylbutyric acids, and their methyl and ethyl esters, have been previously reported in osmeterial secretions from the swallowtail butterfly (*Papilio memnon heronus* Frunstorfer) [37] and of papilionid larvae in the genera *Luehdorfia, Graphium,* and *Atrophaneura* [38], thus a natural origin may be advocated. Aldrich et al. [39] reported that gypsy moth caterpillars (*Lymantria dispar* L.) secretions include biogenic amides (e.g., γ-aminobutyric acid), short-chain hydroxy acids (α-hydroxyisobutyric acid), and Krebs cycle acids (e.g., isocitric acid). In the compounds identified herein, apart from citric acid, other two Krebs cycle intermediates, namely aconitate and itaconate, would be involved [40], thus supporting the natural origin of butyl citrate, dibutyl itaconate, and tributyl aconitate. Regarding the possible role these compounds may play, it is reasonable to assume that they would promote adhesion to pine components such as pine needles, cones, nutshells, etc.

4.3. Comparison with Pine Constituents

In their study, Liu and Xu [41] identified forty-eight volatile chemicals in pine nut shells using GC–MS, of which the following are particularly relevant: α-pinene (12.4%); β-pinene (1.8%); oleic (33.1%), palmitic (13.6%), and stearic (7.2%) acids; 3-carene (1.5%); and 1-methy-4-(1-methylethenyl)-(S)-cyclohexene (7.9%). Pinene was not found in *L. occidentalis* samples, but oleic and palmitic acids were detected in *L. occidentalis* eggs, egg glue, and pine nuts, and stearic acid was present in eggs and pine nuts. The presence of linoleic acid in the eggs is consistent with the findings of Kadri et al. [42], who reported a 30.7% content in *P. pinea* seeds (used as a food source in our experiments), as well as the occurrence of cis-oleic, linolenic, palmitic, and stearic acids. Should the composition of the ovipositional substrate (*P. halepensis* saplings) be considered instead, a recent review paper by El Omari et al. [43] on the phytochemistry of this species showed that the fatty acid profile varies depending on the analyzed plant part (e.g., seeds, cones, branches, and needles [42,44–46]), geographical provenance, and extraction procedure. However, there is general agreement that the main unsaturated fatty acids are linoleic and oleic acids, and that the main saturated fatty acid is palmitic acid, whereas stearic, succinic, azelaic, linolenic, decanoic, myristic, and arachidic acids would be minor constituents.

4.4. Usefulness of the Obtained Results

The combination of reduced penetration through the eggshell (which acts as an excellent barrier to insecticides, fungal pathogens, and some fumigants) and pesticide resistance (given that eggs treated with insecticides repeatedly have been shown to produce high levels of enzymatic activity that break down insecticides) makes eggs an extremely difficult life stage to kill for both agricultural and urban pest species [47].

While the usual management strategy is to wait for the eggs to hatch and then treat the nymphs that emerge because they are easier to kill, the reported data may find application in the development of alternative management strategies, with associated advantages (e.g., avoiding reinfestation due to eggs left behind that were not killed by the treatment, the possibility of using repellants or solvents as non-toxic means of control, etc.).

Hence, based on the reported information regarding the composition of the glue (which consists of oily and proteinaceous components with plasticizers as minority constituents), other possible control strategies (besides spraying with a neonicotinoid/pyrethroid mix) may be envisaged to attack the *L. occidentalis* egg glue system. These include spraying needles with proteases dispersed in a fatty acid medium and/or hydrolyzing the plasticizers using dehydrogenases (given that, according to Bruns and Werners [48], tributyl citrate and tributyl acetyl citrate can be hydrolyzed to citric acid and butanol).

Another relevant topic for follow-up studies would be subjecting the identified material to bioassays to determine its semiochemical properties.

5. Conclusions

The vibrational and GC–MS characterization results suggest that the glue system used by *L. occidentalis* females to agglutinate eggs and adhere to pine components includes plasticizers (such as butyl citrate, dibutyl itaconate, and tributyl aconitate) and thermoplastic elastomer resins with semiochemical properties (such as *cis*-13-docosenamide and *cis*-9-octadecenamide) in an oily matrix (with oleic, palmitic, and stearic acid) and proteinaceous components. Knowledge of this composition paves the way to the development of new *L. occidentalis* control strategies during the egg stage by using proteases dispersed in fatty acid media and/or dehydrogenases to attack the egg glue.

Supplementary Materials: The following supporting information can be downloaded at: https://www.mdpi.com/article/10.3390/insects14040396/s1, Figure S1: GC−MS chromatogram of *L. occidentalis* eggs; Figure S2: GC−MS chromatogram of *L. occidentalis* egg glue; Table S1: Main chemical species identified in *L. occidentalis* eggs by GC–MS; Table S2: Main chemical species identified in *L. occidentalis* egg glue by GC–MS.

Author Contributions: Conceptualization, P.M.-R. and J.J.D.-C.; methodology, J.M.-G. and F.Á.-T.; validation, J.N.-S., S.D.-H. and R.P.-G.; formal analysis: E.S.-H., P.M.-R., J.M.-G. and J.J.D.-C.; investigation, E.S.-H., P.M.-R., J.N.-S., S.D.-H., F.Á.-T., R.P.-G., A.S.-A., J.M.-G. and J.J.D.-C.; resources, J.J.D.-C.; writing—original draft preparation, E.S.-H., P.M.-R., J.M.-G. and J.J.D.-C.; writing—review and editing, E.S.-H., P.M.-R. and J.J.D.-C.; visualization, A.S.-A.; supervision: P.M.-R. and J.J.D.-C.; project administration, J.J.D.-C.; funding acquisition, J.J.D.-C. All authors have read and agreed to the published version of the manuscript.

Funding: This research was funded by LIFE project MYCORESTORE, "Innovative use of mycological resources for resilient and productive Mediterranean forests threatened by climate change", LIFE18 CCA/ES/001110.

Data Availability Statement: The data presented in this study are available on request from the corresponding author.

Acknowledgments: To Pilar Blasco and Pablo Candela, from the Technical Research Services of the University of Alicante, for conducting the GC-MS analyses. Thanks to Laura Ponce-Herrero for providing some of the insect samples for the analyses.

Conflicts of Interest: The authors declare no conflict of interest. The funders had no role in the design of the study, in the collection, analysis, or interpretation of data, in the writing of the manuscript, or in the decision to publish the results.

References

1. Hilker, M.; Meiners, T. *Chemoecology of Insect Eggs and Egg Deposition*; Wiley-Blackwell: Berlin, Germany, 2003; p. 416. [CrossRef]
2. Soldi, R.A.; Rodrigues, M.A.C.M.; Aldrich, J.R.; Zarbin, P.H.G. The male-produced sex pheromone of the true bug, *Phthia picta*, is an unusual hydrocarbon. *J. Chem. Ecol.* **2012**, *38*, 814–824. [CrossRef] [PubMed]
3. Michereff, M.F.F.; Borges, M.; Aquino, M.F.S.; Laumann, R.A.; Mendes Gomes, A.C.M.; Blassioli-Moraes, M.C. The influence of volatile semiochemicals from stink bug eggs and oviposition-damaged plants on the foraging behaviour of the egg parasitoid *Telenomus podisi*. *Bull. Entomol. Res.* **2016**, *106*, 663–671. [CrossRef] [PubMed]
4. Li, D.; Huson, M.G.; Graham, L.D. Proteinaceous adhesive secretions from insects, and in particular the egg attachment glue of *Opodiphthera* sp. moths. *Arch. Insect Biochem. Physiol.* **2008**, *69*, 85–105. [CrossRef]
5. Lei, Y.; Guo, K.; Zhang, Y.; Zhang, X.; Qin, L.; Wang, X.; Zhu, H.; Guo, Y.; Yang, W.; Li, B.; et al. Adhesive property and mechanism of silkworm egg glue protein. *Acta Biomater.* **2021**, *134*, 499–512. [CrossRef] [PubMed]
6. Oliveira Farinha, A.C. *Impact and Ecological Adaptation of Leptoglossus occidentalis (Hemiptera, Coreidae) on Pinus pinea*; Universidade de Lisboa: Lisbon, Portugal, 2019.
7. Connelly, A.E.; Schowalter, T.D. Seed losses to feeding by *Leptoglossus occidentalis* (Heteroptera: Coreidae) during two periods of second-year cone development in western white pine. *J. Econ. Entomol.* **1991**, *84*, 215–217. [CrossRef]
8. Bates, S.L.; Borden, J.H. Life table for *Leptoglossus occidentalis* Heidemann (Heteroptera: Coreidae) and prediction of damage in lodgepole pine seed orchards. *Agric. For. Entomol.* **2005**, *7*, 145–151. [CrossRef]
9. Mutke, S. Leptoglossus, la chinche del piñón en el mundo y cómo afecta al piñón mediterráneo. *Forestalis* **2019**, *32*, 26–29.
10. Ponce Herrero, L.; Domínguez Alonso, J.C. El chinche de las piñas *Leptoglossus occidentalis* Heidemann. *Rev. For.* **2020**, *79*, 90–99.
11. Barta, M. Biology and temperature requirements of the invasive seed bug *Leptoglossus occidentalis* (Heteroptera: *Coreidae*) in Europe. *J. Pest Sci.* **2015**, *89*, 31–44. [CrossRef]
12. Mas, H.; Naya, M.; Pérez-Laorga, E.; Aguado, A.; Marco, M.; Aragoneses, J.; Rodrigo, E. Estudio del ciclo biológico de *Leptoglossus occidentalis* Heidemann, 1910 (Hemiptera, Coreidae) en la Comunitat Valenciana. In Proceedings of the 6th Spanish Forest Congress, Vitoria, Spain, 10–14 June 2013; pp. 10–14.
13. Bernardinelli, I.; Zandigiacomo, P. *Leptoglossus occidentalis* Heidemann (Heteroptera, Coreidae): A conifer seed bug recently found in northern Italy. *J. For. Sci.* **2001**, *47*, 56–58.
14. Naves, P.; Nóbrega, F.; de Sousa, E. An adult *Leptoglossus occidentalis* (Heteroptera: Coreidae) parasitized in Portugal by the native Tachinid fly *Elomya lateralis* (Diptera: Tachinidae). *Ann. Soc. Entomol. Fr. (N.S.)* **2022**, *58*, 287–291. [CrossRef]
15. Strong, W. *BC Cone and Seed Pest Research Program*; Research and Knowledge Management Branch Seminar Series; Ministry of Forests and Range: Vernon, BC, Canada, 2010.
16. Blatt, S. An unusually large aggregation of the western conifer seed bug, *Leptoglossus occidentalis* (Hemiptera: Coreidae), in a man-made structure. *J. Entomol. Soc. B. C.* **1994**, *91*, 71–72.
17. Blatt, S.E.; Borden, J.H. Evidence for a male-produced aggregation pheromone in the western conifer seed bug, *Leptoglossus occidentalis* Heidemann (Hemiptera: Coreidae). *Can. Entomol.* **1996**, *128*, 777–778. [CrossRef]
18. Blatt, S.E.; Borden, J.H.; Pierce, J.H.D.; Gries, R.; Gries, G. Alarm pheromone system of the western conifer seed bug, *Leptoglossus occidentalis*. *J. Chem. Ecol.* **1998**, *24*, 1013–1031. [CrossRef]
19. Millar, J.G.; Zou, Y.; Hall, D.R.; Halloran, S.; Pajares, J.A.; Ponce-Herrero, L.; Shates, T.; Wilson, H.; Daane, K.M. Identification and synthesis of Leptotriene, a unique sesquiterpene hydrocarbon from males of the leaf-footed bugs *Leptoglossus zonatus* and *L. occidentalis*. *J. Nat. Prod.* **2022**, *85*, 2062–2070. [CrossRef]
20. Takács, S.; Bottomley, H.; Andreller, I.; Zaradnik, T.; Schwarz, J.; Bennett, R.; Strong, W.; Gries, G. Infrared radiation from hot cones on cool conifers attracts seed-feeding insects. *Proc. R. Soc. B Biol. Sci.* **2008**, *276*, 649–655. [CrossRef]
21. Franco-Archundia, S.; Gonzaga-Segura, A.; Jiménez-Pérez, A.; Castrejón-Gómez, V. Behavioral response of *Leptoglossus zonatus* (Heteroptera: Coreidae) to stimuli based on colors and its aggregation pheromone. *Insects* **2018**, *9*, 91. [CrossRef]
22. Wilson, H.; Maccaro, J.; Daane, K. Optimizing trap characteristics to monitor the leaf-footed bug *Leptoglossus zonatus* (Heteroptera: Coreidae) in orchards. *Insects* **2020**, *11*, 358. [CrossRef]
23. Inoue, K.; Vidal, D.; Saad, E.; Martins, C.; Zarbin, P. Identification of the alarm and sex pheromones of the leaf-footed bug, *Leptoglossus zonatus* (Heteroptera: Coreidae). *J. Braz. Chem. Soc.* **2018**, *30*, 939–947. [CrossRef]
24. Mitchell, P.L. Leaf-footed bugs (Coreidae). In *Heteroptera of Economic Importance*; Schaefer, C.W., Panizzi, A.R., Eds.; CRC Press: Boca Raton, FL, USA, 2000; pp. 337–403. [CrossRef]
25. Kim, J.H.; Lee, D.E.; Park, S.; Clark, J.M.; Lee, S.H. Characterization of nit sheath protein functions and transglutaminase-mediated cross-linking in the human head louse, *Pediculus humanus capitis*. *Parasites Vectors* **2021**, *14*, 425. [CrossRef]
26. Ma, J.; Ma, L.; Zhang, H.; Zhang, Z.; Wang, Y.; Li, K.; Chen, X. Policosanol fabrication from insect wax and optimization by response surface methodology. *PLoS ONE* **2018**, *13*, e0197343. [CrossRef]
27. Yan, G.; Cao, Z.; Devine, D.; Penning, M.; Gately, N.M. Physical properties of shellac material used for hot melt extrusion with potential application in the pharmaceutical industry. *Polymers* **2021**, *13*, 3723. [CrossRef]

28. Andronie, L.; Coroian, A.; Miresan, V.; Pop, I.; Raducu, C.; Rotaru, A.; Olar, L. Results obtained by investigating saffron ussing FT-IR spectroscopy. *Bull. Univ. Agric. Sci. Vet. Med. Cluj-Napoca Anim. Sci. Biotechnol.* **2016**, *73*, 238. [CrossRef] [PubMed]
29. Mariod, A.A.; Fadul, H. Extraction and characterization of gelatin from two edible Sudanese insects and its applications in ice cream making. *Food Sci. Technol. Int.* **2015**, *21*, 380–391. [CrossRef] [PubMed]
30. Venditti, A. What is and what should never be: Artifacts, improbable phytochemicals, contaminants and natural products. *Nat. Prod. Res.* **2018**, *34*, 1014–1031. [CrossRef] [PubMed]
31. Chaos, A.; Sangroniz, A.; Gonzalez, A.; Iriarte, M.; Sarasua, J.-R.; del Río, J.; Etxeberria, A. Tributyl citrate as an effective plasticizer for biodegradable polymers: Effect of plasticizer on free volume and transport and mechanical properties. *Polym. Int.* **2019**, *68*, 125–133. [CrossRef]
32. Tsochatzis, E.D.; Alberto Lopes, J.; Hoekstra, E.; Emons, H. Development and validation of a multi-analyte GC-MS method for the determination of 84 substances from plastic food contact materials. *Anal. Bioanal. Chem.* **2020**, *412*, 5419–5434. [CrossRef]
33. Yu, Y.; Wang, S.; Yang, Z.; Wang, F.; Deng, L. A novel environment-friendly synthetic technology of dibutyl itaconate. *J. Chem. Technol. Biotechnol.* **2020**, *95*, 2879–2885. [CrossRef]
34. Renthal, R.; Lohmeyer, K.; Borges, L.M.F.; Pérez de León, A.A. Surface lipidome of the lone star tick, *Amblyomma americanum*, provides leads on semiochemicals and lipid metabolism. *Ticks Tick-Borne Dis.* **2019**, *10*, 138–145. [CrossRef]
35. Jiang, X.-C.; Wen-Xia, D.; Chen, B.; Xiao, C.; Gui, F.-R.; Nai-Sheng, Y.A.N.; Qian, L.; Li, Z.-Y. Electrophysiological and oviposition responses of Asian corn borer, *Ostrinia furnacalis* (Lepidoptera: Crambidae), to compounds rinsed from the surfaces of sugarcane and maize leaves. *Eur. J. Entomol.* **2015**, *112*, 295–301. [CrossRef]
36. McAfee, A.; Chapman, A.; Iovinella, I.; Gallagher-Kurtzke, Y.; Collins, T.F.; Higo, H.; Madilao, L.L.; Pelosi, P.; Foster, L.J. A death pheromone, oleic acid, triggers hygienic behavior in honey bees (*Apis mellifera* L.). *Sci. Rep.* **2018**, *8*, 5719. [CrossRef]
37. Chow, Y.S.; Tsai, R.S. Protective chemicals in caterpillar survival. *Experientia* **1989**, *45*, 390–392. [CrossRef]
38. Honda, K. Osmeterial secretions of papilionid larvae in the genera *Luehdorfia*, *Graphium* and *Atrophaneura* (lepidoptera). *Insect Biochem.* **1980**, *10*, 583–588. [CrossRef]
39. Aldrich, J.R.; Schaefer, P.W.; Oliver, J.E.; Puapoomchareon, P.; Lee, C.-J.; Meer, R.K.V. Biochemistry of the exocrine secretion from gypsy moth caterpillars (Lepidoptera: Lymantriidae). *Ann. Entomol. Soc. Am.* **1997**, *90*, 75–82. [CrossRef]
40. Murphy, M.P.; O'Neill, L.A.J. Krebs cycle reimagined: The emerging roles of succinate and itaconate as signal transducers. *Cell* **2018**, *174*, 780–784. [CrossRef] [PubMed]
41. Liu, R.; Xu, B. Characterization of essential oil in pine nut shells from commodity waste in China by steam distillation and GC-MS. *Food Anal. Methods* **2011**, *5*, 435–440. [CrossRef]
42. Kadri, N.; Khettal, B.; Aid, Y.; Kherfellah, S.; Sobhi, W.; Barragan-Montero, V. Some physicochemical characteristics of pinus (*Pinus halepensis* Mill., *Pinus pinea* L., *Pinus pinaster* and *Pinus canariensis*) seeds from North Algeria, their lipid profiles and volatile contents. *Food Chem.* **2015**, *188*, 184–192. [CrossRef] [PubMed]
43. El Omari, N.; Ezzahrae Guaouguaou, F.; El Menyiy, N.; Benali, T.; Aanniz, T.; Chamkhi, I.; Balahbib, A.; Taha, D.; Shariati, M.A.; Zengin, G.; et al. Phytochemical and biological activities of *Pinus halepensis* mill., and their ethnomedicinal use. *J. Ethnopharmacol.* **2021**, *268*, 113661. [CrossRef]
44. Mohamed, A.A.; Behiry, S.I.; Ali, H.M.; El-Hefny, M.; Salem, M.Z.M.; Ashmawy, N.A. Phytochemical compounds of branches from *P. halepensis* oily liquid extract and *S. terebinthifolius* essential oil and their potential antifungal activity. *Processes* **2020**, *8*, 330. [CrossRef]
45. Dhibi, M.; Mechri, B.; Brahmi, F.; Skhiri, F.; Alsaif, M.A.; Hammami, M. Fatty acid profiles, antioxidant compounds and antiradical properties of *Pinus halepensis* Mill. cones and seeds. *J. Sci. Food Agric.* **2012**, *92*, 1702–1708. [CrossRef]
46. Anttonen, S.; Herranen, J.; Peura, P.; Kärenlampi, L. Fatty acids and ultrastructure of ozone-exposed Aleppo pine (*Pinus halepensis* Mill.) needles. *Environ. Pollut.* **1995**, *87*, 235–242. [CrossRef] [PubMed]
47. Campbell, B.E.; Pereira, R.M.; Koehler, P.G. Complications with controlling insect eggs. In *Insecticides Resistance*; Trdan, S., Ed.; InTechOpen: London, UK, 2016; pp. 83–96. [CrossRef]
48. Bruns, F.H.; Werners, P.H. Dehydrogenases: Glucose-6-phosphate dehydrogenase, 6-phosphogluconate dehydrogenase, glutathione reductase, methemoglobin reductase, polyol dehydrogenases. In *Advances in Clinical Chemistry*; Elsevier: Amsterdam, The Netherlands, 1963; Volume 5, pp. 237–294. [CrossRef]

insects

MDPI

Article

Parasitoid Wasp *Acerophagus papayae*: A Promising Solution for the Control of Papaya Mealybug *Paracoccus marginatus* in Cassava Fields in Vietnam

Khac Hoang Le [1], Thi Hoang Dong Tran [2], Dang Hoa Tran [2], Tuan Dat Nguyen [1] and Cong Van Doan [3,*]

1 Entomology Laboratory, Department of Plant Protection, Faculty of Agronomy, Nong Lam University Ho Chi Minh, Ho Chi Minh 700000, Vietnam; lkhoang@hcmuaf.edu.vn (K.H.L.)
2 Faculty of Agronomy, University of Agriculture and Forestry, Hue University, Hue City 530000, Vietnam
3 The Plant Physiology Unit, The Department of Life Sciences and Systems Biology of the University of Turin, Via Accademia Albertina 13/Via Gioacchino Quarello 15/A, 10123 Torino, Italy
* Correspondence: vancong.doan@unito.it

Simple Summary: This study examined the occurrence, biological characteristics, and effectiveness of the parasitoid wasp *Acerophagus papayae* Noyes and Schauff in controlling the papaya mealybug *Paracoccus marginatus* Williams and Granara de Willink, a major pest of cassava crops in Vietnam. The results showed that *A. papayae* naturally occurred more frequently than other species of parasitoid wasps. The honey solution was found to be important for increasing the lifespan of *A. papayae* in the absence of hosts. The second instar of the mealybug was found to be the most suitable host stage for parasitism by *A. papayae*. Female wasps can produce a large number of eggs during their entire lives. These findings suggest that *A. papayae* could be an effective agent for controlling *P. marginatus* and improving crop pest management not only in Vietnam, but in other regions as well.

Abstract: Cassava is a valuable export commodity crop that is often attacked by pests, causing economic losses for this crop. The papaya mealybug *Paracoccus marginatus* has become a major pest of cassava in Vietnam. The parasitoid wasp *Acerophagus papayae* has been demonstrated to be the most efficient parasitoid wasp for controlling *P. marginatus* in many regions. We observed the occurrence of *A. papayae* in Vietnam, studied the biological characteristics of *A. papayae*, and investigated its parasitic activity on *P. marginatus*. The results showed that *A. papayae* occurred more frequently than *Anagyrus loecki*, another known parasitoid of *P. marginatus*. The lifespan of *A. papayae* was approximately 16 days. In the absence of hosts, a 50% honey solution was an essential diet to increase the longevity of both female and male of *A. papayae*. The second instar of *P. marginatus* was a suitable host stage for parasitism by *A. papayae*. Female *A. papayae* laid approximately 60.8 eggs within 17 days, mostly during the first 6 to 7 days. These findings suggest that *A. papayae* has the potential to control *P. marginatus*, and could inform the development of more effective pest management strategies for cassava crops in Vietnam and other regions affected by this pest.

Keywords: parasitoid wasp; mealybugs; *Acerophagus papayae*; *Paracoccus marginatus*; cassava; biocontrol

1. Introduction

Cassava (*Manihot esculenta* Crantz) is a woody shrub in the Euphorbiaceae family, primarily cultivated for its starchy roots. It is widely grown in tropical and subtropical regions of Africa, Asia, and Latin America, with a global production of over 302 million tons per year [1]. Cassava is a staple food for over 500 million people worldwide, particularly across Africa (with 63% of total world production), where it is considered key in combatting food security issues [2–4]. Cassava is also an important export commodity for processing into products such as monosodium glutamate, alcohol, confectioneries, instant noodles, plywood, packaging, biofilm, and pharmaceutical additives. Moreover, cassava is an essential fodder crop in many countries [5–7]. In Vietnam, cassava was introduced in the early 19th century, and has since become a high-value crop, contributing significantly to poverty reduction for farmers in upland, remote, and isolated areas [6,8,9].

The growth period of cassava varies from 6 to 12 months, depending on the variety, growing season, growing area, and intended use [8,10,11]. Cassava plants are vulnerable to numerous pests [12,13]. In South America, nearly 200 species of arthropods feed on cassava as their primary food source, resulting in yield losses of over 50% [14]. In Vietnam, the insect pests of cassava have not been thoroughly studied, and only a few harmful arthropod species have been identified [15]. Mealybugs are insects that cause damage to cassava, including *Ferrisia virgata* Cockerell (Hemiptera: Pseudococcidae), *Paracoccus marginatus* Williams and Granara de Willink (Hemiptera: Pseudococcidae), *Phenacoccus manihoti* Matile-Ferrero (Hemiptera: Pseudococcidae), and *Pseudococcus jackbeardsleyi* Gimpel and Miller (Hemiptera: Pseudococcidae) [16]. In Vietnam, *P. marginatus* was first discovered in Tay Ninh province in 2012, and has since become the primary pest of cassava in 14 provinces and cities throughout the country [16]. According to Williams and Willink [17], *P. marginatus* is native to Belize, Costa Rica, Guatemala, and Mexico, and has spread to some Caribbean islands, Hawaii, and French Guiana [18]. *Paracoccus marginatus* was first discovered on hibiscus plants in Bradenton, Florida, USA, in 1998, and was subsequently found on 18 different plant species in the state [19].

Paracoccus marginatus attacks more than 55 species of host plants, including high-value crops such as papaya, avocado, citrus, mango, strawberry, and pomegranate, as well as hibiscus, cotton, tomato, eggplant, peppers, beans, peas, sweet potatoes, and tapioca [18–22]. This pest is omnivorous and feeds by sucking sap through the epidermis of leaves, fruits, and stems, which leads to leaf wrinkling, yellowing, and wilting, eventually causing plant death [23]. In addition, *P. marginatus* secretes black molds that cover leaves, fruits, and stems, impeding photosynthesis and gas exchange, resulting in stunted growth, deformed leaves, and, ultimately, tree death [21,23,24]. Since it was first reported in Coimbatore in 2007, *P. marginatus* has been identified as a key pest causing severe economic losses to farmers in Coimbatore, Erode, Tirupur, and Salem in India [23]. For example, strawberry farmers suffered losses of over 1500 hectares in Tirupur in 2009, resulting in heavy financial losses [23]. Potential economic losses caused by *P. marginatus* can total between 57% and 75% on papaya trees, with a decrease in fruit quality and a 2–3-fold increase in the cost of production [25,26].

For the long-term management of *P. marginatus*, biological control is the most sustainable and cost-effective option [15,27]. Natural enemies of mealybugs, such as the Australian ladybug *Cryptolaemus montrouzieri* Mulsant (Coleoptera: Coccinellidae), *Scymnus* Kugelann (Coleoptera: Coccinellidae), and some parasitoid wasps [23,27], are present in nature. *Acerophagus papayae* Noyes and Schauff, *Anagyrus loecki* Noyes and Menezes, and *Pseudleptomastix mexicana* Noyes and Schauff (Hymenoptera: Encyrtidae) were regularly introduced to control *P. marginatus*, especially in Guam, Palau, Florida and Hawaii (USA), India, and Sri Lanka [22,24,28,29]. Among the parasitoid wasps, *A. papayae* (Hymenoptera: Encyrtidae) is one of the most effective species in controlling *P. marginatus* [30–33]. Mastoi et al. [34] reported on the natural enemies of *P. marginatus* in the Malaysian Peninsula, and observed *A. papayae* to be a parasitoid wasp on this mealybug. The incidence of mealybugs significantly decreased since the introduction of *A. papayae* into the environment [32,35]. Meyerdirk et al. [36] reported that 46,200 individuals of the parasitoid wasps *A. loecki*, *P. mexicana*, and *A. papayae* had been introduced from Puerto Rico and released in Guam from June to October 2002, resulting in a reduction of more than 99% of *P. marginatus* abundance after 1 year. This has reduced the risk of invasion of *P. marginatus* to neighboring islands. The effectiveness of three parasitoid wasp species (*A. loecki*, *P. mexicana*, and *A. papayae*) against *P. marginatus* was observed in the 4 months after release of the parasitoid wasps in an ecological mulberry garden in Tamil Nadu, India. *Acerophagus papayae* exhibited the highest population growth and parasitism rate, of

75.6–81.7%, followed by *P. mexicana* with 9.3–24.4%; the lowest parasitism was exhibited by *A. loecki* of 0.7–9.0% [37].

Although *P. marginatus* is becoming a key pest, damaging cassava in Vietnam, there have not been many studies evaluating the existence of parasitoid wasps to control this pest in various conditions [27]. Therefore, this study was conducted to (1) determine and identify the presence of the parasitoid wasp *A. papayae* in cassava fields in Vietnam; (2) determine the life cycle of *A. papayae* under specific conditions; and (3) determine the parasitism of *A. papayae* on *P. marginatus* in Vietnam, including its preferred host stage and egg-laying rhythm.

2. Materials and Methods

2.1. Study Sites and Biological Materials

The experiment was conducted during the 2016–2017 cropping season at two locations. The first location was Cam Giang commune, Go Dau district, Tay Ninh province, Vietnam (11°16′10.385″ N and 106°5′10.96″ E), where the frequency of occurrence of the parasitoid wasp *A. papayae* and the papaya mealybug *P. marginatus* were investigated. The second location was the insect laboratory at the Department of Plant Protection, Department of Agronomy, Nong Lam University, Ho Chi Minh City, Vietnam (10°52′17.126″ N and 106°47′31.423″ E), where all the experiments were conducted.

Rearing of *P. marginatus*: *P. marginatus* were collected from cassava fields and reared on a cassava plant in a net cage. When *P. marginatus* laid eggs, all eggs of the same age were transferred to new cassava plants to ensure a source of similarly aged mealybugs for experiments.

Rearing of parasitoid wasps *A. papayae*: Adult male and female specimens of *A. papayae* obtained from the cassava field were paired in collection tubes and provided with a 50% honey + 50% water solution (5–10 pairs of wasps exposed to second-instar *P. marginatus* on cassava in the net cage). After 24 h, the cassava plants with mealybugs were separated and kept in the net cage until *A. papayae* emerged. Emerging adults from infested mealybugs were used to reproduce wasps for the experiments.

2.2. Experimental Designs

2.2.1. Biological Characteristics of *A. papayae*

Determining the Frequency of Parasitoids of the *P. marginatus* in Vietnam

The occurrence of *P. marginatus* parasitoids was recorded once a week during the cassava season. Parasitoids were collected from three cassava fields with high *P. marginatus* damage (>30%) and an area of 0.5–1 ha. Five collection points were randomly selected along two diagonal lines. At each collection point, five adjacent cassava bushes were selected. We cut off the *P. marginatus*-infested part of the cassava plant (randomly picking four points on each tree) and placed them in a net cage to collect the parasitoid wasps. The wasps were identified using Noyes and Schauff's method [38].

The occurrence frequency of parasitoid wasps was assessed as follows:

–	Very low	<5%
+	Low	5–25%
++	Moderate	25–50%
+++	High	50–75%
++++	Very high	>75%

Relative occurrence (%) = $(f/n) \times 100$, where f is the number of individual cassava plants having parasitoid wasps of the papaya mealybug, and n is the total number of cassava observations.

Observing the Mating Activity and Egg-Laying Behavior of *A. papayae* Under Laboratory Conditions

The experiment aimed to observe the mating activity and egg-laying behaviors of *A. papayae* on *P. marginatus*. Ten pairs of newly emerged *A. papayae* were placed in a collection tube and provided with two hundred second-instar *P. marginatus* on cassava leaves. The mating activity and egg-laying behavior of *A. papayae* were observed under a stereo microscope (Olympus SZ51—Japan, 40× magnification). The experiments were repeated for seven days to record the common mating activity and egg-laying behavior of *A. papayae* on *P. marginatus*.

Determining the Immature Developmental Time of the Parasitoid Wasps *A. papayae*

The experiments aimed to characterize the immature developmental time of the parasitoid wasps *A. papayae*, which can inform the optimal timing for releasing the wasps into the environment to control mealybugs. Ten pairs of *A. papayae* (fed with 50% honey) were exposed to 200 individuals of *P. marginatus*, which were fed papaya fruit in a net cage. After 24 h of exposure, the papaya fruit was removed from the cage and placed in another net cage. Every day, ten specimens were dissected and observed under a stereo microscope and electron microscope to determine the life cycle of the wasp. The life cycle of the *A. papayae* was determined from the *P. marginatus* parasitism stage to the adult stage [39].

Effect of Diets on the Longevity of Adult *A. papayae*

In order to provide a diet for *A. papayae* in the absence of hosts, we tested the effect of diets (honey concentration) on the longevity of the parasitoid adults. The experiment was arranged in a completely randomized single-factor design with five treatments (control—distilled water, 10% honey, 30% honey, 50% honey, and 70% honey) with ten replicates. Each replicate included one adult *A. papayae* wasp in a test tube, with the end of test tube covered with a mesh, and cotton wool soaked in honey. After 24 h, the cotton wool was refreshed until the wasp died. The longevity of the parasitoid wasp was recorded as the average lifespan of adult *A. papayae* wasp per treatment.

2.2.2. Parasitism of *A. papayae* on *P. marginatus*

Host Stage Susceptibility and Preference of *A. papayae*

This experiment aimed to identify the host stage susceptibility and preference of *A. papayae*, establish an efficient method for the mass rearing of *A. papayae*, and determine the efficiency of controlling *P. marginatus* under natural conditions. The experiment was conducted in a completely randomized single-factor design with four treatments (second-instar nymphs, third-instar nymphs, non-laying-eggs adults of *P. marginatus*, and egg-laying adults of *P. marginatus*), each with 10 replicates. Each replicate included a female wasp that had mated for three days and was fed with a 50% honey solution (the most optimal concentration of the honey solution to rearing the parasitoid). The female was exposed to 20 *P. marginatus* individuals in each stage. After 24 h of exposure, each treatment was assessed using the following parameters [39]:

$$\text{Parasitism rate (\%)} = (\text{number of infected } P.\ marginatus/\text{total } P.\ marginatus) \times 100$$

$$\text{Emergence rate (\%)} = (\text{number of } A.\ papayae \text{ emerging}/\text{total number of infected } P.\ marginatus) \times 100$$

$$\text{Female rate (\%)} = (\text{number of emerging female}/\text{total number of emerging}) \times 100$$

Parasitoid Fecundity and Progeny of *A. papayae*

The aim of this experiment was to determine the total number of eggs laid by *A. papayae* on *P. marginatus* in the second-instar nymph. One pair of newly emerged *A. papayae* was fed on 50% honey and exposed to 20 *P. marginatus* individuals in the second-instar stage. After 24 h, *P. marginatus* and honey were replaced under the same conditions. Twenty of the exposed *P. marginatus* were dissected to determine the number of eggs laid in the body of the mealybug, as well as the egg-laying ability of *A. papayae*. The experiment was continued until the parasitoid wasps died. The observations were repeated 10 times, each time with a new pair of *A. papayae* individuals. The observation included the total number of eggs laid by one female wasp and the number of eggs laid per day.

The Effect of Parasitism Activity on the Longevity of Female *A. papaya*

The experiment aimed to determine the effect of parasitism activity on the longevity of the female *A. papayae*. The study followed a completely randomized single-factor design with two treatments (nonparasitism activity and parasitism activity), including 10 replicates. Each replicate consisted of a paired *A. papayae* that was fed with 50% honey within 24 h. Then, the wasp was either placed in a test tube without *P. marginatus* + 50% honey or with the second instar of *P. marginatus* and 50% honey. New 50% honey was replaced every 24 h until the parasitoid wasps died. The average longevity of female parasitoid wasp (days) was recorded by observing the number of female wasp deaths every 24 h.

2.3. Data Analysis

Data were entered, processed, and graphed using Excel (2020). For data analysis, the proportional data, such as the parasitism rate (%), emergence (%), and female emergence (%) data, were transformed by arcsin$\sqrt{}$(%) to meet normality [40–42] before analysis. The response variables, such as the parasitism rate, the emergence, the female emergence, and the longevity of female and male *A. papayae*, were assessed by ANOVA analysis of variance IBM SPSS Statistics 28.0 software with the main effect of different host development stages or honey solutions. In addition, Tukey's post hoc test was performed to separate the means at the 95% confidence interval, where ANOVA showed significant differences in the means. The Student's *t*-test was employed to analyze the effect of parasitism activity on the longevity of female *A. papayae*.

3. Results

3.1. Diversity and Abundance of P. marginatus Parasitoids in Vietnam

Only two species of parasitoid wasps belonging to the Encyrtidae family, *Acerophagus papayae* and *Anagyrus loecki*, were recorded on *P. marginatus* in Cam Giang commune, Go Dau district, Tay Ninh province, Vietnam. *A. papayae* occurred at a high frequency of 53.9%, whereas *A. loecki* was observed with a moderate frequency of 34.9% (Table 1).

Table 1. Occurrence frequency of parasitoid wasps of *P. marginatus* in Tay Ninh, Vietnam.

Scientific Name	Occurrence Frequency
A. papayae	+++
A. loecki	++

Note: ++, moderate (25–50%); +++, high (50–75%).

3.2. Mating Activity and Egg-Laying Behavior of A. papayae under Laboratory Conditions

During the mummy stage of the mealybug, the head of *A. papayae* emerges first, and the body segments stretch from the back to the front until the body completely exits the mummy. After emerging, female *A. papayae* can mate and lay eggs within 24 h. The mating process starts when the male courts the female wasp. If the female accepts, the female stands 3 to 4 cm away from the male wasp, moving her antennae continuously and raising her wings slowly. Next, the male jumps onto the back of the female and clings to her back with his feet. The male wasp then bends the end of his abdomen and inserts his genitals into the female wasp's genitals. Mating occurs for approximately 10 s, during which the female remains still. At the end of mating, the male is pushed 4 to 5 cm away by the female. The male stands motionless for 15 to 20 s, then fans his wings to dry the mating parts, and proceeds to pair and mate with other female wasps. If the mating process fails, the male will jump onto the back of the female and mate again.

Female *A. papayae* usually lay eggs in areas where there are many *P. marginatus*. When searching for a host, the wasp's antennae reach forward, similar to two "antennas", to find the host's scent, locating a potential host more quickly. When preparing to lay eggs, the female wasp crawls onto the host's back, finds a suitable location, and uses the egg-laying tray to puncture the skin and lay eggs in the host's body. The egg-laying time of females ranges from 25 to 30 s. After oviposition, the female leaves the parasitized host without any action to mark the host as parasitized. Therefore, if *A. papayae* chooses an already-parasitized host, the *P. marginatus* will respond by flexing its body, rapidly expanding its abdomen to prevent the female from laying eggs.

It was observed that the female *A. papayae* had a habit of feeding on the host's body fluid in addition to laying eggs. The female wasp pierces the skin on the host's dorsal side with a trough, and then turns her head to use her mouth to extract fluids from the hole in the skin. This causes the mealybug's body to become parasitized, mummified, and die. Thus, *A. papayae* is considered to have biological potential for managing *P. marginatus* populations.

3.3. Life Cycle of the *A. papayae* and Developmental Time of Each Phase

The maturation time of the egg was 3.7 ± 0.16 days. The larvae developed in three different stages with two molt processes. The first, second, and third instars had maturation times of 1 to 2 days between each stage. The total maturation time of the larval phase was 4.2 ± 0.22 days. At the end of the third instar stage, *A. papayae* molted for the last time to move into the prepupal phase, which lasted 2.7 ± 0.16 days. After the prepupal phase, *A. papayae* progressed into the pupal phase; the pupal phase was 8.0 ± 0.28 days. The total maturation time of the pupa (including prepupal and pupal phases) was 10.0 ± 0.28 days. The life cycle of *A. papayae* was completed in 16.0 ± 0.32 days, from egg to adult emergence (Supplementary Table S1).

3.4. The Potential of Using Honey Solution as a Diet for *A. papayae* in the Absence of a Host

The 50% honey diet solution influenced the longevity of both male ($p \leq 0.01$, $F = 28.089$) and female *A. papayae* ($p \leq 0.01$, $F = 78.115$). Adult female and male *A. papayae* fed on 50% honey survived for 12.5 ± 0.40 days and 7.3 ± 0.34 days, respectively. Longevity was shortest for those fed on distilled water: 3.2 ± 0.13 and 2.2 ± 0.25 days for females and males, respectively (Table 2).

Table 2. Effect of honey solution on the longevity of female and male *A. papayae*.

Treatments	Longevity of *A. papayae* (Day)	
	Female	Male
	(AVG ± SE)	(AVG ± SE)
Distilled water	3.2 ± 0.13 d	2.2 ± 0.25 c
Honey 10% *	6.5 ± 0.45 c	4.3 ± 0.40 b
Honey 30% *	7.6 ± 0.27 c	5.4 ± 0.43 b
Honey 50% *	12.5 ± 0.40 a	7.3 ± 0.34 a
Honey 70% *	9.6 ± 0.56 b	5.6 ± 0.34 b
p-value	<0.01	<0.01

Notes: AVG, average; SE: standard error; $n = 10$. Within column, means followed by the same lowercase letter do not differ significantly. *: Absence of a host.

3.5. Susceptibility and Preference of Different Host Stages for *A. papayae*

The parasitism rates (%) of *A. papayae* on different host stages of *P. marginatus* were significantly different ($p \leq 0.01$, $F = 143.240$). The highest parasitism rate of 46% was observed when *A. papayae* parasitized the second-instar stage of *P. marginatus*. In contrast, no parasitism was observed on egg-laying adults (Figure 1).

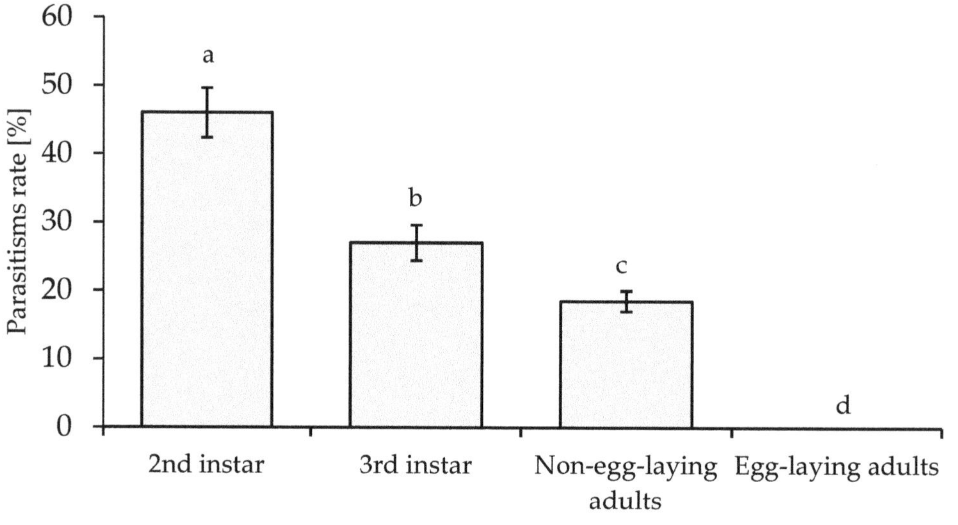

Figure 1. Parasitism rate (%) of *A. papayae* on *P. marginatus*. The data were transformed using arcsin$\sqrt{}$ (%) function prior to statistical analysis. Different letters above columns indicate significant differences. Bar indicates SE value.

Similarly, the emergence values (%) of *A. papayae* from different host stages of *P. marginatus* were significantly different ($p \leq 0.01$, $F = 73.853$). The emergence of *A. papayae* was high on second-instar, third-instar, and non-egg-laying adults. The emergence ranged from 86% to 89%, compared with 0% for egg-laying adults (Figure 2).

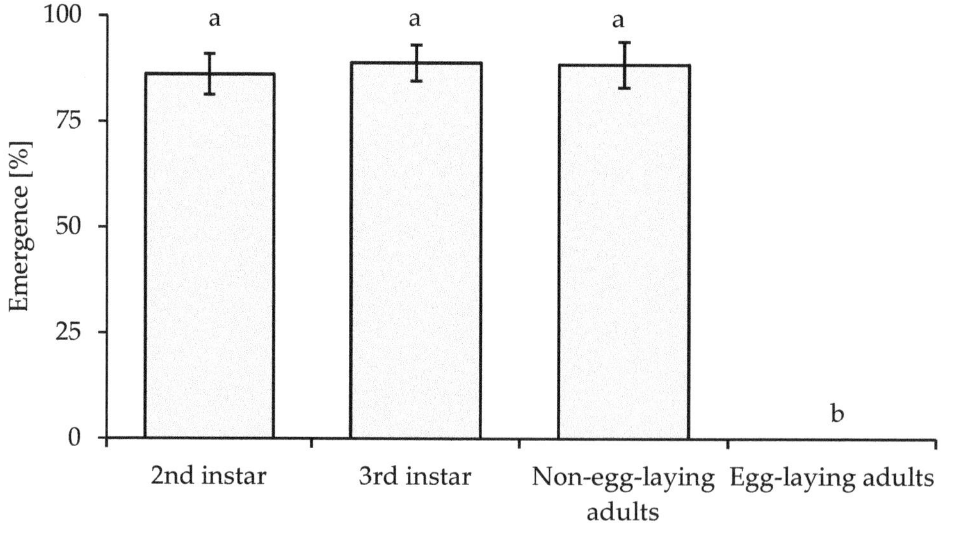

Figure 2. Emergence (%) of *A. papayae* on different development stages of *P. marginatus*. The data were transformed by arcsin$\sqrt{}$(%) before statistical analysis. Different letters above columns indicate significant differences. Bar indicates SE value.

The numbers of female emergences (%) of *A. papayae* parasitizing different stages of *P. marginatus* were significantly different ($p \leq 0.01$, $F = 142.741$). The lowest female emergence rate was observed on egg-laying adults (0%), while the highest female emergence was 74% on non-egg-laying adults and third instars. The second instar showed the most balanced gender ratio of emerging *A. papayae* (Figure 3).

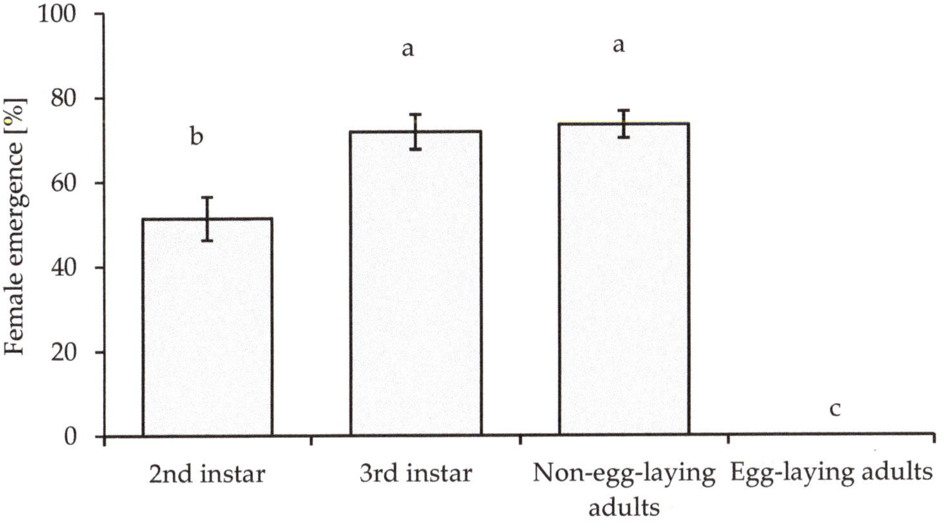

Figure 3. Female emergence (%) of *A. papayae* wasps parasitized on *P. marginatus*. The data were transformed using the arcsin $\sqrt{}$ (%) function prior to statistical analysis. Different letters above columns indicate significant differences. Bar indicates SE value.

3.6. Acerophagus papayae Reproduction on P. marginatus

Female *A. papayae* laid approximately 60.8 ± 1.17 eggs during their life (Supplementary Table S2). The egg-laying rhythm of *A. papayae* decreased with each day. The female wasp laid eggs over 17 days, of which the most was 10 ± 0.54 eggs on the first day; and the highest concentration was laid in the first 6 to 7 days after starting oviposition (Figure 4).

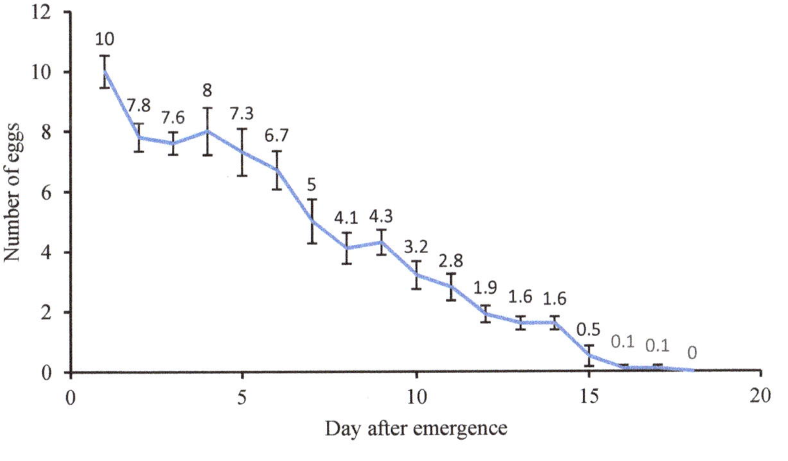

Figure 4. Progeny rhythm of *A. papayae*. Bar indicates SE value.

3.7. Parasitism and Longevity of Female *A. papayae*

Female wasps that exhibited parasitism activity lived longer than the females without parasitoid activity. Females that participated in parasitism activities had an average longevity of 14.2 ± 0.59 days, whereas wasps without any parasitism activity had an average longevity of 12.5 ± 0.40 days (Table 3).

Table 3. Effects of parasitism activity on the longevity of female *A. papayae*.

Treatment + 50% Honey Solution	Longevity of Female *A. papayae* (Days)	
	Min–Max	AVG ± SE
Female without parasitism	11–14	12.5 ± 0.40
Female with parasitism	12–18	14.2 ± 0.59
t-test (*p* value)		0.0288
df (within groups)		18

Notes: AVG, average; SE, standard error; $n = 10$.

4. Discussion

Tay Ninh province was reported to be an invasion area of mealybugs, which cause damage to cassava. During field surveys conducted from 2014 to 2015, four different mealybug species, *Paracoccus marginatus, Phenacoccus manihoti, Pseudococcus jackbeardsleyi*, and *Ferrisia virgata*, were recorded in cassava fields. The abundance of mealybugs ranged from 0 to 698 per cassava tip [27]. *Paracoccus marginatus* is widely expanding across Central and East Africa, Central America, and Asia, because these areas have suitable climatic conditions and diverse host crops [22,43]. The presence of *P. marginatus* may increase the presence of the natural enemies in the field; however, the abundance is normally too low for controlling this pest [34,36]. *Anagyrus lopezi* was found to be parasitizing *P. marginatus* in Tay Ninh province. However, the level of parasitism was not efficient to control *P. marginatus* [27]. In this study, we not only observed *A. loecki*, but also *A. papayae*, parasitizing *P. marginatus* (Table 1). *A. papayae* has been successfully used as biocontrol agent in different regions around the world [36,37]. Our findings indicate a potential new addition to the list of parasitoid wasp species which attack *P. marginatus* in Vietnam [16,24,27,30–32]. However, the abundance of these parasitoids was not efficient for the biocontrol of *P. marginatus* [43]. Consequently, more wasps need to be mass-reared to mitigate against the increasing damage caused by *P. marginatus* to cassava [36,37].

Acerophagus papayae has the potential to be mass-produced for biocontrol programs [44]. To effectively rear *A. papayae* for biocontrol strategies, it is crucial to have a good understanding of their immature developmental time, longevity, and lifetime fertility [45]. Each female parasitoid wasp can produce approximately 60 eggs in total over their lifespan (Figure 4). Female *A. papayae* lay most of their eggs within the first 6 to 7 days after emergence, which is a critical stage to consider when determining the optimal timing to release the wasp into the wild. This enables the wasp to adapt to the natural environment and take full advantage of their egg-laying potential [45].

Determining the appropriate host age is crucial for the efficient multiplication of parasitoid wasp biomass and estimating the optimal time to release the wasps into the environment [39,46]. The preference for a particular host stage is vital for the viability of young *A. papayae*. Adequate and appropriate nutrition are necessary to ensure the complete life cycle development of the parasitoid wasp [47]. The host must have an appropriate body size and sufficient time for the parasitoid wasp to complete its life cycle [48]. *A. papayae* parasitizes *P. marginatus*, and the host responds by stretching its body, which leads to the parasitoid wasp breaking its egg-laying tube and dying. The parasitism rate gradually decreases with the age of the host; older hosts exhibit a stronger response. The second instar of *P. marginatus* is a suitable host for *A. papayae*, with high emergence rates and a well-balanced sex ratio (Figures 1 and 3). This is important for multiplying parasitoid wasp numbers and determining the time to undertake field releases to manage the target pest [39,45]. The *A. papayae* wasp only requires a single *P. marginatus* at the second-instar nymph stage to develop from an egg to adult (See Supplementary Table S1).

A. papayae fed with a honey solution had a significantly higher longevity than when fed with distilled water, as previously described [49,50]. This potentially increases the longevity of female *A. papayae* in the absence of a host. Females exhibiting parasitoid activity had higher longevity than the females without parasitism activity [51]. Females that participated in parasitism activities had an average longevity of 14.2 days, whereas females without parasitism activities had an average longevity of 12.5 days (Table 2).

5. Conclusions

In Vietnam, biological measures for managing *P. marginatus* are not well studied. Our results contribute to the current knowledge about natural enemies that can be used to control *P. marginatus* in cassava cropping systems in Vietnam. To reproduce a large number of *A. papayae* for biocontrol purposes, the parasitoids can be reared on second- or third-instar nymphs of *P. marginatus* as the host. In addition, a 50% honey solution can be provided as a supplementary diet to increase the longevity of the wasps. The appropriate time to release *A. papayae* into the environment is when the majority of the *P. marginatus* population consists of second- or third-instar nymphs, and during the first 6 to 7 days after the initial egg-laying oviposition of the *A. papayae*.

Supplementary Materials: The following supporting information can be downloaded at: https://www.mdpi.com/article/10.3390/insects14060528/s1, Table S1. Development time of each stage and life cycle of *Acerophagus papaya*. Table S2. Egg-laying process of the *Acerophagus papaya*; File S1: original data.

Author Contributions: Conceptualization, K.H.L. and C.V.D.; methodology, K.H.L. and C.V.D.; software, K.H.L.; validation, K.H.L., T.H.D.T., D.H.T., T.D.N. and C.V.D.; formal analysis, K.H.L.; investigation, K.H.L.; resources, K.H.L.; data curation, K.H.L.; writing—original draft preparation, C.V.D.; writing—review and editing, C.V.D., K.H.L., T.H.D.T., D.H.T. and T.D.N.; visualization, K.H.L.; supervision, K.H.L.; project administration, K.H.L.; funding acquisition, K.H.L. and C.V.D. All authors have read and agreed to the published version of the manuscript.

Funding: This research was supported in part by a grant-in-aid by Hue University under the Core Research Program, Grant No. NCM.DHH.2020.14, from the group of Entomology Laboratory, Department of Plant Protection, Faculty of Agronomy, Nong Lam University, and the Projects PNRR M4C2 Initiative 1.2—Funding of projects presented by young researchers—Seal of Excellence, National Recovery and Resilience Plan (NRRP), Italy (Internal code: DOAV_PNRR_YOUR_SOE_22_01; CUP: D18H22001960007).

Data Availability Statement: The original contributions presented in this study are included in the article. Further inquiries can be directed to the corresponding authors.

Acknowledgments: We are very grateful to Ngoc Khoa Nguyen for performing the experiments. We would like to express our sincere thanks to the members of the Department of Plant Protection, Faculty of Agronomy, University of Agriculture and Forestry, and Faculty of Agronomy, University of Agriculture and Forestry, Hue University, for their help and cooperation. We would also like to thank HSC Investment Corporation Company for their contribution to this study.

Conflicts of Interest: The authors declare no conflict of interest.

References

1. FAO. Food and Agriculture Organization of the United Nations Statistics Division. Online Statistical Database: Food Balance. FAOSTAT. 2023. Available online: https://www.fao.org/faostat/en/#data/QCL/visualize (accessed on 20 January 2023).
2. Prudencio, Y.C.; Al-Hassan, R. The food security stabilization roles of cassava in Africa. *Food Policy* **1994**, *19*, 57–64. [CrossRef]
3. Onodu, B.C.; Culas, R.J. The role of Cassava production in improving food security in the delta state of Nigeria. In *Food Security: Threat Factors, Policies and Challenges*; Webb, J., Ed.; Food Science and Technology; Nova Science Publishers: Hauppauge, NY, USA, 2017; pp. 1–34.
4. Tize, I.; Fotso, A.K.; Nukenine, E.N.; Masso, C.; Ngome, F.A.; Suh, C.; Lendzemo, V.W.; Nchoutnji, I.; Manga, G.; Parkes, E.; et al. New cassava germplasm for food and nutritional security in Central Africa. *Sci. Rep.* **2021**, *11*, 7394. [CrossRef] [PubMed]

5. Nguyen, H.H.; Nguyen, T.C.; Tong, Q.A. Cassava leaf production research in Vietnam. In *Cassava Research and Development in Asia: Exploring New Opportunities for an Ancient Crop*; Howeler, R.H., Ed.; Cali CIAT: Valle del Cauca, Colombia, 2007; pp. 494–495. Available online: https://cgspace.cgiar.org/bitstream/handle/10568/56568/cassava_research_development_asia.pdf?sequence=1 (accessed on 20 January 2023).
6. Nguyen, H.H.; Tran, C.K. Achievements in cassava research and development in Vietnam and orientation to 2020. In *Report on Research on Cassava Development 2007–2012*; Institute of Agricultural Science for Southern Vietnam: Ho Chi Minh, Vietnam, 2012; pp. 1–7.
7. Shigaki, T. Cassava: The nature and uses. In *Encyclopedia of Food and Health*; Academic Press: Cambridge, MA, USA, 2016; pp. 687–693. [CrossRef]
8. Hoang, L.; Nguyen, T.T.M.; Nguyen, B.M.; Hoang, K.; Ishitani, M.; Howeler, R. Cassava in Vietnam: Production and research; an overview. In Proceedings of the Paper Presented at Asia Cassava Research Workshop, Hosted by ILCMB-CIAT-VAAS/AGI, Hanoi, Vietnam, 3 November 2014.
9. Hoang, K.; Nguyen, T.T.M.; Nguyen, B.M.; Howeler, R. Cassava Conservation And Sustainable Development in Vietnam. In Proceedings of the Sustainable Cassava Production in Asia for Multiple Uses and for Multiple Markets, the 9th Regional Cassava Workshop, Nanning, China, 27 November–3 December 2011.
10. Safo-Kantanka, O.; Owusu-Nipah, J. Cassava varietal screening for cooking quality: Relationship between dry matter, starch content, mealiness and certain microscopic observations of the raw and cooked tuber. *J. Sci. Food Agric.* **1992**, *60*, 99–104. [CrossRef]
11. Ebah-Djedji, B.C.; Dje, K.M.; Ue, B.N.; Zohouri, G.P.; Amani, N.G. Effect of harvest period on starch yield and dry matter content from the tuberous roots of improved cassava (*Manihot esculenta* Crantz) varieties. *Pak. J. Nutr.* **2012**, *11*, 414–418. [CrossRef]
12. Bellotti, A.C. Cassava Pests and their Management. In *Encyclopedia of Entomology*; Capinera, J.L., Ed.; Springer: Dordrecht, The Netherlands, 2008. [CrossRef]
13. Graziosi, I.; Minato, N.; Alvarez, E.; Ngo, D.T.; Hoat, T.X.; Aye, T.M.; Pardo, J.M.; Wongtiem, P.; Wyckhuys, K.A. Emerging pests and diseases of South-east Asian cassava: A comprehensive evaluation of geographic priorities, management options and research needs. *Pest Manag. Sci.* **2016**, *72*, 1071–1089. [CrossRef]
14. Bellotti, A.C.; Herrera, C.J.; Hernández, M.P.; Ariasm, B.; Guerrero, J.M.; Melo, E.L. Three major cassava pests in Latin America, Africa and Asia. In *A New Future for Cassava in Asia: Its Use as Food, Feed and Fuel to Benefit the Poor*; Howeler, R.H., Ed.; The Nippon Foundation: Tokyo, Japan, 2010; pp. 544–577.
15. Tokunaga, H.; Baba, T.; Ishitani, M.; Ito, K.; Kim, O.-K.; Ham, L.H.; Le, H.K.; Maejima, K.; Natsuaki, K.T.; Van Dong, N.; et al. Sustainable Management of Invasive Cassava Pests in Vietnam, Cambodia, and Thailand. In *Crop Production under Stressful Conditions*; Kokubun, M., Asanuma, S., Eds.; Springer: Singapore, 2018. [CrossRef]
16. Le, T.T.N.; Truong, T.H.L.; La, V.H. Pest composition and natural enemies on cassava in Vietnam. In Proceedings of the 6th National Scientific Conference on Ecology and Biological Resources, Hanoi, Vietnam, 21 October 2015; Natural Science and Technology Publishing House: Hanoi, Vietnam, 2015; pp. 1551–1554.
17. Williams, D.J.; de Willink, M.C.G. *Mealybugs of Central and South America*; CAB International: Wallingford, UK, 1992.
18. Miller, D.R.; Miller, G.L.; Watson, G.W. Invasive species of mealybugs (Hemiptera: Pseudococcidae) and their threat to U.S. agriculture. *Proc. Entomol. Soc. Wash.* **2002**, *104*, 825–836.
19. Walker, A.; Hoy, M.; Meyerdirk, D.E.; Papaya Mealybug. University of Florida: Featured Creatures. 2003. Available online: https://entnemdept.ufl.edu/creatures/fruit/mealybugs/papaya_mealybug.htm (accessed on 22 March 2023).
20. Heu, R.A.; Fukada, M.T.; Conant, P. Papaya mealybug, *Paracoccus marginatus* Williams and Granara de Willink (Hemiptera: Pseudococcidae). New Pest Advisory No. 04-03. Department of Agriculture, Hawaii. *Mol. Evol.* **2007**, *16*, 11–120.
21. Mani, M.; Shylesha, A.N.; Shivaraju, C. First report of the invasive papaya mealybug, *Paracoccus marginatus* Williams & Granara de Willink (Homoptera: Pseudococcidae) in Rajasthan. *Pest Manag. Hortic. Ecosyst.* **2012**, *18*, 234.
22. Finch, E.A.; Beale, T.; Chellappan, M.; Goergen, G.; Gadratagi, B.G.; Khan, M.A.M.; Rehman, A.; Rwomushana, I.; Sarma, A.K.; Wyckhuys, K.A.; et al. The potential global distribution of the papaya mealybug, *Paracoccus marginatus*, a polyphagous pest. *Pest Manag. Sci.* **2021**, *77*, 1361–1370. [CrossRef]
23. Tanwar, R.; Jeyakumar, P.; Vennila, S. *Papaya Mealybug and Its Management Strategies*; NCIPM Technical Bulletin 22; National Centre for Integrated Pest Management: New Delhi, India, 2010.
24. Muniappan, R. *Invasion of Papaya Mealybug in Asia*; Virginia Tech: Blacksburg, VA, USA, 2008.
25. Kansiime, M.K.; Rwomushana, I.; Idah Mugambi, I.; Makale, F.; Julien Lamontagne-Godwin, J.; Chacha, D.; Kibwage, P.; Oluyali, J.; Day, R. Crop losses and economic impact associated with papaya mealybug (*Paracoccus marginatus*) infestation in Kenya. *Int. J. Pest Manag.* **2020**, *69*, 150–163. [CrossRef]
26. Mwanauta, R.W.; Ndakidemi, R.A.; Venkataramana, P.B. Biopesticide efficacy of four plant essential oils against papaya mealybug, *Paracoccus marginatus* Williams and Granara de Willink (Hemiptera: Pseudococcidae). *Heliyon* **2023**, *9*, e14162. [CrossRef]
27. Le, T.T.N.; Graziosi, I.; Cira, T.M.; Gates, M.W.; Parker, L.; Wyckhuys, K.A.G. Landscape context does not constrain biological control of *Phenacoccus manihoti* in intensified cassava systems of southern Vietnam. *Biol. Control* **2018**, *121*, 129–139. [CrossRef]
28. Muniappan, R.; Shepard, B.M.; Watson, G.W.; Carner, G.R.; Sartiami, D.; Rauf, A.; Hammig, M.D. First report of the papaya mealybug, *Paracoccus marginatus* (Hemiptera: Pseudococcidae), in Indonesia and India. *J. Agric. Urban Entomol.* **2008**, *25*, 37–40. [CrossRef]

29. Muniappan, R. Invasion of papaya mealybug, *Paracoccus marginatus* in Asia. In Proceedings of the 57th Annual Meeting of the Entomological Society of America, Indianapolis, IN, USA, 13–16 December 2009.
30. Pham, V.L. Insect and small spiders damage crop plants in Vietnam. *Nong Nghiep J.* **2002**, *1*, 266–267.
31. Nguyen, T.C. Some enemies of scale insects on fruit, industrial and ornamental crops at Ho Chi Minh area and surrounding provinces. In Proceedings of the 6th Vietnam National Conference on Entomology, Hanoi, Vietnam, 9–10 May 2008.
32. Muniappan, R.; Meyerdirk, D.E.; Sengebau, F.M.; Berringer, D.D.; Reddy, G.V.P. Classical biological control of the papaya mealybug, *Paracoccus marginatus* (Hemiptera: Pseudococcidae) in the republic of Palau. *Fla. Entomol.* **2006**, *89*, 212–217. [CrossRef]
33. Muniappan, R. *Papaya Mealybug Parasitoid in Bangladesh*; IAPPS Newsletter; Tamò, M., Ed.; International Association for the Plant Protection Sciences: Cotonou, Republic of Benin, 2014. Available online: https://www.plantprotection.org/iapps-newsletter-3/ (accessed on 22 March 2023).
34. Mastoi, M.I.; Azura, N.; Muhamad, R.; Idris, B.; Solangi, K.; Arfan, G.; Bhatti, I.; Khoso, N. A Report of natural enemies of papaya mealybug, *Paracoccus Marginatus* (Hemiptera: Pseudococcidae) in Peninsular Malaysia. *Sci. Int.* **2016**, *28*, 371–374.
35. Amarasekare, K.G.; Mannion, C.M.; Epsky, N.D. Efficiency and establishment of three introduced parasitoids of the mealybug, *Paracoccus marginatus* (Hemiptera: Pseudococcidae). *Biol. Control* **2009**, *51*, 91–95. [CrossRef]
36. Meyerdirk, D.E.; Muniappan, R.; Warkentin, R.; Bamba, R.; Reddy, G.V.P. Biological control of the papaya mealybug, *Paracoccus marginatus* (Hemiptera: Pseudococcidae) in Guam. *Plant Prot. Q.* **2004**, *19*, 110–114.
37. Sakthivel, N. Effectiveness of three introduced encyrtid parasitic wasps (*Acerophagus papayae*, *Anagyrus loecki* and *Pseudleptomastix mexicana*) against papaya mealybug, *Paracoccus marginatus*, infesting mulberry in Tamil Nadu. *J. Biopestic.* **2013**, *6*, 71–76.
38. Noyes, J.S.; Schauff, M.E. New Encyrtidae (Hymenoptera) from papaya mealybug (*Paracoccus marginatus* | Williams and Granara de Willink) (Hemiptera: Sternorhyncha: Pseudococcidae). *Proc. Entomol. Soc. Wash.* **2003**, *105*, 180–185.
39. Amarasekare, K.G.; Mannion, C.M.; Epsky, N.D. Host instar susceptibility and selection and interspecific competition of three introduced parasitoids of the mealybug *Paracoccus marginatus* (Hemiptera: Pseudococcidae). *Environ. Entomol.* **2010**, *39*, 1506–1512. [CrossRef] [PubMed]
40. Gomez, K.A.; Gomez, A.A. *Statistical Procedures for Agricultural Research*, 2nd ed.; John Wiley and Sons: New York, NY, USA, 1984; p. 680, ISBN 978-0-471-87092-0.
41. Sokal, R.R.; Rohlf, F.J. *Biometry: The Principles and Practice of Statistics in Biological Research*, 3rd ed.; W. H. Freeman and Company: New York, NY, USA, 1995; p. 887.
42. Warton, D.I.; Hui, F.K.C. The arcsine is asinine: The analysis of proportions in ecology. *Ecology* **2011**, *92*, 3–10. [CrossRef] [PubMed]
43. Sharma, A.; Muniappan, R. Ecology and Management of *Paracoccus marginatus* (Papaya Mealybug) (Hemiptera: Pseudococcidae) in the Indian Subcontinent—Achievements, and Lessons. *Indian J. Entomol.* **2021**, *84*, 475–482. [CrossRef]
44. Available online: https://www.cabi.org/news-article/natural-enemy-fight-increased-against-papaya-mealybug-in-kenya/ (accessed on 23 March 2023).
45. Amarasekare, K.G.; Mannion, C.M.; Epsky, N.D. Developmental Time, Longevity, and Lifetime Fertility of Three Introduced Parasitoids of the Mealybug *Paracoccus marginatus* (Hemiptera: Pseudococcidae). *Environ. Entomol.* **2012**, *41*, 1184–1189. [CrossRef] [PubMed]
46. Chong, J.H.; Oetting, R.D. Host stage selection of the mealybug parasitoid *Anagyrus spec.* nov near sinope. *Entomol. Exp. Appl.* **2006**, *121*, 39–50. [CrossRef]
47. Mastoi, A.; Nur Azura, R.; Muhamad, I.M.; Ibrahim, Y. Survey of Papaya Mealybug, *Paracoccus marginatus* (Hemiptera: Pseudococcidae) and its natural enemies in Penninsular Malaysia. *Pak. J. Agric. Agril. Eng. Vet. Sci.* **2014**, *30*, 172–186.
48. Mastoi, M.I.; Nur Azura Adam, N.A.; Muhamad, R.; Idris Abd Ghani, I.A.; Gilal, A.A.; Khan, J.; Bhatti, A.R.; Ahmed Zia, A.; Sahito, J.G.M. Efficiency of *Acerophagus Papayae* on different host stage combinations of Papaya Mealybug, *Paracoccus Marginatus*. *Pak. J. Agri. Sci.* **2018**, *55*, 375–379. [CrossRef]
49. Divya, S.; Kalyanasundaram, M.; Karuppuchamy, P. Effect of Adult nutrition on longevity and parasitisation efficiency of *Acerophagus papayae* Noyes and Schauff (Hymenoptera: Encyrtidae). *J. Biol. Control* **2011**, *25*, 316–319.
50. Wati, M.; Rauf, A.; Pudjianto, P. Aspects of Biology of *Acerophagus Papayae* Noyes & Schauff (Hymenoptera: Encyrtidae), Parasitoid of the Papaya Mealybug. *J. Trop. Plant Pests Dis.* **2019**, *19*, 52–63.
51. Nisha, R.; Kennedy, J.S. Life cycle of Papaya mealybug *Paracoccus marginatus* Williams and Granara de Willink on different host plants vis-à-vis divergent natural selection. *J. Entomol. Zool. Stud.* **2017**, *5*, 91–102.

insects

MDPI

Article

Comparison of Three Attractants for the Effective Capture of *Xylotrechus chinensis* Adults in Multi-Funnel Traps

Nickolas G. Kavallieratos [1,*], Maria C. Boukouvala [1,*], Anna Skourti [1], Spyridon Antonatos [2], Panos V. Petrakis [3], Dimitrios P. Papachristos [2] and Georgios Th. Papadoulis [1]

1 Laboratory of Agricultural Zoology and Entomology, Department of Crop Science, Agricultural University of Athens, 75 Iera Odos Str., 11855 Athens, Greece; annaskourti@aua.gr (A.S.); gpapadoulis@aua.gr (G.T.P.)
2 Laboratory of Agricultural Entomology, Department of Entomology and Agricultural Zoology, Benaki Phytopathological Institute, 8 Stefanou Delta Str., 14561 Kifissia, Greece; s.antonatos@bpi.gr (S.A.); d.papachristos@bpi.gr (D.P.P.)
3 Laboratory of Forest Entomology, Institute of Mediterranean Ecosystems, Hellenic Agricultural Organization—"Dimitra", Terma Alkmanos, 11528 Athens, Greece; pvpetrakis@fria.gr
* Correspondence: nick_kaval@aua.gr (N.G.K.); mbouk@aua.gr (M.C.B.); Tel.: +30-2105294569 (N.G.K.)

Simple Summary: Three blends of attractants and pheromones (i.e., attractant 1, attractant 2, and attractant 3) were compared for monitoring adults of *Xylotrechus chinensis*. The experimental procedure was carried out for a two-year period. The attractants were baited in traps and placed in trees of three heavily infested areas of Athens (Greece). Traps were checked weekly, the captured *X. chinensis* adults were recorded, and they were removed from the traps. Adults of *X. chinensis* start flying in late spring and end in the middle of autumn (end of October). Attractant 3 performed better than attractant 1 and 2, even with low numbers of insects. *Xylotrechus chinensis* did not seem likely to be attracted to the basic part of the blend, which consisted of a blend of pheromones + ethanol. Overall, the inclusion of α-pinene and ipsenol increased the activity of the basic lure.

Abstract: The Asian coleopteran *Xylotrechus chinensis* (Chevrolat) (Cerambycidae: Cerambycinae) is an invasive species in several European countries, attacking mulberry trees. In the current research, we evaluated the performance of three mixtures consisting of pheromones and attractants for the monitoring of *X. chinensis* adults. Attractant 1 (i.e., geranyl acetone, fuscumol acetate, fuscumol, monochamol, 3-hydroxyhexan-2-one, 2-methyl-1-butanol, *anti*-2,3-hexanediol, prionic acid + ethanol), attractant 2 (i.e., geranyl acetone, fuscumol acetate, fuscumol, monochamol, 3-hydroxyhexan-2-one, 2-methyl-1-butanol, *anti*-2,3-hexanediol, prionic acid + α-pinene + ethanol) and attractant 3 (i.e., geranyl acetone, fuscumol acetate, fuscumol, monochamol, 3-hydroxyhexan-2-one, 2-methyl-1-butanol, *anti*-2,3-hexanediol, prionic acid + α-pinene + ipsenol + ethanol) were baited in multi-funnel traps and installed in mulberries for a two-year period in Athens (Greece). The flight activity of *X. chinensis* starts at the end of April and terminates at the end of October. The peaks of *X. chinensis* flight activity were observed on 16 August 2021 and on 6 July 2022. Attractant 3 proved to be the most effective blend, catching 953 adults, followed by attractant 2 (523 adults) and attractant 1 (169 adults), throughout the experimental period. It seems that the pest was not attracted to the basic part of the blend (i.e., pheromones + ethanol). The incorporation of α-pinene and ipsenol resulted in the elevated activity of the base lure. The elevated performance of attractant 3 may be attributed to only the α-pinene and the ipsenol, or possibly the α-pinene, ipsenol, and ethanol, because the pheromone blend did not contain any of the pheromone components of the target species. Overall, attractant 3 could be a useful tool to detect and track *X. chinensis* in new invasive areas, triggering early management strategies against further establishment of this species.

Keywords: blends; monitoring; tiger longicorn beetle; adult flight period; *Morus* spp.

1. Introduction

The invasive species *Xylotrechus chinensis* (Chevrolat) (Coleoptera: Cerambycidae: Cerambycinae) is a destructive pest that feeds and grows on mulberries (*Morus* spp.) [1–3]. This species is native to the Eastern Palaearctic region (i.e., China, South Korea, and Japan) [4–6]. In the last decade, *X. chinensis* has become established in Europe, mainly in countries with a wide presence of *Morus* spp., such as France, Greece, and Spain [1,2,7–10]. Specifically, in Spain, *X. chinensis* was first recorded in 2013 in Catalonia, where in 2020 the infestation had expanded to 12 cities, covering an area of 378.1 km^2 [11]. Meanwhile, a new outbreak of *X. chinensis* occurred in Valencia in 2018 [2]. In the same year, *X. chinensis* was reported for the first time in France (Gironde) [12,13], while in 2019 attacked *Morus* trees were found in the Occitanie region [8,14]. The first detection of this species in Greece was in 2017 in Crete, where approximately 200 mulberries (*Morus* spp.) were found infested, of which 15% were dead [1]. In 2019, an outbreak of *X. chinensis* was recorded in Athens, the capital of Greece, where 300 out of 1300 infested mulberry trees were dead [9,13]. The damage from *X. chinensis* spread further in Athens in the following years, making this species an important pest of mulberry trees throughout the city. In addition, *X. chinensis* has been intercepted twice in Germany, in 2007 (first record in Europe) and 2017, and once in Philadelphia, USA (2011) in containers transporting different wooden objects from China [6,11,15], indicating that wooden materials can be vehicles for its spread.

Early detection of *X. chinensis* is a complex issue, since the immature stages develop inside the tree trunk [8]. The infestation is usually noticed mainly by the exit holes of the adults, as well as by other symptoms that are not so obvious, such as frass and the droppings of the beetles on the trunk [1,2,10]. The xylophagous larvae inhibit the movement of nutrients and water within the infested trees, resulting in their progressive drying [16]. The feeding habit of *X. chinensis* adults is not clear, since it is reported that adults infest leaves and stems of the tops of the mulberry trees [1,2] or they do not feed but drink water [17,18], indicating that the most destructive life stage of this species is larva. For all the above reasons, management of *X. chinensis* is challenging. In recent research efforts in Spain (Barcelona) and Greece (Athens), the trunk injection method was used to control *X. chinensis* larvae [10,11]. During spring 2018 in Barcelona, abamectin was injected into 107 mulberry trees. Almost a third of these treated trees (31 out of 107) showed new infestation symptoms in December, particularly exit holes, indicating a significant reduction in *X. chinensis* presence [11]. In Athens, after two years of application of abamectin, fipronil and imidacloprid via trunk injection to infested *Morus* spp. trees, the emergence holes of *X. chinensis* adults were reduced by 85.6%, 71.8%, and 76.1%, respectively [10]. Chemical control of this species has previously been proposed by spraying the trunks of mulberry trees with contact insecticides, targeting sites where females are resting and/or laying eggs, and newly emerged larvae are present [2]. Both methods were applied during summer [2,10,11]. The trunk injection method is generally safe and can be applied to urban trees [10,19], while trunk spraying is not suggested in urban areas due to concerns related to the exposure of citizens and environment to insecticides [20]. In addition, some preventive measures should be applied to heavily infested trees, including cutting down the affected mulberries, followed by burning or shredding the removed parts of the trees [2,8,12].

Recently, a new device that can monitor trees systematically against wood-boring pests was evaluated [21]. Trees infested by *X. chinensis* and *Rhynchophorus ferrugineus* Olivier (Coleoptera: Curculionidae) were used in the tests. This device automatically records the acoustic emissions generated by the activity of the larvae while feeding inside the trunk of the tree, thus detecting any tree infestation. It showed high accuracy in the detection of wood borer species and provided rapid inspection of many mulberries and palms without human interference, rendering it particularly suitable for trunk inspection in quarantine areas. The vibration recording device potentially contributes to a reduction in unnecessary management treatments and becoming a useful tool for monitoring of *X. chinensis*.

The three ingredients of male pheromones of *X. chinensis*, i.e., 2,3-octanediol, 2-hydroxy-3-octanone, and 3-hydroxy-2-octanone [22,23], have been identified and attract female individuals [24], but they have not yet been developed as a method of detection for this species [8]. Pheromone components could be used for mass trapping of *X. chinensis* females or for mating disruption [8]. Recently, traps containing a mixture of eight kairomones and pheromones, which attracted a wide spectrum of insects [25], were used in Catalonia (Spain), Crete (Greece), and Sète (France) [8,14]. Among the captured species of the Cerambycidae family were *X. chinensis* adult individuals.

In the present study, three mixtures of attractants that can attract a wide range of wood-boring insects were compared to monitor the population of *X. chinensis* in different areas of the Municipality of Athens. It is very important to know whether any of the components that make up each attractant can disrupt the attraction of *X. chinensis* [26]. A search of the global bibliography yielded no data on the seasonal occurrence of *X. chinensis* in Europe. There are sporadic reports on the emergence period of *X. chinensis* adults from the trunks of the mulberries, providing evidence about the beginning of the flight of this pest. In southern Greece, adults of *X. chinensis* appear from the end of spring to the beginning of summer and in northeastern Spain from mid- to late summer [1,2]. Thus, the aim of the current two year (2021–2022) research was to investigate the flight activity of *X. chinensis*, offering a useful monitoring tool for this species.

2. Materials and Methods

2.1. Experimental Areas-Trees

Three experimental areas were selected within the Municipality of Athens: Neos Kosmos (37° 03′ 60.00″ N 22° 25′ 59.99″ E), Kerameikos (37° 58′ 25.19″ N 23° 43′ 4.19″ E), and Sepolia (38° 00′ 9.99″ N 23° 42′ 48.58″ E) (Figure 1). The three selected areas have *Morus* spp. trees (*Morus alba* and *Morus nigra*). Based on the existence of numerous exit holes in mulberry trees, preliminary samplings (through installation of baited traps) indicated that mulberry trees suffered high infestation levels of *X. chinensis* in these areas.

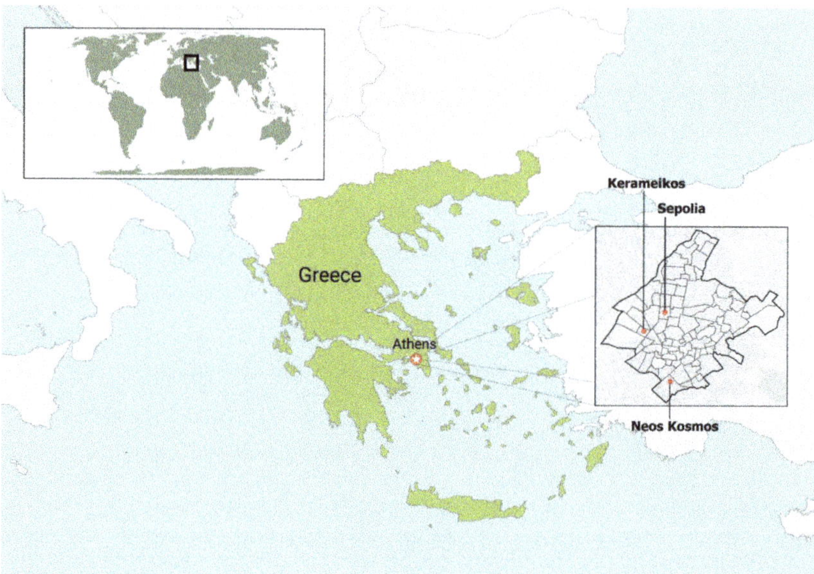

Figure 1. Map of Greece with the three experimental areas: Neos Kosmos, Kerameikos, and Sepolia, in the Municipality of Athens.

2.2. Attractants

The following three blends of pheromones and attractants were used during the experimental period: (i) geranyl acetone, fuscumol acetate, fuscumol, monochamol, 3-hydroxyhexan-2-one, 2-methyl-1-butanol, *anti*-2,3-hexanediol, prionic acid + ethanol (attractant 1); (ii) geranyl acetone, fuscumol acetate, fuscumol, monochamol, 3-hydroxyhexan-2-one, 2-methyl-1-butanol, *anti*-2,3-hexanediol, prionic acid + α-pinene + ethanol (attractant 2); and (iii) geranyl acetone, fuscumol acetate, fuscumol, monochamol, 3-hydroxyhexan-2-one, 2-methyl-1-butanol, *anti*-2,3-hexanediol, prionic acid + α-pinene + ipsenol (2-methyl-6-methylene-7-octen-4-ol) + ethanol (attractant 3). The three attractants had components in common of geranyl acetone, fuscumol acetate, fuscumol and monochamol, which are components of the pheromones of various species of the cerambycid subfamilies Spondylidinae and Lamiinae [27–29], 3-hydroxyhexan-2-one, 2-methyl-1-butanol, and *anti*-2,3-hexanediol, which are components of the pheromones of various species of the subfamily Cerambycinae [30–32], prionic acid, which is the sex pheromone of various species of the genus *Prionus* (Coleoptera: Cerambycidae: Prioninae) [33], and ethanol, which is released from infested trees [34–36], enhancing the attraction of wood-boring insects [37,38]. Attractant 3 additionally contained the components α-pinene, which is the kairomonic component released by host trees [39] and ipsenol, which is an aggregation pheromone of wood-boring insects of the subfamily Scolytinae (Curculionidae) [40,41]. Attractant 2 contained only α-pinene as an additional component. We assume that the blend of pheromones and ethanol was enhanced by α-pinene and ipsenol [42–44].

The preparation of the common part of the three attractants was carried out at the French National Institute of Agriculture, Food and the Environment (INRAE) by Dr Alain Roques. This blend consisted of 25 mg of geranyl acetone, 50 mg of fuscumol acetate, 50 mg of fuscumol, 50 mg of monochamol, 50 mg of 3-hydroxyhexan-2-one, 50 mg of 2-methyl-1-butanol, 50 mg of *anti*-2,3-hexanediol, and 1 mg of prionic acid dissolved in isopropanol (carrier) to a final volume of 1 mL. Aliquots of 1 mL of this mixture were poured into 1.5 mL glass vials with plastic lids and kept at 4 °C until the beginning of the experiment (Figure 2a). The extra components α-pinene (12 mL) (+ 16 mL ethanol) of attractant 2 or α-pinene (12 mL) and ipsenol (20 mg) (+ 16 mL ethanol) of attractant 3 are commercially available and contained in vials of 28 mL volume each with a constant evaporation rate of average 150 mg/day, lasting for 100 days, as measured by the manufacturer (NovAgrica, Athens, Greece). Ethanol alone as part of the first attractant is contained in the same type of vials (28 mL volume). The vials with ethanol alone were inspected frequently and filled if it was necessary. Therefore, per attractant two vials were used: attractant 1, a 1.5 mL vial containing the common part and a 28 mL vial containing ethanol (Figure 2b); attractant 2, a 1.5 mL vial containing the common part and a 28 mL vial containing α-pinene + ethanol (Figure 2c); and attractant 3, a 1.5 mL vial containing the common part and a 28 mL vial containing α-pinene + ipsenol + ethanol (Figure 2d). Upon installation of traps, the caps of the 28 mL vials were removed and the vials left open.

2.3. Experimental Design

The commercially available multi-funnel traps, which are suitable for catching species of cerambycids [45], were supplied by NovAgrica (Athens, Greece) and were used in the experiment. The multi-funnel trap consists of six black funnels with a removable white collection cup that facilitates the counting of the captures (dry trapping) (Figure 2e). A killing agent (transfluthrin 0.4% *w*/*w* strip) was applied in the collection cup to prevent captured insects from escaping. The traps were hung from trees with their bottoms 2.5 m above the ground. In the areas of Neos Kosmos and Kerameikos, 6 traps were placed for each attractant (i.e., 6 traps × 3 attractants × 2 areas = 36 traps), while in Sepolia, 3 traps were placed for each attractant (i.e., 3 traps × 3 attractants x 1 area = 9 traps). In each area, traps were suspended on main roads in blocks consisting of the three attractants and were 100 m apart (i.e., 6 blocks for Neos Kosmos and Kerameikos, 3 blocks in Sepolia). The distance between the blocks was 100 m. Before the suspension of the traps on mulberry trees, each vial of 1.5 mL containing the common part of the attractants was placed without the lid in a 65 mm × 95 mm, 100 µm thickness, resealable polyethylene bag (Minigrip bags, Avantor, The Metropolis, Singapore) with a cylindrical cotton wad (3.5 cm × 1 cm) [46] (Figure 2f). Attractants were placed in the middle funnel (Figure 2g) of the trap and replaced every three months. Trap inspections were conducted weekly from 19 July 2021 to 31 December 2022. At each trap check, traps were clockwise rotated to avoid the effect of a particular sampling location [47]. At each check, captures of *X. chinensis* adults were recorded. Then, all captured individuals were removed from the traps.

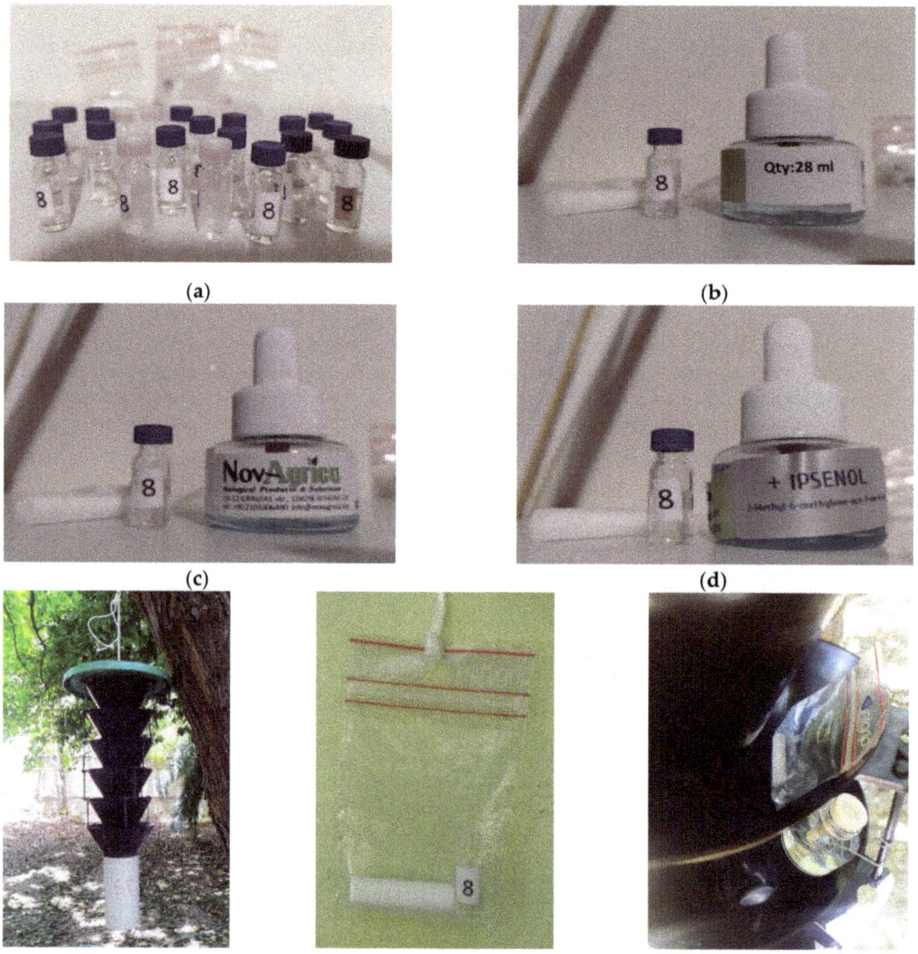

Figure 2. (**a**) The common part of the three attractants in glass vials. (**b**) Attractant 1. (**c**) Attractant 2. (**d**) Attractant 3. (**e**) Multi-funnel trap. (**f**) The common part of the three attractants in a glass vial in a polyethylene bag with cotton wad. (**g**) Placement of attractant in the middle funnel of the multi-trap.

2.4. Statistical Analysis

Before the data of captures were analyzed, they were log (x + 1)-transformed to normalize variances and standardize means [10,48]. Data were analyzed using a three-way ANOVA [49]. The response variable was the number of captured adults. Main effects were attractant, date and experimental area. Their interactions (area x attractant, area x date, attractant × date, and area × attractant × date) were also considered in the analysis. Year was not considered in the analysis, since 2021 included fewer checks than 2022. Means were separated by the Tukey–Kramer HSD test, $p = 0.05$ [50]. All dates with zero captures in any treatment were dropped from the analysis. The analyses were carried out with the statistical package JMP 16.2 [51].

3. Results

The main effects area and the interactions area x attractant and area x attractant x date were not significant (Table 1). The total number of *X. chinensis* adults captured in the traps for the entire experimental period was 1645, of which 169, 523, and 953 individuals were recorded in attractants 1, 2, and 3, respectively (Figure 3). Within each experimental area, significant differences were observed among the three attractants ($F = 51.5$; DF = 8, 1709; $p < 0.01$) (Figure 3). In Neos Kosmos, the most effective was attractant 3, catching a total of 372 adults, followed by attractant 2 and 1 with 213 and 63 individuals captured, respectively. The same trend was observed in Kerameikos and Sepolia, where the most catches were noted for attractant 3 (399 and 182 adults in Kerameikos and Sepolia, respectively) and the least for attractant 1 (71 and 35 adults in Kerameikos and Sepolia, respectively). No significant differences were recorded for each attractant among the three experimental areas, indicating that there was no influence of area on the performance of attractants.

Table 1. ANOVA parameters for main effects and their interactions for *X. chinensis* adults captured weekly in each trap for attractants in three experimental areas during 2021–2022 (total DF = 1709).

Source	DF	*F*	*p*
Area	2	0.2	0.84
Attractant	2	294.2	<0.01
Date	37	20.6	<0.01
Area × attractant	4	0.4	0.83
Area × date	74	1.6	0.01
Attractant × date	74	3.3	<0.01
Area × attractant × date	148	1.0	0.54

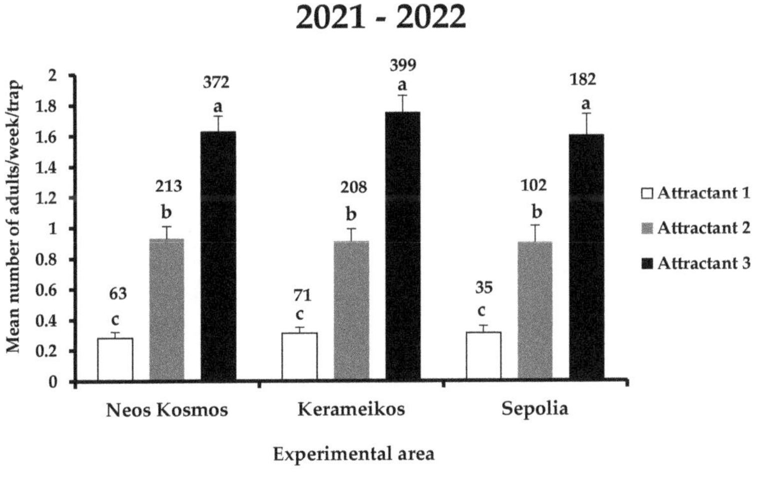

Figure 3. Mean number (+ standard error) of *X. chinensis* adults captured weekly in traps baited with each attractant in the three experimental areas (i.e., Neos Kosmos, Kerameikos, and Sepolia) during the entire experimental period (2021-2022). Different letters indicate significant differences. Numbers above columns indicate the total number of captures. All dates when all treatments had zero catches were dropped.

A view of the flight activity of *X. chinensis* recorded in 2021 and 2022 in Athens is shown in Figures 4 and 5. The peaks of captures for 2021 and 2022 were observed in August and July, respectively. Specifically, the mean numbers of captures for August 2021 were 2.47, 1.15, and 0.75 for attractants 3, 2, and 1, respectively. In July 2022, the mean numbers of captures were 3.03, 2.53, and 0.60 for attractants 3, 2, and 1, respectively. During 2021, significant differences were noted among captures in traps baited with the three tested attractants ($F = 51.6$; DF = 2, 584; $p < 0.01$). Traps baited with attractant 3 (i.e., 303 adults) caught significantly more adults than traps baited with attractant 1 (i.e., 89 adults) and 2 (i.e., 148 adults) (Figure 4). The highest mean number of *X. chinensis* adults captured from the beginning of observations until 13 September 2021 was recorded for attractant 3, followed by captures corresponding to attractant 2 and attractant 1 (Figure 4). The maximum mean number of catches was recorded on 16 August 2021, where 2.9 adults/week/trap were captured in traps baited with attractant 3. From 22 September 2021 onwards, a reduction in captures of *X. chinensis* was observed for all three tested attractants. From 29 October 2021 until 26 April 2022, there were zero captures (Figures 4 and 5). A similar trend was recorded in 2022. There were significant differences in captures when traps were baited with attractants 1, 2, and 3 ($F = 160.6$; DF = 2, 1124; $p < 0.01$). Traps containing attractant 3 caught significantly more adults vs. traps containing attractant 2 vs. traps containing attractant 1 (Figure 5). The highest mean number of adults captured from the beginning to the end of *X. chinensis* flight activity was recorded for attractant 3, followed by captures corresponding to attractant 2 and attractant 1 (Figure 5). The first captures for attractants 2 and 3 occurred on 6 May 2022 (0.33 and 0.07 adults/week/trap for attractants 3 and 2, respectively), while traps baited with attractant 1 caught the first adults (0.07 adults/week/trap) one week later. The highest mean numbers of captured *X. chinensis* adults were noted on 6 July 2022 for attractants 2 and 3 (2.86 and 3.13 adults/week/trap for attractants 2 and 3, respectively), while for attractant 1 this was observed on 18 July (0.73 adults/week/trap). The last catches were observed on September 1, 2022 for attractant 1 (0.2 adults/week/trap), 11 October for attractant 2 (0.07 adults/week/trap), and 18 October for attractant 3 (0.2 adults/week/trap). No captures were recorded from 25 October to the end of December 2022.

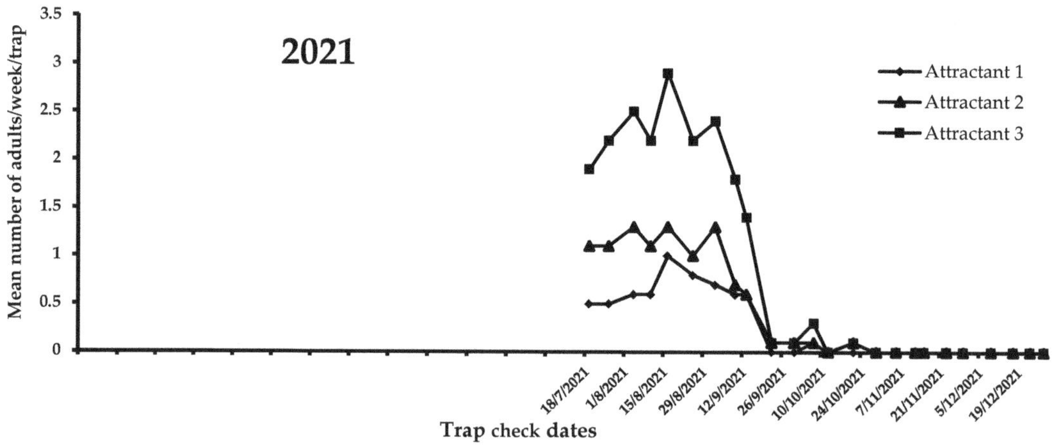

Figure 4. Mean number of *X. chinensis* adults captured weekly in traps baited with attractants 1, 2, and 3 during 2021.

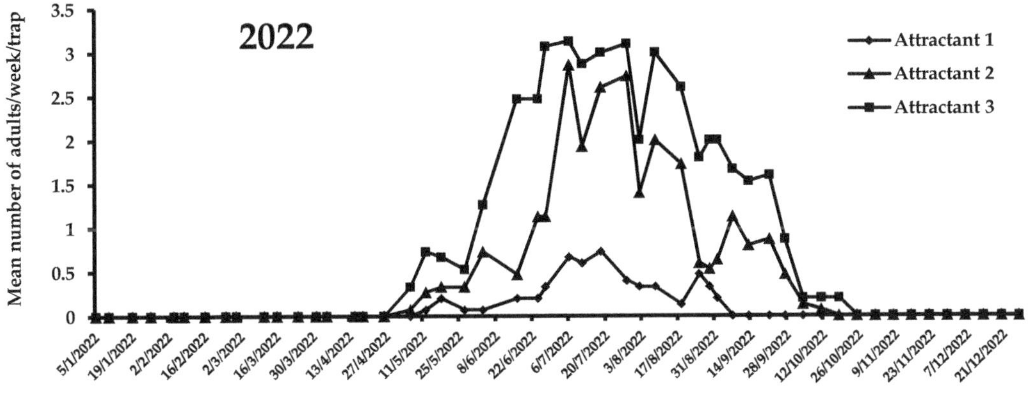

Figure 5. Mean number of *X. chinensis* adults captured weekly in traps baited with attractants 1, 2, and 3 during 2022.

During the first week of the observations in 2021 (i.e., end of July), the mean weekly temperature was 28.5 °C (Figure 6), which increased in the following three weeks, ranging from 30.3 to 33.1 °C. A gradual reduction in the average weekly temperature was noted from middle August (i.e., 28.7 °C) to the end of 2021 (i.e., 7.1 °C). By the end of the flight period in 2021, the mean weekly temperature was <20 °C. In the first two weeks of 2022, the average weekly temperature increased, reaching ~13 °C, while for the following three weeks it remained <10 °C (Figure 7). The same trend was observed over the next 7 weeks, ranging from 7.6 to 13.8 °C. A gradual increase in the mean weekly temperature was observed between the end of March till 26 April, reaching 17.8 °C. During the first two weeks of May, where the first captures were recorded, the mean weekly temperature increased further, reaching 20.2 °C. In the following weeks until the beginning of August, the mean weekly temperature ranged from 23.1 to 30.5 °C, while it gradually decreased after the second week of August. Similarly to 2021, the mean weekly temperature dropped below 20 °C by the end of the flight period of *X. chinensis*.

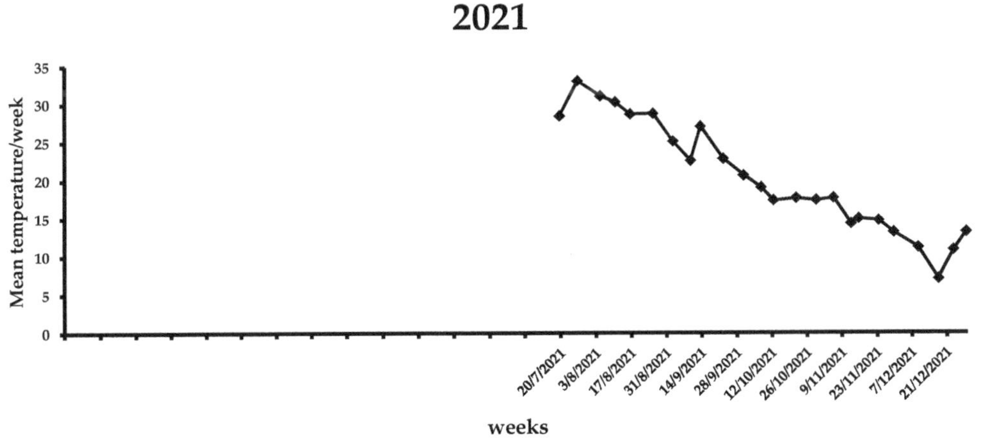

Figure 6. Mean weekly temperature (°C) in Municipality of Athens during the 2021 experimental period. Data obtained from freemeteo.gr.

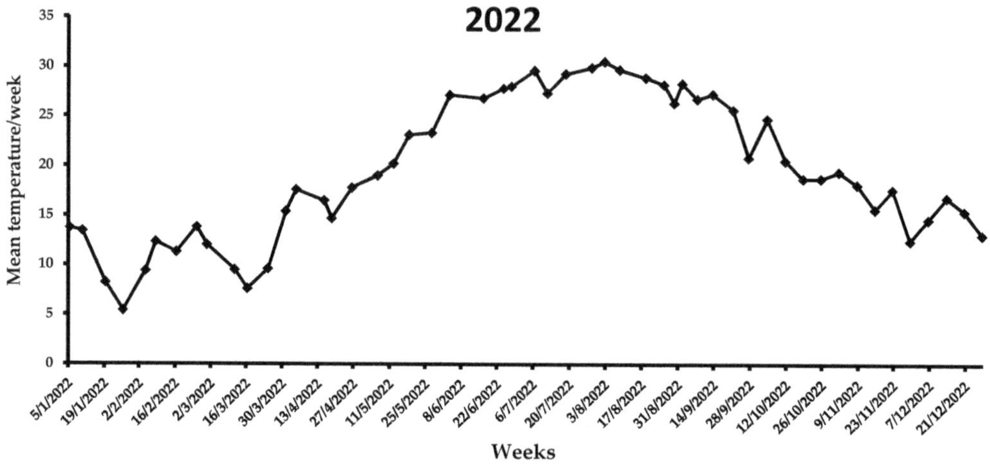

Figure 7. Mean weekly temperature (°C) in Municipality of Athens during the 2022 experimental period. Data obtained from AccuWeather.com.

4. Discussion

The findings of the current study revealed that the most effective attractant mixture was attractant 3, which contained the two additional components α-pinene and ipsenol (+ethanol). When ipsenol was absent (attractant 2) or when α-pinene and ipsenol were absent (attractant 1), captures were significantly fewer than those corresponding to attractant 2. Captures corresponding to attractant 2, which contained α-pinene (+ethanol), were significantly higher than the captures corresponding to attractant 1 (+ethanol). Therefore, the combination of α-pinene and ipsenol with the common part of all attractants and ethanol resulted in the increased attraction of *X. chinensis* adults compared with the presence of α-pinene with the common part of the attractants + ethanol. Ethanol and α-pinene perhaps act synergistically to attract many species of wood-boring and bark beetles, including species of the family Cerambycidae, issues that have classified them as broad-spectrum combinational attractants [43,52,53]. According to Rodríguez-González et al. [42], the addition of ethanol to the compound 3-hydroxy-2-hexanone, which is produced by the males of *Xylotrechus arvicola* (Olivier) (Coleoptera: Cerambycidae: Cerambycinae), increased the attractiveness of this compound to both sexes of this species. In previous studies, it has been noted that the kairomones ethanol and α-pinene increased the attractiveness of monochamol, which is a component of *Monochamus galloprovincialis* Olivier (Coleoptera: Cerambycidae: Lamiinae) pheromone [54,55]. Hoch et al. [46] reported that the inclusion of α-pinene and ethanol in a mixture of geranyl acetone, fuscumol acetate, fuscumol, monochamol, 3-hydroxyhexan-2-one, 2-methyl-1-butanol, *anti*-2,3-hexanediol, and prionic acid significantly increased the number of captured individuals of species of the family Cerambycidae compared to the captures recorded in the mixture without the addition of kairomones. In our study, the addition of ipsenol to attractant 3 further enhanced its effectiveness. Similarly, Miller et al. [44] reported that the number of captured individuals of *Monochamus scutellatus* (Say) (Coleoptera: Cerambycidae: Lamiinae) in traps that had a mixture of ipsenol and α-pinene was higher than the number of captures in traps that had α-pinene alone. The authors also found that ipsenol significantly increased the number of captures of *Acanthocinus princeps* (Walker), *Acanthocinus obliquus* (LeConte), *Acanthocinus nodosus* (F.), *Acanthocinus obsoletus* (Olivier) (Coleoptera: Cerambycidae: Lamiinae), and *Rhagium inquisitor* (L.) (Coleoptera: Cerambycidae: Lepturinae) when α-pinene was added to the traps.

One other important finding of the present study was the depiction of the flight activity of *X. chinensis*, revealing valuable data on the ecology of this pest that could be useful for its successful management. Specifically, during the monitoring of the traps, it was found that the flight of *X. chinensis* begins around the end of April and terminates around the end of October. This stands in agreement with the suggestion of Leivadara et al. [1] for the emergence of *X. chinensis* in Crete (Greece), which takes place between May and June. In Spain (Catalonia), Sarto i Monteys and Torras i Tutusaus [2], after two years of observations of specimens kept in cages under semi-field environment conditions, reported that adults emerge later, in July and August, exactly the same period as adults of *X. chinensis* emerge in Japan, which is one of the countries of origin of this species [22]. The authors also found that in both years of their observations, the emergence of adults began after ten or seven consecutive days with mean daily temperature >20 °C, for 2016 and 2017, respectively [2]. In Athens, the first captures of traps were recorded at the beginning of May 2022. In the time between the previous check (late April) of the traps and the check of the first captures, it was observed that the average daily temperature in Athens was ~20 °C. New adults of *X. chinensis* after eclosion have been reported to remain inside the trees for a period between 12 and 40 days before leaving the trees [17,18], probably awaiting stabilization of the temperature above 20 °C. Hence, considering our results and those obtained from Spain, we assume that emergence of *X. chinensis* adults from infested trees requires certain temperature conditions. Additional research is required to verify this hypothesis.

Concerning the termination of the flight period of *X. chinensis*, in both years of the study, the last captures were recorded between 18 and 21 October. Although the life span of *X. chinensis* adults is not known, considering the period of adults' emergence and the flight activity of *X. chinensis* obtained in our experiments, we assume that the last adults that emerged from the trees correspond to the last individuals captured in the traps. This issue should be further investigated. Regarding the performance of each attractant, attractant 1 ceased to be active in early September. The early reduction in the captures of attractant 1 compared to attractants 2 and 3 is probably due to the low performance exhibited since the beginning of *X. chinensis* flight activity. It should be noted that the last captures in 2022 were recorded in traps containing attractant 3, although in 2021 the last captures were recorded in traps containing either attractant 2 or attractant 3. Having expanded 334 km^2 in Catalonia in two years, *X. chinensis* is expected to spread rapidly to nearby areas [8,11,55] and through wooden boxes that are used for the transportation of goods [16]. Therefore, attractant 3 could be used to detect early arrivals of the pest. Precise records of *X. chinensis* in new areas could trigger management measurements to restrict further dispersal of this species.

5. Conclusions

To conclude, the most effective attractant was found to be attractant 3, which can help the successful monitoring of *X. chinensis*. The combination of α-pinene and ipsenol enhanced the activity of the basic part of the blend (i.e., pheromones + ethanol), while the role of ethanol is rather uncertain. It does not seem likely that *X. chinensis* was attracted to the basic part of the blend, because it was the least active attractant. Thus, it seems more likely that the pest was attracted to the blend of kairomones, with the data showing that α-pinene and ipsenol were attractive. Further investigation is needed by testing other blends of attractants. In addition, the role of α-pinene, ipsenol, and ethanol should be evaluated for the attraction of *X. chinensis* to make a direct comparison to attractant 3. The precise monitoring of populations of noxious insect populations is crucial, since it provides timely and detailed information on pest densities, resulting in a reduction in synthetic insecticides and the prompt application of ecologically friendly methods, such as mass trapping. The next step of investigation deals with the assessment of new trap devices to optimize the attractant-based trapping protocol of this noxious species.

Author Contributions: Conceptualization, N.G.K., M.C.B., P.V.P., D.P.P. and G.T.P.; methodology, N.G.K., M.C.B., S.A., A.S., P.V.P., D.P.P. and G.T.P.; validation, N.G.K., M.C.B., A.S., S.A., P.V.P., D.P.P. and G.T.P.; formal analysis, N.G.K. and M.C.B.; investigation, N.G.K., M.C.B., A.S., S.A., P.V.P., D.P.P. and G.T.P.; data curation, N.G.K., M.C.B., A.S., S.A., P.V.P., D.P.P. and G.T.P.; writing—original draft preparation, N.G.K., M.C.B. and A.S.; writing—review and editing, N.G.K., M.C.B., A.S., S.A., P.V.P., D.P.P. and G.T.P.; visualization, N.G.K., M.C.B., A.S., S.A., P.V.P., D.P.P. and G.T.P.; supervision, N.G.K., P.V.P., D.P.P. and G.T.P.; project administration; N.G.K., M.C.B., P.V.P., D.P.P. and G.T.P.; resources, N.G.K., P.V.P., D.P.P. and G.T.P.; funding acquisition, N.G.K., P.V.P., D.P.P. and G.T.P. All authors have read and agreed to the published version of the manuscript.

Funding: This study was supported by the project 33.0256 "Management of *Xyloterchus chinensis* (Coleoptera: Cerambycidae) with pheromones and insecticides in Municipality of Athens" (Municipality of Athens).

Data Availability Statement: Data are contained within the article.

Acknowledgments: We thank E.P. Nika for technical assistance.

Conflicts of Interest: The authors declare no conflict of interest.

References

1. Leivadara, E.; Leivadaras, I.; Vontas, I.; Trichas, A.; Simoglou, K.; Roditakis, E.; Avtzis, D.N. First record of *Xylotrechus chinensis* (Coleoptera, Cerambycidae) in Greece and in the EPPO region. *EPPO Bull.* **2018**, *48*, 277–280. [CrossRef]
2. Sarto i Monteys, V.; Torras i Tutusaus, G. A new alien invasive longhorn beetle, *Xylotrechus chinensis* (Cerambycidae), is infesting mulberries in Catalonia (Spain). *Insects* **2018**, *9*, 52. [CrossRef]
3. Korakaki, E.; Legakis, A.; Katsanevakis, S.; Koulelis, P.P.; Avramidou, E.V.; Soulioti, N.; Petrakis, P.V. Invasive Alien Species of Greece. In *Invasive Alien Species*; Pullaiah, T., Ielmini, M.R., Eds.; Wiley-Blackwell: Hoboken, NJ, USA, 2021; Volume 3, pp. 124–189.
4. Han, Y.; Lyu, D. Taxonomic review of the genus *Xylotrechus* (Coleoptera: Cerambycidae: Cerambycinae) in Korea with a newly recorded species. *Korean J. Appl. Entomol.* **2010**, *49*, 69–82. [CrossRef]
5. Sama, G.; Löbl, I. Cerambycinae. In *Catalogue of Palaearctic Coleoptera. Chrysomeloidea*; Löbl, I., Smetana, A., Eds.; Apollo Books: Stenstrup, Denmark, 2010; pp. 143–207.
6. Schrader, G. Express—PRA zu *Xylotrechus chinensis*. Available online: https://pflanzengesundheit.julius-uehn.de/dokumente/upload/b75e3_xylotrechus-chinensis_express-pra.pdf (accessed on 2 July 2023).
7. Valladares, L.; Brustel, H.; Cocquempot, C. *Xylotrechus chinensis*: Nouvelle Espèce Invasive Arrivée d'Asie. Available online: https://passion-entomologie.fr/xylotrechus-chinensisnouvelle-especeinvasive/ (accessed on 2 July 2023).
8. Bragard, C.; Baptista, P.; Chatzivassiliou, E.; Di Serio, F.; Gonthier, P.; Jaques Miret, J.A.; Justesen, A.F.; Magnusson, C.S.; Milonas, P.; Navas-Cortes, J.A. Pest categorisation of *Xylotrechus chinensis*. *EFSA J.* **2021**, *19*, e07022.
9. Demetriou, J.; Kalaentzis, K.; Kazilas, C.; Koutsoukos, E.; Avtzis, D.; Georgiadis, C. Revisiting the non-native insect fauna of Greece: Current trends and an updated checklist. *NeoBiota* **2021**, *65*, 93–108. [CrossRef]
10. Kavallieratos, N.G.; Boukouvala, M.C.; Skourti, A.; Nika, E.P.; Papadoulis, G.T. Trunk injection with insecticides manages *Xylotrechus chinensis* (Chevrolat) (Coleoptera: Cerambycidae). *Insects* **2022**, *13*, 1106. [CrossRef]
11. Sarto i Monteys, V.; Ribes, A.C.; Savin, I. The invasive longhorn beetle *Xylotrechus chinensis*, pest of mulberries, in Europe: Study on its local spread and efficacy of abamectin control. *PLoS ONE* **2021**, *16*, e0245527. [CrossRef]
12. Cocquempot, C.; Desbles, F.; Mouttet, R.; Valladares, L. *Xylotrechus chinensis* (Chevrolat, 1852), nouvelle espèce invasive pour la France métropolitaine (Coleoptera, Cerambycidae, Clytini). *Bull. Soc. Entomol. Fr.* **2019**, *124*, 27–32. [CrossRef]
13. EPPO (European and Mediterranean Plant Protection Organization). New data on quarantine pests and pests of the EPPO alert list. *EPPO Rep. Serv.* **2020**, *91*, 5.
14. EPPO (European and Mediterranean Plant Protection Organization). Invasive alien wood borers: Trapping studies in France. *EPPO Rep. Serv.* **2021**, *157*, 7.
15. Philadelphia, U.S.; Customs and Border Protection. Philly CBP Intercepts Nation's First *Xylotrechus chinensis* Beetle. Available online: https://www.cbp.gov/newsroom/local-media-release/philly-cbpintercepts-nations-first-xylotrechus-chinensis-beetle (accessed on 2 July 2023).
16. EPPO (European and Mediterranean Plant Protection Organization). Alert list—*Xylotrechus chinensis* (Coleoptera: Cerambycidae). Available online: https://www.eppo.int/ACTIVITIES/plant_quarantine/alert_list_insects/xylotrechus_chinensis (accessed on 2 July 2023).
17. Murakami, M. Study on biology and control of *Xylotrechus chinensis* Chevrolat. *Tech. Bull. Seric. Exp. Stat.* **1960**, *77*, 1–23.
18. Togashi, K.; Sasaki, M.; Fujii, T. Relationship between body size and ovariole number in three cerambycids inhabiting mulberry orchards. *Appl. Entomol. Zool.* **2020**, *55*, 9–16. [CrossRef]

19. Fettig, C.J.; Grosman, D.M.; Munson, A.S. Efficacy of abamectin and tebuconazole injections to protect lodgepole pine from mortality attributed to mountain pine beetle (Coleoptera: Curculionidae) attack and progression of blue stain fungi. *J. Entomol. Sci.* **2013**, *48*, 270–278. [CrossRef]
20. Nicolopoulou-Stamati, P.; Maipas, S.; Kotampasi, C.; Stamatis, P.; Hens, L. Chemical pesticides and human health: The urgent need for a new concept in agriculture. *Front. Public Health* **2016**, *4*, 148. [CrossRef]
21. Potamitis, I.; Rigakis, I.; Tatlas, N.A.; Potirakis, S. In-vivo vibroacoustic surveillance of trees in the context of the IoT. *Sensors* **2019**, *19*, 1366. [CrossRef]
22. Kuwahara, Y.; Matsuyama, S.; Suzuki, T. Identification of 2,3-octanediol, 2-hydroxy-3-octanone, and 3-hydroxy-2-octanone from male *Xylotrechus chinensis* Chevrolat as possible sex pheromones (Coleoptera: Cerambycidae). *Appl. Entomol. Zool.* **1987**, *22*, 25–28. [CrossRef]
23. Iwabuchi, K.; Takahashi, J.; Sakai, T. Occurrence of 2, 3-octanediol and 2-hydroxy-3-octanone, possible male sex pheromone in *Xylotrechus chinensis* Chevrolat (Coleoptera: Cerambycidae). *Appl. Entomol. Zool.* **1987**, *22*, 110–111. [CrossRef]
24. EPPO (European and Mediterranean Plant Protection Organization). *Xylotrechus chinensis* (Coleoptera: Cerambycidae): Addition to the EPPO alert list. *EPPO Rep. Serv.* **2018**, *157*, 7–9.
25. Fan, J.T.; Denux, O.; Courtin, C.; Bernard, A.; Javal, M.; Millar, J.G.; Hanks, L.M.; Roques, A. Multi-component blends for trapping native and exotic longhorn beetles at potential points-of-entry and in forests. *J. Pest Sci.* **2019**, *92*, 281–297. [CrossRef]
26. Miller, D.R.; Crowe, C.M.; Mayo, P.D.; Reid, L.S.; Silk, P.J.; Sweeney, J.D. Interactions between ethanol, *syn*-2,3-hexanediol, 3-hydroxyhexan-2-one, and 3-hydroxyoctan-2-one lures on trap catches of hardwood longhorn beetles in southeastern United States. *J. Econ. Entomol.* **2017**, *110*, 2119–2128. [CrossRef]
27. Sweeney, J.D.; Silk, P.J.; Gutowski, J.M.; Wu, J.; Lemay, M.A.; Mayo, P.D.; Magee, D.I. Effect of chirality, release rate, and host volatiles on response of *Tetropium fuscum* (F.), *Tetropium cinnamopterum* Kirby, and *Tetropium castaneum* (L.) to the aggregation pheromone, fuscumol. *J. Chem. Ecol.* **2010**, *36*, 1309–1321. [CrossRef]
28. Pajares, J.A.; Alvarez, G.; Ibeas, F.; Gallego, D.; Hall, D.R.; Fahman, D.I. Identification and field activity of a male-produced aggregation pheromone in the pine sawyer beetle, *Monochamus galloprovincialis*. *J. Chem. Ecol.* **2010**, *36*, 570–583. [CrossRef]
29. Mitchell, R.F.; Graham, E.E.; Wong, J.C.H.; Reagel, P.F.; Striman, B.L.; Hughes, G.P.; Paschen, M.A.; Ginzel, M.D.; Millar, J.G.; Hanks, L.M. Fuscumol and fuscumol acetate are general attractants for many species of cerambycid beetles in the subfamily Lamiinae. *Entomol. Exp. Appl.* **2011**, *141*, 71–77. [CrossRef]
30. Hanks, L.M.; Millar, J.G.; Moreira, J.A.; Barbour, J.D.; Lacey, E.S.; McElfresh, J.S.; Reuter, F.R.; Ray, A.M. Using generic pheromone lures to expedite identification of aggregation pheromones for the cerambycid beetles *Xylotrechus nauticus*, *Phymatodes lecontei*, and *Neoclytus modestus modestus*. *J. Chem. Ecol.* **2007**, *33*, 889–907. [CrossRef]
31. Hanks, L.M.; Millar, J.G.; Mongold-Diers, J.A.; Wong, J.C.H.; Meier, L.R.; Reagel, P.F.; Mitchell, R.F. Using blends of cerambycid beetle pheromones and host volatiles to simultaneously attract a diversity of cerambycid species. *Can. J. For. Res.* **2012**, *42*, 1050–1059. [CrossRef]
32. Millar, J.G.; Hanks, L.M. Chemical Ecology of Cerambycids. In *Cerambycidae of the World: Biology and Pest Management*; Wang, Q., Ed.; CRC Press/Taylor and Francis Group: Boca Raton, FL, USA, 2017; pp. 161–208.
33. Barbour, J.D.; Millar, J.G.; Rodstein, J.; Ray, A.M.; Alston, D.G.; Rejzek, M.; Dutcher, J.D.; Hanks, L.M. Synthetic 3,5-dimethyldodecanoic acid serves as a general attractant for multiple species of *Prionus* (Coleoptera: Cerambycidae). *Ann. Entomol. Soc. Am.* **2011**, *104*, 588–593. [CrossRef]
34. Kelsey, R.G.; Joseph, G. Ethanol in ponderosa pine as an indicator of physiological injury from fire and its relationship to secondary beetles. *Can. J. For. Res.* **2003**, *33*, 870–884. [CrossRef]
35. Ranger, C.M.; Reding, M.E.; Persad, A.B.; Herms, D.A. Ability of stress-related volatiles to attract and induce attacks by *Xylosandrus germanus* and other ambrosia beetles. *Agric. For. Entomol.* **2010**, *12*, 177–185. [CrossRef]
36. Kelsey, R.G.; Gallego, D.; Sánchez-García, F.J.; Pajares, J.A. Ethanol accumulation during severe drought may signal tree vulnerability to detection and attack by bark beetles. *Can. J. For. Res.* **2014**, *44*, 554–561. [CrossRef]
37. Hanks, L.M.; Millar, J.G. Field bioassays of cerambycid pheromones reveal widespread parsimony of pheromone structures, enhancement by host plant volatiles, and antagonism by components from heterospecifics. *Chemoecology* **2013**, *23*, 21–44. [CrossRef]
38. Rodríguez-González, A.; Sánchez-Maíllo, E.; Peláez, H.J.; Mayo, S.; González-López, O.; Carro-Huerga, G.; Casquero, P.A. Evaluation of commercial and prototype traps for *Xylotrechus arvicola* (Coleoptera: Cerambycidae), an insect pest in Spanish vineyards. *Aust. J. Grape Wine Res.* **2018**, *24*, 190–196. [CrossRef]
39. Sukovata, L.; Dziuk, A.; Plewa, R.; Jaworski, T. The effect of trap color on catches of *Monochamus galloprovincialis* and three most numerous non-target insect species. *Insects* **2022**, *13*, 220. [CrossRef]
40. Costello, S.L.; Negrón, J.F.; Jacobi, W.R. Traps and attractants for wood-boring insects in ponderosa pine stands in the Black Hills, South Dakota. *J. Econ. Entomol.* **2008**, *101*, 409–420. [CrossRef]
41. Blomquist, G.J.; Figueroa-Teran, R.; Aw, M.; Song, M.; Gorzalski, A.; Abbott, N.L.; Chang, E.; Tittiger, C. Pheromone production in bark beetles. *Insect Biochem. Mol. Biol.* **2010**, *40*, 699–712. [CrossRef]
42. Rodríguez-González, A.; Sánchez-Maillo, E.; Peláez, H.J.; González-Núñez, M.; Hall, D.R.; Casquero, P.A. Field evaluation of 3-hydroxy-2-hexanone and ethanol as attractants for the cerambycid beetle pest of vineyards, *Xylotrechus arvicola*. *Pest Manag. Sci.* **2017**, *73*, 1598–1603. [CrossRef]

43. Miller, D.R.; Rabaglia, R.J. Ethanol and (−)-α-pinene: Attractant kairomones for bark and ambrosia beetles in the southeastern U.S. *J. Chem. Ecol.* **2009**, *35*, 435–448. [CrossRef]
44. Miller, D.R.; Allison, J.D.; Crowe, C.M.; Dickinson, D.M.; Eglitis, A.; Hofstetter, R.W.; Munson, A.S.; Poland, T.M.; Reid, L.S.; Steed, B.E.; et al. Pine sawyers (Coleoptera: Cerambycidae) attracted to α-pinene, monochamol, and ipsenol in North America. *J. Econ. Entomol.* **2016**, *109*, 1205–1214. [CrossRef]
45. Dodds, K.J.; DiGirolomo, M.F. Effect of cleaning multiple-funnel traps on captures of bark and woodboring beetles in northeastern United States. *Insects* **2020**, *11*, 702. [CrossRef]
46. Hoch, G.; Connell, J.; Roques, A. Testing multi-lure traps for surveillance of native and alien longhorn beetles (Coleoptera, Cerambycidae) at ports of entry and in forests in Austria. *Manag. Biol. Invasions* **2020**, *11*, 677–688. [CrossRef]
47. Kavallieratos, N.G.; Athanassiou, C.G.; Balotis, G.N.; Tatsi, G.T.; Mazomenos, B.E. Factors affecting male *Prays oleae* (Lepidoptera: Yponomeutidae) captures in pheromone-baited traps in olive orchards. *J. Econ. Entomol.* **2005**, *98*, 1499–1505. [CrossRef]
48. Zar, J.H. *Biostatistical Analysis*; Pearson: Harlow, UK, 2014.
49. Sall, J.; Lehman, A.; Creighton, L. JMP start statistics. In *A Guide to Statistics and Data Analysis Using JMP and JMP in Software*; Duxbury Press: Belmont, ON, Canada, 2001.
50. Sokal, R.R.; Rohlf, F.J. *Biometry*; Freeman & Company: New York, NY, USA, 1995.
51. SAS Institute Inc. *Using JMP 16.2*; SAS Institute Inc.: Cary, NC, USA, 2021.
52. Brockerhoff, E.G.; Jones, D.C.; Kimberley, M.O.; Suckling, D.M.; Donaldson, T. Nationwide survey for invasive wood-boring and bark beetles (Coleoptera) using traps baited with pheromones and kairomones. *For. Ecol. Manag.* **2006**, *228*, 234–240. [CrossRef]
53. Dodds, K.J.; Dubois, G.D.; Hoebeke, R.R. Trap type, lure placement, and habitat effects on Cerambycidae and Scolytinae (Coleoptera) catches in the Northeastern United States. *J. Econ. Entomol.* **2010**, *103*, 698–707. [CrossRef] [PubMed]
54. Pajares, J.; Álvarez, G.; Hall, D.; Ibarra, N.; Hoch, G.; Halbig, P.; Cocoş, D.; Johansson, H.; Schroeder, M. Attractants for management of the pine sawyer beetle *Monochamus sutor*, a potential vector of *Bursaphelenchus xylophilus*. *J. Appl. Entomol.* **2017**, *141*, 97–111. [CrossRef]
55. EPPO (European and Mediterranean Plant Protection Organization). Available online: https://gd.eppo.int/taxon/XYLOCH (accessed on 19 July 2023).

insects

Article

GIS-Based Tool for Pest Specific Area-Wide Planning of Crop Rotation Distance with Land Use Data

Manuela Schieler [1,2,*], Natalia Riemer [3], Paolo Racca [1], Benno Kleinhenz [1], Helmut Saucke [3], Michael Veith [2] and Bernd Meese [1]

1 Central Institute for Decision Support Systems in Crop Protection, Rüdesheimer Str. 60-68, D-55545 Bad Kreuznach, Germany; racca@zepp.info (P.R.); kleinhenz@zepp.info (B.K.)
2 Department of Biogeography, Faculty of Regional and Environmental Sciences, Trier University, Universitätsring 15, D-54286 Trier, Germany; veith@uni-trier.de
3 Section of Ecological Plant Protection, Faculty of Organic Agricultural Sciences, University of Kassel, Nordbahnhofstr. 1a, D-37213 Witzenhausen, Germany; natalia.riemer@uni-kassel.de (N.R.); hsaucke@uni-kassel.de (H.S.)
* Correspondence: schieler@zepp.info

Simple Summary: Today's society demands a better balance between the needs of agricultural production and environmental protection, particularly in reducing the application of pesticides. Compared to the past, fewer pesticide active ingredients are allowed and available in the European Union. One pest that is controlled by pesticides is our model pest, the pea moth (*Cydia nigricana*), which can cause damage to pea seeds (*Pisum sativum*). The aim of this study was to develop alternative and preventive strategies that can be used as crop protection in the shape of a GIS-based planning tool for crop rotation with a minimum distance to avoid specific pests. In our tool, the distance between current and previous pea sites was important, because insect pests follow their host plants. The hypothesis is that the greater the distance, the less infestation. We developed a web tool to create buffers with different distances around the previous pea sites to calculate the infestation risk for future pea sites. The web tool can be adapted in terms of distance and risk classes for other pests. Consequently, our tool helps to avoid pest infestation, and therefore, farmers can reduce pesticide applications, which protects the environment and strengthens biodiversity.

Abstract: Crop rotation is an important strategy for pest reduction. For mono-, or oligophagous pests that overwinter at a previously infested site, crop rotation means that the pests must find new host crop sites in the following year, and it is more efficient if a pest-specific distance is applied. Here, we report the development of a GIS-based tool for efficient cultivation planning using the example of the pest complex pea moth (*Cydia nigricana*) and grain and green peas (*Pisum sativum*). Monitoring data for four consecutive years (2016–2019) from 513 sites were used. Infestation of pea seeds and the distance to the previous year's pea sites were recorded. An adjustable Python script was developed by means of infestation–distance–correlation as a pest and crop-specific minimum migration distance (MD). The output of the tool is a risk map as decision support for cultivation planning. It shows different risk buffers with distances from 1261 m to 1825 m, depending on the cultivation type. The web tool is easily adjustable to other pests and crops anywhere in the world. The tool helps to prevent damages caused by agricultural, mono-, or oligophagous insect pests and consequently reduces pesticide applications for the benefit of the environment and biodiversity.

Keywords: area-wide pest management; crop rotation distance; integrated pest management; Python script; *Cydia nigricana*

1. Introduction

Crop rotation is a proven and valuable technique for ecosystems and the reduction of soil-borne pests and diseases [1–4]. For sustainable crop protection, it is better to avoid pests in advance than the obligation to have to control them with pesticides. For example, Wright [2] studied crop rotation effects on Colorado potato beetle (*Leptinotarsa decemlineata*) infestation and found out that crop rotation is the best way to control pest species with few hosts and little migration distances. Area-wide crop rotation has an even greater effect on pest populations.

Further reasons for crop rotation are limited approaches for pest control in organic farming, the availability of allowed pesticide active ingredients, insecticide resistances, and a lack of specific pesticides against target species. First, in organic farming, the efficacy of organic pesticides is often lower compared to conventional pesticides [5]. Second, fewer pesticide active ingredients are allowed to be used in the European Union, pesticides like some neonicotinoids can no longer be used even though they have been effective in insect pest control [6]. The third important factor, insecticide resistance, makes it harder to control some pest species. For Colorado potato beetles, insecticide resistance is well known for neonicotinoids [7,8], pyrethroids, nereistoxins [9], and novaluron [10]. The fourth reason is that there are some new pest species, for example, the cixiid planthopper *Pentastiridius leporinus*, without pesticide control strategies so far. Furthermore, there is the "National action plan for a sustainable application of pesticides" in Germany [11], which claims to use pesticides responsibly, meaning not only to protect crop plants but also to protect human and animal health. In summary, there are many reasons to concentrate on crop rotation distance as an avoidance strategy for insect pests.

There are pests that overwinter site-specifically or at least in close proximity to previous host plant sites from where they have to migrate to find their host in the next season. The pest-specific distances, which they have to overcome to reach the new host plants, are important for a more efficient crop rotation strategy. Furthermore, there are several studies on crop rotation distance of Colorado potato beetles [12–16], which cause damage by defoliating potato plants [7]. For example, with a distance > 100 m, the infestation risk is reduced [16], and with a distance > 400 m, the effect is even better [14,16,17]. Additionally, with a rotation distance of 300–900 m, the insecticide necessity was reduced by 50% versus a non-rotated field [12]. Consequently, crop rotation will be more efficient with distances adapted to the pests and digitalized procedures, like a web tool. Therefore, it is necessary to research pest-specific crop rotation distances [15,17] for pests that overwinter site-specifically in harvested sites or at least close to them, as well.

The use of geographic information systems (GIS) within decision support systems (DSS) in agriculture can be very effective and useful, and many different applications already exist [18,19]. If input data are available, they are area-wide applicable. When developing a web tool, using a Python script is convenient, because it is easily adjustable with respect to crop rotation distances and different pests as soon as needed parameters are available. Additionally, automated calculations save a lot of time. For Colorado potato beetles, Weisz et al. [13] developed models to calculate infestation as a function of migratory distance. With one of the models, they created a risk map for planned potato fields, as an interpolated heat map, which shows the infestation risk compared to a non-rotated field. Thus, a risk map according to the crop rotation distance could reduce or avoid infestation by pests in advance.

In our study, we used the pea moth (*Cydia nigricana*, Lepidoptera: Tortricidae) as a model species. Pea moths can cause great damage to pea seeds (*Pisum sativum*, Fabaceae), which leads to yield and quality loss [20,21]. The model species overwinters as the fifth instar in the soil of previous pea sites. In spring, the adults emerge and search for new pea sites, lured by the odor of pea flowers [22]. Former studies focused on the minimum crop rotation distance (MD) of pea moths, which describes the distance between infested pea sites of the last year and the ones of the current season. For example, Thöming et al. [23] discovered that an MD of 500 m for green peas had a positive effect on a decreasing infestation by pea moths. Huusela-Veistola and Jauhiainen [24] found that seed and pod infestation decreased exponentially with the distance to a previous grain pea field of >1000–3000 m, and similar results are shown by Riemer et al. [25].

We show how to prevent damage to crops by insect pests that overwinter site-specifically by introducing a planning tool as an area-wide application, which was developed by the Central Institute for Decision Support Systems in Crop Protection (German acronym: ZEPP) as part of the project CYDNIGPRO (CYDia NIGricana PROgnosis). The infestation of pea fields by pea moths as well as the minimum distance to previously infested pea fields were analyzed and implemented in a risk map in order to reduce and avoid damage to peas. The results of the project are partly used as input for a risk map for improved crop rotation distances. Such a risk map has not been established for practical use in Germany so far. The planning tool can be used and adapted to other pests worldwide if input data are available.

2. Materials and Methods

Field data for the development of the risk map were acquired between 2016 and 2019 in three representative German pea-growing areas, which are located in the federal states Saxony-Anhalt (ST), Saxony (SN), and Hesse (HE) (Table 1). In Saxony-Anhalt, northeast of the Harz Mountains, in the proximity of Quedlinburg (51.93–51.74 N, 10.98–11.29 E, WGS84, Figure 1), the cultivation of peas has a long history, especially for seed production. The "Lommatzscher Pflege" (51.40–51.18 N, 13.20–13.53 E) in Saxony is a traditional green pea-growing area, a major one for frozen green pea products in Germany. The cultivation of peas in Hesse (51.38–51.14 N, 9.75–10.12 E) is not as intense and commercial as in Saxony-Anhalt and Saxony, the pea sites are much smaller, and most of the peas are forage crops. Each model area had a diameter of approx. 30 km. The sample sites were cultivated by local farmers or agricultural cooperatives. Some of the farmers used insecticides, others did not. The data were used in three different groups: all grain peas, and grain peas with and without insecticide treatment. For the outcome, the data grain peas with insecticide treatment are not usable, because the goal is to use satellite data, with which those differences are not detectable.

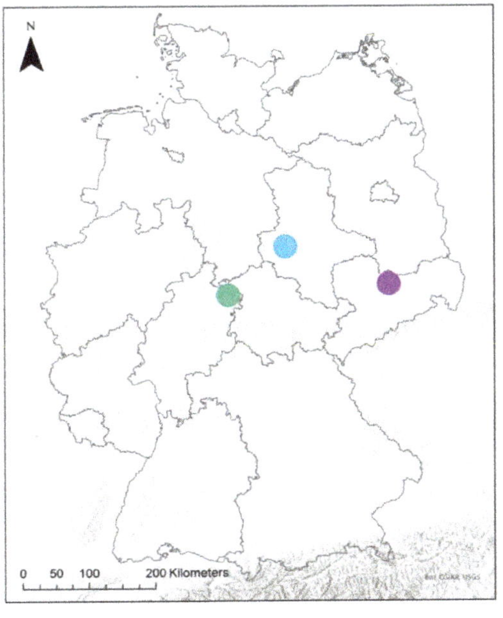

Figure 1. Map of Germany with the location of the model areas, ST = blue, SN = purple, HE = green (map source: Esri, CGIAR, USGS).

Table 1. Number of sample sites per federal state in Germany, separated by with/without insecticides.

	Federal State/Year	ST	SN	HE		Federal State/Year	SN
Grain peas	2016	57/1	7/21	3/31	Green peas	2016	33/19
	2017	50/3	4/15	14/44		2017	13/24
	2018	32/9	2/9	11/34		2018	20/32
	2019		2/6	0/15		2019	0/2
	Subtotal	139/13	15/51	28/124			66/77
	Total	152	66	152			143

2.1. Data Acquisition

2.1.1. Infestation Data

To determine the correlation between infestation and minimum crop rotation distance, the sampling of the infestation was required. Therefore, at the end of the growing season, at each pea site, a sample of 100 pods was taken to record percentage infestation by pea moths that is easily recognizable from nutritional traces, mostly around the seeds. We collected the pods along the tractor tracks, starting after the headland, for approx. 50 m. The number of seeds infested by pea moths was counted and converted in percentage.

2.1.2. Pea Site Locations

There are different possibilities to obtain information about the location of pea sites. Surveying farmers in a greater area is too labor-intensive. Consequently, a larger dimension to obtain area-wide locations of pea sites is needed. In Germany, one possible format is anonymous InVeKoS (Integriertes Verwaltungs- und Kontrollsystem) data. These data are available in a vector format (polygon shapefiles) and are provided by governmental agencies that are responsible for agricultural subsidies related to the common agricultural policy (CAP) of the EU (Figure 2, Appendix A: Figures A1 and A2). Examples of these programs are "ecological focus areas" or "crop diversification" within the greening direct payments [26,27]. In our study, these geographical locations were used as sample sites and for all further calculations. The data we used were from the years 2015–2019, whereas the data from the year 2015 only served as previous pea site locations, not as sample sites. In the maps of the model areas, we show the examples for the years 2015–2017 to obtain an overview of how close the pea sites are from one to the next year.

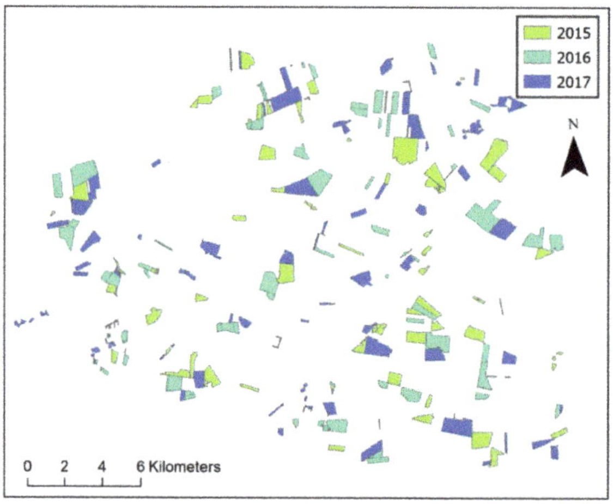

Figure 2. Map of pea sites in model area ST for the years 2015–2017 as an example. Current pea sites and those of the previous year are often right next to each other, see the appendix for SN and HE.

2.2. *Infestation–Distance–Correlation*

In spring, pea moths that had overwintered in pea sites have to overcome the distances to new sites with their host plants. Therefore, we correlated the distance and infestation data to provide basic information about the minimum crop rotation distance (MD). We used the provided polygon shapefiles to calculate the distance between the current pea sites and those of the previous year, from field edge to field edge. Only the nearest pea site was used. For the calculation of infestation as a function of MD, we used the exponential decay function of Weisz et al. [12], Thöming et al. [23] and Huusela-Veistola and Jauhiainen [24]:

$$y = a \times e^{\left(-\frac{MD}{b}\right)} \tag{1}$$

where a is the y-intercept, b is the inflection point of the slope, MD is the minimum distance in meters to the closest pea field of the previous year, and y is the estimated infestation in percent. The correlation of infestation data and minimum distance to the closest previous pea site serve as background information for the risk classes provided by the risk maps as described in the next section. The influence of multiple previous pea fields is explained in section "overlapping buffer zones".

2.3. *Geographical Implications*

Crop site locations of the previous year are the base of the calculation. Only the fact that there used to be a cultivated site is taken as information, not the absolute infestation of each site, due to the effort to gather the infestation data of each site each year. The calculations were conducted with the ArcGIS Pro 3.1.0 (Esri) and Python 3.6.

2.3.1. Python Script

We provide a universal script for crop rotation options for different crops. Therefore, several parameters are adjustable, such as distance, the number of risk classes, and the value of the risk classes. The risk classes can be linear, exponential, or individually adapted. In this paper, the proceedings are presented using the example of pea moths and our study data. In Figure 3, the general workflow of the Python script is visualized.

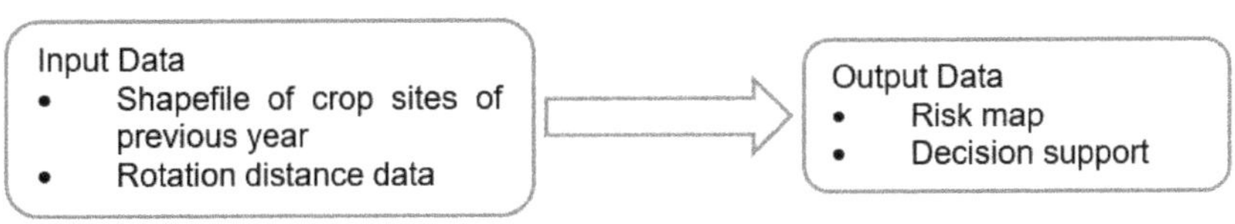

Figure 3. Workflow of the Python script.

2.3.2. Buffer Zones

The first step is to build buffer zones with different distances. They are calculated around the input polygons, which are the crop sites of the previous year. It is possible to create one or more buffer zones with desirable distances depending on the pest species data. In case of more than one buffer zone, there is always a ring-shaped buffer built around the next inner buffer.

2.3.3. Risk Classes

Each buffer zone gets a specific risk class. In case of more than one buffer zone, risk classes decrease with an increase in distance to previous year's crops. Risk classes may have linear or exponential numerical order, be in percent, or other individual classes may be used.

2.3.4. Overlapping Buffer Zones

The next step unites overlapping buffer zones. If two or more previous fields are close to each other, buffer zones may overlap. In this case, the overlapping layers are automatically deleted except for one. The polygon that is left is automatically set to the next highest risk class because sources of infestation come from at least two directions (Figure 4).

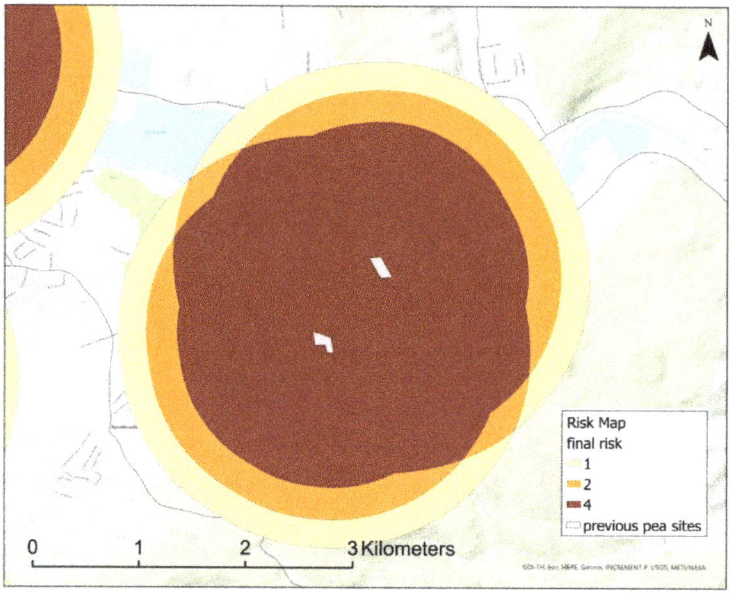

Figure 4. Risk classes of overlapping areas are automatically raised to the next highest risk class (here exponential). In this case, the risk comes from two pea sites of the previous year: white polygons (map source: GDI-TH, Esri, HERE, Garmin, INCREMENT P: USGS, METI/NASA).

In Figure 5, a pseudo code of the crop rotation distance tool is written. It shows the stylized single steps of the calculation, which are also described above.

```
Crop Rotation Distance Tool (pseudo code)
Data:    shape file of prev. year crop locations
         risk class distances (d1, ..., dmax)
Result: risk zone shape file
For location in prev. year crop locations:
    i = 1
    While i ≤ length(risk class distances):
        risk(i) = length(risk class distances) - i + 1
        risk_zone(i) = buffer(location, di)
        assign risk(i) to buffer
        i +=1
union risk class zones(all locations)
If risk class zones overlap:
    use highest occurring risk class of overlapped area
    increase risk by number of overlapping risk zones
Return risk zone shape file
```

Figure 5. Pseudo code of crop rotation distance tool.

3. Results

3.1. *Infestation–Distance–Correlation*

In the following, the result of the infestation–distance–correlation is described, which results in the recommended crop rotation distances. In Figure 6, seed infestation is plotted against MD. The observed data are shown with the fitted model curves for all grain peas (red), grain peas without insecticide treatment (green), and grain peas with insecticide treatment (blue). For example, for all grain peas (red), the infestation rate is 9.84% at 0 m, declining to 4.32% at 1000 m and 2.86% at 1500 m; the latter is less than a third of the infestation rate at 0 m.

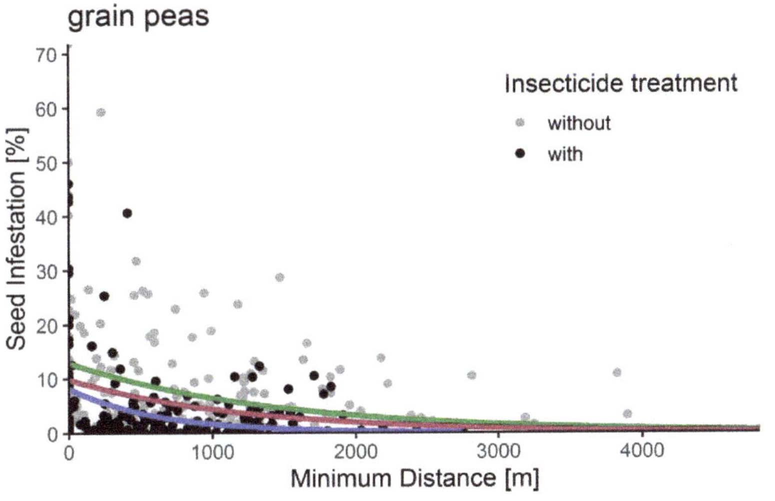

Figure 6. Seed infestation in percent as a function of minimum distance in meters. Each dot depicts one sample site. The gray dots symbolize the sites without insecticide treatment, and the black dots are the ones with insecticide treatment. The red curve shows the exponential decay function of all grain peas, the green curve indicates sites without insecticide treatment (n = 188), and the blue curve indicates sites with insecticide treatment (n = 182).

In Table 2, the parameters of the model equations are described. For all grain peas, the effective crop rotation distance is 1261 m ± 299 m, with an RMSE goodness of fit parameter of the exponential decay function of 9.1. Additionally, the calculations were also conducted for grain peas with and without insecticide treatment. For grain peas with insecticide treatment, the effective crop rotation distance is 619 m ± 187 m, with an RMSE of 7.41 and an R^2 of 0.07. For grain peas without insecticides, the effective crop rotation distance is 1426 m ± 399 m, with an RMSE of 9.98 and an R^2 of 0.08. The poor R^2 of all three calculations reflects the wide variation in the data in Figure 6. However, the significant estimates prove that the tendency of the curves is significant. Consequently, we decided to offer the following risk buffers:

- 1261 m = inner risk buffer, which is parameter b for all grain peas;
- 1560 m = middle risk buffer, which is parameter b for all grain peas plus standard error;
- 1825 m = outer risk buffer, which is parameter b for grain peas without insecticides plus standard error.

The decision on which risk buffer should be used depends on the cultivation type, such as organic or conventional, green or grain peas. Nevertheless, we highly recommend using at least the inner safety buffer and not parameter b of grain peas with insecticides, since we do not know if a previous pea site was treated or not. In the case of organic pea cultivation, we recommend using the outer risk buffer.

The same calculation was not possible for green peas because the mean infestation was very low (0.11%); therefore, we decided to use a safety buffer of 500 m as Thöming et al. [23] recommended.

Table 2. Parameters and estimates of the infestation–distance–correlation for grain peas, the significance of the parameters is as follows: <0.001 = ***, <0.01 = **; SE = standard error.

	Parameter	a	b	RMSE	$R^2_{adjusted}$
All grain peas	Estimate SE	9.79 *** 0.92	1260.70 *** 299.45	9.10	0.05
Without insecticide treatment	Estimate SE	12.85 *** 1.45	1426.23 ** 398.92	9.98	0.08
With insecticide treatment	Estimate SE	8.13 *** 1.09	619.05 ** 187.02	7.41	0.07

3.2. Geographical Implication

3.2.1. Risk Map

The first part of the calculations results in a risk map, which shows the buffer zones according to risk classes around the crop sites of the previous year (Figure 7). If the infestation risk comes from multiple directions, buffer zones overlap and risk classes are increased to the next highest class, here exponential from two to four.

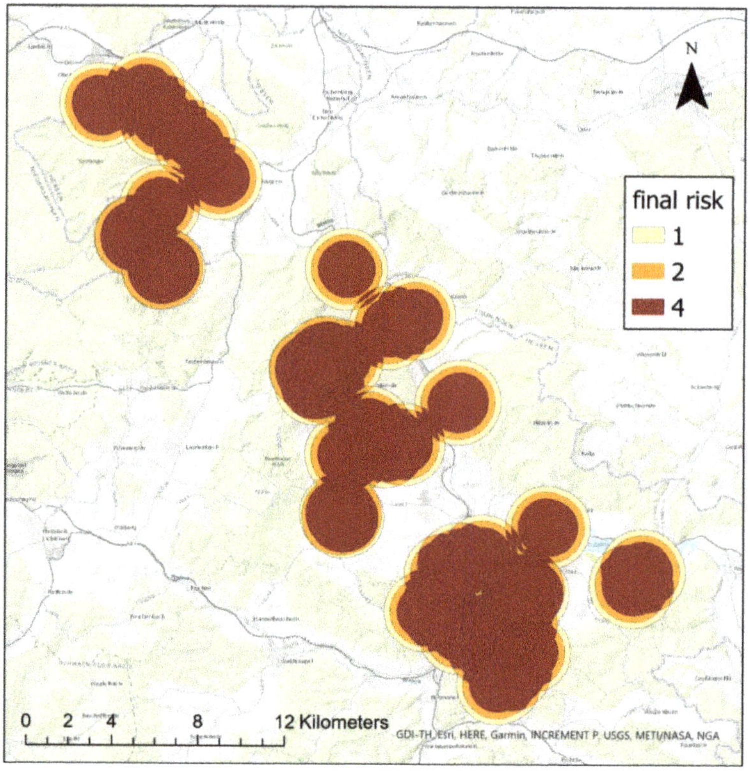

Figure 7. Risk map with buffers around potential infestation sites of the previous year (map source: GDI-TH, Esri, HERE, Garmin, INCREMENT P: USGS, METI/NASA, NGA).

3.2.2. Planning Tool

In the second part of the calculations, the user defines the new pea site of interest (risk query). The planning tool then cuts out a circular zone around the queried coordinates (Figure 8A, blue circle) and provides it as the site-specific risk map (Figure 8B). The user can now decide if the planned site is suitable or not. In addition, it is possible to identify directions of risk sources. If there is an option to grow peas on an alternative site, the user can now choose a site in a less risky area.

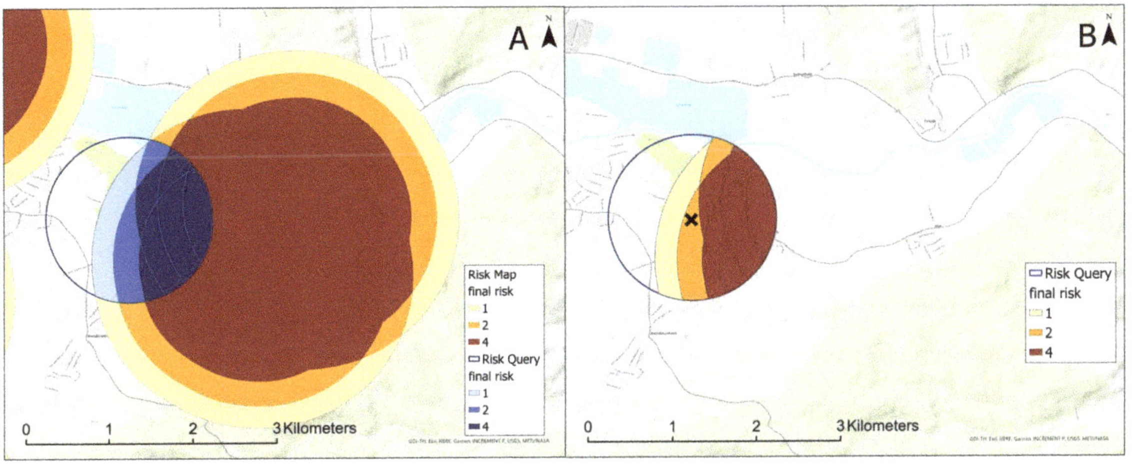

Figure 8. (**A**): For the user's query (blue), a buffer around geographical coordinates is clipped out of the risk map. (**B**): The displayed user output is the clipped region of interest, the '×' marks the queried geographical coordinates. The chosen new pea site has a risk potential, apparently coming from two previous pea sites (map source: GDI-TH, Esri, HERE, Garmin, INCREMENT P: USGS, METI/NASA).

4. Discussion

We here show how to mitigate crop damage by site-specifically overwintering insect pests and how this information can be implemented in an area-wide, commonly available web tool. We digitalized and automatized the calculation of infestation risk as a function of minimum crop rotation distance (MD). A key factor for obtaining effective pest reduction is the pest-specific MD.

4.1. Infestation–Distance–Correlation

The infestation–distance–correlation showed that the greater the distance to the previous year's pea site, the less infestation by pea moths (Figure 6, Table 2). For grain peas, the effect was apparent, and the effective rotation distance for all grain peas is 1261 m ± 299 m. We used these distances for the inner (1261 m) and the middle risk buffer (1560 m). The outer risk buffer (1825 m), which is the worst-case scenario for grain peas without insecticides, should be used if organic peas are grown or if it is known that the previous year's infestation was very high. It cannot be recommended to use a smaller buffer, since with the use of satellite data, we do not know if a previous year's pea site was treated with insecticides or not. Even though the function almost seems to be linear within the examined range, the underlying function has been proven to be an exponential decay function by Huusela-Veistola and Jauhiainen [24], Thöming et al. [23], Weisz et al. [12], and Riemer et al. [25] in their studies. The MD for grain peas of 1–3 km was effective in reducing the risk of pea moth infestation, depending on the year and cropping system [24]. They recommend a distance of 1.5 km to previous pea fields, which is similar to our recommendation. As a consequence of the calculation, insecticides could be reduced, in terms of fewer applications, or left out, if the determined distance to the previous year's crops is maintained.

Green peas, which are harvested around BBCH 79 in a green, immature, and tender condition, do not implicate a big risk for infestation in the following year. Most of the pea moths are not able to complete their life cycle, due to early harvest. Hence, the infestation pressure in dense and intensive green pea cultivation areas does not get really high. Therefore, we decided to use a comparably small risk buffer of 500 m around the previous year's green pea sites. In organic green peas, an MD of 500 m resulted in a significant reduction in larval infestation [23].

Similar studies concerning insect pests have been conducted in the past. For example, an MD of more than 400 m was effective in reducing Colorado potato beetle infestation in the following year [12,14,16], with increasing distances further reducing infestation. Carrière et al. [28] studied *Lygus hesperus*, the western tarnished plant bug, and its maximum migration distance. They found that at a maximum distance of more than 1500 m, there is no additional spread of *L. hesperus* from alfalfa into cotton, which is important for farmers.

4.2. Geographical Calculations

Our new GIS-based web tool generates a risk map and provides decision support for applicants (Figure 8). Although only the nearest pea site was used as input for the infestation–distance–correlation, the risk map calculations include all previous pea sites in the vicinity of a potential new site by overlapping risk buffers (Figure 4). Interestingly, we found no correlation between infestation intensity and the sizes of pea sites (Appendix A, Figure A3). While one might expect a higher infestation probability with increasing site sizes due to a more intense flower odor, in our study we found numerous small sites with high infestations as well as large sites with low infestation. Therefore, we assume that the number and the distance of fields in the vicinity are the main factors that affect infestation in the following year.

In the past, different approaches to mitigate infestation by insect pests were developed. For example, Weisz et al. [13] developed an "all source model" for Colorado potato beetles by calculating one risk zone or interpolated hotspots with different risk classes for an area, so the new fields can be placed outside the risk zone. In our risk map, there are risk buffers around each previous field, which means that the provided risk areas are much more differentiated. Beckler et al. [29] also created an interpolated abundance map, which shows hot spots of corn rootworms (*Diabrotica barberi* and *Diabrotica virgifera virgifera*) emergence. Compared to our approach, both approaches require annual area-wide monitoring, so to reduce the labor required, we decided to use the location of the pea sites as the main input factor.

In the future, continuous annual input data will be required for each crop. In this regard, the imminent availability of processed satellite data will be a huge enhancement. Waldhoff et al. [30] classified major crops to analyze crop rotation sequences in the Rur catchment in Germany. Meroni et al. [31] also classified major crops or crop groups and their phenology as BBCH stages all over Europe [32]. By using phenological timelines, it can be easier to classify the different pea-growing types, like grain peas, green peas, and green peas for propagation. Maps can be made available by different providers (e.g., www.map.onesoil.ai, [33]) by the classification of satellite images during vegetation periods, however by now mostly for main cultures such as wheat, corn, and rape seed or combined groups such as vegetables. Some German research institutes are still in the development phase to create maps with classified crops for permanent use, such as Thünen Institute [34], and FERN.Lab with Habitat Sampler and Minimal Sample Classifier (MiSa.C) [35,36]. The advantage of classified satellite images is that information on cultivated crops is area-wide, freely available, and does not depend on the application for agricultural subsidies. However, long-lasting cloud cover or difficult-to-distinguish combined cultures can pose a problem regarding these not-yet-available data sets.

4.3. Limitations

Parameters such as soil type, soil temperature, and parasitoids—currently not included in our model—may significantly affect infestation probability and overwintering. Finding out all relevant site effects might improve further infestation predictions but was not within the scope of this study. In our study, only the parameters distance and infestation were included. Nevertheless, other landscape structures such as urban areas and forests can serve as barriers. Since we do not have any study results for our model pest, we were not able to include such landscape structures in our calculations.

In some cases, the infestation can be high despite assumed large distances due to the federal borders, because InVeKoS data were only available for the federal states with model areas. Moreover, InVeKoS data are not in all cases crop-specifically classified, especially for our use. For example, a site might be labeled as bee pasture, but actually, it can be a pure pea site or include pea plants. Ecological focus areas like bee pastures could not be taken into account in our calculations.

Our tool cannot be used for polyphagous insects, such as some aphids, especially *Myzus persicae,* because it is impossible to detect their primary host *Prunus* spp. (holocycle) as well as all of the secondary hosts (anholocycle) with satellite data. Insects that change their host multiple times during the year are not suitable for the tool because the geographic position of the previous source is unknown. In addition, there are insects, here also aphis, for example, that can be spread by wind because of their light weight as aerial plankton. So, the real source of these aphids is in both cases not detectable.

4.4. Outlook

We here used pea moths and pea crops in three German growing areas as a model system to develop a GIS-based planning tool for crop rotation distance in order to minimize pea moth infestation rates. However, the tool can easily be adjusted to other insect pests, such as Colorado potato beetles on potatoes, Cixiid planthoppers on sugar beets, and corn rootworms on corn. Also, infestation with other pests such as fungi, nematodes, or pathogens can survive in the soil for several years, so in the next step numerous different pre-crops could be modeled. When more than one pre-crop are host plants for pests, they can be included. In addition, pest species-specific dispersion barriers can be implemented.

The use of crop rotation distance to ensure proper buffers between crops planted in one year and the next is best implemented through a system of area-wide management. Wright [2], Weisz et al. [13], and Sexson and Wyman [14] already recommended rotating crops through the cooperation of large groups of farmers in an area to ensure that host crops are not planted in high-risk areas. The goal should be to digitalize the cooperation of farmers for better area-wide communication and area-wide pest management.

Currently, our tool is only available for registered members on the Internet platform www.isip.de (information system for integrated plant production, [37]), a plant protection service provider that can be used by advisors and farmers in Germany. By the time of publication, the tool is completely programmed and implemented in the platform with an interim solution, where users must provide pea sites of the previous year manually, at least until satellite data are available. Even though the input mask and calculation are currently limited to members, the tool's availability can be extended to other countries as well. The tool depends on the availability of input data only and is not limited to a country or region. With the development of more precise processing and classification of Earth observation data, the availability of such data will be possible in the future.

5. Conclusions

We developed a new web tool that allows better planning of crop rotation distance to last year's pea sites by estimating the risk of infestation of mono- or oligophagous insect pests, who overwinter site-specifically. We recommend cultivating peas outside the risky areas, which is at least 1261 to 1825 m from last year's pea sites, depending on the cultivation type. Currently, it is available throughout Germany, though use may be extended in the future to other countries as well. If the land use data are available, the tool is very convenient to plan crop rotation distance. This may lead to a reduced application of insecticides [11]. A major advantage of this tool is its adaptability to other pests and crops.

Author Contributions: Conceptualization, M.S., B.M., N.R., B.K., P.R. and H.S.; methodology, M.S. and B.M.; software, M.S. and B.M.; writing—original draft preparation, M.S., B.M. and N.R.; writing—review and editing, M.S., B.M., N.R., B.K., P.R. and M.V.; visualization, M.S. and B.M.; supervision, B.M., B.K., H.S. and M.V.; project administration, H.S. and B.K.; funding acquisition, H.S. and B.K. All authors have read and agreed to the published version of the manuscript.

Funding: The project CYDNIGPRO (Project No. 2812OE034) was supported by funds of the Federal Ministry of Food and Agriculture (BMEL) based on a decision of the parliament of the Federal Republic of Germany via the Federal Office for Agriculture and Food (BLE) under the Federal Program for Ecological Farming and Other Forms of Sustainable Agriculture.

Data Availability Statement: The data presented in this study are available on request from the corresponding author.

Acknowledgments: We would like to thank Charlotte Clemenz, Niklas Jacob, Ulf Müller, Markus Hammer-Weis, Gerd Schrage, and Kristin Schwabe for their work on the sample sites. We thank the local farmers for letting us monitor their pea sites; Jean Fred Fontaine and Lisa Bartsch for their help in R and Stephan Estel for proofreading the manuscript.

Conflicts of Interest: The authors declare no conflicts of interest. The funders had no role in the design of the study; in the collection, analyses, or interpretation of data; in the writing of the manuscript; or in the decision to publish the results.

Appendix A

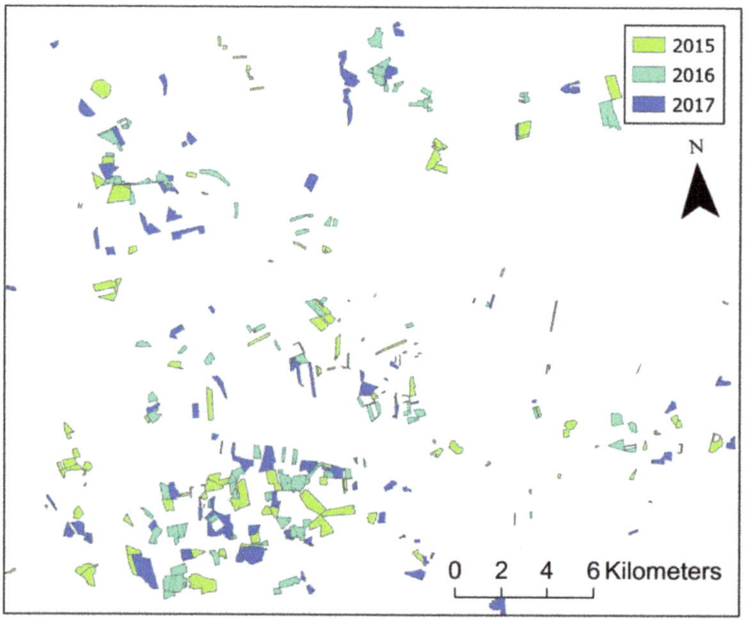

Figure A1. Map of pea sites in model area SN for the years 2015–2017. Current pea sites and those of the previous year are often right next to each other.

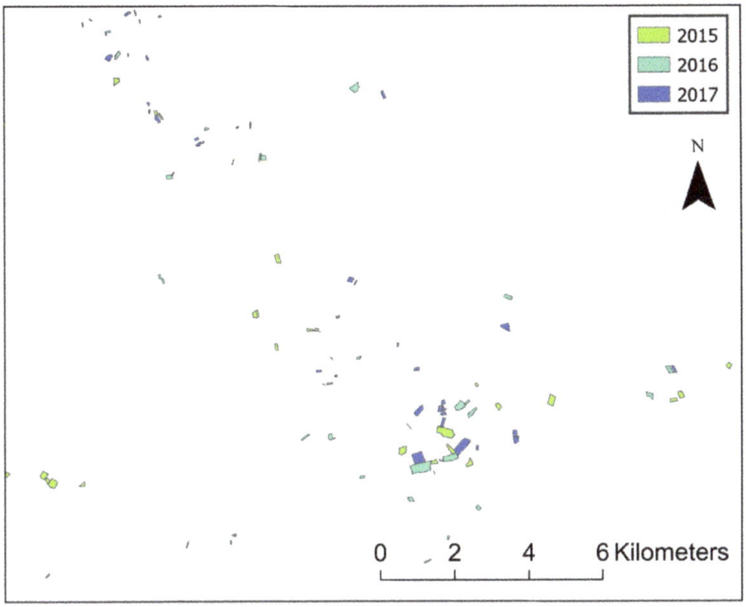

Figure A2. Map of pea sites in model area HE for the years 2015–2017. Current pea sites and those of the previous year are often right next to each other.

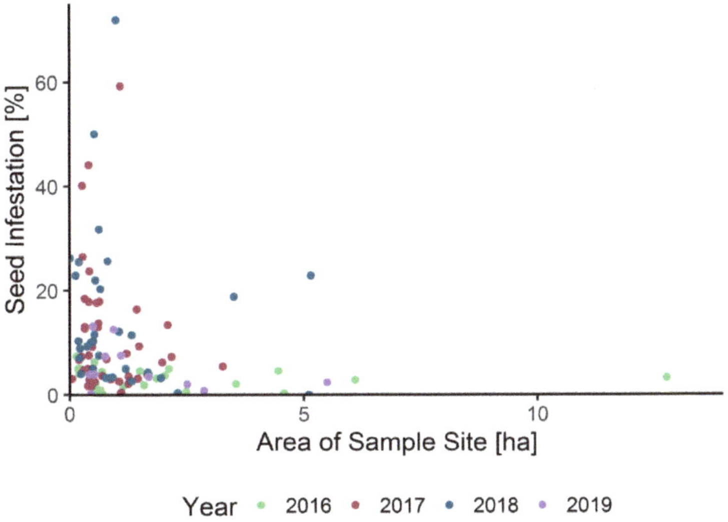

Figure A3. Seed infestation in percent as a function of sample site area in hectares. In the example are sites from Hesse that were not sprayed with insecticides. Small sites can have a high and low infestation by pea moths; $R^2 = 0.02$, linear regression. Peas that are cultivated for seed production must have a very low infestation level for seed approval.

References

1. Reddy, P.P. *Agro-Ecological Approaches to Pest Management for Sustainable Agriculture*; Springer Singapore: Singapore, 2017; ISBN 978-981-10-4324-6.
2. Wright, R.J. Evaluation of Crop Rotation for Control of Colorado Potato Beetles (Coleoptera: Chrysomelidae) in Commercial Potato Fields on Long Island. *J. Econ. Entomol.* **1984**, *77*, 1254–1259. [CrossRef]

3. Stoner, K.A. Management of Insect Pests with Crop Rotation and Field Layout. In *Crop Rotation on Organic Farms: A Planning Manual*; Johnson, S.E., Mohler, C.L., Eds.; NRAES: Ithaca, NY, USA, 2009; pp. 41–43, ISBN 978-1-933395-21-0.
4. Farooq, M.; Nawaz, A.; Nadeem, F. Crop Rotations, Fallowing, and Associated Environmental Benefits. In *Oxford Research Encyclopedia of Environmental Science*; Nadeem, F., Nawaz, A., Farooq, M., Eds.; Oxford University Press: Oxford, UK, 2019; pp. 1–49, ISBN 9780199389414.
5. Bahlai, C.A.; Xue, Y.; McCreary, C.M.; Schaafsma, A.W.; Hallett, R.H. Choosing Organic Pesticides over Synthetic Pesticides May Not Effectively Mitigate Environmental Risk in Soybeans. *PLoS ONE* **2010**, *5*, e11250. [CrossRef]
6. Wood, T.J.; Goulson, D. The environmental risks of neonicotinoid pesticides: A review of the evidence post 2013. *Environ. Sci. Pollut. Res. Int.* **2017**, *24*, 17285–17325. [CrossRef]
7. Ferro, D.; Logan, J.; Voss, R.; Elkinton, J. Colorado potato beetle (Coleoptera: Chrysomelidae) temperature-dependent growth and feeding rates. *Environ. Entomol.* **1985**, *14*, 343–348. [CrossRef]
8. Alyokhin, A.; Baker, M.; Mota-Sanchez, D.; Dively, G.; Grafius, E. Colorado potato beetle resistance to insecticides. *Am. J. Pot Res* **2008**, *85*, 395–413. [CrossRef]
9. Stanković, S.; Kostic, M.; Sivcev, I.; Jankovic, S.; Kljajic, P.; Todorovic, G.; Radosav, J. Resistance of Colorado potato beetle (Coleoptera: Chrysomelidae) to neonicotinoids, pyrethroids and nereistoxins in Serbia. *Rom. Biotechnol. Lett.* **2012**, *17*, 7599–7609.
10. Cutler, G.C.; Tolman, J.H.; Scott-dupree, C.D.; Harris, C.R. Resistance Potential of Colorado Potato Beetle (Coleoptera: Chrysomelidae) to Novaluron. *J. Econ. Entomol.* **2005**, *98*, 1685–1693. [CrossRef] [PubMed]
11. BMEL. *Nationaler Aktionsplan zur Nachhaltigen Anwendung von Pflanzenschutzmitteln*; BMEL: Bonn, Germany, 2017.
12. Weisz, R.; Smilowitz, Z.; Christ, B. Distance, Rotation, and Border Crops Affect Colorado Potato Beetle (Coleoptera: Chrysomelidae) Colonization and Population Density and Early Blight (Alternaria Solani) Severity in Rotated Potato Fields. *J. Econ. Entomol.* **1994**, *87*, 723–729. [CrossRef]
13. Weisz, R.; Smilowitz, Z.; Fleischer, S. Evaluating Risk of Colorado Potato Beetle (Coleoptera: Chrysomelidae) Infestation as a Function of Migratory Distance. *J. Econ. Entomol.* **1996**, *89*, 435–441. [CrossRef]
14. Sexson, D.L.; Wyman, J.A. Effect of Crop Rotation Distance on Population of Colorado Potato Beetle (Coleoptera: Chrysomelidae): Development of Areawide Colorado Potato Beetle Pest Management Strategies. *J. Econ. Entomol.* **2005**, *98*, 716–724. [CrossRef] [PubMed]
15. French II, N.; Follett, P.; Nault, B.; Kennedy, G. Colonization of potato fields in eastern North Carolina by Colorado potato beetle. *Entomol. Exp. Appl.* **1993**, *68*, 247–256. [CrossRef]
16. Boiteau, G.; PICKA, J.D.; Watmough, J. Potato Field Colonization by Low-Density Populations of Colorado Potato Beetle as a Function of Crop Rotation Distance. *J. Econ. Entomol.* **2008**, *101*, 1575–1583. [CrossRef] [PubMed]
17. Hough-Goldstein, J.; Whalen, J. Relationship Between Crop Rotation Distance From Previous Potatoes and Colonization and Population Density of Colorado Potato Beetle. *J. Agric. Urban Entomol.* **1996**, *13*, 293–300.
18. Racca, P.; Kleinhenz, B.; Zeuner, T.; Keil, B.; Tschöpe, B.; Jung, J. Decision Support Systems in Agriculture: Administration of Weather Data, Use of Geographic Information Systems (GIS) and Validation Methods in Crop Protection Warning Service. In *Efficient Decision Support Systems: Practice and Challenges-From Current to Future/Book 1*; Jao, C., Ed.; InTech: London, UK, 2011; pp. 331–354, ISBN 978-953-308-65-7.
19. *GIS Applications in Agriculture, Volume Three: Invasive Species*; Clay, S., Ed.; CRC Press: Boca Raton, FL, USA, 2011; ISBN 9781420078800.
20. Capinera, J.L. *Handbook of Vegetable Pests*; Academic Press: San Diego, CA, USA, 2001; ISBN 9780080533261.
21. Hoffmann, G.M.; Schmutterer, H. *Parasitäre Krankheiten und Schädlinge an Landwirtschaftlichen Kulturpflanzen*, 2nd ed.; Verlag Eugen Ulmer GmbH & Co.: Stuttgart, Germany, 1999; ISBN 9783800132072.
22. Thöming, G.; Norli, H.R.; Saucke, H.; Knudsen, G.K. Pea plant volatiles guide host location behaviour in the pea moth. *Arthropod-Plant Interact.* **2014**, *8*, 109–122. [CrossRef]
23. Thöming, G.; Pölitz, B.; Kühne, A.; Saucke, H. Risk assessment of pea moth *Cydia nigricana* infestation in organic green peas based on spatio-temporal distribution and phenology of the host plant. *Agric. For. Entomol.* **2011**, *13*, 121–130. [CrossRef]
24. Huusela-Veistola, E.; Jauhiainen, L. Expansion of pea cropping increases the risk of pea moth (*Cydia nigricana*; Lep., Tortricidae) infestation. *J. Appl. Entomol.* **2006**, *130*, 142–149. [CrossRef]
25. Riemer, N.; Schieler, M.; Saucke, H. The efficacy of spatio-temporal predictors in forecasting the risk of *Cydia nigricana* infestation. *Entomol. Exp. Appl.* **2024**, *00*, 1–10. [CrossRef]
26. *Umsetzung der EU-Agrarreform in Deutschland*, BMEL, Ed.; Bundesministerium für Ernährung und Landwirtschaft: Berlin, Germany, 2015.
27. European Commission. CAP Explained—Direct Payments for Farmers 2015–2020. 2017. Available online: https://ec.europa.eu/agriculture/direct-support/direct-payments_en (accessed on 9 November 2021).
28. Carrière, Y.; Ellswoth, P.C.; Dutilleul, P.; Ellers-Kirk, C.; Barkley, V.; Antilla, L. A GIS-based apporach for areawide pest management: The scales of Lygus hesperus movements to cotton from alfalfa, weeds, and cotton. *Entomol. Exp. Appl.* **2006**, *118*, 203–210. [CrossRef]
29. Beckler, A.A.; French, B.; Wade; Chandler, L.D. Using GIS in Areawide Pest Management: A Case Study in South Dakota. *Trans. GIS* **2005**, *9*, 109–127. [CrossRef]

30. Waldhoff, G.; Lussem, U.; Bareth, G. Multi-Data Approach for remote sensing-based regional crop rotation mapping: A case study for the Rur catchment, Germany. *Int. J. Appl. Earth Obs. Geoinf.* **2017**, *61*, 55–69. [CrossRef]
31. Meroni, M.; d'Andrimont, R.; Vrieling, A.; Fasbender, D.; Lemoine, G.; Rembold, F.; Seguini, L.; Verhegghen, A. Comparing land surface phenology of major European crops as derived from SAR and multispectral data of Sentinel-1 and -2. *Remote Sens. Environ.* **2021**, *253*, 112232. [CrossRef]
32. Meier, U. *Growth Stages of Mono-and Dicotyledonous Plants*, 2nd ed.; Federal Biological Research Centre for Agriculture and Forestry: Braunschweig, Germany, 2001.
33. OneSoil. OneSoil Global Analytics. Available online: www.map.onesoil.ai (accessed on 11 November 2023).
34. Blickensdörfer, L.; Schwieder, M.; Pflugmacher, D.; Nendel, C.; Erasmi, S.; Hostert, P. Mapping of crop types and crop sequences with combined time series of Sentinel-1, Sentinel-2 and Landsat 8 data for Germany. *Remote Sens. Environ.* **2022**, *269*, 112831. [CrossRef]
35. Beamish, A.L.; Anbuhl, L.; Behling, R.; Goncalves, R.; Lingenfelser, A.; Neelmeijer, J.; Rabe, D.; Scheffler, D.; Thiele, M.; Spengler, D. FERN.Lab: Bridging the gap between remote sensing academic research and society. *Remote Sens. Appl. Soc. Environ.* **2021**, *24*, 100641. [CrossRef]
36. Neumann, C. Habitat sampler—A sampling algorithm for habitat type delineation in remote sensing imagery. *Divers. Distrib.* **2020**, *26*, 1752–1766. [CrossRef]
37. ISIP e. V. ISIP—Informationssystem für die Integrierte Pflanzenproduktion. Available online: www.isip.de (accessed on 11 November 2023).

insects

MDPI

Article

High Trunk Truncation as a Potential Sustainable Management Option for Asian Longhorned Beetle on *Salix babylonica*

Chen Huang [1,†], Hualing Wang [1,2,†], Xiaoxia Hai [1], Zhigang Wang [1] and Fei Lyu [1,*]

[1] College of Forestry, Hebei Agricultural University, Baoding 071000, China; huangchenlil@163.com (C.H.); wang_hual@126.com (H.W.); moqi.hai@foxmail.com (X.H.); wzhg432@163.com (Z.W.)
[2] Hebei Urban Forest Health Technology Innovation Center, Baoding 071000, China
* Correspondence: haimolv@foxmail.com; Tel.: +86-03127520216
† These authors contributed equally to this work.

Simple Summary: The Asian longhorned beetle (ALB), *Anoplophora glabripennis* Motschulsky, is a serious wood borer of hardwood trees in North America, Europe, and China, causing substantial economic and ecological losses. Several ALB management strategies have been applied; however, certain drawbacks have been identified with these strategies. To explore effective and environmentally sustainable management options, we investigated the biological characteristics of ALB, including the distribution of frass and emergence holes on trees and preferred branches for ALB foraging and ovipositing. The results showed that 76.31–78.88% of frass holes and 85.08–87.93% of emergence holes were located above a height of 200 cm. Adults showed a preference for feeding on branches 2–3 cm in diameter, and eggs were predominantly laid on 5 cm branches, primarily located above a height of 200 cm. These findings suggest a correlation between the number of ALBs and the height of the tree crown. To test this hypothesis, we investigated whether the number of ALBs decreased when the tree crown was removed above 200 cm (high trunk truncation, HTT). The results revealed a significant decrease in the number after the implementation of HTT. Based on our results we advocate for HTT as an effective ALB management strategy, offering valuable insights into the development of a sustainable approach to controlling the number of ALBs.

Abstract: The Asian longhorned beetle (ALB) causes substantial economic and ecological losses, thus, an environmentally friendly management strategy is needed. Here, we propose high trunk truncation (HTT), the removal of the above 200 cm portion of trees, as a sustainable management strategy to control ALB. To examine the hypothesis, an initial step involved the assessment of various biological characteristics of ALB. Subsequently, a controlled field experiment was carried out utilizing HTT. Finally, HTT was applied in two additional ALB infestation regions. The results of the study of the biological characteristics of ALB showed that 76.31–78.88% of frass holes and 85.08–87.93% of emergence holes were located on branches above 200 cm. Adults preferred to feed on branches 2–3 cm in diameter, ALB eggs were predominantly laid on 5 cm branches, and both were primarily located above 200 cm. These results revealed a correlation between the number of ALBs and the tree crown height. The controlled field experiment showed that the number of ALBs was significantly decreased when the HTT strategy was implemented: approximately 90% of frass holes and 95% of adults were eradicated by HTT compared with the control. Different field surveys involving HTT have shown similar results. These findings provide valuable insights into a sustainable and efficient management strategy for reducing the number of ALBs.

Keywords: Asian longhorn beetle; invasive insects; high trunk truncation; eradication

1. Introduction

Forests and urban trees provide multiple ecosystem services for city residents, including the provision of food and water resources, as well as timber and nontimber products [1,2]. In these ecosystems, insects, as the most abundant organisms, play a key role in maintaining ecosystem health. However, certain insect species can have detrimental effects on trees, leading to reduced ecological and economic function [1]. This issue is further exacerbated by the widespread distribution of non-native insect species on a global scale [3,4]. The Asian longhorn beetle (ALB), *Anoplophora glabripennis* (Motschulsky), is one of the top 100 most destructive invasive species worldwide and has successfully established itself in various locations across North America and Europe [5,6]. Its capability to kill healthy trees results in significant economic losses for urban parks and rural forests within both native and non-native ALB ranges [5,7]. To date, the strategies employed to combat the ALB have included the eradication of infested trees [7], the application of insecticides to both infested and healthy trees [8,9], and the implementation of biological control through the utilization of natural enemies [10–12].

Between 2010 and 2021, more than 45% of eradication plans were successful, but these campaigns were labor-intensive, expensive, and time-consuming [4]. ALB infestations occurring in the Veneto region, specifically in the municipality of Cornuda, northeastern Italy, were initially identified in June 2009 [7]. In response, an extensive and intensive monitoring and eradication program was implemented immediately. From 2009 to 2020, this program involved the establishment of buffer zones covering approximately 7600 hectares, visual inspections of infested host plants or plants within the clear-cut radius, trapping protocols within buffer zones, and public awareness campaigns [7]. In contrast, a substantial 288.4 square miles of infestation area are currently under quarantine for the ALB in various locations, including Worcester County, Long Island, Clermont County, Charleston, and Dorchester counties in the United States [13]. The extensive spread of the infection poses a formidable challenge to achieving successful eradication [7]. In addition, the costs associated with eradication escalate significantly as the area expands, potentially rendering the eradication strategy economically unfeasible [7].

Trunk or soil injections and pesticide sprays with imidacloprid have been applied in China, the United States, and Japan to mitigate the population density of ALB [5,14,15]. In Beijing, China, emamectin benzoate has been injected into the trunks of trees in an infested willow forest as an ALB control measure [8]. Additionally, interventions such as inserting wooden sticks with aluminum phosphide into larval galleries and injecting organophosphate insecticides (methamidophos) into trunks have been utilized to eliminate ALB larvae [9,14]. However, many investigations have highlighted the adverse consequences of pesticide use. One notable example is the reduction in regional insect biodiversity observed in both freshwater and terrestrial systems [16]. Furthermore, the presence of toxic substances within the food web may cause bio-amplification, posing potential risks to human and animal health [16–18] and adversely impacting insect pollinators adversely [19,20].

Potential biological control agents for ALB include a variety of organisms, including entomopathogenic fungi and bacteria, parasitic nematodes, parasitoids, and predators [21]. In Japan, the entomopathogenic fungi *Beauveria brongniartii* Petch and *Metarhizium brunneum* Petch have already been developed into commercial products, while the latter, *M. brunneum*, is also available for commercial purposes within the United States [21]. These fungi exhibit a high mortality rate against ALB, and the effect of *M. brunneum* fungal infection on accelerated host death is enhanced by neonicotinoid insecticides, as demonstrated in laboratory experiments [22]. Despite their efficacy, the virulence of fungi is constrained by specific environmental temperature conditions [23,24]. To date, fourteen larval parasitoid species have been identified in China and Korea, along with two ectoparasitoids (*Spathius erythrocephalus* Wesmael and *Trigonoderus princeps* Westwood) in Europe and seven braconid parasitoids in North America, that are associated with the ALB in their native and nonnative ranges [6,21,25]. Some larval parasitoids are extreme generalists [10], attacking not only ALB and *A. chinensis* (citrus longhorn beetle, CLB) but also other woodboring pests. However, their mass release as a biocontrol agent in North America has not been considered to date [26], such as might occur with the biocontrol agent *Dastarcus helophoroides*

(Fairmaire) [27]. Overall, the abovementioned strategies for the management of ALB have certain drawbacks. Consequently, further investigations into environmentally friendly and economically viable control strategies are critical for effectively managing this pest [6].

This study aimed to develop an environmentally friendly and economically viable control strategy. Understanding the biological characteristics of insect pests is crucial for enhancing insect pest management strategies. Therefore, we examined a variety of biological characteristics, including the distribution of frass and emergence holes on the upper trunks and main branches of *Salix babylonica*, and the ALB foraging and ovipositing preference on branches of different diameters. According to the observed characteristics, we concluded that pruning tree crowns above a height of 200 cm (high trunk truncation treatment) might decrease the number of ALBs. To validate this hypothesis, we conducted a controlled field experiment to analyze the amount of ALB captured across high trunk truncation and no-high trunk truncation treatments in Mancheng, Hebei Province. To validate the efficacy of the control measures, the methodologies were implemented across various regions, and the number of ALBs was assessed by comparing the impacts of high trunk truncation and no-high trunk truncation interventions. The results supported the initial hypothesis, that high trunk truncation treatment could decrease the number of ALBs. We thus propose that high trunk truncation could serve as a viable approach for sustainably managing the ALB on *S. babylonica*.

2. Materials and Methods

2.1. Survey Sites

The survey conducted in this study included a total of three sites, each of which had pure stands of *S. babylonica*, including Mancheng and Gaoyang in Baoding City, and Lingshou in Shijiazhuang City, Hebei Province, China. In Mancheng, *S. babylonica* served as a windbreak and was located adjacent to agricultural fields in BeiXinzhuang (114.43 N, 38.43 E). In Gaoyang city (115.71 N, 38.72 E), *S. babylonica* was used as a landscape tree within a park setting, and in Lingshou (114.44 N, 38.29 E), *S. babylonica* was used as a roadway tree along the sides of roads. The diameter at breast height (DBH) of the sample trees was approximately 18 cm, and the height of the trees was about 600–700 cm.

2.2. Insects

Wild ALB males and females were obtained from *S. babylonica* trees in Mancheng, in Baoding, Hebei Province, China. The adults were maintained in rearing cages (diameter = 24 cm; height = 22 cm) at densities of six insects (three females and three males) per cage. Insects were provided with fresh *S. babylonica* branches, which were replaced every 2 d. The cages were maintained in the laboratory under controlled conditions at 25 ± 1 °C and 60 ± 10% relative humidity. A natural light/dark cycle of approximately 14 h light/10 h dark (GMT + 8) was used.

2.3. Biological Characteristics of ALB

2.3.1. Spatial Distributions of Frass and Emergence Holes

To understand the biological characteristics of the ALB, the heights of frass and emergence holes were surveyed in Gaoyang (Zhuanxu Park) and Mancheng (Beixinzhuang), which are located in Baoding City, Hebei Province, China. The data collection was conducted during the spring and summer seasons, specifically from 22 May to 4 July. Three adult observers identified together the fresh frass holes and emergence holes from the ground. The heights of the holes were measured using a telemeter rod (measurement range: 5 m; Hebei Zhufeng Instrument Equipment Co., Ltd., Handan, China) in four orientations (east, west, south, and north).

2.3.2. Bark Consumption by ALB Adults on Branches with Different Diameters

To determine foraging preferences among branches with different diameters, individual female and male beetles were provided with single branch segments of *S. babylonica* after a 12 h starvation period. Bark consumption experiments were performed in an insect cage (length = 20 cm × height = 8 cm × width = 10 cm) under controlled conditions (25 ± 1 °C and 60 ± 10% relative humidity) (Figure 1A). Branches with different diameters (0.5 cm, 1.0 cm, 2.0 cm, 3.0 cm, and 4.0 cm) were cut into 5 cm-long segments immediately before the experiment and placed in the middle of the cages containing adults. After 24 h, sulfuric acid paper was used to trace the outline of the adult bite marks on the cut branches with a pencil. Subsequently, the tracings of the bite scars were placed over scale paper to calculate the area of the bite scars, representing the area of bark consumed [28].

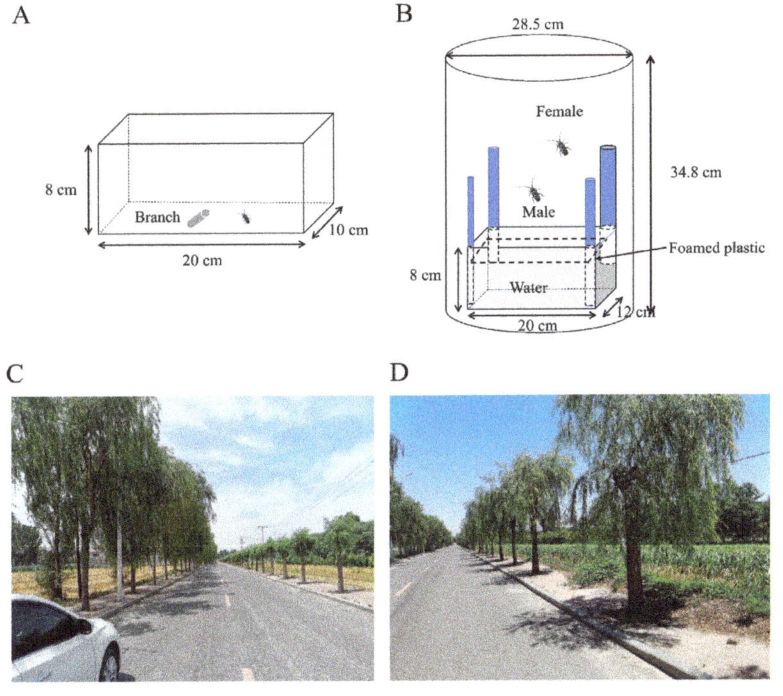

Figure 1. Schematic of the ALB adult feeding (**A**) and oviposition (**B**) behavior bioassay chamber, and the high trunk truncation for *S. babylonica* trees after 1 year (**C**) and 2 years (**D**) in Mancheng, Baoding City, Hebei Province.

2.3.3. Oviposition Selection on Branches with Different Diameters

Oviposition selection experiments were performed in a cylindrical cage (height = 34.8 cm × diameter = 28.5 cm) under controlled conditions (25 ± 1 °C and 60 ± 10% relative humidity). In this setup, four branches with different diameters (1.0 cm, 2.0 cm, 3.0 cm, and 4.0 cm) and (3.0 cm, 4.0 cm, 5.0 cm, and 6.0 cm) were randomly placed in the four corners of the rectangular acrylic box (length = 20 cm × height = 8 cm × width = 12 cm), and then the rectangular acrylic box was placed inside the cylindrical cage (Figure 1B). To secure the branches, foamed plastic was used to affix the branches in the corners of the rectangular acrylic box. Water was placed in the bottom half of the box to provide a water source for the branches, and foamed plastic obstructed the water to prevent the beetles from falling in. A mated pair of ALBs was introduced into the cage, and a 16–20 mesh nylon gauze net was placed over the cage to prevent insect escape. After 10 days, the bark of the branches was removed, and the numbers of eggs on branches with different diameters were recorded. Ten replicates were measured, respectively.

2.4. Control Effectiveness of High Trunk Truncation for ALB in a Controlled Field Experiment

To validate the hypothesis that high trunk truncation treatment could decrease the number of ALBs, approximately 80 *S. babylonica* tree crowns were removed on one side of the road in spring in Mancheng (Figure 1C,D). Conversely, the *S. babylonica* trees on the opposite side of the road were left untreated to serve as the control group (Figure 1C,D). Approximately 30 trees were randomly selected to investigate the number of frass holes per tree within the treatment and control groups in May, one year after high trunk truncation treatments. Additionally, the number of adults per tree was examined on three separate occasions, separated by seven-day intervals, between June and July. The detection of ALB adults was carried out by three adult observers on the ground. The beetles located on the branches below 200 cm were manually captured, and those on branches above 200 cm were poked with fishing poles.

2.5. Control Effectiveness of High Trunk Truncation for ALB in Different Areas

To provide additional evidence regarding the efficiency of high trunk truncation treatment in controlling ALB, the practice of high trunk truncation treatment was implemented in another two regions affected by ALB infestation. These regions were Lingshou and Gaoyang, Hebei Province. The total captured beetles before and after high trunk truncation treatments were analyzed in 2014–2018 in Lingshou, Hebei Province. In the spring of 2016, the Landscape Management Bureau in Lingshou, Hebei Province conducted a high trunk truncation procedure on *S. babylonica* trees. In 2014–2018, the total number of beetles captured per collection was recorded during the collection of test insects, with six collection times per year for both 2014 and 2015 and five collection times per year for 2016–2018. Twenty adjacent sample trees were compiled into a group, and then four groups were selected randomly and surveyed each year. The distance between the two groups was at least 20 m. Insects were collected from *S. babylonica* trees at 7–15-day intervals in June–August. In Gaoyang, approximately 40 *S. babylonica* trees were randomly selected to investigate the number of adults per tree in the high trunk truncation treatment and no-high trunk truncation treatment (control) on 29 July and 4 August 2020, respectively.

2.6. Statistical Analysis

All the statistical analyses were performed using SPSS Statistics v. 21.0 (IBM Corp., Armonk, NY, USA) for Windows. To assess the efficacy of high trunk truncation in *S. babylonica* trees as a control measure for the ALB, the total number of captured ALB adults per time at Lingshou in 2014–2018 was assessed with a Generalized linear model (GLM) with a Poisson distribution and a log link function, followed by Bonferroni's test ($\alpha = 0.05$). The numbers of frass holes and adults per tree at Mancheng and Gaoyang were sqrt(1 + x)-transformed to meet the assumptions of Gaussian distribution and homoscedasticity, followed by comparisons using independent-sample *t*-tests. Each measured height was regarded as a data point to calculate the distribution percentages of different height regions in the total number of frass and emergence holes. Then, the distribution percentages of the different height regions in each orientation were regarded as a sample for analyzing the spatial distribution difference of frass and emergence holes using one-way analysis of variance (ANOVA), followed by Bonferroni's test ($\alpha = 0.05$). The feeding areas in branches with different diameters were log(1 + x)-transformed to meet the assumptions of Gaussian distribution and homoscedasticity and then compared using ANOVA. The numbers of eggs in branches with different diameters were compared with GLM with a Poisson distribution and a log link function, followed by Bonferroni's test ($\alpha = 0.05$).

3. Results

3.1. Biological Characteristics of ALB

3.1.1. Spatial Distributions of Frass and Emergence Holes

To elucidate the spatial distribution of frass and emergence holes on *S. babylonica* trees, we examined 1047 frass holes and 521 emergence holes in no-pruning tree crowns in

Gaoyang and Mancheng, Hebei Province, China (Figures 2 and 3). The frass holes of the ALB larvae were mainly distributed in the *S. babylonica* trees above 200 cm (Figure 2A,B), accounting for approximately 76.30% and 78.88% of the total at Gaoyang and Mancheng, respectively (Figure 2C,D). Within this range, approximately 37.62% and 45.64% of the frass holes were distributed between 201 and 300 cm at Gaoyang and Mancheng, respectively. Frass holes were also present between 301 and 400 cm, making up approximately 32.17% and 29.36% at Gaoyang and Mancheng, respectively (Figure 2C; F = 23.469, $df = 4$, $p < 0.001$ and Figure 2D, F = 46.231, $df = 4$, $p < 0.001$).

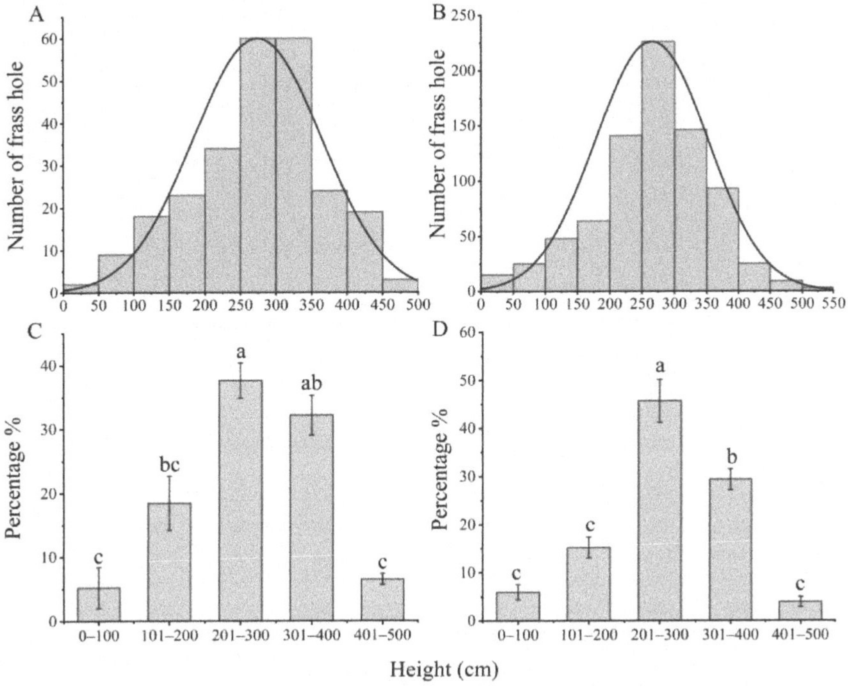

Figure 2. Height distribution conditions of the ALB frass holes in Gaoyang (**A**,**C**) and Mancheng (**B**,**D**), in Hebei Province, China. Count statistics of frass holes at Gaoyang (**A**) and Mancheng (**B**) at different heights. Distribution percentages of frass holes at Gaoyang (**C**) and Mancheng (**D**) at different heights; the percentage indicates the proportion of frass holes at different height ranges of the total number of frass holes. The sample sizes at Gaoyang and Mancheng were 252 and 795, respectively. Differences in the frass holes found in branches of varying diameters were assessed with a one-way analysis of variance, followed by posthoc Bonferroni's test, $\alpha = 0.05$. Different letters in the bar indicate that the data are significantly different at $\alpha < 0.05$.

ALB emergence holes were predominantly located in *S. babylonica* trees above a height of 200 cm (Figure 3A,B), accounting for approximately 87.93% and 85.08% of the total holes at Gaoyang and Mancheng, respectively (Figure 3C,D). The majority of emergence holes, about 73.84% and 33.08%, were distributed between 200 and 300 cm at Gaoyang and Mancheng, respectively. Moreover, approximately 14.09% and 34.09% of the emergence holes were distributed in the 301–400 cm layer (Figure 3C; F = 140.591, $df = 3$, $p < 0.001$ and Figure 3D, F = 27.526, $df = 5$, $p < 0.001$).

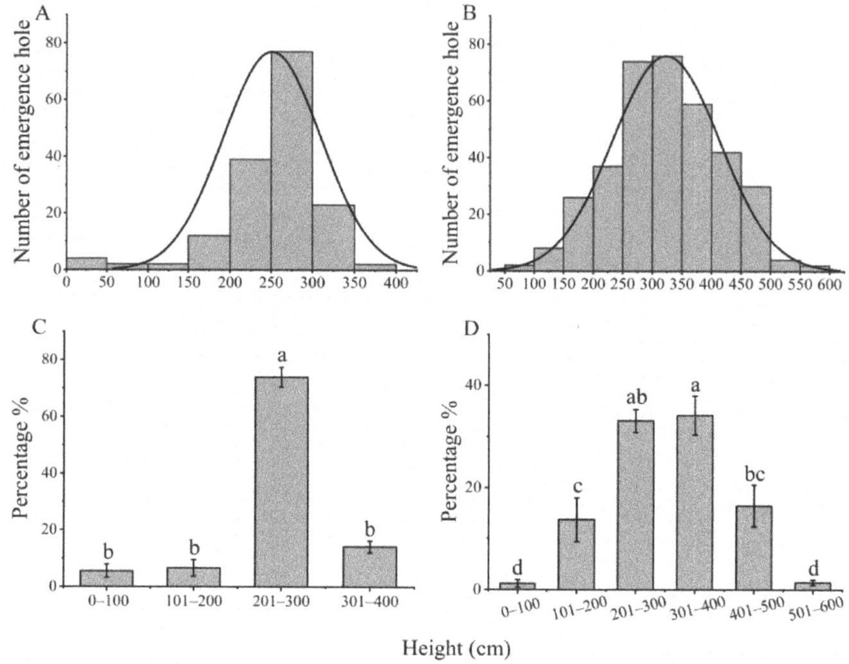

Figure 3. Height distributions of the ALB emergence holes in Gaoyang (**A**,**C**) and Mancheng (**B**,**D**), Hebei Province, China. Count statistics of emergence holes at Gaoyang (**A**) and Mancheng (**B**) at different heights. Distribution percentages of emergence holes at Gaoyang (**C**) and Mancheng (**D**) at different heights. The percentage indicates the relative distribution of emergence holes across various height ranges in relation to the overall number of emergence holes. The sample sizes were 161 and 360 at Gaoyang and Mancheng, respectively. Differences in the emergence holes in branches of different diameters were assessed with a one-way analysis of variance, followed by posthoc Bonferroni's test, $\alpha = 0.05$. The different letters in the bar indicate that the data are significantly different $\alpha < 0.05$.

3.1.2. Bark Consumption by ALB Adults on Branches with Different Diameters

After being incubated with the branches of *S. babylonica* for a period of 24 h, both the male and female feeding areas displayed significantly larger dimensions, specifically measuring 2.0 cm and 3.0 cm in diameter, in comparison to the diameters of other branches (Figure 4; $F_{male} = 6.311$, $p < 0.001$; $F_{female} = 13.693$, $p < 0.001$; all $df = 4$). There were no significant differences in the feeding areas of either males or females among *S. babylonica* branches with diameters of 2.0 cm, 3.0 cm, or 4.0 cm, but the feeding area was significantly greater than that of branches with diameters of 0.5 cm or 1.0 cm.

3.1.3. Oviposition Selection on Branches with Different Diameters

Significant differences were observed in the average number of eggs on branches with different diameters (Figure 5A: $X^2 = 58.000$, $df = 3$, $p < 0.001$; Figure 5B: $X^2 = 32.795$, $df = 3$, $p < 0.001$). The average numbers of eggs per 10 days were 0, 0.70 ± 0.37, 1.00 ± 0.37, and 4.1 ± 0.35 on branches with diameters of 1 cm, 2 cm, 3 cm, and 4 cm, respectively. The number of eggs laid per female was found to be larger on the branches with a diameter of 4 cm compared to on the branches with other diameters. However, no significant difference was observed in the number of eggs between branches with diameters of 2 cm and 3 cm (Figure 5A). On the branches with 3 cm, 4 cm, 5 cm, and 6 cm diameters, the mean numbers of eggs were 0.50 ± 0.22, 1.30 ± 0.36, 4.10 ± 0.64, and 1.90 ± 0.44, respectively. The number of eggs per female on branches with a 5 cm diameter was also greater than that on branches of other diameters (Figure 5B).

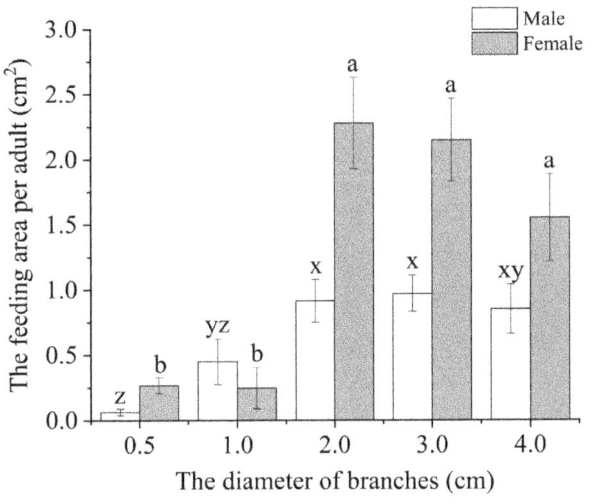

Figure 4. Feeding areas of individual male and female adults per 24 h period on single *S. babylonica* branches. The sample sizes for males and females were 19, 19, 21, 20, and 20 and 19, 19, 18, 21, and 18 for *S. babylonica* branches with diameters of 0.5, 1.0, 2.0, 3.0, and 4.0 cm, respectively. The data are presented as the means ± SEs. Differences in the feeding areas of adults in branches of different diameters were assessed with a one-way analysis of variance, followed by posthoc Bonferroni's test, $\alpha = 0.05$. The different letters in the bar indicate that the data are significantly different.

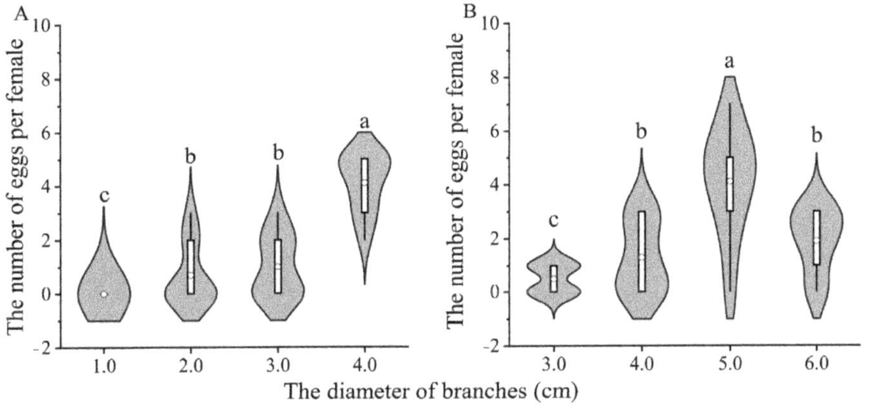

Figure 5. Number of eggs per pair of adults on *S. babylonica* branches in different treatment groups, 1.0, 2.0, 3.0 and 4.0 cm diameter of branches (**A**) and 3.0, 4.0, 5.0 and 6.0 cm diameter of branches (**B**). Differences in the number of eggs laid by a female ALB on branches with different diameters were assessed with a GLM link passion distribution, and pairwise comparisons were conducted with Bonferroni's test, $\alpha = 0.05$. Different letters in the violin figure indicate that the data are significantly different. The number of samples was 10.

Based on the aforementioned characteristics, it was deduced that pruning the tree crown beyond 200 cm could lead to a reduction in the number of ALBs. In order to verify this hypothesis, a series of field experiments were conducted.

3.2. Control Effectiveness of High Trunk Truncation for ALB in a Controlled Field Experiment

To validate the hypothesis that high trunk truncation treatment could decrease the number of ALBs, a controlled field experiment was conducted to compare the number of frass holes and adults between the high trunk and non-high trunk truncation groups in Mancheng (Figures 1C,D and 6A). The results showed that the average number of frass holes per tree was approximately 0.53 ± 0.12 when the tree crown underwent high trunk truncation, in contrast to 5.70 ± 0.25 per tree in the absence of high trunk truncation; approximately 90% [(5.70 − 0.53)/5.70] of the frass holes were eradicated by high trunk truncation treatment compared with control. Significantly, the number of frass holes per tree differed between the high trunk and non-high trunk truncation groups (Figure 6B; $t = 20.208$, $df = 58$, $p < 0.001$). Additionally, the number of adults per tree exhibited a significant difference between the high trunk and noon-high trunk truncation treatments (Figure 6C; $5.129 < t < 7.691$, $df = 62$, $p < 0.001$). Specifically, the average number of adults per tree ranged from approximately 0.06 ± 0.06 to 0.09 ± 0.05 when the tree crown underwent high trunk truncation, while it varied from 1.25 ± 0.24 to 1.72 ± 0.23 per tree in the non-high trunk truncation treatment (Figure 6C); approximately 95% [(1.72 − 0.09)/1.72] of the adults specimens were eradicated by high trunk truncation treatment compared with the control.

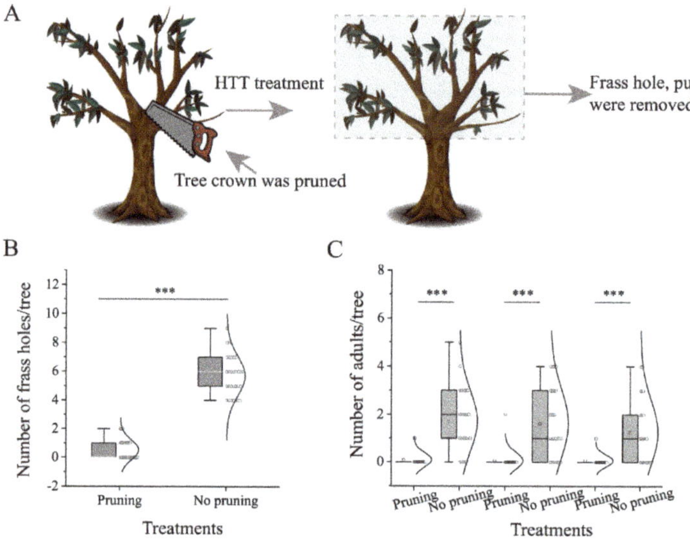

Figure 6. Schematic of high trunk truncation for *S. babylonica* (**A**), number of fresh frass holes (**B**), and number of adults (**C**) per tree in the high trunk truncation and non-high trunk truncation treatments in Mancheng. HTT: high trunk truncation. The upper and lower end points of the vertical lines and the lines of the upper and lower edges of the box indicate the maximum and minimum and 25% and 75%, respectively; the square indicates the mean, and the line indicates the median in the box, the same as below. *** $p < 0.001$ by independent-samples t-test. Thirty trees per treatment were investigated in the different treatments in the frass hole experiments, and 32 trees per treatment were investigated in the different treatments in the number of adult experiments.

3.3. Control Effectiveness of High Trunk Truncation for ALB in Different Areas

To ascertain the effectiveness of the control measures, the methodologies were implemented in different regions (Lingshou and Gaoyang), and the number of ALBs was evaluated through a comparison of the impacts of high trunk truncation and non-high trunk truncation interventions. In the years 2014–2015, the recorded total number of captured ALB adults in Lingshou, Hebei Province, was 132.83 ± 2.76 and 131.67 ± 7.71 per 80 trees, respectively (Figure 7A). Following the high trunk truncation of *S. babylonica*, a significant reduction in the number of captured beetles occurred, with 28.40 ± 2.46, 37.75 ± 3.78, and 37.00 ± 1.14 beetles captured per 80 trees in 2016, 2017, and 2018, respectively (Figure 7A). Significant differences were observed in the total number of captured beetles across different years, with the total number in 2014 and 2015 exceeding that in 2016–2018 (Figure 7A; $X^2 = 734.055$, $df = 4$, $p < 0.001$). In Gaoyang, the average number of adults per tree was approximately 0.37 ± 0.13–0.43 ± 0.13 when the tree crown underwent high trunk truncation, whereas it was 0.93 ± 0.15–1.15 ± 0.17 per tree in the absence of high trunk truncation. The number of adults per tree differed significantly between the high-trunk and non-high-trunk truncation treatments (Figure 7B; $2.714 < t < 3.795$, $df = 78$, $p < 0.008$).

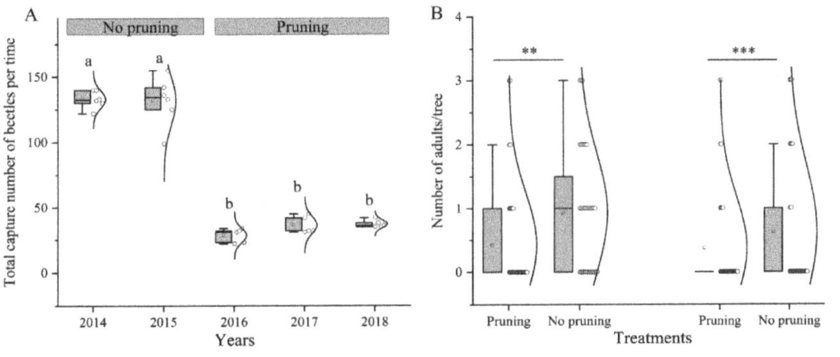

Figure 7. Total number of captured ALBs per time in Lingshou in 2014–2018 (**A**) and the number of adults per tree in Gaoyang in 2020 (**B**). Differences in the number of ALBs were assessed with a GLM linked to a Poisson distribution, followed by Bonferroni's test, $\alpha = 0.05$ in Lingshou in 2014–2018. The same letter above the box indicates no significant difference in the number of captured beetles among the different years. Test samples were collected six times per year in both 2014 and 2015, and five times per year in 2016, 2017, and 2018. ** $p < 0.01$, *** $p < 0.001$ by independent-samples *t*-test (n = 40).

4. Discussion

The ALB is the most serious nonnative invasive species in Europe and North America [5]. Its threat is considerable, as it can target 209 species or cultivars of healthy trees, out of which 101 species exhibit a higher susceptibility, resulting in substantial economic and ecological losses [4,6,15,29,30]. Although eradication programs have achieved a success rate of more than 45%, the task of addressing larger infestation areas continues to present a significant challenge [4]. In our current investigation, a noteworthy revelation emerged: the application of high trunk truncation treatments led to the removal of 90% of the frass holes and 95% of the adults in the initial year, compared to the non-high trunk truncation treatment (Figure 6B,C). This discovery has promising implications, suggesting that the management of the ALB may not rely solely on complete eradication but could also involve the strategic pruning of the tree crowns of trees that tolerate it and will regrow. Such an approach has the potential to reduce the number of ALBs.

The successful eradication of the ALB involves a multifaceted approach, including the early detection and identification of infested plants [7,31]. However, challenges are encountered in terms of the efficacy of these eradication measures, especially in larger infestation areas. The costs associated with eradication efforts escalate as the eradication area expands, rendering the operation economically unfeasible in extensively infested regions, particularly when dealing with insects exhibiting pronounced polyphagia [7,32]. For example, a region exceeding 20,000 ha in Worcester (Massachusetts, United States) was identified as being infested, prompting the monitoring of over 5 million trees and the subsequent removal of approximately 34,000 trees by 2015 [7,33]. Therefore, successful eradication faces a significant challenge when confronting extensive areas of infestation.

To develop an environmentally friendly and economically viable control strategy, it is essential to understand the biological characteristics of insect pests. Over the past decade, the primary focus in combating wood-boring pests has consistently been on eradication methods, particularly through the removal of their preferred hosts based on their biological characteristics [4,7,34]. For example, the approach of removing preferred hosts has been used for the management of ALB and CLB, which rank among the most serious invasive species that pose significant threats to forests and urban trees in both North America and Europe. Nevertheless, significant differences can be observed in the biological characteristics of these two *Anoplophora* species. Notably, CLB exhibits a different pattern, with oviposition primarily occurring on the lower trunk, root collar region, and exposed roots, and larval development concentrated in the lower trunk and roots [4]. Thus, the complete removal of infected trees is warranted for CLB, given the prevalence of oviposition and larval development on the lower trunk and in the root collar region. In contrast, the ALB typically engages in oviposition and larval development on the upper trunk and main branches [4]. This finding is consistent with our findings, which indicate that the frass and emergence holes, as well as the foraging and oviposition preference of the ALB, are predominantly concentrated in tree trunks above a height of 200 cm (Figures 2–5). Therefore, a viable alternative to complete plant removal could be the removal of the upper 200 cm portion of the tree (High trunk truncation) for ALB management. This recommendation is supported by our findings, which indicated that 76.31–78.88% of the frass holes and 85.08–87.93% of the emergence holes were located on the trunks of trees above 200 cm (Figures 2 and 3).

When emamectin benzoate trunk injections were used to control ALB in an infested willow forest in China, an impressive 89% reduction in the ALB larval population was observed during the first spring after application [8]. Similarly, the implementation of high trunk truncation targeting the control ALB demonstrated significant outcomes, as evidenced by the elimination of 90% of the frass holes and 95% of the adults within the initial year following treatment, in contrast to the absence of pruning intervention (Figure 6). The effectiveness of chemical and high trunk truncation is equivalent in terms of the reduction in the number of adults following treatment. Therefore, the high trunk truncation strategy could serve as an effective alternative for ALB, potentially replacing chemical pesticide applications, as suggested by Wang [35]. Furthermore, the previous study showed that ALB is primarily observed within windbreaks adjacent to agricultural landscapes [4,5]. Reducing chemical pesticide use in windbreaks adjacent to agricultural landscapes can significantly increase the abundance of *Propylea japonica* (Thunberg) and *Harmonia axyridis* (Pallas), which are natural enemies and so enhance the biocontrol function [36]. Honeydew, a carbohydrate-rich secretion produced by Hemipteran insects, serves as a significant source of carbohydrates for beneficial insects. However, the presence of insecticides in honeydew leads to a decrease in the number of numerous beneficial insects [17]. Therefore, reducing the use of insecticides to control the ALB may also maintain beneficial insects, increasing the biodiversity of farmland and the biocontrol efficiency of natural enemies.

Furthermore, the presence of Allee effects presents a substantial and persistent limitation for populations with low-density [37]. The control experiments showed that 90% of the frass holes and 95% of the adults were eliminated by high trunk truncation within the initial year following treatment. As a result, management strategies that prove effective may prioritize reducing the population below the Allee threshold instead of attempting to eliminate every individual [31]. According to Branco et al. [4], certain countries may prioritize the implementation of containment measures if eradication efforts continue to yield success over the next decade. Consequently, the adoption of the high trunk truncation management strategy could potentially emerge as a prominent choice for controlling the ALB in the coming years.

In summary, the high trunk truncation strategy significantly reduced the number of ALBs on the tree *S. babylonica* in three different experiment sites. Compared to alternative management methods, this approach offers several advantages for ALB control. First, the number of ALBs showed a substantial decrease in the case of the high trunk truncation treatment compared to the control. Notably, after one year of high trunk truncation in Mancheng, 90% of the frass holes and 95% of the adults were removed, surpassing the control treatment group. Second, high trunk truncation acts as a preventive measure against strong winds breaking or uprooting ALB-infested trees, mitigating the severe damage caused by fallen trees. Third, the biomass of trees can rapidly recover within approximately 3–5 years, as exemplified in fast-growing trees such as *Salix*, *Populus*, and other cultivated species. Moreover, the trees that have undergone high trunk truncation treatment in China include *Platanus orientalis*, *Poplar*, *Willow*, *Acer negundo*, *Platanus orientalis*, and *Sophora japonica* (Figure S1). The high trunk truncation strategy may be useful for managing ALB on the above-mentioned trees. Finally, the high trunk truncation strategy might minimally impact the habitats of other arthropods while concurrently reducing the use of pesticides. Given these advantages, we recommend the adoption of the high trunk truncation strategy as a practical approach for controlling the ALB population in other tree species or regions.

Supplementary Materials: The following supporting information can be downloaded at: https://www.mdpi.com/article/10.3390/insects15040278/s1, Figure S1: Different tree species managed by high trunk truncation in China. A: Pruned crown of *Platanus orientalis*. B: Pruned crown of *Poplar*. C: Pruned crown of *Willow*. D: *A. negundo* 3 years after high trunk truncation. E: *P. orientalis* after high trunk truncation in the summer of the current year. F: *S. babylonica* 7 years after high trunk truncation. G: *S. babylonica* were managed by high trunk truncation in the Mancheng. H: Pruned crown of *Sophora japonica*.; Table S1: The bark thickness of different diameter of *S. babylonica*.

Author Contributions: Conceptualization, F.L., H.W. and Z.W.; methodology, F.L.; formal analysis, C.H., F.L. and X.H.; investigation, C.H., F.L. and X.H. writing—original draft preparation, C.H., F.L., X.H. and H.W.; writing—review and editing, C.H., F.L., X.H. and H.W.; visualization, F.L., X.H. and H.W.; supervision, F.L.; project administration, F.L.; funding acquisition, F.L. and Z.W. All authors have read and agreed to the published version of the manuscript.

Funding: This research was funded by the Natural Science Foundation of China (No. 32171800), the Natural Science Foundation of Hebei Province of China, grant number C2019204276, the Basic Scientific Research of Universities in Hebei Province of China, grant number KY202008; KY2022048.

Data Availability Statement: All data supporting the findings of this study are available within the article. Source data are provided with this paper after publication.

Acknowledgments: We thank Hebei Agricultural University (Wang Pengwen, Feng Zhao, Kaixing Hai, Xiaoling Kang, and Yanjun He) for field assistance and insect collection. We also thank the Natural Science Foundation of China (No. 32171800), the Hebei Province Natural Science Foundation of China (No. C2019204276), and the Basic Scientific Research of Universities in Hebei Province of China (No. KY202008; KY2022048).

Conflicts of Interest: The authors declare no conflicts of interest. The funders had no role in the design of the study; in the collection, analyses, or interpretation of data; in the writing of the manuscript; or in the decision to publish the results.

References

1. Allison, J.D.; Redak, R.A. The impact of trap type and design features on survey and detection of bark and woodboring beetles and their associates: A review and meta-analysis. *Annu. Rev. Entomol.* **2017**, *62*, 127–146. [CrossRef]
2. Elisabeth, C.B.; Erik, G.B. Ecosystem services from urban forests: The case of Oslomarka, Norway. *Ecosystem Serv.* **2021**, *51*. [CrossRef]
3. Rogosch, J.S.; Olden, J.D. Comparing opportunistic and strategic removal efforts to manage invasive fish species using a dynamic multi-state occupancy model. *J. Appl. Ecol.* **2021**, *58*, 2797–2809. [CrossRef]
4. Branco, S.; Faccoli, M.; Brockerhoff, E.G.; Roux, G.; Jactel, H.; Desneux, N.; Gachet, E.; Mouttet, R.; Streito, J.C.; Branco, M. Preventing invasions of Asian longhorn beetle and citrus longhorn beetle: Are we on the right track? *J. Pest. Sci.* **2022**, *95*, 41–66. [CrossRef]
5. Haack, R.A.; Herard, F.; Sun, J.; Turgeon, J.J. Managing invasive populations of Asian longhorned beetle and citrus longhorned beetle: A worldwide perspective. *Annu. Rev. Entomol.* **2010**, *55*, 521–546. [CrossRef]
6. Golec, J.R.; Li, F.; Cao, L.; Wang, X.; Duan, J.J. Mortality factors of *Anoplophora glabripennis* (Coleoptera: Cerambycidae) infesting *Salix* and *Populus* in central, northwest, and northeast China. *Biol. Control* **2018**, *126*, 198–208. [CrossRef]

7. Marchioro, M.; Faccoli, M. Successful eradication of the Asian longhorn beetle, *Anoplophora glabripennis*, from North-Eastern Italy: Protocol, techniques and results. *Insects* **2021**, *12*, 877. [CrossRef]
8. Wang, J.; Che, S.; Qiu, L.; Li, G.; Shao, J.; Zhong, L.; Zhang, G.; Xu, H. Efficacy of emamectin benzoate trunk injection against the Asian Long-Horned Beetle [*Anoplophora glabripennis* (Coleoptera: Cerambycidae)]. *J. Econ. Entomol.* **2020**, *113*, 340–347. [CrossRef]
9. Wang, B.D.; Gao, R.T.; Mastro, V.C.; Reardon, R.C. Toxicity of four systemic neonicotinoids to adults of *Anoplophora glabripennis* (Coleoptera: Cerambycidae). *J. Econ. Entomol.* **2005**, *98*, 2292–2300. [CrossRef]
10. Wang, X.; Wang, X.Y.; Kenis, M.; Cao, L.M.; Hoelmer, K.A. Exploring the potential for novel associations of generalist parasitoids for biological control of invasive woodboring beetles. *BioControl* **2021**, *66*, 97–112. [CrossRef]
11. Wang, X.; Aparicio, E.M.; Murphy, T.C.; Duan, J.J.; Elkinton, J.S.; Gould, J.R. Assessing the host range of the North American parasitoid *Ontsira mellipes*: Potential for biological control of Asian longhorned beetle. *Biol. Control* **2019**, *137*, 104028. [CrossRef]
12. Li, F.; Zhang, Y.; Wang, X.; Cao, L.; Yang, Z.; Gould, J.R.; Duan, J.J. Discovery of parasitoids of *Anoplophora glabripennis* (Coleoptera: Cerambycidae) and their seasonal abundance in China using sentinel host eggs and larvae. *J. Econ. Entomol.* **2020**, *113*, 1656–1665. [CrossRef] [PubMed]
13. APHIS. USDA Announces Plans for Eradicating Asian Longhorned Beetle in Massachusetts, New York, Ohio, and South Carolina. Available online: https://content.govdelivery.com/accounts/USDAAPHIS/bulletins/319935b (accessed on 26 May 2022).
14. Hu, J.; Angeli, S.; Schuetz, S.; Luo, Y.; Hajek, A.E. Ecology and management of exotic and endemic Asian longhorned beetle *Anoplophora glabripennis*. *Agric. Forest Entomol.* **2009**, *11*, 359–375. [CrossRef]
15. Wang, L.; Li, C.; Luo, Y.; Wang, G.; Dou, Z.; Haq, I.U.; Shang, S.; Cui, M. Current and future control of the wood-boring pest *Anoplophora glabripennis*. *Insect Sci.* **2023**, *30*, 1534–1551. [CrossRef]
16. Beketov, M.A.; Kefford, B.J.; Schafer, R.B.; Liess, M. Pesticides reduce regional biodiversity of stream invertebrates. *Proc. Natl. Acad. Sci. USA* **2013**, *110*, 11039–11043. [CrossRef]
17. Calvo-Agudo, M.; Tooker, J.; Dicke, M.; Tena, A. Insecticide-contaminated honeydew: Risks for beneficial insects. *Biol. Rev.* **2021**, *97*, 664–678. [CrossRef]
18. Lykogianni, M.; Bempelou, E.; Karamaouna, F.; Aliferis, K.A. Do pesticides promote or hinder sustainability in agriculture? The challenge of sustainable use of pesticides in modern agriculture. *Sci. Total Environ.* **2021**, *795*, 148625. [CrossRef] [PubMed]
19. Basset, Y.; Lamarre, G. Toward a world that values insects. *Science* **2019**, *364*, 1230–1231. [CrossRef] [PubMed]
20. Siviter, H.; Bailes, E.J.; Martin, C.D.; Oliver, T.R.; Koricheva, J.; Leadbeater, E.; Brown, M.J.F. Agrochemicals interact synergistically to increase bee mortality. *Nature* **2021**, *596*, 389–392. [CrossRef]
21. Brabbs, T.; Collins, D.; Herard, F.; Maspero, M.; Eyre, D. Prospects for the use of biological control agents against *Anoplophora* in Europe. *Pest. Manag. Sci.* **2015**, *71*, 7–14. [CrossRef]
22. Fisher, J.J.; Castrillo, L.A.; Donzelli, B.G.G.; Hajek, A.E. Starvation and imidacloprid exposure influence immune response by *Anoplophora glabripennis* (Coleoptera: Cerambycidae) to a fungal pathogen. *J. Econ. Entomol.* **2017**, *110*, 1451–1459. [CrossRef] [PubMed]
23. Fisher, J.J.; Hajek, A.E. Thermoregulatory behavior and fungal infection of *Anoplophora glabripennis* (Coleoptera: Cerambycidae). *Environ. Entomol.* **2014**, *43*, 384–392. [CrossRef] [PubMed]
24. Clifton, E.H.; Jaronski, S.T.; Hajek, A.E. Virulence of commercialized fungal entomopathogens against Asian Longhorned Beetle (Coleoptera: Cerambycidae). *J. Insect Sci.* **2020**, *20*, 1–6. [CrossRef] [PubMed]
25. Wang, X.; Wang, X.Y.; Gould, J.R.; Duan, J.J.; Lee, S.; Lee, S. An update on biological control of Asian Longhorned Beetle. In Proceedings of the ESA's Virtual Annual Meeting, Helsinki, Finland, 14 April 2020; pp. 370–382.
26. Gould, J.R.; Aflague, B.; Murphy, T.C.; McCartin, L.; Elkinton, J.S.; Rim, K.; Duan, J.J. Collecting nontarget wood-boring insects for host-specificity testing of natural enemies of Cerambycids: A case Study of *Dastarcus helophoroides* (Coleoptera: Bothrideridae), a parasitoid of the Asian Longhorned Beetle (Coleoptera: Cerambycidae). *Environ. Entomol.* **2018**, *47*, 1440–1450. [CrossRef] [PubMed]
27. Lyu, F.; Hai, X.X.; Wang, Z.G.; Liu, B.X.; Yan, A.H.; Bi, Y.G. Research progress in *Dastarcus helophoroides* Fairmaire (Bothrideridae), an important natural enemy of the longhorn beetle pests. *Sci. Seric.* **2014**, *40*, 1107–1113. [CrossRef]
28. Lyu, F.; Hai, X.X.; Wang, Z.G. Green-colored paperboard enhances the Asian longhorned beetle response to host plant odor cues. *J. Pest. Sci.* **2021**, *94*, 1–11. [CrossRef]
29. Lyu, F.; Hai, X.; Wang, Z. A review of the host plant location and recognition mechanisms of Asian longhorn beetle. *Insects* **2023**, *14*, 292. [CrossRef] [PubMed]
30. Pedlar, J.H.; Mckenney, D.W.; Denys, Y.; Hope, E.S. Potential economic impacts of the Asian Longhorned Beetle (Coleoptera: Cerambycidae) in Eastern Canada. *J. Econ. Entomol.* **2020**, *113*, 839–850. [CrossRef] [PubMed]
31. Liebhold, A.M.; Berec, L.; Brockerhoff, E.G.; Epanchin-Niell, R.S.; Hastings, A.; Herms, D.A.; Kean, J.M.; McCullough, D.G.; Suckling, D.M.; Tobin, P.C.; et al. Eradication of invading insect populations: From concepts to applications. *Annu. Rev. Entomol.* **2016**, *61*, 335–352. [CrossRef]
32. Tobin, P.C.; Kean, J.M.; Suckling, D.M.; McCullough, D.G.; Herms, D.A.; Stringer, L.D. Determinants of successful arthropod eradication programs. *Biol. Invasions* **2014**, *16*, 401–414. [CrossRef]
33. Rhonda, S.; Suzanne, B. USDA Announces 2021 Plans for Asian Longhorned Beetle Eradication Efforts in Massachusetts, New York, Ohio and South Carolina. Available online: https://www.aphis.usda.gov/aphis/newsroom/news/sa_by_date/sa-2021/alb-eradication-efforts (accessed on 2 September 2021).

34. Eyre, D.; Barbrook, J. The eradication of Asian longhorned beetle at Paddock Wood, UK. *CABI Agric. Biosci.* **2021**, *2*, 12. [CrossRef]
35. Wang, Q. *Ceramycidae of the World: Biology and Pest Management*; CRC Press: Taylorand Francis, FL, USA, 2017; pp. 306–309.
36. Dong, Z.; Ouyang, F.; Lyu, F.; Ge, F. Shelterbelts in agricultural landscapes enhance ladybeetle abundance in spillover from cropland to adjacent habitats. *BioControl* **2015**, *60*, 351–361. [CrossRef]
37. Stephens, P.A.; Sutherland, W.J. Consequences of the Allee effect for behaviour, ecology and conservation. *Trends Ecol. Evol.* **1999**, *14*, 401–405. [CrossRef] [PubMed]

insects

MDPI

Review

Impact of Abiotic and Biotic Environmental Conditions on the Development and Infectivity of Entomopathogenic Nematodes in Agricultural Soils

Joanna Matuska-Łyżwa *, Sandra Duda, Dominika Nowak and Wiesław Kaca

Department of Microbiology, Institute of Biology, Faculty of Natural Sciences, Jan Kochanowski University in Kielce, 7 Uniwersytecka St., 25-406 Kielce, Poland; sandraduda24@onet.pl (S.D.); domi334@o2.pl (D.N.); wieslaw.kaca@ujk.edu.pl (W.K.)

* Correspondence: joanna.matuska-lyzwa@ujk.edu.pl

Simple Summary: Entomopathogenic nematodes (EPNs) are beneficial soil fauna that participate in the food chain. Natural populations of these animals keep insect numbers at a certain level and provide food for other groups of invertebrates. EPNs are also used as biological pest control agents. The life and effectiveness of EPNs are affected by many biotic and abiotic factors of the soil environment. Our review describes characteristics and directions of the effects of various environmental factors on the biology of EPNs.

Abstract: Many organisms, including beneficial entomopathogenic nematodes (EPNs), are commonly found in the soil environment. EPNs are used as biopesticides for pest control. They have many positive characteristics and are able to survive at sites of application for a long time, producing new generations of individuals. The occurrence of populations depends on many environmental parameters, such as temperature, moisture, soil texture, and pH. Extreme temperatures result in a decrease in the survival rate and infectivity of EPNs. Both high humidity and acidic soil pH reduce populations and disrupt the biological activity of EPNs. Nematodes are also exposed to anthropogenic agents, such as heavy metals, oil, gasoline, and even essential oils. These limit their ability to move in the soil, thereby reducing their chances of successfully finding a host. Commonly used fertilizers and chemical pesticides are also a challenge. They reduce the pathogenicity of EPNs and negatively affect their reproduction, which reduces the population size. Biotic factors also influence nematode biology. Fungi and competition limit the reproduction and survival of EPNs in the soil. Host availability enables survival and affects infectivity. Knowledge of the influence of environmental factors on the biology of EPNs will allow more effective use of the insecticidal capacity of these organisms.

Keywords: parasites of insects; soil environment; biopesticides

1. Introduction

The continuing growth of industry worldwide has resulted in an increase in environmental pollution. These developments have had a highly negative effect on beneficial soil organisms. Research carried out so far shows that it is mainly heavy metals, petroleum derivatives, and insecticides that affect these organisms. At high concentrations and after a long exposure time, they can adversely affect beneficial soil organisms.

Within the soil environment, various forms of co-existence occur among different organisms. One of them is parasitism, which is an antagonistic form of dependence. This means that one of the species that participates in the interaction loses while the other gains. We recognize as parasites species that have an intimate long-lived relationship with their host and which benefit from food absorbed from their host, which provides the place of existence for the parasite and its development, which often adversely affects the condition (including the reproductive potential) of the host. Parasitism is common in nature, and parasites include those species that are considered pathogens to plants and animals and to humans [1,2].

Entomopathogenic nematodes (EPNs) have been shown to be highly effective biological control agents with unique advantages over chemical and other biological pesticides. EPNs can kill insect pests within several days of infestation. Tens to hundreds of thousands of new EPNs can be produced from a single infested insect host and then travel to infest new hosts, thus theoretically allowing EPNs to persist in the field throughout the growing season [3,4]. EPNs are compatible with most other agrochemicals, with which they can be co-applied [5–11], and have minimal undesirable side effects on nontarget organisms. This is the case with chemical pesticides, which lead to soil contamination and bioaccumulation [4,12]. Despite these advantages, EPNs are still rarely used in agriculture due to costs and the sensitivity of EPNs to environmental factors [13].

The article aims to characterize the use of entomopathogenic nematodes (EPNs) for biological pest control, present the development and parasitism strategies of EPNs, describe the optimal environmental conditions for EPNs and their interspecies relationships, and present the effects of soil contaminants on EPNs (heavy metals, petroleum substances, pesticides, insecticides, and anthropogenic activities).

2. Entomopathogenic Nematodes (EPNs) and Biological Pest Control

Entomopathogenic nematodes are a group of parasitic organisms that are part of the natural soil mesofauna and are a factor that naturally controls the density of insect populations. EPNs are also an example of another interesting relationship between organisms—mutualism—that is beneficial to both coexisting species, which are dependent on each other to such an extent that, under natural conditions, they could not exist separately. EPNs exist in this particular kind of symbiosis with the bacteria to which they owe their ability to quickly kill an infested insect [14]. Using the right species of nematodes, pests of various crops can be controlled. This method is an alternative to the use of chemical plant protection products, enabling a reduction in the use of hazardous substances in agriculture or horticulture. Moreover, the isolation of highly virulent EPN strains from local soils removes the need for the introduction of foreign organisms, thus avoiding potential regulatory constraints [15,16]. In addition, EPNs pose no threat to warm-blooded organisms and nontarget organisms; they only attack target organisms [17,18].

Currently, preparations based on insecticidal nematodes are directed against thrips (Thysanoptera), diptera (Diptera), beetles (Coleoptera), and butterflies (Lepidoptera). Research on the effectiveness of EPNs against various pest species has been conducted under field and laboratory conditions [19–27]. One of the advantages of using preparations containing these living organisms is that, following a single application, EPN populations can survive at the application site for a long time, producing new generations. Sources say that the duration of action of the first generation introduced into the soil is 6 to as long as 16 weeks. Due to the short development cycle, this is enough time to produce a significant number of offspring. Also important is the fact that, during the process of preparing a biopreparation for spraying, nematode mortality is negligible. As a result, the product does not lose quality, which increases the chances of effective control of crop pests [28–30]. Unfortunately, EPN preparations have shortcomings in terms of biological pest control. The main ones are, primarily, the high level of host selectivity exhibited by EPNs, as well as the possibility of their encountering natural enemies in the soil, such as other nematodes or fungi [31]. Commercially available EPN preparations containing infective juveniles (modified third-stage free-living juveniles) are produced in bioreactors, and biopreparations are placed in the soil using tractor or knapsack sprayers, although nematodes can also be introduced into the soil by manual watering for home garden use. Infective juveniles can be passed through spray tubes with a diameter of at least 500 μm and are capable of withstanding pressure levels of up to 2000 kPa [14]. Commercial preparations of nematode juveniles also contain components that improve the persistence of these organisms. An example is the addition of gelling substances that protect nematodes from drying out. Due to these additives, EPNs can perform a protective function for plants for much longer. Gel formulations allow the application of doses of EPNs lower than those commonly used to control underground pests. EPN–gel applications increase crop yields [5,32–36].

3. Development and Strategies of Entomopathogenic Nematode Parasitism

Entomopathogenic nematodes probably evolved from free-living nematodes which fed on bacteria [37]. By 2020, more than 100 species of nematodes had been assigned to the *Steinernema* genus (Rhabditida: Steinermatidae), while 21 species had been assigned to the *Heterorhabditis* genus (Rhabditida: Heterorhabditidae) [38]. Each species of nematode coexists with only one species of bacteria, but the same bacterial species can live in symbiosis with different species of EPNs. Each *Steinernema* species is mutualistic with only one *Xenorhabdus* species, but some *Xenorhabdus* species are symbionts of multiple *Steinernema* species. *Steinernema feltiae* Filipjev (Steinernema: Steinermatidae) is related to *Xenorhabdus bovienii* Akhurst and Boemare (Enterobacterales: Morganellaceae), while *X. bovienii* lives in symbiosis with *Steinernema affine* Bovien (Steinernema: Steinermatidae), *Steinernema kraussei* Steiner (Steinernema: Steinermatidae), and *Steinernema intermedium* Poinar (Steinernema: Steinermatidae) [39,40].

Of the different stages of EPN development, only the third-stage infective juvenile (IJ) can live in the external environment, leaving the body of the infested insect host. Therefore, juveniles at this stage are adapted to the conditions they encounter in the soil. These juveniles have a thick and resistant cuticle, and the mouth and intestines are closed [41]. Inside their body, there are fat granules, which act as food reserves, because the IJs do not absorb nutrients from the external environment [28]. Depending on the amount of accumulated fat reserves and the general condition of the juveniles, the search for a host can last for several days. The hosts of EPNs are usually insect larvae, much less often adults, in which the development of nematodes is slower [17,42,43]. The susceptibility of *Tuta absoluta* Meyrick (Lepidoptera: Gelechiidae) adults has been shown by Batalla-Carrera et al. (2010) in preliminary experiments with continuous exposure to EPNs in laboratory conditions [44]. The juveniles at this stage are called infective because they actively search for a host, and, when they encounter a suitable insect host, they penetrate inside it via either the cuticle or natural orifices (e.g., the mouth, anus, or spiracles) of the body [45].

The finding and infestation of an insect occurs thanks to a specialized system of chemoreceptors (amphids) developed in the form of lateral organs in the area of the lips. Infective juveniles react to the presence of a host by moving towards an increasing concentration of chemical signals (chemotaxis) emitted by it, such as CO_2, components of insect feces (e.g., ammonia, uric acid, and arginine), and other specific compounds [46–48]. In addition to chemical signals, IJs also respond to insect movements in the soil (thigmotaxis). Studies have shown that IJs of the genus *Heterorhabditis* show a greater chemotactic capacity than IJs belonging to the genus *Steinernema*, which show a greater thigmotactic capacity [49].

Some EPN species are not very active, staying close to the soil surface and using an "ambush" strategy in which they stand on their tails and wait for passing insects. This is the strategy used by *Steinernema carpocapsae* Weiser (Steinernema: Steinermatidae). It is one of many tactics used by insect nematodes. Other species penetrate more deeply into the soil matrix and use an "active hiking" strategy to locate and infest sedentary insects. "Cruiser" behavior is used by *Steinernema glaseri* Steiner (Steinernema: Steinermatidae), *Steinernema arenarium* Artyukhovsky (Steinernema: Steinermatidae), *Heterorhabditis bacteriophora* Poinar (Rhabditida: Heterorhabditidae), and *Heterorhabditis megidis* Poinar, Jackson and Klein (Rhabditida: Heterorhabditidae), using the "trap" method. EPN species which use both methods of host searching include *S. feltiae* and *Steinernema riobrave* Cabanillas, Poinar and Raulston (Steinernema: Steinermatidae) [50].

The mutualistic relationship between bacteria and EPNs is based on several interactions. Nematodes carry symbiotic bacteria in the intestines of their bodies, allowing them to spread and infest new insects. In turn, bacteria kill the insect quickly, thanks to the production of enzymes and toxins. Bacteria producing these compounds also deter saprophytes from colonizing an infested insect cadaver, making it an unattractive food source for other organisms, as well as providing the nematodes with nutrients [51,52].

In the intestines of infective juveniles there are cells of symbiotic bacteria [53,54]. Bacteria of the genus *Xenorhabdus* are mutualistically related to the nematode genus *Steinernema* and are found in specialized intestinal alveoli [55,56]. *Heterorhabditis* nematodes form a symbiotic system with *Photorhabdus* bacteria, which are stored throughout the intestines of the nematodes [57–59]. Before the release of the bacteria, *Steinernema* spp. produce toxins (e.g., *Xpts*, *Txp40*, and *Xax*) that block the immune response of the insect, creating better conditions for the bacteria which will be released later. These toxins also enhance the pathogenic effects of the symbiotic bacteria [45,60–63]. In the nematode family Heterorhabditidae, no proteins with similar effects have been reported [56].

After penetration by an infective juvenile into the body of an insect host, bacteria are released from the intestines of nematodes into the hemocoel of the insect. Depending on the nematode family concerned, the bacteria are released in one of two ways: members of the Steinernematidae release their bacteria by defecation [64], whereas members of the Heterorhabditidae release bacteria through their mouths [57].

In the hemolymph of a host insect and its internal organs, the bacteria destroy the hemocytes (causing septicemia), multiply, and begin to produce exoenzymes as well as the toxins that destroy the immune system of the insect. These bacterial products cause biochemical changes in the body of the insect host, which leads to its death within 24–48 h [47,56,60,65–68]. In addition, they produce antimicrobial bacteriocins [69] and antibiotics that inhibit the growth of fungi, yeasts, putrefactive bacteria, and other pathogens (e.g., *Shigella sonnei* Levine (Enterobacteriales: Enterobacteriaceae) and *Flavobacterium* sp. (Flavobacteriales: Flavobacteriaceae) [70–72].

As a result of the action of antimicrobial substances produced by the symbiotic bacteria, the body of an infested insect does not decompose. In the initial period of infestation, the body of the host becomes rubbery, darkens over time, becomes stickier, and, finally, disintegrates. Depending on the species of nematode and the bacteria associated with it, the body of an insect after infestation can take on a different color [58,73–76].

Research on the nematode–bacterium system indicates that, in addition to the main EPN bacterial endosymbionts (*Xenorhabdus* and *Photorhabdus*), there is a frequently associated microbiota (FAM) consisting of several members of the Proteobacteria phylum. The endosymbionts and the accompanying microbiota determine the pathobiome of EPNs [77–79].

Infective juveniles (J_3) that penetrate into the host insect molt to the fourth juvenile stage (J_4), which is capable of taking in nutrition from the host [17,80]. The developing nematodes feed on dead bacteria, substances produced by the bacteria, as well as insect tissues, altered by the influence of bacteria [74,81]. J_4-stage nematodes reach adulthood within 2 or 3 days when cultured in *Galleria mellonella* Linnaeus (Lepidoptera: Pyralidae) at 23 °C [82]. In the Steinernematidae, at the J_4 stage, male and female adults develop, whereas in the Heterorhabditidae, hermaphroditic individuals are formed (in subsequent generations, there are individuals of both sexes). Individuals in the first generation of adults are called giants, due to their larger body size compared to the individuals of generations which arise later. Giant females (Figure 1) reach over 6 mm in length, while males may slightly exceed 3 mm in length [28,81,83]. The number of generations that develop in the body of an insect depends mainly on the amount of food available. Different ways of development can occur depending on the density of nematodes inside the body of the host insect. If the density is low, the females reach larger individual body sizes and lay a large number of eggs, with a proportion of the eggs developing in the body of the female. First-stage juveniles (J_1), emerging from the eggshells, initially feed on the reproductive system of the female and, after its rupture, on the pseudocoel, leading to her death. This phenomenon of intrauterine birth is called endotokia matricida or facultative vivipary. During this time, the juveniles develop rapidly from J_1 to J_4 and from J_4 to adulthood. When the density of nematodes inside the body of the host insect is high, however, individual females are smaller, and most eggs develop within their bodies [84].

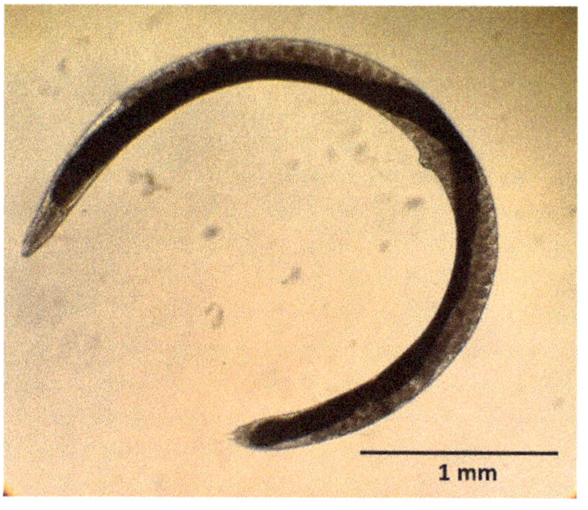

Figure 1. First-generation female of *Steinernema feltiae*.

In *H. bacteriophora*, an individual in the hermaphrodite generation lays about 1000 eggs; in the case of *S. carpocapsae*, however, a first-generation female lays about 3000–4000 eggs. Individuals of the next generation of *H. bacteriophora* each lay 3000–5000 eggs, whereas those of *S. carpocapsae* lay 12,000–14,000 eggs [82]. The final number of infective juveniles is similar in both species; in *H. bacteriophora*, −1% of the eggs in hermaphroditic individuals develop only to the J_3 stage, which then leave the body of the insect host. From the second generation, in which there are females and males, 5–10% of the eggs develop to the J_3 stage, which also leave the body of the insect host. From the next generation, all eggs develop into J_3s, which leave the host and enter the external environment. In *S. carpocapsae*, the J_3 juveniles that leave the body of the insect come from the second and third generations of adults [81]. A multistage population transforms into a monostadial population (infective juveniles) when food begins to run out. At this time, nematodes end their development as a generation of infective juveniles which take into their digestive tract cells of the symbiotic bacteria which multiplied in the body of the dead insect host [58,85]. In these juveniles, the opening anal orifice closes, the mouth regresses, and bacteria are stored in the intestine. Juveniles at this stage are capable of migrating to the external environment and infesting another insect [86]. The length of the development cycle, from the penetration of juveniles into the body of the insect to the exit of the next generation, in the Heterorhabditidae is 10 to 16 days, and in the Steinernematidae it is 10 to 14 days [82].

If suitable conditions prevail outside the body of the host, the juveniles leave the body. However, if the humidity or temperature is too low, the juveniles can stop migrating to the external environment for up to 50 days. The juveniles also have the ability to enter diapause, in which they can remain for up to 3 years [87,88]. During the free-living phase of the nematodes, their symbiotic bacteria stay active at all times. The greater the metabolic activity of the bacteria, the shorter the period for which the infective juveniles of the nematode can live in the external environment. This is due to the limited amount of nutrients accumulated in the body of the nematode. However, when a host is infested by the nematode, the high activity of bacteria has positive effects because they lead to faster insect death. Because of the possibility of the premature death of the nematode in the absence of a host, it is beneficial to maintain metabolic activity at an optimal, moderate level [89–91].

Insects have all kinds of defense mechanisms. The first of these is behavioral protection, e.g., intensive defecation in beetle larvae significantly hinders the penetration of nematode juveniles through the opening in the beetle larvae [92]. Another strategy is mechanical defense in the form of a protective cuticle on the surface of the body. However, if a nematode juvenile gets inside a host larva, the mechanisms of the host's immune response are activated. The cellular response involves hemocytes, immune cells that are responsible for phagocytosis but which can also perform other complex functions, such as surrounding and encapsulating parasites like the nematode. The humoral response includes the production of antibacterial peptides. One of the functions of the peptides is to interact with hemocytes in the activation of phenoloxidase, an enzyme involved in the activation of metabolic pathways that lead to the neutralization of pathogens. This enzyme induces the production of toxins and products involved in encapsulation. Other defense strategies include the production of reactive oxygen and nitrogen species [51,93–96].

EPNs have adapted in response to the harsh conditions they encounter in the bodies of attacked host insects. They have developed mechanisms that make it possible to bypass or even neutralize the host's defense factors. It is possible, in principle, to achieve the biochemical masking of the nematode from the insect's immune system. In turn, on the surface of the cuticles of nematodes, there are molecules which block the secretion of antibacterial peptides from the host. These nematodes also secrete substances that kill hemocytes. An example is *S. feltiae*, which has lipopolysaccharides (LPSs) in its cuticle that inhibit the production of the phenoloxidase enzyme [94,97].

Entomopathogenic nematodes exhibit intraspecies variability, which is related to their location and adaptation to a particular habitat. This diversity can be manifested in differences in developmental cycles, reproductive strategies, and virulence toward host insects. Examples of intraspecific diversity could be seen in *S. feltiae* isolates collected from different habitats in northern Spain. Not only did the nematodes themselves show differences, but the associated bacteria did also. Their functions differed between populations of different origins [98–100].

Another example of intraspecies variability is the differential tolerance to high temperature and low humidity in representatives of the genus *Heterorhabditis*. During the study, it was observed that, within the same species, there are individuals that are more or less tolerant to stress conditions. Genetic mechanisms are responsible for this diversity, which allows populations inhabiting hot and dry areas to develop adaptations necessary for living in such environments [101].

4. Composition of Soil and Its Effect on Nematode Populations

Entomopathogenic nematodes occur in both natural and anthropogenic ecosystems, where they shape biological diversity. The occurrence of certain nematode species depends on their nutritional preferences and, consequently, on the presence of specific host insect species in these locations [102].

Research on the distribution of EPNs showed that the presence of nematodes of the Steinernematidae family is commonly reported in forested areas [102–106]. *S. kraussei* and *S. intermedium* are most commonly isolated from forest areas, with the former begin found mainly in coniferous forest soils of Europe and North America [105,106]. Areas transformed by humans also create suitable living conditions for these organisms. *S. feltiae* has been shown to be a recorded species in meadows, forests, and agricultural fields [105,107,108], while *S. affine* and *S. carpocapsae* have also been found in cultivated soils [105,107].

Nematodes of the family Heterorhabditidae are often associated with sandy coastal soils and grasslands [104,107,109,110], the most commonly isolated species being *H. megidis* [107,111–113].

The distribution of EPN populations in a particular area depends on many environmental factors, such as soil moisture, temperature, structure, and pH [114]; the availability of hosts; and the presence of competitive or predatory organisms [115–117] (Figure 2).

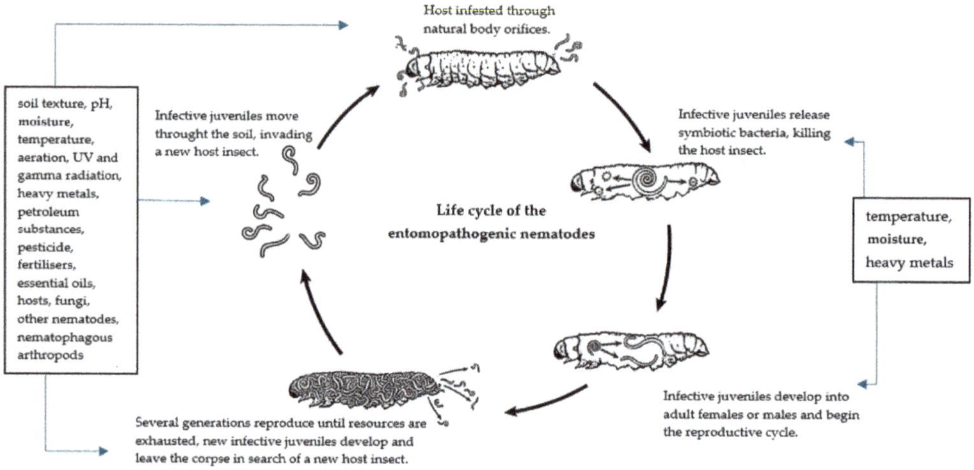

Figure 2. Factors influencing the development of entomopathogenic nematodes.

4.1. Soil Texture

The mechanical composition of soil, e.g., grain size, affects soil organisms and other conditions of the soil environment, such as the degree of soil moisture and aeration. Lighter soils are better aerated and are more easily permeable to lower layers. Heavier soils have a greater water-carrying capacity, but air circulation is poorer. Individual species of EPNs differ in their requirements in terms of air–water relations in the soil environment. Soil structure influences, among other things, the movement, infectivity, and survival of nematodes [118–122]. Depending on the size of the soil fraction, the rate of movement and the frequency of undulating movements changes. In heavier soils, infectivity and movement are hindered, whereas in light loam soils with similar percentages of silt and clay, they are facilitated [48,99,121,123,124]. Other research has shown that, in soil composed of about 40% sand, 33% silt, and 26% clay, nematodes moved at a rate of 4.35 cm/day and covered a distance of 46 cm in 14 days. In sandy soil with a mean grain diameter of 0.2–0.5 μm, infective juveniles covered a distance of 2 cm (87%) within 48 h; 0.5% of the infective juveniles covered a distance of 12–14 cm from the site of release [125].

Research into vertical migration showed that it decreases in response to the increase in clay and silt content, with sandy soil making it easier for nematodes to move in this direction [126]. Most nematode species prefer to migrate upward rather than downward. One example is the species *H. bacteriophora*, which infects insects above the EPN application site [127]. Other species, such as *S. carpocapsae*, prefer to migrate downward; from the site of application of the infective juveniles, they move deep into the soil, where they effectively infest host insects [128]. The reason for such different directions of movement is the difference in the rate of penetration of signal chemicals secreted by hosts that attract the nematodes [129].

The structure of the soil also affects the infectivity of EPNs [130]. Juveniles have been shown to penetrate hosts more quickly and willingly in sandy and sandy loam soils [48]. The highest infestation levels were obtained in clay–sandy soils with low moisture and in sand with high relative moisture, close to saturation. It was shown that, as the clay content of the soil increased, the infectivity of *S. glaseri* decreased [131].

Soil texture is also one of the most important factors influencing nematode survival [123,132]. In research on *S. carpocapsae* and *S. glaseri*, infective juveniles of these species placed on sterile sandy, sandy loam, or loam soils for 16 weeks showed different degrees of survival. The lowest survival rate was associated with the loam soil because the low porosity and poor aeration of this type of soil limited the survival rate of infective juveniles. Similar conditions prevailed in clay soils, which also reduced the effectiveness of the EPNs. On the other hand, the highest survival rate was recorded in sandy loam and sandy soils. It was shown that different EPN isolates can show different adaptations to the textural classes of soil. Native isolates may show better performance in the environment from which they originate [48,121,133,134].

4.2. Moisture

The level of moisture in the environment in which EPNs are located affects their survival rate, with the appropriate moisture level making it easier for them to move around and search for a host. The movement of nematodes within soil pores is enabled by a thin water capsule around the body of the nematode [14,135–137].

The type and rate of nematode movement depend on the thickness of the water capsule: the thicker the capsule, the greater the distance the infective form travels [129]. In a dry environment, movement practically stops, and individuals either enter a state of anabiosis or die [124].

High soil moisture also adversely affects the ability of nematodes to move [138]. This is due to the lack of surface tension necessary for movement [139]. Research on infective juveniles has shown that, by making pendulum movements on the soil surface at a moisture content of 25–40%, juveniles have the most favorable conditions for invasion of the host. When the soil moisture content exceeds 50%, the biological activity of infective juveniles is limited [48]. The infectivity of nematodes is also reduced when the soil environment is too dry [140].

Soil moisture also affects nematodes living inside a host insect's body. In dry soil, a small number of juveniles were observed leaving the host body. At low moisture, *S. glaseri* and *S. carpocapsae* juveniles leave the body of the host insect the earliest, whereas *S. riobrave* juveniles leave the insect body the latest [141]. The reason for this phenomenon may be the inability of the nematodes to physiologically adapt to the low moisture conditions of the soil environment after abruptly leaving the host organism. Under such unfavorable conditions, the juveniles can survive for between a few hours and several days, as they are exposed to rapid drying [142]. On the other hand, when the moisture is low, the juveniles cannot stay in the body of the host for too long, as the cuticle of the insect begins to dry out and harden, making it difficult for the juveniles to leave the body of the insect [141].

4.3. Temperature

Another factor influencing the biology of EPNs is temperature. The degree of temperature tolerance varies from species to species and even between strains of the same species [110,143]. A common feature of temperature response is the tendency for nematode survival to decrease with increasing temperature. Nematodes are less sensitive to low temperatures than to high temperatures [143–145]. Nematodes that are in anabiosis are more tolerant of extreme temperatures than nematodes that are active under these unfavorable conditions [124]. Nematode strains that are used in the biocontrol of pests of living plants are active at temperatures between 10 °C and 30 °C [146–148].

The optimal temperature for infestation by different species of the Steinernematidae is in the range of 13–20 °C [149]. Studies have shown that the optimal temperature for *S. carpocapsae* is in the range of 20–25 °C. These studies also showed that, at 10 °C, only fourth-stage juveniles (J_4) were present in the dead host (*G. mellonella*, the greater wax moth). The full development of the nematodes occurred at a temperature of 15 °C, but no migration of the juveniles to the external environment was found at that temperature. When the temperature rose above the optimal value (30 °C), it was shown that only first-stage juveniles were present in the insect host [150]. Research on temperature-dependent infectivity showed that *S. glaseri* and *S. feltiae* showed a greater infestation of insects at 25 °C than at 15 °C, although no significant differences was observed for *S. arenarium* [149,151, 152]. Similar studies were conducted for *H. megidis*, where the highest infectivity and fastest development were recorded at 25 °C [153], although soil moisture had to be acceptable. The activity of *H. megidis* juveniles decreased when the ambient temperature fell to 15 °C.

As the temperature increased, these juveniles regained their ability to attack and kill the insect host [124].

In the case of extreme low temperatures, nematodes have been shown to survive temperatures below 0 °C. In the case of unencapsulated nematodes, juveniles freeze at −6 °C, while *Heterorhabditis zelandica* Poinar (Rhabditida: Heterorhabditidae) is encapsulated and can survive at −32 °C [154]. Other species also exhibit high tolerance to low temperatures. Examples include *S. feltiae*, *S. arenarium*, and *H. bacteriophora*, which can survive at −22 °C, −14 °C, and −19 °C, respectively [155].

High temperatures above 32 °C also adversely affect the infectivity, reproduction, growth, and survival of nematodes [146,156]. It has been shown that storing *S. feltiae* at 35 °C for one hour caused a dramatic decrease in activity, whereas storing them at 37 °C for 16 h led to the death of all infective juveniles [157,158].

Some strains of EPN species from the Heterorhabditidae are also tolerant to high temperatures. This is evidenced by the isolation of these nematodes from hot and dry areas. Such examples have been isolated from Egypt, Sri Lanka, Israel, and Iran [159–162].

4.4. Soil pH

Soil pH is a key element that affects many biological and physicochemical processes in soils. The degree of soil acidification largely determines the mobility and bioavailability of heavy metals, as well as ionic organic pollutants. With an increase in acidification of the soil environment, the bioavailability of heavy metals increases. The optimal pH range of soil for the growth and reproduction of most plant species and soil organisms is considered to be between 5.5 and 7.2 [163].

In the case of EPNs, acidic soil pH levels cause decreases in juvenile infectivity and survival. The survival of *S. carpocapsae* and *S. glaseri* was reported to be similar at pH 4, 6, and 8 for the first 4 weeks. Storage of these species for 16 weeks in an environment with a pH of 10 showed that the survival rate of *S. carpocapsae* was significantly greater than that of *S. glaseri* [164]. *S. feltiae* and *H. bacteriophora* showed greater mobility, activity, and pathogenicity at pH 6.8 and 8 than at pH 5.5 [165].

4.5. Radiation

An important factor influencing EPNs is ultraviolet (UV) and gamma radiation. Understanding the effects of UV radiation is important when using nematodes to control pests found on the above-ground parts of plants. It has been shown that 30 min of exposure of nonencapsulated nematodes to sunlight reduced their reproduction and increased their mortality [166]; encapsulated nematodes were not affected as much. The researchers also showed that exposure to long-wavelength (366 nm) UV radiation for up to 8 h did not have a negative effect on infective juveniles, unlike the response to short-wavelength UV (254 nm). Exposure to short-wavelength UV radiation for 1.75 min reduced the pathogenicity of infective juveniles by 19%, and exposure for 7 min reduced pathogenicity by 100% [166]. In viability assessments, *Steinernema* species, especially *S. carpocapsae*, showed greater tolerance to UV radiation than *Heterorhabditis* species, *H. megidis* and *H. bacteriophora* being the most sensitive [167,168].

Research on the effect of gamma radiation on entomopathogenic nematodes showed that infective juveniles of *Steinernema scapterisci* Nguyen and Smart (Steinernema: Steinernematidae) and *H. bacteriophora* irradiated with 2 Gy gamma radiation were more effective in biocontrol of Egyptian cotton leafworm, *Spodoptera littoralis* Boisduval (Lepidoptera: Noctuidae), than were nonirradiated nematodes. Similar results were obtained for *S. carpocapsae* BA2 [169,170].

4.6. Soil Aeration

Soil aeration and compaction also have a major impact on the development of EPNs. The oxygen and carbon dioxide concentrations of the water molecules in which the infective juveniles are present depend on aeration. EPNs are aerobic organisms, and a low oxygen content may reduce their survival in the soil environment [171,172]. The air content of the soil is variable and depends on the soil type. Heavy moist soils are conducive to anaerobic conditions, which, in most cases, are harmful to nematodes. Factors that are negatively related to the oxygen content of the soil environment include the contents of clay and organic matter [48]. The percentage composition of gases in dry sandy soil is similar to that in the atmosphere, so such soils create conditions conducive to nematodes [124]. Despite oxygen being present in soil clods, nematodes only use the oxygen dissolved in the water surrounding the clods. The amount of oxygen depends on the spaces between the clods, the type of soil, and the degree of water saturation [124].

Research shows that the respiration rate of *S. carpocapsae* depends on the ambient temperature [173]. It was also shown that low O_2 concentrations interfered with pathogenicity and led to the death of nematode juveniles. The survival rate of *S. carpocapsae* and *S. glaseri* infective juveniles in sandy loam soils decreased in response to a decrease in oxygen content from 20% to 1% [164,173]. Infective juvenile forms of EPNs use elevated carbon dioxide concentrations as signals with which to locate a potential host [48].

5. Soil Contaminants Affecting Entomopathogenic Nematodes

Soil is a natural habitat for the development of many organisms that condition the decomposition and production of biomass. The soil environment filters groundwater and also ensures the circulation of matter and the flow of energy within the ecosystem. Soil also buffers, filters, and accumulates harmful substances. Many factors contribute to increases in soil pollutants, such as industry, transport, and improper sewage and waste management [174].

5.1. Metal Ions

Heavy metals are potential contaminants resulting from anthropogenic pollution. They enter the environment from various sources [175,176]. These pollutants are very harmful to the biologically active layer of the soil; because the metals are not susceptible to leaching, they persist in the soil environment for a long time, limiting the suitability of the soils for agriculture. Excessive accumulation of heavy metals in the soil negatively affects the organisms present. In many cases, these compounds accumulate in the organisms, leading to disorders in metabolism and changes in the structure and function of cells and tissues [177–179]. The toxicity of heavy metals and their high bioavailability depend on many factors, such as the amount of contamination and the chemical form, as well as on the properties of the soil environment: temperature, pH, presence of other metals, and oxidation–reduction potential [180,181].

In the soil environment, infective EPN juveniles will be directly exposed to any heavy metals found there. These compounds affect the occurrence, survival, infectivity, and reproductive capacity of nematodes. Natural concentrations of heavy metals in the habitats of nematodes are not the cause of their mortality. On the other hand, they are the reason for the reduced infectivity of EPNs. The most adverse effects were observed with nickel (Ni(II)) and lead (Pb(II)) ions [182,183]. Laboratory studies on the effect of heavy metals on EPNs have shown that heavy metal ions can adversely affect the pathogenicity of EPNs [107,180,182–190].

In the first days of contact with infective *S. feltiae* juveniles, low concentrations of Pb(II) and cadmium Cd(II) ions in the sand cause an increase in the intensity of invasion of the host. This is caused by a stress response that stimulates the nematodes to quickly invade the body of the host. In soils with a higher concentration of heavy metal ions, negative parameters, such as lower infectivity, increased mortality of nematodes inside the body of the insect host, and increased mortality of individuals migrating in the external environment, have already been shown to occur in a metal ion-concentration-dependent manner [187,191].

Heavy metals also hinder the movement of nematodes. Studies on the effects of metal salts showed that the presence of these compounds in soil prevented nematodes from moving, finding, and infesting hosts. Chloride salts have been found to be particularly toxic [192]. In subsequent analyses, it was determined that the addition of magnesium salts to solutions in which nematodes were stored had a protective effect against heavy metal ions [191,193,194].

Research into the effects of combinations of several metals on *S. carpocapsae* showed that these combinations reduced the pathogenicity of EPNs. The presence of lead, cadmium, and copper in the soil resulted in the nematodes killing a smaller number of host insects compared to the control samples (in natural, unadulterated soil). The addition of magnesium to soil contaminated with heavy metal ions significantly increased the number of nematodes entering hosts [190]. Different combinations of metal ion triads were also created in the study, based on Cd(II), cobalt (Co(II)), copper (Cu(II)), magnesium (Mg(II)), manganese (Mn(II)), Pb(II), and zinc (Zn(II)). In most cases, the nematodes tolerated the toxicity of heavy metals. On the other hand, Cd(II) and Pb(II) ions always killed the nematodes. None of the added metal ions reduced the toxicity of Pb(II), while the toxicity of Cd(II), together with Cu(II), Zn(II), and Co(II), was reduced by the addition of Mg(II) but not Mn(II). The toxicity toward *S. carpocapsae* of aqueous solutions of selected triads of metal ions depended on the method of introduction of these ions into the nematode environment. Mg(II) ions showed a protective effect over a long period of time or induced rapid death of nematodes. Mn(II) ions, on the other hand, provided short-term protection, with generally low toxicity [180].

Insect nutrients contaminated with Pb, Cd, or Cu ions also had an adverse effect on infectivity, nematode development in the host, and the viability of juveniles migrating from the host into the environment. These ions accumulate in the tissues of the insect host, which indirectly affects the nematodes [195–197].

Heavy metals affect not only EPNs, but also the associated bacteria present in their digestive tracts. It was shown that copper, manganese, and nickel ions were most harmful to these microorganisms. In contrast, lead was not toxic to nematode-associated bacteria [198].

The characteristics of the soil environment in which EPNs occur are also influenced by human activities. These include chemical or physical changes to the soil caused by harrowing or application of fertilizers, pesticides, insecticides, or fungicides [199].

5.2. Petroleum

Petroleum derivatives found in agricultural soil destabilize the biological properties of soils, as well as the biotic community, and finally affect the quality and fertility of the soil. Petroleum substances that may adversely affect the development and function of EPNs are mainly crude oil, diesel oil, engine oil, and gasoline ("petrol"). Of the above-mentioned compounds, gasoline is the least harmful, and engine and diesel oils are the most harmful [200]. All these substances reduce the pathogenicity and reproductive capacity of EPNs. The longer the contaminant is active, the higher the mortality of the nematodes. Mortality is also related to the occurrence of EPNs in land contaminated with petroleum substances. EPNs could not be isolated from soil contaminated with petroleum derivatives [201]. In addition, these compounds inhibit the ability of EPNs to find a host [202] and reduce their infectivity and reproduction [203,204].

5.3. Pesticides

Among the pesticides used in plant protection, products containing oxamyl, sulfur, mancozeb, and fenitrothion turn out to be highly toxic to nematodes. Trifluralin causes a high mortality rate in EPNs. Pesticide concentration has also been shown to have a major impact on the infectivity of EPNs [205].

In research into the effects of various pesticides on EPNs, it was shown that most fungicides, pyrethroid insecticides, and chlorinated hydrocarbons were nontoxic or showed low toxicity toward *S. feltiae* larvae. These compounds also did not reduce the infectivity of the EPNs. In the case of urea and carbamine herbicides, the viability of infective juveniles decreased significantly, and those that survived contact with herbicides were characterized by low infectivity. The ability to find and infect a host was also impaired. In an environment in which herbicides were present, female EPN reproduction was limited. Within 5 months of the application of triazine, diazine, and uracil herbicides, there were no differences in the numbers of nematode juveniles harvested from the soil and capable of infesting insects [206].

Similar effects were observed for organophosphate compounds, chlorinated hydrocarbons, diflubenzuron (an insecticide from the group of acylurea compounds), and synthetic pyrethroids. These agents adversely affected nematodes in in vitro culture, while methomyl reduced the reproduction of *S. carpocapsae* [207,208].

Insecticides are also used as protection products against pests in animals. Such an agent is fipronil, which controls fleas and ticks in dogs, poultry red mites, and ants, cockroaches, flies, and mosquitoes. Another example is malathion, an agent effective in the treatment of head lice, scabies, and poultry lice. These insecticides were found to have no adverse effects on the biology of *S. carpocapsae* [209].

Insecticides can also be isolated from plants. An example is azadirachtin, extracted from *Azadirachta indica* A. Juss. (Sapindales: Meliaceae) (the neem tree). It is effective against mealybugs (Hemiptera: Pseudococcidae), aphids (Hemiptera: Aphidomorpha), and *L. decemlineata* and is safe for birds and beneficial insects. It was shown that this agent does not adversely affect EPNs, is toxic only at doses higher than recommended, and can be used in combination with EPN-based insecticides [210,211].

Insecticides are chemical agents directed against plant insect pests. They are used as stand-alone preparations or in combination with biological methods in Integrated Pest Management (IPM). However, they can affect the biology of EPNs. One group of insecticides which can be used together with EPNs are the neonicotinoids [212]. Imidacloprid was shown to have no adverse effect on EPN pathogenicity or the infectivity of juveniles and can be used in combination with EPN-based preparations [209,212–214]. However, at high doses, similarly to acetamiprid, it causes a decrease in fertility and an increase in the mortality of EPNs [215]. Acetamiprid and thiamethoxam show weaker synergism with EPNs [216]. Chlorpyrifos is an organophosphate insecticide that is more toxic than acetamiprid toward *S. feltiae* [217]. It was also demonstrated that this EPN species was more sensitive to insecticides than were other species, e.g., *H. bacteriophora*, *S. arenarium*, and *S. kraussei* [205,217]. An EPN species that was not adversely affected by chlorpyrifos and the pyrethroid insecticide cypermethrin was *S. carpocapsae* [209]. Another group of insecticides are the carbamates, among which are carbosulfan and carbofuran. These agents did not affect the pathogenicity of the nematodes studied, but carbofuran killed more nematodes than did carbosulfan [218]. Either of the above-mentioned carbamates and pirimicarb can be used in IPM together with EPNs [210,218]. Macrocyclic lactones include abamectin. Like lufenuron, a benzoylurea insecticide also used to control fleas, fungi, and nematodes in mammals, abamectin adversely affects *H. bacteriophora* [210]. Another agent from the macrocyclic lactone insecticides is emamectin benzoate, which is effective against the spotted wing fly (*D. suzukii*) in berry crops. This insecticide has also exhibited adverse effects on the infectivity of *S. carpocapsae* [209].

5.4. Fertilizers

Fertilizer usage is also one of the agrochemical treatments which may impact EPNs. Excessive use of inorganic fertilizers contributes to increased soil salinity and accumulation of heavy metals and nitrates [219,220]. In saline soil, EPN movement is hindered and their ability to find and identify hosts is limited [221].

Long-term exposure to high concentrations of inorganic fertilizers inhibited the infectivity and reproduction of nematodes, whereas 1-day exposure of EPNs to the fertilizer increased infectivity. Interspecific differences in EPNs' sensitivity to fertilizers were also observed. *H. bacteriophora* was more sensitive to negative impacts of fertilizers than were *S. feltiae* and *S. arenarium*. In field trials, an organic fertilizer increased *S. feltiae* density, whereas NPK fertilizer decreased nematode density regardless of manure applications [222].

Analysis of the effect of various fertilizers on EPNs showed that all the fertilizers tested led to the death of the studied nematodes. On the other hand, the highest toxicity to *H. bacteriophora* and *S. feltiae* was achieved by nitrogen–phosphorus and diammonium phosphate fertilizers [223]. In other research, it was proven that the infectivity of EPNs to insects was lower in soil treated with phosphorus fertilizer than in soil treated with potassium, NPK, or nitrogen fertilizers [224]. The combined application of inorganic fertilizers with NPK and Mikrovit (MgO, 2.33%; Cu, 0.25%; Mn, 0.17%; Zn, 0.18%; B, 0.015%; Fe, 0.23%; Mo, 0.01%), against a background of liming was conducive to greater activity of nematodes in infesting insects, while it adversely affected the reproduction of EPNs [225]. The reproduction of nematodes was also depressed by NPK fertilizer in a time-dependent manner. It was shown that, after 20 and 40 days of exposure to NPK fertilizer, there was a strong decrease in the intensity of infestation. The use of higher doses of mineral fertilizers resulted in a decrease in the reproduction of nematodes developing within the host, which led to a decrease in the number of nematodes in the soil. The exceptions were potassium nitrate and manure, which stimulated the reproduction of the nematodes [129].

5.5. Essential Oils

Essential oils, due to their broad spectrum biological of activities (bactericidal, virucidal, fungicidal, antiparasitic, insecticidal, therapeutic, and cosmetic), are currently used in the pharmaceutical, sanitary, cosmetic, food, and agricultural industries [226–230]. In plant protection, they are increasingly used as natural pesticides and as a rational element of IPM [231].

In research on the toxicity toward EPNs of essential oils, thyme, cinnamon, clove, and garlic oils have been shown to exert the most negative effects, causing 100% mortality of *S. feltiae* and *H. bacteriophora* at a 1% concentration [231]. Pine oil led to mortality of 50% of *H. bacteriophora*, while lemongrass oil killed more *S. feltiae* individuals than *H. bacteriophora* juveniles. Spearmint, cedarwood, eucalyptus, and rosemary oils had no significant effect on nematode mortality [231]. The most promising oil seems to be eucalyptus oil. It is also an effective pesticide against plant pests [232]. These properties indicate that this oil can be used in IPM in combination with EPN-based preparations [231].

In order to reduce EPN stress caused by the drying out or UV irradiation of surfactants applied to the above-ground parts of plants, the possibility of using coconut oil and olive oil as adjuvants in EPN preparations has been investigated [233].

6. Interspecies Relationships

For EPNs, hosts are essential biotic factors. The morphological, anatomical, and physiological properties of the insect enable nematodes to find, infest, and develop in its body cavity. For example, an increasing concentration of host excretions, which is a signal of the nearby presence of the host, causes an increase in the activity of infective juveniles [234].

EPNs have to share soil with a huge diversity of living organisms and microorganisms in addition to the pests against which they are targeted. Numerous positive and negative interactions between EPNs and soil organisms other than their hosts have been documented [235,236].

6.1. Hosts

Natural invasions of EPNs in the environment in the form of nematode-infested insects in ecosystems are rarely observed. This is due to the rapid development of the nematode population in the host and the rapid destruction of its body. Of the dozens of beetle species in which the larvae are sensitive to the nematode *S. carpocapsae*, only a few have been successfully studied with respect to natural invasion, namely, *Zabrus tenebrioides*

Goeze (Coleoptera: Carabidae) (the wheat ground beetle) [237], *Agriotes lineatus* Linnaeus (Coleoptera: Elateridae) (the lined click beetle) [238], *Melolontha melolontha* Linnaeus (Coleoptera: Scarabaeidae) (the cockchafer), *Drosophila suzukii* Matsumura (Diptera: Drosophilidae), and *Melolontha hippocastani* Fabricius (Coleoptera: Scarabaeidae) (the forest cockchafer) [10,239]. It has also been possible to isolate *S. feltiae* from flies [240] and *Heterorhabditis heliothidis* Khan, Brooks and Hirschmann (Rhabditida: Heterorhabditidae) from *Helicoverpa zea* Boddie (Lepidoptera: Noctuoidea) (the corn earworm) [241]. *H. bacteriophora* has also been found in *Helicoverpa punctigera* Wallengren (Lepidoptera: Noctuoidea) (the native budworm) [237] and *Ceratitis capitata* Wiedemann (Diptera: Tephritidae). It has also been shown that *S. kraussei* infested *Cephalcia abietis* Linnaeus (Hymenoptera: Pamphiliidae) (the false spruce webworm) under natural conditions [242,243].

The wide host range of EPNs allows for the biological control of many phytophagous species. To date, EPNs have been used in the control of ground flies (Diptera: Sciaridae) in greenhouse crops [244] and mushroom crops [245]; weevils (Coleoptera: Curculionidae) in greenhouses and nurseries of ornamental and forest trees [246–248]; beetles (Coleoptera: Scarabaeidae) on urban lawns, golf courses, and forest nurseries [249]; and geometer moths (Lepidoptera: Geometridae), *Leptinotarsa decemlineata* Say (Coleoptera: Chrysomelidae) (the Colorado potato beetle), and sawflies (Hymenoptera: Symphyta), as well as in the control of a number of tree pests that pupate in the soil [250,251]. However, certain EPNs are specialists that usually target one type of insect [108]. For example, *Steinernema diaprepesi* Nguyen and Duncan (Rhabditida: Steinernematidae) targets the larval stages of the *Diaprepes abbreviatus* Linnaeus (Coleoptera: Curculionidae) (the root weevil) [252,253] and *Steinernema neocurtillae* Nguyen and Smart (Rhabditida: Steinernematidae) targets only *Neocurtilla hexadactyla* Perty (Ortoptera: Gryllotalpidae) (mole crickets) [70,83].

Despite the association of EPN hosts with particular insect orders, as shown above, different sensitivity to EPNs is commonly shown. The larvae of moths or butterflies (Lepidoptera) are most often infected. Beetles (Coleoptera) (especially adults) and flies (Diptera: Muscidae) are more resistant to EPNs [254]. The larvae of ladybirds *Hippodamia convergens* Crotch (Coleoptera: Coccinellidae) (the convergent ladybird) and golden-eyed beetles *Hemerobius* spp. (Neuroptera: Hemerobiidae) and *Chrysopa* spp. (Neuroptera: Chrysopidae) were shown to be resistant to EPN infestation [237]. The development of nematodes in a host's body is also influenced by the developmental stage of the insect. Nematodes thrive best in the larval stages, less so in pupal and mature imago forms [255,256]. In addition to the developmental stage of the insect, the development of EPNs is also influenced by the level of nutrition of the host. It has been shown that in starved host insects, the development of nematodes slows down [257]. It has also been found that as the density of nematodes in the host increases, the number of generations decreases [258].

Among the morphological features of insects that facilitate the invasion of nematodes are the size of the natural openings (the mouth, anus, and spiracles) and the structure of the outer sheath [259,260]. The ability of infective juveniles to penetrate an insect also depends on the period of contact of the nematodes with the host or its feces. The longer the duration of contact, the lower the chance of penetrating the insect [261,262]. The development of nematodes and their density in the environment depend, among other things, on the number of potential insect hosts in the environment: the more insects available, the better the chances of survival of EPN populations [115]. The reproduction rate of EPNs also depends on the host's body weight. Nematode reproduction is much greater in a host with a higher body weight [263]. The reproductive capacity of nematodes is also affected by the origin of the infective form. It was found that the reproductive rate was higher after several passages trough the same host species than with the use of juveniles where the previous generations had undergone development in another host species [256,264].

6.2. Mites

It has been observed that a large number of nematophagous arthropods attack and eat infective juveniles, whereas smaller arthropod larvae only damage them. The introduction of EPNs into the soil stimulates the development of populations of nematophagous arthropods, which, in turn, effectively reduce the number of nematodes. It was shown that certain species of mites (*Gamasellodes vermivorax* Walter (Mesostigmata: Ascidae) and *Alycus roseus* Koch (Trombidiformes: Alycidae)) feed on IJ EPNs [265]. Mites that do not eat EPNs have been observed too. Studies have shown that *Sancassania polyphyllae* Zachvatkin (Acari: Acaridae) do not consume infective juveniles in the soil [266].

There are also mites in the soil which are killed by the symbiotic EPN bacterium *Photorhabdus luminescens* (Thomas and Poinar) Boemare, Akhurst and Mourant (Enterobacterales: Morganellaceae). *P. luminescens* resulted in a significantly higher mortality rate of adult *Tetranychus truncatus* Ehara (Acari: Tetranychidae) than *Pseudomonas aeruginosa* (Schroeter) Migula (Pseudomonadales: Pseudomonadaceae) [267].

6.3. Springtails

Springtails (Collembola) are organisms that mainly inhabit the soil and litter layers, which are also the environment of IJs of EPNs. These arthropods pose a major threat to them. It was shown that springtail species—*Sinella curviseta* Brook (Entomobryomorpha: Entomobryidae), *Folsomia candida* Willem (Entomobrymorpha: Isotomidae), and *Hypogastrura scotti* Yosii (Poduromorpha: Hypogastruridae)—consumed significant numbers of IJs [265,266]. Moreover, the presence of EPNs in the soil increases the abundance of springtails [268]. The ability of *F. candida* and *Sinella caeca* Zhao (Entomobryomorpha: Entomobryidae) to eat three species of entomopathogenic nematodes (*S. carpocapsae*, *S. feltiae*, and *S. glaseri*) and to reduce the infectiveness of nematode applications against insects was investigated. It was shown that as few as five springtails are enough to reduce the invasiveness of EPNs against the insect [269].

6.4. Nematodes

Competition is a natural factor in the struggle for existence in the soil. Organisms that inhabit an ecosystem often have to compete in many ways for food, territory, or survival. The biological activity of nematodes is also influenced by both interspecific competition among nematodes and the attractiveness of the host. Scavenger free-living nematodes such as *Oscheius* spp. (Rhabditida: Rhabditidae) are able to displace EPNs from their place of occurrence [270]. By studying the simultaneous introduction of infective juveniles of both *S. feltiae* and *H. bacteriophora* into the same environment, it was shown that, in the absence of insect hosts in the soil, dispersal is characterized very largely by randomness. On the other hand, after the introduction of potential hosts (*G. mellonella*, *Tenebrio molitor* Linnaeus (Coleoptera: Tenebrionidae), and *Tribolium confusum* Jacquelin du Val (Coleoptera: Tenebrionidae)) into the environment, there was a change in the speed and direction of movement of the nematode juveniles. In response to the introduction of one host insect species into the soil environment, an increase in the activity of both *S. feltiae* and *H. bacteriophora* was demonstrated. Both nematode species showed the ability to recognize and select the more attractive host. *S. feltiae* and *H. bacteriophora* accumulated considerably in the vicinity of *G. mellonella*, less so in the vicinity of *T. molitor*, and least near *T. confusum*. Simultaneous introduction of the two nematode species into the soil resulted in the faster movement of *S. feltiae* than *H. bacteriophora* toward the hosts. As the contact time increased, the number of infective juveniles near the insect increased. In cross-inoculation, *S. feltiae* showed the same food preference as in single-species inoculation. In cross-inoculation, *H. bacteriophora* was directed to the regions of the host that were less attractive to and therefore less affected by the competitor species (*S. feltiae*) [271].

The density of nematodes in the soil can be affected by the presence of other soil fauna. Among the nematodes found in the soil environment, there are nematophagous nematodes, which reduce the number of EPNs. It was found that, during the 8-week period after the introduction of infective *S. carpocapsae* juveniles into the soil, their number systematically decreased, while the number of predatory nematodes (Mononchida) and saprophagous nematodes (Rhabditida) definitely increased. This showed that EPNs provide a food base for other trophic groups of nematodes [144]. Other research showed that *Mononchoides longicaudatus* Khera (Rhabditida: Diplogastridae) feed on the infective juveniles of *S. carpocapsae* and *H. bacteriophora* [272].

There are also nematodes in the soil environment that are pests of plants, for instance, *Meloidogyne* spp. (Tylenchida: Heteroderidae). Plant-pathogenic nematodes suppress entomopathogenic nematodes and factors from EPNs can repel plant-pathogenic nematodes [273,274].

6.5. Fungi

Another group of organisms that reduce EPN populations are fungi. More than 150 species of predatory and endoparasitic fungi have been shown to reduce the population of EPNs in the natural soil environment [275]. Predatory fungi form two types of traps (sticky and ring traps). Sticky traps of *Arthrobotrys oligospora* have hyphae which form a network of sticky meshes. EPN infective juveniles adhere to the mesh of the web and form a food base for the fungus. The ring trap of *Dactylaria candida* is made up of several mycelial cells. The ring is sensitive to touch and movement and can tighten when a nematode is inside. Traps of this type can also be adapted to the diameters of nematodes; as a nematode passes through the ring, it constricts when the nematode touches the inner walls of the ring [124]. The study showed high infectivity of *Arthrobotrys musiformis* and *A. oliogospora* against EPNs. *A. musiformis* is more dangerous because it killed more nematode species used in the experiment [276].

Research on predatory fungi (*Monacrosporium ellipsosporum* and *A. oligospora*) showed that, within 72 h, these fungi attacked and destroyed nematodes in vitro. In the case of the endoparasitic fungus *Drechmeria coniospora*, it was observed that the spores of the fungus infected EPN juveniles in the soil, colonized them, but did not infect individuals of the first generation of nematodes developing within the host [277].

Among fungi, there are also species that can act similarly to EPNs in their activity toward pests. Synergism was shown to exist between these two groups [142]. Research used entomopathogenic fungi (*Beauveria bassiana* and *Paecilomyces farinosus*) and EPNs (*S. carpocapsae* and *H. bacteriophora*) to protect peas from the pea leaf weevil, *Sitona lineatus* Linnaeus (Coleoptera: Curculionidae). It was shown that there were significantly fewer live larvae of the pest on plots with both fungi and nematodes present [278].

6.6. Bacteria

The life processes of EPNs are also affected by interactions with soil bacteria. Studies have shown that *Serratia marcescens* Bizio (Enterobacterales: Yersiniaceae) adversely affects nematodes' penetration into the host's body, limits their growth, and inhibits reproduction [279]. After entering the host, the nematodes are exposed to its intestinal microflora, which is unfavorable to symbiotic bacteria EPNs, reducing the infectivity of the parasite [280]. Other bacteria associated with nematodes are *Paenibacillus* sp. (Bacillales: Paenibacillaceae). These bacteria adhere to the nematode's cuticle and reproduce in the body of the insect infected with the nematode. A phoretic relationship between *Paenibacillus nematophilus* and EPNs (*Heterorhabditis* spp.) has been documented. It depends on the fact that the bacteria use nematodes as a means of transport into the insect's body without doing them any harm [235,281,282].

7. Summary

Entomopathogenic nematodes are naturally occurring organisms in the environment. A wide host circle and adaptability to a diverse range of soil environments make EPNs excellent agents for biological control of insect pests. Like other organisms, EPNs are affected in various ways by environmental soil pollutants in the form of heavy metals, petroleum substances, pesticides, fertilizers, and even essential oils, in response to which these organisms will struggle. Continuous development of plant protection methods allows for even more effective use of the insecticidal and adaptive abilities of EPNs (Table 1).

Table 1. Environmental factors influencing entomopathogenic nematodes.

Environmental Factor	Value	Significance for EPNs
	Abiotic Factors	
Soil texture	Textural classes (e.g., sandy, loam, and sandy loam)	Movement, infectivity, and survivability [48,99,118–127,129–132]
Moisture	25–40%	Under optimal conditions, nematodes move actively in the soil [48]
	>50%	Significant decreases in nematode motility and survival occur [48]
Temperature	13–25 °C	Optimal temperature range for development and biological activity [149]
	<−6 °C (nonencapsulated nematodes)	Below these values, death of the organisms occurs [154]
	<−32 °C (capsulated nematodes)	
	>32 °C	Above this value, biological activity and survival decrease significantly [146,156]
	>37 °C	Above this value, there is death of the organisms [157,158]
Soil pH	5.5–10	Optimal value range for survival and infectivity [163,164]
	<4	Significant deterioration in survival and infectivity [164]
Radiation	Wavelength of 366 nm	No noticeable changes [166,167]
	Wavelength of 254 nm, with an exposure time of >1.75 min	Reduced pathogenicity and reproduction and increased mortality [166,167]
	Gamma radiation	Improved biological effectiveness against plant pests [169,170]
Soil aeration	1–20%	High survivability and density [164,173]
	Low oxygen content	Significant increase in mortality [171,172]
Metal ions	Cd(II) and Pb(II)	Low concentrations of these metals result in an increase in infectivity, whereas higher concentrations result in a decrease in infectivity and speed of movement and an increase in mortality [187,191]
	Co(II), Cu(II), Mg(II), and Zn(II)	Toxic effects in combination with Cd(II) and Pb(II) [180]
	Mg(II)	Reduction in metal triad toxicity from Cd(II) or high nematode mortality [180]
	Mn(II)	Exhibits the lowest toxicity toward EPNs of all the metals tested [180]
Petroleum	Crude oil, diesel, engine oil, and gasoline	Reduction in pathogenicity and reproduction of EPNs [200,202–204]
Pesticides	E.g., thiamethoxam, chlorpyrifos, and carbosulfan	Reduced pathogenicity, reproduction, and invasiveness [205,207,208]

Table 1. *Cont.*

Environmental Factor	Value	Significance for EPNs
Fertilizers	Nitrates and phosphates	Cause an increase in mortality and reduce reproduction, infectivity, and speed of movement in soil [129,221,224,225]
Essential oils	Thyme oil, cinnamon oil, clove oil, and garlic oil	Significant increase in mortality [231]
	Spearmint oil, cedarwood oil, eucalyptus oil, and rosemary oil	No significant effect on nematode mortality [231,232]
	Biotic Factors	
Hosts	Increased concentration of insect excreta in the environment	Improved host localization [115,255,256]
	Host larval stage	Greatest infectivity [255,256]
	Pupal or imago stage	Significantly reduced infectivity [255,256]
	Resistance to EPNs	The more resistant the insect host species, the lower the chance of the EPN population expanding [237,254]
Mites	*Gamasellodes vermivorax* and *Alycus roseus*	Decreased numbers of EPNs in the soil [265,266]
Springtails	*Sinella curviseta*, *Folsomia candida*, and *Hypogastrura scotti*	Decreased numbers of EPNs in the soil and reduction in infectivity [265,266,269]
Nematodes	*Oscheius* spp.	Displace EPNs form the place of occurrence [270]
	Other species of EPNs	Increased biological activity [271]
	Nematophagous nematodes	Reduced the number of EPNs [144,272]
	Plant-pathogenic nematodes	Suppressed EPNs and antagonistic effect [273,274]
Fungi	Predatory fungi	Reduced populations of nematodes in the soil [124,275–277]
	Entomopathogenic fungi	Synergistic effect [142,278]
Bacteria	*Serratia marcescens*	Limits growth and reproduction [279]
	Host microbiome	Reduced infectivity [280]
	Paenibacillus nematophilus	Phoretic relationship [235,281,282]

Author Contributions: Conceptualization, J.M.-Ł. and W.K.; investigation, J.M.-Ł.; writing—original draft preparation, J.M.-Ł., S.D. and D.N.; writing—review and editing, J.M.-Ł., S.D., D.N. and W.K.; visualization, J.M.-Ł., S.D. and D.N.; supervision, J.M.-Ł.; project administration, J.M.-Ł. and W.K.; funding acquisition, J.M.-Ł. and W.K. All authors have read and agreed to the published version of the manuscript.

Funding: This study was supported by grant number SUPB.RN.23.236, awarded by W.K.

Institutional Review Board Statement: Not applicable.

Data Availability Statement: Dataset is available on reasonable request from the corresponding author.

Conflicts of Interest: The authors declare no conflicts of interest.

References

1. Lincicome, D.R. *The Goodness of Parasitism: A New Hypothesis. [W:] Aspect of the Biology of Symbiosis*; Cheng, T.C., Ed.; University Park Press: Baltimore, MD, USA, 1971; pp. 139–227.
2. Baer, J.G. *Ecology of Animal Parasites*; The University of Illinois Press: Urbana, IL, USA, 1951; pp. 1–224; ISBN 978-0-598-35258-3.
3. Kurtz, B.; Toepfer, S.; Ehlers, R.U.; Kuhlmann, U. Assessment of Establishment and Persistence of Entomopathogenic Nematodes for Biological Control of Western Corn Rootworm. *J. Appl. Entomol.* **2007**, *131*, 420–425. [CrossRef]
4. Lewis, E.E.; Campbell, J.; Griffin, C.T.; Kaya, H.K.; Peters, A.; Campbell, J.; Griffin, C.; Kaya, H.; Peters, A. Behavioral Ecology of Entomopathogenic Nematodes. *Biol. Control* **2006**, *38*, 66–79. [CrossRef]

5. Grewal, P.S. Formulation and Application Technology. In *Entomopathogenic Nematology*; Gaugler, R., Ed.; CABI Publishing: Wallingford, UK, 2002; pp. 265–287.
6. Imperiali, N.; Chiriboga, X.; Schlaeppi, K.; Fesselet, M.; Villacrés, D.; Jaffuel, G.; Bender, S.F.; Dennert, F.; Blanco-Pérez, R.; Van Der Heijden, M.G.A.; et al. Combined Field Inoculations of Pseudomonas Bacteria, Arbuscular Mycorrhizal Fungi, and Entomopathogenic Nematodes and Their Effects on Wheat Performance. *Front. Plant Sci.* **2017**, *8*, 1809. [CrossRef] [PubMed]
7. Shapiro-Ilan, D.I.; Han, R.; Dolinksi, C. Entomopathogenic Nematode Production and Application Technology. *J. Nematol.* **2012**, *44*, 206–217. [PubMed]
8. Shapiro-Ilan, D.I.; Jackson, M.; Reilly, C.C.; Hotchkiss, M.W. Effects of Combining an Entomopathogenic Fungi or Bacterium with Entomopathogenic Nematodes on Mortality of *Curculio caryae* (Coleoptera: Curculionidae). *Biol. Control* **2004**, *30*, 119–126. [CrossRef]
9. Aioub, A.A.A.; El-Ashry, R.M.; Hashem, A.S.; Elesawy, A.E.; Elsobki, A.E.A. Compatibility of Entomopathogenic Nematodes with Insecticides against the Cabbage White Butterfly, *Pieris rapae* L. (Lepidoptera: Pieridae). *Egypt J. Biol. Pest. Control* **2021**, *31*, 153. [CrossRef]
10. Dias, S.D.C.; De Brida, A.L.; Jean-Baptiste, M.C.; Leite, L.G.; Ovruski, S.M.; Lee, J.C.; Garcia, F.R.M. Compatibility of Entomopathogenic Nematodes with Chemical Insecticides for the Control of *Drosophila suzukii* (Diptera: Drosophilidae). *Plants* **2024**, *13*, 632. [CrossRef] [PubMed]
11. Jean-Baptiste, M.C.; De Brida, A.L.; Bernardi, D.; Da Costa Dias, S.; De Bastos Pazini, J.; Leite, L.G.; Wilcken, S.R.S.; Garcia, F.R.M. Effectiveness of Entomopathogenic Nematodes against *Ceratitis capitata* (Diptera: Tephritidae) Pupae and Nematode Compatibility with Chemical Insecticides. *J. Econ. Entomol.* **2021**, *114*, 248–256. [CrossRef]
12. Bathon, H. Impact of Entomopathogenic Nematodes on Non-Target Hosts. *Biocontrol. Sci. Technol.* **1996**, *6*, 421–434. [CrossRef]
13. Georgis, R.; Koppenhöfer, A.M.; Lacey, L.A.; Bélair, G.; Duncan, L.W.; Grewal, P.S.; Samish, M.; Tan, L. Successes and Failures in the Use of Parasitic Nematodes for Pest Control. *Biol. Control* **2006**, *38*, 103–123. [CrossRef]
14. Kaya, H.K.; Gaugler, R. Entomopathogenic Nematodes. *Annu. Rev. Entomol.* **1993**, *38*, 181–206. [CrossRef]
15. Fallet, P.; Machado, R.A.R.; Toepfer, S.; Ye, W.; Kajuga, J.; Waweru, B.; Bazagwira, D.; Turlings, T.C.J. A Rwandan Survey of Entomopathogenic Nematodes That Can Potentially Be Used to Control the Fall Armyworm. *IOBC-WPRS Bull.* **2020**, *150*, 87–90.
16. Fallet, P.; De Gianni, L.; Machado, R.A.R.; Bruno, P.; Bernal, J.S.; Karangwa, P.; Kajuga, J.; Waweru, B.; Bazagwira, D.; Degen, T.; et al. Comparative Screening of Mexican, Rwandan and Commercial Entomopathogenic Nematodes to Be Used against Invasive Fall Armyworm, *Spodoptera frugiperda*. *Insects* **2022**, *13*, 205. [CrossRef] [PubMed]
17. Ehlers, R.U. Mass Production of Entomopathogenic Nematodes for Plant Protection. *Appl. Microbiol. Biotechnol.* **2001**, *56*, 623–663. [CrossRef] [PubMed]
18. Ehlers, R.U.; Hokkanen, H.M.T. Insect Biocontrol with Non-Endemic Entomopathogenic Nematodes (*Steinernema* and *Heterorhabditis* spp.): Conclusions and Recommendations of a Combined OECD and COST Workshop on Scientific and Regulatory Policy Issues. *Biocontrol Sci. Technol.* **1996**, *6*, 295–302. [CrossRef]
19. Matuska-Łyżwa, J.; Kaca, W.; Żarnowiec, P. Biological Activity Wild Isolates of Entomopathogenic Nematodes to Horse-Chestnut Leaf Miner (*Cameraria ohridella*). *Pol. J. Environ. Stud.* **2015**, *24*, 1181–1184. [CrossRef] [PubMed]
20. Matuska-Łyżwa, J.; Żarnowiec, P.; Kaca, W. Comparison of Biological Activity of Field Isolates of *Steinernema feltiae* with a Commercial *S. Feltiae* Biopesticide Product. *Insects* **2021**, *12*, 816. [CrossRef] [PubMed]
21. Matuska-Łyżwa, J.; Wodecka, B.; Kaca, W. Characterization of *Steinernema feltiae* (Rhabditida: Steinernematidae) Isolates in Terms of Efficacy against Cereal Ground Beetle *Zabrus tenebrioides* (Coleoptera: Carabidae): Morphometry and Principal Component Analysis. *Insects* **2023**, *14*, 150. [CrossRef] [PubMed]
22. Taşkesen, Y.E.; Yüksel, E.; Canhilal, R. Field Performance of Entomopathogenic Nematodes against the Larvae of *Zabrus* spp. Clairville, 1806 (Coleoptera: Carabidae). *Uluslararası Tarım Yaban Hayatı Bilim. Derg.* **2021**, *7*, 429–437. [CrossRef]
23. Lacey, L.A.; Bettencourt, R.; Garrett, F.J.; Simões, N.J.; Gaugler, R.H. Factors Influencing Parasitism of Adult Japanese Beetles, *Polillia japonica* (Col.: Scarabaeidae) by Entomopathogenic Nematodes. *Entomophaga* **1993**, *38*, 501–509. [CrossRef]
24. Kumari, B.; Kumar, A.; Sujata, S.; Sharma, L. In Vitro Efficacy of Entomopathogenic Nematodes (EPNS) against Economically Important Insect-Pests of Cauliflower. *Int. J. Environ. Agric. Biotechnol.* **2024**, *9*, 50–58. [CrossRef]
25. Zeinolabedin Fard, N.; Abbasipour, H.; Saeedizadeh, A.; Karimi, J. Laboratory Assay of Entomopathogenic Nematodes against the Elm Leaf Beetle, *Xanthogaleruca luteola* Müller (Col.: Chrysomelidae). *J. For. Sci.* **2020**, *66*, 524–531. [CrossRef]
26. Vargas, G.; Velazquez-Hernandez, Y.; Daniel Greene, A. Entomopathogenic Nematodes to Control the Hibiscus Bud Weevil *Anthonomus testaceosquamosus* (Coleoptera: Curculionidae), above Ground and on Soil Surface. *BioControl* **2024**, *69*, 91–101. [CrossRef]
27. War, W.A.; Askary, T.H.; Ahad, I.; Mir, R.R.; Wani, F.J.; Bhat, M.A.; Rasool, J. Biocidal Efficacy of a New and Native Species of Entomopathogenic Nematode against Gram Pod Borer, *Helicoverpa armigera* (Hubner). *Int. J. Environ. Clim. Chang.* **2024**, *14*, 906–913. [CrossRef]
28. Poinar, G.O. *Biology and Taxonomy of Steinernematidae and Heterorhabditidae, Entomopathogenic Nematodes in Biological Control*; Gaugler, R., Kaya, H.K., Eds.; CRC Press: Boca Raton, FL, USA, 1990; pp. 23–61.
29. Smits, P.H. Post-Application Persistence of Entomopathogenic Nematodes. *Biocontrol Sci. Technol.* **1996**, *6*, 379–388. [CrossRef]
30. Kung, S.P.; Gaugler, R.; Kaya, H.K. Soil Type and Entomopathogenic Nematode Persistence. *J. Invertebr. Pathol.* **1990**, *55*, 401–406. [CrossRef]

31. Timper, P.; Davies, K.G. Biotic Interactions. In *Nematode Behaviour*; Gaugler, R., Bilgrami, A.L., Eds.; CABI Publishing: Wallingford, UK, 2004; pp. 277–307; ISBN 0-85199-818-6.
32. Shapiro-Ilan, D.I.; Gouge, D.H.; Piggott, S.J.; Fife, J.P. Application Technology and Environmental Considerations for Use of Entomopathogenic Nematodes in Biological Control. *Biol. Control* **2006**, *38*, 124–133. [CrossRef]
33. Malinowski, H. Stan badań nad wykorzystaniem czynników biologicznych do ograniczania populacji owadów żerujących na korzeniach. *Pr. Inst. Badaw. Leśn. Ser. A* **1998**, *860*, 104–129.
34. Tomalak, M. *Ekspertyza. Czynniki Biologiczne Dostępne w Ochronie Upraw Ekologicznych Przed Szkodnikami*; Państwowy Instytut Badawczy: Poznań, Poland, 2009; p. 28.
35. Boczek, J.; Brzeski, M.; Korpczyńska-Linkiewicz, D. *Wybrane Działy Zoologii*; PWN: Warszawa, Poland, 2000; p. 343.
36. Fallet, P.; Bazagwira, D.; Ruzzante, L.; Ingabire, G.; Levivier, S.; Bustos-Segura, C.; Kajuga, J.; Toepfer, S.; Turlings, T.C.J. Entomopathogenic Nematodes as an Effective and Sustainable Alternative to Control the Fall Armyworm in Africa. *PNAS Nexus* **2024**, *3*, 122. [CrossRef] [PubMed]
37. Dillman, A.R.; Chaston, J.M.; Adams, B.J.; Ciche, T.A.; Goodrich-Blair, H.; Stock, S.P.; Sternberg, P.W. An Entomopathogenic Nematode by Any Other Name. *PLoS Pathog.* **2012**, *8*, e1002527. [CrossRef]
38. Bhat, A.H.; Chaubey, A.K.; Askary, T.H. Global Distribution of Entomopathogenic Nematodes, *Steinernema* and *Heterorhabditis*. *Egypt J. Biol. Pest Control* **2020**, *30*, 31. [CrossRef]
39. Campos, R.; Barbercheck, M.; Hoy, C.W.; Stock, S.P. Entomopathogenic Nematodes as a Model System for Advancing the Frontiers of Ecology. *J. Nematol.* **2012**, *44*, 162–176.
40. Koppenhöfer, H.S. Bacterial Symbionts of *Steinernema* and *Heterorhabditis*. *Entomopathog. Nematodes Syst. Phylogeny Bact. Symbionts* **2007**, *5*, 735–808.
41. Campbell, J.; Gaugler, R.H. Role of the Sheath in Desiccation Tolerance of Two Entomopathogenic Nematodes. *Nematologica* **1991**, *37*, 324–332. [CrossRef]
42. Ciche, T.A.; Darby, C.; Ehlers, R.U.; Forst, S.; Goodrich-Blair, H. Dangerous Liaisons: The Symbiosis of Entomopathogenic Nematodes and Bacteria. *Biol. Cont.* **2006**, *38*, 22–46. [CrossRef]
43. Wouts, W.M. The Biology and Life Cycle of a New Zealand Population of *Heterorhabditis heliothidis* (Heterorhabditidae). *Nematologica* **1979**, *25*, 191–202. [CrossRef]
44. Batalla-Carrera, L.; Morton, A.; García-del-Pino, F. Efficacy of Entomopathogenic Nematodes against the Tomato Leafminer *Tuta absoluta* in Laboratory and Greenhouse Conditions. *BioControl* **2010**, *55*, 523–530. [CrossRef]
45. Wang, Y.; Gaugler, R. Host and Penetration Site Location by Entomopathogenic Nematodes against Japanese Beetle Larvae. *J. Invertebr. Pathol.* **1998**, *72*, 313–318. [CrossRef] [PubMed]
46. Banerjee, N.; Hallem, E.A. The Role of Carbon Dioxide in Nematode Behaviour and Physiology. *Parasitology* **2020**, *147*, 841–854. [CrossRef] [PubMed]
47. Dunphy, G.B.; Webster, J.M. The Monoxenic Culture of Neoaplectana Carpocapsae DD 136 and *Heterorhabditis heliothidis*. *Rev. Nematol.* **1989**, *12*, 113–123.
48. Kaya, H.K. Soil Ecology. In *Entomopathogenic Nematodes in Biological Control*; Gaugler, R., Kaya, H.K., Eds.; CRC Press: Boca Raton, FL, USA, 1990; pp. 93–115.
49. Torr, P.; Heritage, S.; Wilson, M.J. Vibrations as Anovel Signal for Host Location by Parasitic Nematodes. *Int. J. Parasitol.* **2004**, *34*, 997–999. [CrossRef]
50. Koppenhöfer, A.M. Nematodes. In *Field Manual of Techniques for the Application and Evaluation of Entomopathogens*; Lacey, L., Kaya, H., Eds.; Kluwer: Dordrecht, The Netherlands, 2000; pp. 249–264.
51. Goodrich-Blair, H.; Clarke, D.J. Mutualism and Pathogenesis in Xenorhabdus and Photorhabdus: Two Roads to the Same Destination. *Mol. Microbiol.* **2007**, *64*, 260–268. [CrossRef]
52. Nielsen-LeRoux, C.; Gaudriault, S.; Ramarao, N.; Lereclus, D.; Givaudan, A. How the Insect Pathogen Bacteria *Bacillus thuringiensis* and Xenorhabdus/Photorhabdus Occupy Their Hosts. *Curr. Opin. Microbiol.* **2012**, *15*, 220–231. [CrossRef]
53. Endo, B.Y.; Nickle, W.R. Ultrastructure of the Buccal Cavity Region and Oesophagus of the Insect Parasitic Nematode, *Heterorhabditis bacteriophora*. *Nematologica* **1994**, *40*, 379–398. [CrossRef]
54. Spiridonov, S.E.; Akhmedov, E.N.; Belostotskaya, F.N. Proliferation of Symbiotic Bacteria in the Intestinal Vesicles of Invasive Larvae of *Neoplectana* spp. (Nematoda, Steinernematidae). *Helminthol* **1991**, *28*, 141–142.
55. Bird, A.; Akhurst, R.J. The Nature of the Intestinal Vesicle in Nematodes of the Family Steinernematidae. *Int. J. Parasitol.* **1983**, *13*, 599–606. [CrossRef]
56. Forst, S.; Clarke, D.J. Bacteria-Nematode Symbiosis. In *Entomopathogenic Nematology*; Gaugler, R., Ed.; CAB International: Wallingford, UK, 2002; pp. 57–77.
57. Ciche, T.A.; Ensign, J.C. For the Insect Pathogen *Photorhabdus luminescens*, Which End of a Nematode Is Out? *Appl. Environ. Microbiol.* **2003**, *69*, 1890–1897. [CrossRef]
58. Adams, B.J.; Fodor, A.; Koppenhöfer, H.S.; Stackebrandt, E.; Patricia Stock, S.; Klein, M.G. Reprint of "Biodiversity and Systematics of Nematode–Bacterium Entomopathogens" [Biol. Control 37 (2006) 32–49]. *Biol. Control* **2006**, *38*, 4–21. [CrossRef]
59. Hominick, W.M. Entomopathogenic Rhabditid Nematodes and Pest Control. *Parasitol. Today* **1990**, *6*, 148–152. [CrossRef]
60. Han, R.C.; Ehlers, R.U. Pathogenicity, Development, and Reproduction of *Heterorhabditis bacteriophora* and *Steinernema carpocapsae* under Axenic In Vivo Conditions. *J. Invertebr. Pathol.* **2000**, *75*, 55–58. [CrossRef]

61. Kinkar, O.U.; Prashar, A.; Kumar, A.; Hadapad, A.B.; Hire, R.S.; Makde, R.D. Txp40, an Insecticidal Toxin Protein from *Xenorhabdus nematophila*: Purification, Toxicity Assessment and Biophysical Characterization. *Toxicon* **2022**, *218*, 40–46. [CrossRef] [PubMed]
62. Cowles, K.N.; Goodrich-Blair, H. Expression and Activity of a Xenorhabdus Nematophila Haemolysin Required for Full Virulence towards *Manduca sexta* Insects. *Cell Microbiol.* **2005**, *7*, 209–219. [CrossRef] [PubMed]
63. Morgan, J.A.W.; Sergeant, M.; Ellis, D.; Ousley, M.; Jarrett, P. Sequence Analysis of Insecticidal Genes from *Xenorhabdus nematophilus* PMFI296. *Appl. Environ. Microbiol.* **2001**, *67*, 2062–2069. [CrossRef]
64. Martens, E.C.; Heungens, K.; Goodrich-Blair, H. Early Colonization Events in the Mutualistic Association between *Steinernema carpocapsae* Nematodes and *Xenorhabdus nematophila* Bacteria. *J. Bacteriol.* **2003**, *185*, 3147–3154. [CrossRef] [PubMed]
65. Matha, V.; Mraček, Z. Changes in Haemocyte Counts in *Galleria mellonella* (L.) (Lepidoptera: Gallerüdae) Larvae Infected with *Steinernema* Sp. (Nematoda: Steinernematidae). *Nematology* **1984**, *30*, 86–89. [CrossRef]
66. Bartyzel, M.; Jarosz, J. Bakterie Nematofilne z Rodzaju Xenorhabdus Związane Mutualistycznie z Owadobójczymi Nicieniami. *Postępy Mikrobiol.* **1987**, *26*, 55–63.
67. Simoes, N.; Rosa, J.S. Pathogenicity and Host Specificity of Entomopathogenic Nematodes. *Biocontrol Sci. Technol.* **1996**, *6*, 403–412. [CrossRef]
68. Bowen, D.; Rocheleau, T.A.; Blackburn, M.; Andreev, O.; Golubeva, E.; Bhartia, R.; French-Constant, O. Insecticidal Toxins from the Bacterium *Photorhabdus luminescens*. *Science* **1998**, *280*, 2129–2132. [CrossRef] [PubMed]
69. Sharma, S.; Waterfield, N.; Bowen, D.; Rocheleau, T.; Holland, L.; James, R.; Ffrench-Constant, R. The Lumicins: Novel Bacteriocins from *Photorhabdus luminescens* with Similarity to the Uropathogenic-Specific Protein (USP) from Uropathogenic *Escherichia coli*. *FEMS Microbiol. Lett.* **2002**, *214*, 241–249. [CrossRef]
70. Nguyen, K.; Smart, G.C.J. Life Cycle of Steinernema Scapterisci. *J. Nematol.* **1992**, *24*, 160–169.
71. Akhurst, R.J. Antibiotic Activity of *Xenorhabdus* spp., Bacteria Symbiotically Associated with Insect Pathogenic Nematodes of the Families Heterorhabditidae and Steinernematidae. *Microbiology* **1982**, *128*, 3061–3065. [CrossRef]
72. Boemare, N.; Givaudan, A.; Brehelin, M.; Laumond, C. Symbiosis and Pathogenicity of Nematode-Bacterium Complexes. *Symbiosis* **1997**, *22*, 21–45.
73. Thaler, J.O.; Baghdiguian, S.; Boemare, N. Purification and Characterization of Xenorhabdicin, a Phage Tail-like Bacteriocin, from the Lysogenic Strain F1 of *Xenorhabdus nematophilus*. *Appl. Environ. Microbiol.* **1995**, *61*, 2049–2052. [CrossRef] [PubMed]
74. Jarosz, J. Strategie Obrony Przeciwzakaźnej i Sposoby Ucieczki Entomopatogenów Spod Kontroli Immunologicznej Owadów. *Wiad. Parazyt.* **1996**, *42*, 3–37. [PubMed]
75. Bode, H.B. Entomopathogenic Bacteria as a Source of Secondary Me-Tabolites. *Curr. Opin. Chem. Biol.* **2009**, *13*, 224–2230. [CrossRef]
76. Gulcu, B.; Hazir, S.; Kaya, H.K. Scavenger Deterrent Factor (SDF) from Symbiotic Bacteria of Entomopathogenic Nematodes. *J. Invertebr. Pathol.* **2012**, *110*, 326–333. [CrossRef]
77. Ogier, J.C.; Pagès, S.; Frayssinet, M.; Gaudriault, S. Entomopathogenic Nematode-Associated Microbiota: From Monoxenic Paradigm to Pathobiome. *Microbiome* **2020**, *8*, 25. [CrossRef]
78. Ogier, J.C.; Akhurst, R.; Boemare, N.; Gaudriault, S. The Endosymbiont and the Second Bacterial Circle of Entomopathogenic Nematodes. *Trends Microbiol.* **2023**, *31*, 629–642. [CrossRef]
79. Lechowicz, I.; Chrapek, M.; Czerwonka, G.; Korzeniowska-Kowal, A.; Tobiasz, A.; Urbaniak, M.; Matuska-Łyżwa, J.; Kaca, W. Detection of Ureolytic Activity of Bacterial Strains Isolated from Entomopathogenic Nematodes Using Infrared Spectroscopy. *J. Basic Microbiol.* **2016**, *56*, 922–928. [CrossRef] [PubMed]
80. Croll, N.A. *Pasożytnictwo i Inne Związki*; Wyd. Naukowe PWN: Warszawa, Polska, 1977; p. 193.
81. Burnell, A.; Stock, S.P. *Heterorhabditis*, *Steinernema* and Their Bacterial Symbionts—Lethal Pathogens of Insects. *Nematol* **2000**, *2*, 31–42. [CrossRef]
82. Wang, J.; Bedding, R.A. Population Development of *Heterorhabditis bacteriophora* and *Steinernema carpocapsae* in the Larvae of *Galleria mellonella*. *Fundam. Appl. Nematol.* **1996**, *19*, 363–367.
83. Nguyen, K.B. Methodology, Morphology and Identification. In *Entomopathogenic Nematodes: Systematics, Phylogeny and Bacterial Symbionts*; Nguyen, K.B., Hunt, D., Eds.; Brill: Leiden, The Netherland, 2007; pp. 59–119.
84. Johnigk, S.A.; Ehlers, R.U. Endotokia Matricida in Hermaphrodites of *Heterorhabditis* spp. and the Effect of the Food Supply. *Nematology* **1999**, *1*, 717–726. [CrossRef]
85. Popiel, I.; Friedman, M.J.; Grove, D.L. Infective Juvenile Formation in the Insect Parasitic Nematode *Steinernema feltiae*. *Parasitology* **1989**, *99*, 77–81. [CrossRef]
86. Sandner, H. *Pasożyty w Służbie Człowieka*; PWN: Warszawa, Polska, 1990; p. 168.
87. Brown, I.M.; Gaugler, R. Temperature and Humidity Influence Emergence and Survival of Entomopathogenic Nematodes. *Nematologica* **1997**, *43*, 363–375.
88. Tomalak, M. Rejestracja biologicznych środków ochrony roślin w Europie—Nowe perspektywy. *Prog. Plant Prot.* **2007**, *47*, 233–240.

89. Sicard, M.; Brugirard-Ricaud, K.; Lanois, A.; Boemare, N.E.; Pagès, S.; Brehe, M.; Givaudan, A. Stages of Infection during the Tripartite Interaction between *Xenorhabdus nematophila*, Its Nematode Vector, and Insect Hosts. *Appl. Environ. Mi-Crobiol.* **2004**, *70*, 6473–6480. [CrossRef] [PubMed]
90. Emelianoff, V.; Sicard, M.; Le Brun, N.; Moulia, C.; Ferdy, J.-B. Effect of Bacterial Symbionts *Xenorhabdus* on Mortality of Infective Juveniles of Two Steinernema Species. *Parasitol. Res.* **2007**, *100*, 657–659. [CrossRef] [PubMed]
91. Chapuis, É.; Arnal, A.; Ferdy, J.-B. Trade-Offs Shape the Evolution of the Vector-Borne Insect Pathogen *Xenorhabdus nematophila*. *Proc. R. Soc. B* **2012**, *279*, 2672–2680. [CrossRef] [PubMed]
92. Gaugler, R.; Wang, Y.I.; Campbell, J.F. Aggressive and Evasive Behaviors in *Popillia japonica* (Coleoptera: Scarabaeidae) Larvae: Defenses against Entomopathogenic Nematode Attack. *J. Invertebr. Pathol.* **1994**, *64*, 193–199. [CrossRef]
93. Lavine, M.D.; Strand, M.R. Insect Hemocytes and Their Role in Immunity. *Insect Biochem. Mol. Biol.* **2002**, *32*, 1295–1309. [CrossRef]
94. Castillo, J.C.; Reynolds, S.E.; Eleftherianos, I. Insect Immune Responses to Nematode Parasites. *Trends Parasitol.* **2011**, *27*, 537–547. [CrossRef] [PubMed]
95. Lackie, A.M. Immune Recognition in Insects. *Der. Comp. Immunol.* **1981**, *5*, 191–204. [CrossRef] [PubMed]
96. Söderhäll, K. Prophendoxidase Activating System and Melanization: A Recognition Mechanism of Arthropodas? A review. *Dev. Comp. Immunol.* **1982**, *6*, 601–611. [PubMed]
97. Brivio, M.F.; Mastore, M.; Moro, M. The Role of Steinernema Feltiae Body-Surface Lipids in Host–Parasite Immunological Interactions. *Mol. Biochem. Parasitol.* **2004**, *135*, 111–121. [CrossRef] [PubMed]
98. Campos-Herrera, R.; Gómez-Ros, J.M.; Escuer, M.; Cuadra, L.; Barrios, L.; Gutiérrez, C. Diversity, Occurrence, and Life Characteristics of Natural Entomopathogenic Nematode Populations from La Rioja (Northern Spain) under Different Agricultural Management and Their Relationships with Soil Factors. *Soil Biol. Biochem.* **2008**, *40*, 1474–1484. [CrossRef]
99. Campos-Herrera, R.; Gutierrez, C. Screening Spanish Isolates of Steinernematid Nematodes for Use as Biological Control Agents through Laboratory and Greenhouse Microcosm Studies. *J. Invertebr. Pathol.* **2009**, *100*, 100–105. [CrossRef]
100. Campos-Herrera, R.; Escuer, M.; Labrador, S.; Robertson, L.; Barrios, L.; Gutierrez, C. Distribution of the Entomopathogenic Nematodes from La Rioja (Northern Spain). *J. Invertebr. Pathol.* **2007**, *95*, 125–139. [CrossRef] [PubMed]
101. Levy, N.; Faigenboima, A.; Salame, L.; Molina, C.; Ehlers, R.U.; Glazer, I.; Ment, D. Characterization of the Phenotypic and Genotypic Tolerance to Abiotic Stresses of Natural Populations of *Heterorhabditis bacteriophora*. *Sci. Rep.* **2020**, *10*, 10500. [CrossRef] [PubMed]
102. Hominick, W.M.; Reid, A.P.; Bohan, D.A.; Briscoe, B.R. Entomopathogenic Nematodes: Biodiversity, Geographical Distribution and the Convention on Biological Diversity. *Biocontrol Sci. Technol.* **1996**, *6*, 317–332. [CrossRef]
103. Hominick, W.M.; Reid, A.P.; Briscoe, B.R. Prevalence and Habitat Specificity of Steinernematid and Heterorhabditid Nematodes Isolated during Surveys of the UK and The Netherlands. *J. Helminthol.* **1995**, *69*, 27–32. [CrossRef]
104. Stock, S.P.; Pryor, B.M.; Kaya, H.K. Distribution of Entomopathogenic Nematodes (Steinernematidae and Heterorhabditidae) in Natural Habitats in California, USA. *Biodivers. Conserv.* **1999**, *8*, 535–549. [CrossRef]
105. Sturhan, D. Prevalence and Habitat Specificity of Entomopathogenic Nematodes in Germany. In *COST 819 Entomopathogenic Nematodes. Application and Persistence of Entomopathogenic Nematodes*; Gwynn, R.L., Smits, P.H., Griffin, C., Ehlers, R.U., Boemare, N., Masson, J.P., Eds.; European Commission EUR 18873: Luxembourg, 1999; Volume DG XII, pp. 123–132.
106. Sturhan, D.; Liskova, A.M. Occurrence and Distribution of Entomopathogenic Nematodes in the Slovak Republic. *Nematology* **1999**, *1*, 273–277. [CrossRef]
107. Matuska-Łyżwa, J. *Dzikie Izolaty Steinernematidae i Heterorhabditidae z Różnych Siedlisk Ekologicznych Polski, Wpływ Niektórych Elementów Środowiska Na Zdrowie i Funkcjonowanie Człowieka. Konieczność Koegzystencji w Obcowaniu z Przyrodą*; IOŚ-PIB: Warszawa, Polska, 2017; pp. 87–104. ISBN 978-83-60312-23-0.
108. Hominick, W.M. Biogeography. In *Entomopathogenic Nematology*; Gaugler, R., Ed.; CABI Publishing: Oxon, UK, 2002; pp. 115–143.
109. Griffin, C.T.; Joyce, S.A.; Dix, I.; Burnell, A.M.; Downes, M.J. Characterisation of the Entomopathogenic Nematode *Heterorhabditis* (Nematoda: Heterorhabditidae) from Ireland and Britain by Molecular and Cross-Breeding Techniques, and the Occurrence of the Genus in These Islands. *Fundam. Appl. Nematol.* **1994**, *17*, 245–254.
110. Yoshida, M.; Reid, A.P.; Briscoe, B.R.; Hominick, W.M. Survey of Entomopathogenic Nematodes (Rhabditida: Steinernematidae and Heterorhabditidae) in Japan. *Fundam. Appl. Nematol.* **1998**, *21*, 185–198.
111. Constant, P.; Fischer-Le-Saux, M.; Briand-Panoma, S.; Mauléon, H. Natural Occurrence of Entomopathogenic Nematodes (Rhabditida: Steinernematidae and Heterorhabditidae) in Guadeloupe Islands. *Fundam. Appl. Nematol.* **1998**, *21*, 667–672.
112. Stock, S.P.; Strong, D.; Gardner, S.L. Identification of *Heterorhabditis* (Nematoda: Heterorhabditidae) from California with a New Species Isolated from the Larvae of the Ghost Moth *Hepialis californicus* (Lepidoptera: Hepialidae) from the Bodega Bay Natural Reserve. *Fundam. Appl. Nematol.* **1996**, *19*, 585–592.
113. Phan, K.L.; Subbotin, S.A.; Nguyen, N.C.; Moens, M. *Heterorhabditis baujardi* sp. n. (Rhabditida: Heterorhabditidae) from Vietnam and Morphometric Data for *H. indica* Populations. *Nematology* **2003**, *5*, 367–382.
114. Hominick, W.M.; Briscoe, B.R. Occurrence of Entomopathogenic Nematodes (Rhabditida; Steinernematidae; Heterorhabditidae) in British Soil. *Parasitology* **1990**, *100*, 295–302. [CrossRef]
115. Ridgeway, R.L.; Vilson, S.B. *Biological Control by Augmentation of Natural Enemies: Insect and Mite Control with Parasites and Predators*; Plenum Press: New York, NY, USA, 1977; p. 479.

116. Kaya, H.K. Natural Enemies and Other Antagonists. In *Entomopathogenic Nematology*; Gaugler, R., Ed.; CABI Publishing: Wallingford, UK, 2002; pp. 189–202.
117. Timper, P.; Kaya, H.K. Impact of a Nematode-Parasitic Fungus on the Effectiveness of Entomopathogenic Nematodes. *J. Nematol.* **1992**, *24*, 1–8.
118. Kapranas, A.; Maher, A.M.D.; Griffin, C.T. The Influence of Organic Matter Content and Media Compaction on the Dispersal of Entomopathogenic Nematodes with Different Foraging Strategies. *Parasitology* **2017**, *144*, 1956–1963. [CrossRef]
119. Choo, H.Y.; Kaya, H.K. Influence of Soil Texture and Presence of Roots on Host Finding by *Heterorhabditis bacteriophora*. *J. Invertebr. Pathol.* **1991**, *58*, 279–280. [CrossRef]
120. Toledo, J.; Williams, T.; Pérez, C.; Liedo, P.; Valle, J.F.; Ibarra, J.E. Abiotic Factors Affecting the Infectivity of *Steinernema carpocapsae* (Rhabditida: Steinernematidae) on Larvae of *Anastrepha obliqua* (Diptera: Tephritidae). *Biocontrol Sci. Technol.* **2009**, *19*, 887–898. [CrossRef]
121. Shapiro, D.I.; McCoy, C.W.; Fares, A.; Obreza, T.; Dou, H. Effects of Soil Type on Virulence and Persistence of Entomopathogenic Nematodes in Relation to Control of *Diaprepes* (Coleoptera: Curculionidae). *Environ. Entomol.* **2000**, *29*, 1083–1087. [CrossRef]
122. Koppenhöfer, A.M.; Fuzy, E.M. Effect of Soil Type on Infectivity and Persistence of the Entomopathogenic Nematodes *Steinernema scarabaei*, *Steinernema glaseri*, *Heterorhabditis zealandica*, and *Heterorhabditis bacteriophora*. *J. Invertebr. Pathol.* **2006**, *92*, 11–22. [CrossRef] [PubMed]
123. Hazir, S.; Kaya, H.K.; Stock, S.; Keskin, N. Entomopathogenic Nematodes (Steinernematidae and Heterorhabditidae) for Biological Control of Soil Pests. *Turk. J. Biol.* **2003**, *27*, 181–202.
124. Brzeski, M.; Sandner, H. *Zarys Nematologii*; PWN: Warszawa, Poland, 1974; pp. 18–256.
125. Moyle, P.L.; Kaya, H.K. Dispersal and Infectivity of the Entomogenous Nematode, *Neoaplectana carpocapsae* Weiser (Rhabditida: Steinernematidae), in Sand. *J. Nematol.* **1981**, *13*, 295–300.
126. Georgis, R.; Poinar, G.O. Effect of Soil Texture on the Distribution and Infectivity of *Neoaplectana glaseri* (Nematoda: Steinernematidae). *J. Nematol.* **1983**, *15*, 329–332. [PubMed]
127. Sandner, H. Movement of Invasive Larvae of *Heterorhabditis bacteriophora* Poinar in Soil. *Ann. Wars. Agricult. Univ. SGGW-AR Anim.* **1986**, *20*, 45–47.
128. Sandner, H. Movement of Invasive Larvae of *Steinernema feltiae* (Filipjev) in Soil. *Ann. Wars. Agricult. Univ. SGGW-AR Anim. Sc.* **1986**, *20*, 41–44.
129. Bednarek, A. *Ekologiczne Uwarunkowania Aktywności Biologicznej Nicieni Entomofilnych w Środowisku Glebowym Agrocenoz*; Rozprawy Naukowe i Monografie, SGGW: Warszawa, Polska, 1990; p. 108. ISBN 83-00-02646-0.
130. Sanchez, W.; Shapiro, D.; Williams, G.; Lawrence, K. Entomopathogenic Nematode Management of Small Hive Beetles (*Aethina tumida*) in Three Native Alabama Soils under Low Moisture Conditions. *J. Nematol.* **2021**, *53*, e2021-63. [CrossRef]
131. Molyneux, A.S.; Bedding, R.A. Influence of Soil Texture and Moisture on the Infectivity of *Heterorhabditis* sp. D1 and *Steinernema glaseri* for Larvae of the Sheep Blowfly, *Lucilia cuprina*. *Nematologica* **1984**, *30*, 358–365. [CrossRef]
132. Susurluk, A.; Bütüner, A.K. The Role of Drought in the Efficacy of Some Entomopathogenic Nematodes. *Türk. Entomol. Derg.* **2024**, *48*, 103–110. [CrossRef]
133. Toepfer, S.; Kurtz, B.; Kuhlmann, U. Influence of Soil on the Efficacy of Entomopathogenic Nematodes in Reducing *Diabrotica virgifera virgifera* in Maize. *J. Pest Sci.* **2010**, *83*, 257–264. [CrossRef] [PubMed]
134. Lankin, G.; Vidal-Retes, G.; Allende, G.; Castaneda-Alvarez, C.; San-Blas, E.; Aballay, E. Soil Texture, Infective Juvenile Concentration, and Soil Organic Matter Influence the Efficacy of *Steinernema feltiae* Isolate Lican Ray. *J. Nematol.* **2020**, *52*, 1–11. [CrossRef] [PubMed]
135. Shapiro-Ilan, D.I.; Gaugler, R. Production Technology for Entomopathogenic Nematodes and Their Bacterial Symbionts. *J. Ind. Microbiol. Biotechnol.* **2002**, *28*, 137–146. [CrossRef] [PubMed]
136. Norton, D.C. *Ecology of Plant Parasitic Nematodes*; John Wiley & Sons: New York, NY, USA, 1978; p. 284.
137. Glazer, I. *Survival Biology [w:] Entomopathogenic Nematology*; Gaugler, R., Ed.; CABI Publishing: Wallingford, UK, 2002; pp. 169–187.
138. Kulkarni, N.; Mishra, V.K.; Daksh, S.; Paunikar, S.D. Response of Native Entomopathogenic Nematode, *Steinernema* spp. (TFRIEPN-57) Isolated from Central India to Variation in Temperature and Soil Moisture. *J. Environ. Biol.* **2016**, *37*, 399–406.
139. Wallace, H.R. The Dynamics of Nematode Movement. *Annu. Rev. Phytopathol.* **1968**, *6*, 91–114. [CrossRef]
140. Molyneux, A.S. Survival of Infective Juveniles of *Heterorhabditis* spp., and *Steinernema* spp. (Nematoda: Rhabditida) at Various Temperatures and Their Subsequent Infectivity for Insects. *Rev. Nétnatol.* **1985**, *8*, 165–170.
141. Koppenhöfer, A.M.; Baur, M.E.; Stock, S.P.; Choo, H.Y.; Chinnasri, B.; Kaya, H.K. Survival of Entomopathogenic Nematodes within Host Cadavers in Dry Soil. *Appl. Soil Ecol.* **1997**, *6*, 231–240. [CrossRef]
142. Kamionek, M.; Maslana, I.; Sandner, H. The Survival of Invasive Larvae of *Neoaplectana carpocapsae* Weiser in a Waterless Environment under Various Conditions of Temperature and Humidity. *Zesz. Probl. Post. Nauk Roln.* **1974**, *134*, 409–412.
143. Geden, C.J.; Axtell, R.C. Effect of Temperature on Nematode (*Steinernema feltiae* [Nematoda: Steinernematidae]) Treatment of Soil for Control of Lesser Mealworm (Coleoptera: Tenebrionidae) in Turkey Houses. *J. Econ. Entomol.* **1988**, *81*, 800–803. [CrossRef]
144. Ishibashi, N.; Kondo, E. *Steinernema feltiae* (DD-136) and *S. glaseri*: Persistence in Soil and Bark Compost and Their Influence on Native Nematodes. *J. Nematol.* **1986**, *18*, 310–316. [PubMed]
145. Kung, S.P.; Gaugler, R.; Kaya, H.K. Effects of Soil Temperature, Moisture, and Relative Humidity on Entomopathogenic Nematode Persistence. *J. Invertebr. Pathol.* **1991**, *57*, 242–249. [CrossRef]

146. Grewal, P.S.; Selvan, S.; Gaugler, R. Thermal Adaptation of Entomopathogenic Nematodes: Niche Breadth for Infection, Establishment, and Reproduction. *J. Therm. Biol.* **1994**, *19*, 245–253. [CrossRef]
147. Georgis, R.; Manweiler, S.A. Entomopathogenic Nematodes: A Developing Biological Control Technology. *Agric. Zool. Rev.* **1994**, *6*, 63–94.
148. Aryal, S.; Nielsen, U.N.; Sumaya, N.H.; Wilson, C.; Riegler, M. Effect of Temperature on Survival of Australian Entomopathogenic Nematodes and Their Virulence against the Queensland Fruit Fly, *Bactrocera tryoni*. *BioControl* **2022**, *67*, 617–628. [CrossRef]
149. Danilov, L.G.; Vasil'eva, S.O.; Gogolev, A.N. Influence of Temperature on the Infective Activity of Entomopathogenic Nematodes from the Families Steinernematidae and Heterorhabditidae. *Parazitol* **1994**, *28*, 495–500.
150. Kaya, H.K. Development of the DD-136 Strain of *Neoaplectana carpocapsae* at Constant Temperatures. *J. Nematol.* **1977**, *9*, 346–349. [PubMed]
151. Jagdale, G.B.; Gordon, R.; Gordon, R. Effect of Recycling Temperature on the Infectivity of Entomopathogenic Nematodes. *Can. J. Zool.* **1997**, *75*, 2137–2141. [CrossRef]
152. Kowalska, J. Effect of Three Temperatures on Infectivity of Selected Entomopathogenic Nematodes. *Chem. Inż. Ekol.* **2005**, *12*, 411–414.
153. Trdan, S.; Valič, N.; Urek, G.; Milevoj, L. Concentration of Suspension and Temperature as Factors of Pathogenicity of Entomopathogenic Nematodes for the Control of Granary Weevil, *Sitophilus granarius* (L.) (Coleoptera: Curculionidae). *Acta Agric. Slov.* **2005**, *85*, 117–124. [CrossRef]
154. Wharton, D.A.; Surrey, M.R. Cold Tolerance Mechanisms of the Infective Larvae of the Insect Parasitic Nematode, *Heterorhabditis zealandica* Poinar. *Cryo Lett.* **1994**, *15*, 353–360.
155. Brown, I.M.; Gaugler, R. Cold Tolerance of Steinernematid and Heterorhabditid Nematodes. *J. Therm. Biol.* **1996**, *21*, 115–121. [CrossRef]
156. Zervos, S.; Johnson, S.C.; Webster, J.M. Effect of Temperature and Inoculum Size on Reproduction and Development of *Heterorhabditis heliothidis* and *Steinernema glaseri* (Nematoda: Rhabditoidea) in Galleria Mellonella. *Can. J. Zool.* **1991**, *69*, 1261–1264. [CrossRef]
157. Wouts, W.M. *Steinernema (Neoaplectana) and Heterorhabditis Species. Manual of Agricultural Nematology*; Nickle, W.R., Ed.; Marcel Dekker: New York, NY, USA, 1991; pp. 855–897.
158. Sanda, N.B. Effect of Temperatures on the Pathogenicity of Entomopathogenic Nematodes (*Steinernema carpocapsae* and *Heterorhabditis bacteriophora*) for Biocontrol of *Octodonta nipae* (Maulik). *Niger. J. Entomol.* **2020**, *36*, 88–95. [CrossRef]
159. Shamseldean, M.M.; Abd-Elgawad, M.M. Natural Occurrence of Insect Pathogenic Nematodes (Rhabditida: Heterorhabditidae) in Egyptian Soils. *Afro Asian J. Nematol.* **1994**, *4*, 151–154.
160. Amarasinghe, L.D.; Hominick, W.M.; Briscoe, B.R.; Reid, A.P. Occurrence and Distribution of Entomopathogenic Nematodes in Sri Lanka. *J. Helminthol.* **1994**, *68*, 277–286. [CrossRef]
161. Glazer, I.; Liran, N.; Poinar, G.O.; Smits, P.H. Identification and Biological Activity of Newly Isolated Heterorhabditid Populations from Israel. *Fundam. Appl. Nematol.* **1993**, *16*, 467–472.
162. Karimi, J.; Alvadni, J.; Moravej, G.H. Survey on Natural Enemies of White Grub Larvae, *Polyphylla adspersa* (Coleoptera: Melolonthidae) from Razavi Khorasan Province, Iran. *Entomol. News* **2019**, *128*, 410. [CrossRef]
163. Stuczynski, T.; Siebielec, G.; Maliszewska-Kordybach, B.; Smreczak, B.; Gawrysiak, L. *Wyznaczanie Obszarów, na Których Przekroczone są Standardy Jakości Gleb*; Biblioteka Monitoringu Środowiska: Warszawa, Poland, 2004; p. 103. ISBN 83-7217-228-5.
164. Kung, S.P.; Gaugler, R.; Kaya, H.K. Influence of Soil pH and Oxygen on Persistence of *Steinernema* spp. *J. Nematol.* **1990**, *22*, 440–445.
165. Jaworska, M.; Dudek, B. Występowanie Owadobójczych Nicieni w Glebach Wybranych Upraw. *Zesz. Nauk. Akad. Rolniczej. Kraków.* **1992**, *20*, 131–147.
166. Gaugler, R.; Boush, G.M. Effects of Ultraviolet Radiation and Sunlight on the Entomogenous Nematode, *Neoaplectana carpocapsae*. *J. Invertebr. Pathol.* **1978**, *32*, 291–296. [CrossRef]
167. Shapiro-Ilan, D.I.; Hazir, S.; Lete, L. Viability and Virulence of Entomopathogenic Nematodes Exposed to Ultraviolet Radiation. *J. Nematol.* **2015**, *47*, 184–189. [PubMed]
168. Gaugler, R.; Bednarek, A.; Campbell, J. Ultraviolet Inactivation of Heterorhabditid and Steinernematid Nematodes. *J. Inv. Pathol.* **1992**, *59*, 155–160. [CrossRef]
169. Sayed, R.M.; Shairra, S.A. Enhancing the Efficacy of Entomopathogenic Nematodes by Gamma Radiation in Controlling *Spodoptera littoralis* Larvae. *J. Radiat. Res. Appl. Sci.* **2017**, *10*, 97–102. [CrossRef]
170. Sayed, R.M.; Khidr, A.-A.A.-E.; Moustafa, H.Z. Changes in Defense Mechanism Related to Controlling *Galleria mellonella* Larvae by Gamma Irradiated *Steinernema carpocapsae* BA2. *J. Entomol. Rese.* **2015**, *39*, 287–292. [CrossRef]
171. Evans, A.A.F.; Perry, R.N. Survival Strategies in Nematodes. In *The Organization of Nematodes*; Croll, N.A., Ed.; Academic Press: London, UK, 1976; pp. 383–424.
172. Wharton, D.A. *A Functional Biology of Nematodes*; The Johns Hopkins University Press: Baltimore, MD, USA, 1986.
173. Lindegren, J.E.; Rij, R.E.; Ross, S.R.; Fouse, D.C. Respiration Rate of *Steinernema feltiae* Infective Juveniles at Several Constant Temperatures. *J. Nematol.* **1986**, *18*, 221–224.
174. Järup, L. Hazards of Heavy Metal Contamination. *Br. Med. Bull.* **2003**, *68*, 167–182. [CrossRef]

175. Baran, S.; Turski, R. *Degradacja Ochrona i Rekultywacja Gleb*; Akademia Rolnicza w Lublinie: Lublin, Polska, 1996; p. 223; ISBN 83-86761-48-2.
176. Khan, A.; Scullion, J. Effects of Metal (Cd, Cu, Ni, Pb or Zn) Enrichment of Sewage-Sludge on Soil Micro-Organisms and Their Activities. *Appl. Soil Ecol.* **2002**, *20*, 145–155. [CrossRef]
177. Gruca-Królikowska, S.; Wacławek, W. Metale w Środowisku Cz. II. Wpływ Metali Ciężkich Na Rośliny. *Chem. Dydakt. Ekol. Metrol.* **2006**, *11*, 41.
178. He, C.Q.; Tan, G.E.; Liang, X.; Du, W.; Chen, Y.L.; Zhi, G.Y.; Xhu, Y. Effect of Zn-Tolerant Bacterial Strains on Growth and Zn Accumulation in *Orychophragmus violaceus*. *Appl. Soil Ecol.* **2010**, *44*, 1–5. [CrossRef]
179. Kijeński, J.; Polaczek, J. Recykling Tworzyw Sztucznych Według Koncepcji Zielonej Chemii. *Polimery* **2010**, *10*, 667–744.
180. Jaworska, M.; Gorczyca, A.; Sepioł, J.; Tomasik, P. Oddziaływania Metal—Metal w Układach Biologicznych. Synergizm i Antagonizm Triad Jonów Metali w Owadobójczych Nicieniach *Steinernema carpocapsae*. *Chem. Inż. Ekol.* **2000**, *7*, 313–326.
181. Słaba, M.; Długoński, J. Mikrobiologiczne Usuwanie i Odzyskiwanie Metali Ciężkich. *Post. Mikrobiol.* **2002**, *41*, 168–183.
182. Matuska-Łyżwa, J.; Żeber-Dzikowska, I.; Chmielewski, J. Nicienie Jako Wskaźnik Zanieczyszczenia Gleb Ołowiem Na Terenach Uprzemysłowionych. *Przemysł Chem.* **2019**, *6*, 980–985. [CrossRef]
183. Jaworska, M.; Gorczyca, A.; Sepioł, J.; Tomasik, P. Effect of Metal Ions on the Entomopathogenic Nematode *Heterorhabditis bacteriophora* Poinar (Nematoda: Heterohabditidae) under Laboratory Conditions. *Water Air Soil Pollut.* **1997**, *93*, 157–166. [CrossRef]
184. Matuska, J. Inwazyjność i reprodukcja nicieni entomopatogennych *Heterorhabditis megidis* (Poinar, Jackson i Klein 1987) po kontakcie z jonami ołowiu(II). *Proc. ECOpole* **2009**, *3*, 95–99.
185. Matuska, J.; Kamionek, M. Effect of Lead(II) Ions on the Intensity and Extent of Infection. *Chem. I Inż. Ekol.* **2006**, *6*, 553–557.
186. Sun, Y.; Bai, G.; Wang, Y.; Zhang, Y.; Pan, J.; Cheng, W.; Feng, X.; Li, H.; Ma, C.; Ruan, W.; et al. The Impact of Cu, Zn and Cr Salts on the Relationship between Insect and Plant Parasitic Nematodes: A Reduction in Biocontrol Efficacy. *Appl. Soil Ecol.* **2016**, *107*, 108–115. [CrossRef]
187. Jarmuł, J.; Kamionek, M. Przeżywalność i Zdolność Do Migracji Nicieni Entomopatogenicznych *Steinernema feltiae* Filipjev w Środowisku Skażonym Jonami Ołowiu. *Chem. Inż. Ekol.* **2003**, *10*, 281–284.
188. Jarmuł, J.; Kamionek, M. Wpływ Związków Ołowiu Na Intensywność i Ekstensywność Porażania Gąsienic *Gallerii mellonella* L. Przez Nicienie Entomopatogeniczne *Steinernema feltiae* Filipjev. *Chem. Inż. Ekol.* **2001**, *8*, 601–607.
189. Jarmuł, J.; Kamionek, M. Wpływ Jonów Ołowiu (Pb^{2+}) Na Śmiertelność IJs *Steinernema feltiae* Filipjev i *Heterorhabditis megidis* Poinar. *Chem. Inż. Ekol.* **2000**, *7*, 1023–1029.
190. Jaworska, M.; Ropek, D.; Gorczyca, A. Żywotność i Patogeniczność Owadobójczych Nicieni w Glebie Skażonej Metalami Ciężkimi. *Chem. Inż. Ekol.* **1999**, *6*, 469–475.
191. Jaworska, M.; Gorczyca, A.; Antonkiewicz, J.; Jasiewicz, C. Wpływ Skażenia Gleby Metalami Ciężkimi Na Patogeniczność Nicieni *Steinernema feltiae* Filipjev. *Chem. Inż. Ekol.* **1998**, *5*, 719–725.
192. Thurston, G.S.; Ni, Y.; Kaya, H.K. Influence of Salinity on Survival and Infectivity of Entomopathogenic Nematodes. *J. Nematol.* **1994**, *26*, 345–351. [PubMed]
193. Jaworska, M. The Role of Magnesium in the Protection of Entomopathogenic Nematodes from Soil Pollution with Oil Derivatives. *J. Elem.* **2014**, *19*, 673–682. [CrossRef]
194. Jaworska, M.; Gospodarek, J. Effect of Magnesium on Beneficial Organisms. *J. Elem.* **2009**, *2*, 257–263. [CrossRef]
195. Matuska, J.; Kamionek, M. Effect of Lead Nitrate [$Pb(NO_3)_2$] on Development. *Ecol. Chem. Eng.* **2008**, *15*, 89–92.
196. Jarmuł, J.; Kamionek, M. Effect of Lead Nitrate on a Population of the Nematodes *Steinernema feltiae* Filipjev Developing in the Body of Its Host *Galleria mellonella* L. *Chem. Inż. Ekol.* **2002**, *9*, 367–371.
197. Jarmuł, J.; Kamionek, M. Wpływ Jonów Ołowiu Na Gąsienice *Galleria mellonella* L. Oraz Na Nicienie Entomopatogeniczne *Steinernema feltiae* Filipjev. *Chem. Inż. Ekol.* **2004**, *11*, 151–156.
198. Ropek, D.; Jaworska, M.; Grzyb, J. Effect of Heavy Metals on the Bacterial Symbionts of the Entomopathogenic Nematodes *Steinernema carpocapsae*. *Chem. I Inż. Ekol.* **2002**, *4*, 455–459.
199. Stuart, R.J.; Barbercheck, M.E.; Grewal, P.S.; Taylor, R.A.J.; Hoy, C.W. Population Biology of Entomopathogenic Nematodes: Concepts, Issues, and Models. *Biol. Control* **2006**, *38*, 80–102. [CrossRef]
200. Ropek, D.R.; Gospodarek, J. Entomopathogenic Nematode *Steinernema feltiae* as an Indicator of Soil Pollution with Oil Derivatives in Bioremediation Process. *Agriculture* **2022**, *12*, 2033. [CrossRef]
201. Ropek, D.; Gospodarek, J. Effect of Soil Pollution with Oil—Derivatives on the Occurrence of Entomopathogenic Nematodes. *Ecol. Chem. Eng. A* **2013**, *20*, 157–166. [CrossRef]
202. Ropek, D.; Gospodarek, J. The Effect of Oil Derivatives on The Ability of Entomopathogenic Nematode *Steinernema feltiae* to Find Host. *Ecol. Chem. Eng. A* **2013**, *20*, 857–865. [CrossRef] [PubMed]
203. Jaworska, M.; Gospodarek, J. Organizmy glebowe i ich aktywność jako wskaźnik stanu środowiska w warunkach zanieczyszczenia gleby substancjami ropo-pochodnymi. *Zesz. Probl. Post. Nauk Roln.* **2004**, *501*, 189–194.
204. Ropek, D.; Gospodarek, J. Wpływ ropopochodnych na zdolność do migracji nicienia owadobójczego *Steinernema feltiae*. *Proceeding ECOpole* **2013**, *7*, 237–240. [CrossRef]
205. Nermuť, J.; Mrácek, Z. The Influence of Pesticides on the Viability and Infectivity of Entomopathogenic Nematodes (Nematoda: Steinernematidae). *Russ. J. Nematol.* **2010**, *18*, 141–148.

206. Kamionek, M. Wpływ Pestycydów Na *Steinernema feltiae* (Filipjev) i Inne Nicienie Entomofilne. *Rozp. Nauk. I Monogr. Wyd. SGGW* **1992**, *63*, 231–236.
207. Hara, A.H.; Kaya, H.K. Effects of Selected Insecticides and Nematicides on the In Vitro Development of the Entomogenous Nematode *Neoaplectana carpocapsae*. *J. Nematol.* **1982**, *14*, 486–491.
208. Hara, A.H.; Kaya, H.K. Development of the Entomogenous Nematode, *Neoaplectana carpocapsae* (Rhabditida: Steinernematidae), in Insecticide-Killed Beet Armyworm (Lepidoptera: Noctuidae). *J. Econ. Entomol.* **1983**, *76*, 423–426. [CrossRef]
209. Yan, X.; Moens, M.; Han, R.; Chen, S.; De Clercq, P. Effects of Selected Insecticides on Osmotically Treated Entomopathogenic Nematodes. *J. Plant Dis. Prot.* **2012**, *119*, 152–158. [CrossRef]
210. Laznik, Ž.; Trdan, S. The Influence of Insecticides on the Viability of Entomopathogenic Nematodes (Rhabditida: Steinernematidae and Heterorhabditidae) under Laboratory Conditions. *Pest Manag. Sci.* **2014**, *70*, 784–789. [CrossRef] [PubMed]
211. Stark, J.D. Entomopathogenic Nematodes (Rhabditida: Steinernematidae): Toxicity of Neem. *J. Econ. Entomol.* **1996**, *89*, 68–73. [CrossRef]
212. Koppenhöfer, A.M.; Grewal, P.S.; Kaya, H.K. Synergism of Imidacloprid and Entomopathogenic Nematodes against White Grubs: The Mechanism. *Entomol. Exp. Appl.* **2000**, *94*, 283–293. [CrossRef]
213. Koppenhöfer, A.M.; Cowles, R.S.; Cowles, E.A.; Fuzy, E.M.; Kaya, H.K. Effect of Neonicotinoid Synergists on Entomopathogenic Nematode Fitness. *Entomol. Exp. Appl.* **2003**, *106*, 7–18. [CrossRef]
214. Özdemir, E.; İnak, E.; Evlice, E.; Laznik, Z. Compatibility of Entomopathogenic Nematodes with Pesticides Registered in Vegetable Crops under Laboratory Conditions. *J. Plant Dis. Prot.* **2020**, *127*, 529–535. [CrossRef]
215. Dzięgielewska, M.; Kaup, G. Wpływ Preparatów Układowych Aplikowanych Doglebowo Na Aktywność Biologiczną Nicieni Owadobójczych *Steinernema feltiae* Filipjev i *Heterorhabditis megidis* Poinar, Jackson. *Chem. I Inż. Ekol.* **2007**, *14*, 1167–1173.
216. Koppenhöfer, A.M.; Cowles, R.S.; Cowles, E.A.; Fuzy, E.M.; Baumgartner, L. Comparison of Neonicotinoid Insecticides as Synergists for Entomopathogenic Nematodes. *Biol. Control* **2002**, *24*, 90–97. [CrossRef]
217. Kruk, K.; Dzięgielewska, M. Wpływ acetamiprydu i chloropiryfosu na aktywność biologiczną nicieni owadobójczych (Steinernematidae, Heterorhabditidae). *Prog. Plant Prot.* **2020**, *60*, 179–185. [CrossRef]
218. Bednarek, A.; Popowska-Nowak, E.; Pezowicz, E.; Kamionek, M. Sensitivity of Two Species of Entomopathogenic Fungi: *Beauveria bassiana* and *B. brongniartii* and of Three Species of Entomopathogenic Nematodes: *Steinernema glaseri*, *S. feltiae* and *Heterorhabditis megidis* to Low Doses of Two Insecticides Carbosulfan and Carbofuran. *Pol. J. Ecol.* **2004**, *52*, 223–228.
219. Sönmez, İ.; Kaplan, M.; Sönmez, S. Effect of chemical fertilizers on environmental pollution and its prevention methods. *Hortic. Stud.* **2008**, *25*, 24–34.
220. Campos-Herrera, R.; Piedra-Buena, A.; Escuer, M.; Montalba, B.; Gutierrez, C. Effect of Seasonality and Agricultural Practices on Occurrence of Entomopathogenic Nematodes and Soil Characteristics in La Rioja (Northern Spain). *Pedobiologia* **2010**, *53*, 253–258. [CrossRef]
221. Nielsen, A.L.; Spence, K.O.; Nakatani, J.; Lewis, E.E. Effect of Soil Salinity on Entomopathogenic Nematode Survival and Behaviour. *Nematol* **2011**, *13*, 859–867. [CrossRef]
222. Bednarek, A.; Gaugler, R. Compatibility of Soil Amendments with Entomopathogenic Nematodes. *Nematol. J.* **1997**, *29*, 220–227.
223. Şahin, Y.; Susurluk, A. Effects of Some Inorganic Fertilizers on Entomopathogenic Nematodes, *Steinernema feltiae* (Tur-S3) and *Heterorhabditis bacteriophora* (Hb.H). *Türk. Biyolojik Mücadele Derg.* **2018**, *9*, 102–109. [CrossRef]
224. Shehata, I.E.; Hammam, M.M.A.; Abd-Elgawad, M.M.M. Effects of Inorganic Fertilizers on Virulence of the Entomopathogenic Nematode *Steinernema glaseri* and Peanut Germination under Field Conditions. *Agronomy* **2021**, *11*, 945. [CrossRef]
225. Gondek, K.; Ropek, D. Wybrane właściwości gleby pod trwałym użytkiem zielonym oraz aktywność owadobójczych nicieni i grzybów. *Acta Agrophys.* **2007**, *10*, 327–340.
226. Bakkali, F.; Averbeck, S.; Averbeck, D.; Idaomer, M. Biological Effects of Essential Oils—A Review. Biological Effects of Essential Oils—A Review. *Food Chem. Toxicol.* **2008**, *46*, 446–475. [CrossRef]
227. Burt, S. Essential Oils: Their Antibacterial Properties and Potential Applications in Foods—A Review. *Int. J. Food Microbiol.* **2004**, *94*, 223–253. [CrossRef]
228. Chaieb, K.; Hajlaoui, H.; Zmantar, T.; Kahla-Nakbi, A.B.; Rouabhia, M.; Mahdouani, K.; Bakhrouf, A. The Chemical Composition and Biological Activity of Clove Essential Oil, *Eugenia caryophyllata* (*Syzigium aromaticum* L. Myrtaceae): A Short Review. *Phytother. Res.* **2007**, *21*, 501–506. [CrossRef] [PubMed]
229. Górski, R.; Szklarz, M.; Kaniewski, R. Skuteczność Olejku Eterycznego z Konopi Siewnej w Zwalczaniu Mszycy Jabłoniowo-Babkowej (*Dysaphis plantaginea* Pass.) Występującej Na Jabłoni. *Prog. Plant Prot./Post. Ochr. Roślin* **2009**, *49*, 2013–2016.
230. Dutka, A. Application of essential oils in plant protection against pest insects—Paper review. *Prog. Plant Prot.* **2013**, *53*, 36–42. [CrossRef]
231. Barua, A.; McDonald-Howard, K.L.; Mc Donnell, R.J.; Rae, R.; Williams, C.D. Toxicity of Essential Oils to Slug Parasitic and Entomopathogenic Nematodes. *J. Pest Sci.* **2020**, *93*, 1411–1419. [CrossRef]
232. Gupta, A.; Sharma, S.; Naik, S.N. Bio Pesticidal Value of Selected Essential Oils against Pathogenic Fungus, Termites, and Nematodes. *Int. Biodeter. Biodegr.* **2011**, *65*, 703–707. [CrossRef]
233. Castruita-Esparza, G.; Bueno-Pallero, F.Á.; Blanco-Pérez, R.; Dionísio, L.; Aquino-Bolaños, T.; Campos-Herrera, R. Activity of *Steinernema colombiense* in Plant-Based Oils. *J. Nematol.* **2020**, *52*, e2020-72. [CrossRef] [PubMed]

234. Schmidt, J.; All, J.N. Chemical Attraction of *Neoaplectana carpocapsae* (Nematoda: Steinernematidae) to Insect Larvae. *Environ. Entomol.* **1978**, *7*, 605–607. [CrossRef]
235. El-Borai, F.E.; Duncan, L.W.; Preston, J.F. Bionomics of a Phoretic Association between *Paenibacillus* sp. and the Entomopathogenic Nematode *Steinernema diaprepesi*. *J. Nematol.* **2005**, *37*, 18–25.
236. Eng, M.S.; Preisser, E.L.; Strong, D.R. Phoresy of the Entomopathogenic Nematode *Heterorhabditis marelatus* by a Non-Host Organism, the Isopod Porcellio Scaber. *J. Invertebr. Pathol.* **2005**, *88*, 173–176. [CrossRef]
237. Poinar, G.O. *Nematodes for Biological Control of Insects*; CRC Press: Boca Raton, FL, USA, 1979; p. 289.
238. Poinar, G.O.; Veremchuk, G.V. A New Strain of Nematodes Pathogenic to Insects and the Geographical Distribution of *Neoaplectana carpocapsae* Weiser (Rhabditida, Steinernematidae). *Zool. Zhurnal.* **1970**, *49*, 966–969.
239. Poinar, G.O.; Thomas, G.M.; Veremtschuk, G.V.; Pinnock, D.E. Further Characterization of *Achromobacter nematophilus* from American and Soviet Populations of the Nematode *Neoaplectana carpocapsae* Weiser. *Int. J. Syst. Bacteriol.* **1971**, *21*, 78–82. [CrossRef]
240. Bovien, P. *Some Types of Association between Nematodes and Insects*; Videnskabelige Meddelelser Fra Dansk Naturhistorisk Forening; Kobenhavn: Reitzel, Denmark, 1937; Volume 101, p. 114.
241. Khan, A.; Brooks, W.M.; Hirschmann, H. *Chromonema heliothidis* n. Gen., n. sp. (Steinernematidae, Nematoda), a Parasite of *Heliothis zea* (Noctuidae, Lepidoptera), and Other Insects. *J. Nematol.* **1976**, *8*, 159–168.
242. Bednarek, A.; Mraček, Z. The Incidence of Nematodes of the Family Steinernematidae in *Cephalcia falleni* Dalm. (Hymenoptera: Pamphilidae) Habitat after an Outbreak of the Pest. *J. Appl. Entomol.-Z. Angew. Entomol.* **1986**, *102*, 527–530. [CrossRef]
243. Mokrini, F.; Laasli, S.; Benseddik, Y.; Joutei, A.B.; Blenzar, A.; Lakhal, H.; Sbaghi, M.; Imren, M.; Özer, G.; Paulitz, T.; et al. Potential of Moroccan Entomopathogenic Nematodes for the Control of the Mediterranean Fruit Fy *Ceratitis capitata* Wiedemann (Diptera: Tephritidae). *Sci. Rep.* **2020**, *10*, 19204. [CrossRef]
244. Buxton, J.H. Control of Sciarid Fly with Parasitic Nematodes. *Bull. OILB SROP* **1993**, *16*, 23–25.
245. Grewal, P.; Richardson, P.; Collins, G.; Edmondson, R. Comparative Effects of *Steinernema feltiae* (Nematoda: Steinernematidae) and Insecticides on Yield and Cropping of the Mushroom Agaricus Bisporus. *Ann. Appl. Biol.* **2008**, *121*, 511–520. [CrossRef]
246. Armendariz, I.; Downes, M.J.; Griffin, C.T.; Smits, P.H. Pine Weevil Biocontrol: Compatibility of Entomopathogenic Nematodes and Stump-Degrading Fungi in Laboratory Conditions. *Bull. OILB SROP* **2000**, *23*, 47–50.
247. Ekanayakei, H.M.R.K.; Abeysinghei, A.M.C.P.; Toida, Y. Potential of Entomopathogenic Nematodes as Bio-Control Agents of Sweet Potato Weevil, *Cylas formicarius* (FABRICIUS) (Coleoptera: Brenthidae). *Jpn. J. Nematol.* **2001**, *31*, 19–25. [CrossRef]
248. Fitters, P.F.L.; Dunne, R.; Griffin, C.T. Vine Weevil Control in Ireland with Entomopathogenic Nematodes: Optimal Time of Application. *Ir. J. Agric. Food Res.* **2001**, *40*, 199–213.
249. Kowalska, J. Próba Zastosowania Nicieni Owadobójczych Oraz Metody Integrowanej w Zwalczaniu Pędraków Chrabąszcza Majowego *Melolontha melolontha* L. w Uprawie Leśnej. *Sylwan* **2001**, *145*, 89–95.
250. Tomalak, M. Możliwości Rozszerzenia Zakresu Stosowania Metod Biologicznych w Zwalczaniu Szkodników Parków i Lasów Miejskich. *Prog. Plant Prot.* **2002**, *42*, 53–58.
251. Tomalak, M. Infectivity of Entomopathogenic Nematodes to Soil-Dwelling Developmental Stages of the Tree Leaf Beetles *Altica quercetorum* and *Agelastica alni*. *Entom. Exp. Appl.* **2004**, *110*, 125–133. [CrossRef]
252. Ali, J.G.; Alborn, H.T.; Stelinski, L.L. Subterranean Herbivore-Induced Volatiles Released by Citrus Roots upon Feeding by Diaprepes Abbreviatus Recruit Entomopathogenic Nematodes. *Chem. Ecol.* **2010**, *36*, 361–368. [CrossRef] [PubMed]
253. Renkema, J.M.; Krey, K.L.; Lahiri, S. Control of *Diaprepes abbreviatus* (Coleoptera: Curculionidae) with *Steinernema riobrave* (Rhabditida: Steinernematidae) in Plasticulture Florida Strawberry. *Fla Entomol.* **2021**, *104*, 124–130. [CrossRef]
254. Dutky, S.R. Insect Microbiology. *Adv. Appl. Microbiol.* **1959**, *1*, 175–198. [CrossRef] [PubMed]
255. Bélair, G.; Boivin, G. Susceptibility of the Carrot Weevil (Coleoptera: Curculionidae) to *Steinernema feltiae*, *S. bibionis* and *Heterorhabditis heliothidis*. *J. Nematol.* **1985**, *17*, 363–366.
256. Shapiro, M.; Poinar, G.O., Jr.; Lindegren, J.E. Suitability of *Lymantria dispar* (Lepidoptera: Lymantriidae) as a Host for the Entomogenous Nematode, *Steinernema feltiae* (Rhabditida: Steinernematidae). *J. Econ. Entomol.* **1985**, *78*, 342–345. [CrossRef]
257. Wójcik, W.F. Influence of Structure and Dynamics of the Population of the *Neoaplectana carpocapsae* Weiser, 1955 Nematodes. *Ann. Wars. Agricult. Univ. SGGW-AR Anim. Sc.* **1986**, *20*, 97–197.
258. Bednarek, A.; Nowicki, T.; Wójcik, W.F. Sex Structure in Population of *Steinernema feltiae*. *Zesz. Probl. Post. Nauk. Roln.* **1986**, *323*, 213–223.
259. Poinar, G.O. Insect Immunity to Parasitic. Nematodes. In *Contemporary Topics in Immunology*; Plenum Press: New York, NY, USA, 1974; Volume 4, pp. 167–178.
260. Bedding, R.A.; Molyneux, A.S. Penetration of Insect Cuticle By Infective Juveniles of *Heterorhabditis* spp. (Heterorhabditidae: Nematoda). *Nematol* **1982**, *28*, 354–359. [CrossRef]
261. Bednarek, A.; Nowicki, T. Effect of Intrapopulation Factors in the Nematodes *Steinernema feltiae* (Steinernematidae) on Intensity of Insect Infestation. *Zesz. Probl. Post. Nauk Roln.* **1986**, *323*, 199–212.
262. Alikhan, M.A.; Bednarek, A.; Grabiec, S. The Physiological and Morphological Characteristics of *Neoaplectana carpocapsae* (Nematoda: Steinernematidae) in Two Host. *J. Invert. Pathol.* **1985**, *45*, 168–173. [CrossRef]
263. Boff, M.I.C.; Wiegers, G.L.; Smits, P.H. Influences of host size and host species on the infectivity and development of *Heterorhabditis megidis* (strain NLH-E 87.3). *BioControl* **2000**, *45*, 469–482. [CrossRef]

264. Matuska, J.; Kamionek, M. The Effect of Host Species on the Biology of *Heterorhabditis megidis* (Poinar, Jackson and Klein 1987). *Prog. Plant Prot.* **2005**, *42*, 893–897.
265. Epsky, N.D.; Walter, D.E.; Capinera, J.L. Potential Role of Nematophagous Microarthropods as Biotic Mortality Factors of Entomogenous Nematodes (Rhabditida: Steinernematidae, Heterorhabditidae). *J. Econ. Entomol.* **1988**, *81*, 821–825. [CrossRef]
266. Ulug, D.; Hazir, S.; Kaya, H.K.; Lewis, E. Natural Enemies of Natural Enemies: The Potential Top-down Impact of Predators on Entomopathogenic Nematode Populations. *Ecol. Entomol.* **2014**, *39*, 462–469. [CrossRef]
267. Ashwini, M.N.; Bhaskar, H.; Mathew, D.; Shylaja, M.R.; Girija, D. Isolation and Evaluation of Bacteria Associated with Entomopathogenic Nematode *Heterorhabditis* spp. against the Spider Mite, *Tetranychus truncatus* Ehara (Acari: Tetranychidae). *Egypt J. Biol. Pest Control* **2022**, *32*, 87. [CrossRef]
268. Hodson, A.K.; Siegel, J.P.; Lewis, E.E. Ecological Influence of the Entomopathogenic Nematode, *Steinernema carpocapsae*, on Pistachio Orchard Soil Arthropods. *Pedobiologia* **2012**, *55*, 51–58. [CrossRef]
269. Gilmore, S.K.; Potter, D.A. Potential Role of Collembola as Biotic Mortality Agents for Entomopathogenic Nematodes. *Pedobiologia* **1993**, *37*, 30–38. [CrossRef]
270. Campos-Herrera, R.; Půža, V.; Jaffuel, G.; Blanco-Pérez, R.; Čepulytė-Rakauskienė, R.; Turlings, T.C.J. Unraveling the Intraguild Competition between *Oscheius* spp. Nematodes and Entomopathogenic Nematodes: Implications for Their Natural Distribution in Swiss Agricultural Soils. *J. Invertebr. Pathol.* **2015**, *132*, 216–227. [CrossRef]
271. Kreft, A.; Skrzypek, H. The Role of the Host's Biomass in Their Localization and Recognition by Enthomopathogenic Nematodes (Rhabditida: Steinernematidae and Heterorhabditidae). *Ann. UMCS Sect. C* **2003**, *58*, 41–52.
272. Bilgrami, A.L.; Brey, C. Potential of Predatory Nematodes. In *Nematodes as Biocontrol Agents*; Grewal, P.S., Ehlers, R.U., Shapiro-Ilan, D.I., Eds.; CABI Publishing: Wallingford, UK, 2005; pp. 447–464. ISBN 978-0-85199-017-0.
273. Kenney, E.; Eleftherianos, J. Entomopathogenic and Plant Pathogenic Nematodes as Opposing Forces in Agriculture. *Int. J. Parasitol.* **2016**, *46*, 13–19. [CrossRef] [PubMed]
274. Felicitas, E.F.A.; Caoili, B.L.; Latina, R.A. Antagonistic Effect of *Steinernema abbasi* and *Heterorhabditis indica* Philippine Isolates on Root Penetration and Development of *Meloidogyne incognita* in Tomato. *Sci. Technol.* **2021**, *31*, 865–876. [CrossRef]
275. Barron, G.L. Nematode-Destroying Fungi. In *Experimental Microbial Ecology*; Burns, R.H., Slater, J.H., Eds.; Blackwell: Oxford, UK, 1982; pp. 533–552.
276. Arêdes, A.; Ribeiro, L.P.; Dias, V.M.; Dolinski, C. Studies on the Interaction between Tropical Nematophagous Fungi and Entomopathogenic Nematodes (Rhabditida) under Laboratory Conditions. *Nematoda* **2017**, *4*, e112016. [CrossRef]
277. Poinar, G.O., Jr.; Jansson, H.B. Susceptibility of *Neoaplectana* spp. and *Heterorhabditis heliothidis* to the Endoparasitic Fungus *Drechmeria coniospora*. *J. Nematol.* **1986**, *18*, 225–230. [PubMed]
278. Jaworska, M.; Ropek, D. Stymulacja Chemiczna Aktywności Owadobójczej Grzybów i Nicieni Aplikowanych Łącznie Na Owady Testowe. In Proceedings of the Materiały Konferencyjne II Sympozjum Naukowego, Kraków, Polska, 8 June 1997; pp. 39–45.
279. Darissa, O.M.; Iraki, N.M. In Vivo and in Vitro Interactions between Two Entomopathogens: The Bacterium *Serratia marcescens* and the Nematode *Heterorhabditis indica*. *Russ. J. Nematol.* **2013**, *21*, 83–91.
280. Skowronek, M.; Sajnaga, E.; Kazimierczak, W.; Lis, M.; Wiater, A. Screening and Molecular Identification of Bacteria from the Midgut of *Amphimallon solstitiale* Larvae Exhibiting Antagonistic Activity against Bacterial Symbionts of Entomopathogenic Nematodes. *Int. J. Mol. Sci.* **2021**, *22*, 12005. [CrossRef]
281. Enright, M.R.; Finnegan, M.M.; Mclnerney, J.O.; O'Laighleis, M.; Griffin, C.T. *Specific Adherence of Sporangia of a Paenibacillus sp. Bacterium to Heterorhabditis spp. Nematodes. Hitching a Ride to Lunch?* Griffin, C.T., Burnell, A.M., Downes, M.J., Mulder, R., Eds.; Development of Entomopathogenic Nematode/Bacteria Research; European Commission: Brussels, Belgium, 2001; pp. 301–306. ISBN 9289404671.
282. Enright, M.R.; Mclnerney, J.O.; Griffin, C.T. Characterization of Endospore-Forming Bacteria Associated with Entomopathogenic Nematodes, *Heterorhabditis* spp., and Description of *Paenibacillus nematophilus* sp. Nov. *Int. J. Syst. Evol. Microbiol.* **2003**, *53*, 435–441. [CrossRef]

insects

MDPI

Article

Unconventional Strategies for Aphid Management in Sorghum

Ivan Grijalva [1,*], Qing Kang [2], Daniel Flippo [3], Ajay Sharda [3] and Brian McCornack [1]

1 Department of Entomology, Kansas State University, Manhattan, KS 66506, USA; mccornac@ksu.edu
2 Department of Statistics, Kansas State University, Manhattan, KS 66506, USA; qkang@ksu.edu
3 Department of Biological and Agricultural Engineering, Kansas State University, Manhattan, KS 66506, USA; dkflippo@ksu.edu (D.F.); asharda@ksu.edu (A.S.)
* Correspondence: grijalva@ksu.edu

Simple Summary: Sorghum holds significant economic importance across both the northern and southern regions of the U.S. Since 2013, the presence of the sorghum aphid *Melanaphis sorghi* (Theobald) has led to -yield reduction in unmanaged sorghum. Various management strategies have been critical in controlling aphid infestations, including cultural practices, host–plant resistance, and insecticide use. Typically, the common practice is applying insecticides at a whole-field level once aphids reach they reach a certain economic threshold. However, this traditional approach often includes treating plants that are not yet infested. To minimize insecticide usage in sorghum, we suggested selectively spraying individual plants only when aphids are present or not, as opposed to conventional spraying based on economic thresholds. Through field experiments conducted over two summer seasons, our findings demonstrated that spraying individual plants can manage aphid populations while reducing overall insecticide usage. This approach holds promise for advancing targeted insecticide applications in sorghum.

Abstract: Since the invasion of the sorghum aphid *Melanaphis sorghi* (Theobald), farmers in the sorghum (*Sorghum bicolor* L. Moench) production region in the Great Plains of the U.S. have faced significant crop damage and reduced yields. One widely used practice to aid in managing sorghum aphids is pest monitoring, which often results in field-level insecticide applications when an economic threshold is reached. However, relying on this traditional management practice includes the application of insecticides to non-infested plants. To reduce insecticide usage in sorghum, we proposed spraying individual plants when aphids are present or absent compared to traditional spraying based on a standard economic threshold using field replicate plots over two summer seasons. The experimental results of this study indicated fewer aphids in plots managed with an economic threshold, followed by randomly sprayed and plant-specific treatments compared with the untreated control treatment. Therefore, compared with traditional management, those treatments can be alternative strategies for managing aphids on sorghum within our field plot study.

Keywords: field crops; sorghum; integrated pest management; insecticides; sorghum aphid

1. Introduction

Agricultural pests represent 37% of annual crop losses, 13% of which are directly caused by insects [1]. Advantageously integrated pest management was promoted in the early 1970s as a strategy for pest control that promotes sustainable agriculture with a strong ecological basis [2]. Different strategies have been used in integrated pest management to manage insect pests in agriculture, including the application of insecticides, cultural practices, host–plant resistance, and mechanical and biological management methods [3]. One of the most widely used strategies for managing insect pests is the use of insecticides; however, there is much concern about their use for pest control due to their cumulative adverse effects on the environment [4–6].

Primary concerns include the effects of insecticides on non-target or beneficial organisms, including natural enemies and pollinators [7,8], and the potential for developing insecticide resistance [9,10]. Additionally, the use of insecticides can create indirect costs, including negative impacts on the human health of growers or consumers (e.g., poisoning), as well as long-term effects on the environment (e.g., bioaccumulation, biodiversity, etc.) [11,12]. Therefore, given the various negative impacts of insecticides and the reliance on single strategies to manage key pests, there is a need to find alternatives to reduce the use of insecticides to promote sustainability in field crops [4].

Sorghum (*Sorghum bicolor* (L.) Moench) is a field crop that relies on insecticide application to manage pests. In recent years, fields infested by sorghum aphid (*Melanaphis sorghi* (Theobald)) [13] have significantly impacted production in the Southern Great Plains, which has increased the use of seed-applied and foliar insecticides [14]. Kansas is one of the major sorghum-producing states in this region, and sorghum aphid infestations have reduced yields since 2015 [15]. The use of resistant hybrids and the management of aphid populations early in the southern range of its distribution have reduced the overall impact of this pest on sorghum. However, sorghum aphids reach damaging levels in sorghum production areas across the state.

Damage caused by sorghum aphids includes leaf chlorosis, a reduction in plant nutrients, and decreased photosynthesis due to honeydew secretion and sooty mold growth [16,17]. During pre-flowering, aphid infestations can decrease the number of heads, seed weights, and plant maturity, causing a reduction in yield [15]. Due to their high reproductive rates and ability to disperse over great distances, usually during their alate stage with the wind, monitoring changes in aphid populations is imperative for the timely and warranted use of insecticides, which are applied based on established economic thresholds [17].

Traditional aphid management, which depends on published economic thresholds, suggests the application of insecticides when 20–30% of sorghum plants are infested with an estimated 25–125 sorghum aphids/wingless/leaf threshold [18]. Others suggest the use of a 40 sorghum aphid wingless/leaf threshold or "tally threshold" for insecticide treatment [19]. Usually, thresholds and related sampling plans rely on mathematical probabilities that unsampled plants will require treatment based on aphid densities using subsamples of the entire plant population. For example, areas or unsampled plants with low or no aphid presence can be sprayed unnecessarily, resulting in a needless increase in the use of insecticides. Whole-field insecticide applications require high amounts of insecticides to treat a crop's area, resulting in significant production costs and an increased environmental impact [20]. Additionally, spraying an entire field with insecticides could increase the risk of aphid resistance [14] and reduce the number of predators and parasitoids contributing to natural predation by sorghum aphid populations.

Therefore, to reduce the use of insecticides on unnecessary areas or non-infested plants with aphids, we propose a field plot study in which individual plants are sprayed with or without aphids compared to traditional spraying based on standard economic thresholds. Sorghum was used as a crop model to evaluate the effect of plant-level applications on sorghum aphid populations and yield differences. We suggest that our approach could be used to manage sorghum aphid populations while reducing insecticide use by monitoring aphid pests and only applying insecticides at the plant level when aphids are present or absent.

2. Materials and Methods

2.1. Site Location

This research study was conducted in fields located in Manhattan, KS, with coordinates of (39°07′33.6″ N, 96°38′09.6″ W) and (39°12′50.4″ N, 96°35′49.2″ W). The fields were planted with 0.20 ha of sorghum using susceptible seeds of varieties 84P80 and 84G62 (Pioneer, Johnston, IA, USA) at a seeding rate of 60,000 seeds/acre in 2021 and 2022, respectively. The aphid tolerance of the seeds was rated as 2 out of 10, which is considered "poor" according to the company. The sorghum plants were spaced at 0.76 m between rows using no-till practices and were planted in June during the 2021 and 2022 growing seasons. Several herbicides were sprayed to manage weeds using a boom sprayer, including Brawl II (2046 mL/ha) (Tenkoz, Alpharetta, GA, USA), Atrazine 4L (3069.27 mL/ha) (Syngenta, Greensboro, NC, USA), Explorer (394.62 mL/ha) (Syngenta, Greensboro, NC, USA), and Warrant (5846.22 mL/ha) (Bayer CropScience, St. Louis, MO, USA). No fertilizers were applied during this study, and insecticide was applied to manage sorghum aphid populations.

2.2. Insecticide Applications

The insecticide Sivanto^{TM} prime (Flupyradifurone) (Bayer CropScience, St Louis, MO, USA) was applied at a standard commercial rate (4 fl oz/acre) of 292.31 mL/ha for 4 treatments (Table 1). This was completed using a backpack pump sprayer with dimensions of 22.86 cm × 39.37 cm × 52.07 cm that was equipped with a Teejet 110015 nozzle. The untreated control plots did not receive insecticide during the duration of this study. The economic threshold plots were sprayed with a broad-spectrum insecticide on all plants in the two middle rows when they reached an economic threshold of ≥40 mean wingless aphids per leaf on 12 evaluated plants. In contrast, the randomly sprayed and plant-specific plots were treated based on the presence or absence of aphids on individual plants, simulating the actions of a ground robotic vehicle. Finally, the tally threshold plots were sprayed based on an economic threshold of ≥40 wingless sorghum aphids per plant on 12 evaluated plants.

Table 1. Abbreviations of the treatments along with their descriptions and the cumulative number of sprayed plants (CSP) over both years.

Treatment Abbreviation	Treatment Name	Treatment Description and CSP
UC	Untreated control	No insecticide applications were made. No plants were sprayed.
ET	Economic threshold	All plants in the two middle rows of the plot were sprayed based on an economic threshold of ≥40 mean sorghum aphid wingless/leaf on 12 randomly evaluated plants. A CSP of 2588 was sprayed.
RS	Randomly sprayed	Consisted of the same number of plants treated as in PS but randomly sprayed plants regardless of sorghum aphid wingless presence. A CSP of 612 was sprayed.
PS	Plant-specific	Consisted of specific plants sprayed upon visually confirming the presence of wingless (≥1 sorghum aphid wingless/leaf) on 12 randomly evaluated plants. A CSP of 612 was sprayed.
TT	Tally threshold	Consisted of individual plants sprayed based on an economic threshold per plant basis of ≥40 sorghum aphid wingless/plant on 12 randomly evaluated plants. A CSP of 214 was sprayed.

Various spray applications on individual plants were conducted at different phases for the randomly sprayed, plant-specific, and tally threshold treatments. However, in 2021 and 2022, individual plots were sprayed only once at different phases because the plot replicates did not simultaneously reach the economic threshold, due to typical variations in aphid populations. The insecticide was applied on a weekly basis according to the treatments outlined in Table 1.

2.3. Field Design and Treatments

In this study, we used a randomized complete block design to assess the effects of individual plant insecticide applications on the overall dynamics of aphid populations within plots, compared to standard whole-plot insecticide applications based on established economic thresholds. The sorghum fields were divided into 30 plots. Each plot was 4 rows wide by 12 m long, and each block was spaced 3 m apart from the other. The field experiments in both years consisted of 5 randomized treatments within each of the 6 blocks.

The location of the plots in the field served as the blocking factor. The 5 treatments were randomly assigned to a plot within each block. Each treatment was represented by six plots. The counts of wingless aphids were visually assessed at 15 sampling days, including 3 days at baseline, 1 day (1D), and 6 days (6D) after insecticide applications, and they were grouped into evaluation phases or weeks (6 in total) to observe the efficacy of the treatment over time. The goal was to see if randomly sprayed, and plant-specific are feasible treatments compared with the economic threshold, as these treatments simulate the feasibility of deploying a ground robotic vehicle capable of detecting or not aphids and spraying individual plants.

Sorghum aphids were counted from 2 leaves of 12 randomly selected plants on each sampling day, with different plants being selected randomly for each sampling day. This same experimental design was used for the 2021 and 2022 summer seasons.

2.4. Non-Natural Infestations of Sorghum Aphid Populations

Based on preliminary field experiments conducted in the summer of 2020, various sorghum plots did not reach sorghum aphid economic threshold levels for insecticide applications. For this reason, we manually infested 8 random, individual plants with approximately 200 wingless aphids in the two middle rows of each individual plots when the sorghum plants had 7–10 leaves (stage 3) during the growing seasons of 2021 and 2022, respectively (Figure 1). Sorghum aphids were obtained from a research colony maintained at the Entomology Department of Kansas State University, which was established in 2020 from field-collected aphid populations. After randomly selected sorghum plants were infested, each plant was covered with sleeve cages. Each cage was 50 cm long by 20 cm wide with two openings at the ends built with noseeum mesh long enough to allow for plant growth development while protecting the aphids from natural enemies, especially coccinellids (Figure 2). All sleeve cages were removed two weeks before visual assessments started, which allowed sorghum aphids to spread through the center two rows of each replicate plot.

Figure 1. Sorghum plants manually infested with wingless aphids.

Figure 2. Sorghum plants infested with wingless aphids were covered with a sleeve cage.

2.5. Insect Sampling

Two months after sorghum was planted (stage 4: the flag leaf was visible [15]), visual counts were made of all wingless aphids, which is the visually observed life stage used for management decisions. Density estimates were performed by randomly selecting 12 plants within the two middle rows of each plot. From each plant, we visually assessed two leaves, the most recent completely unfurled leaf below the flag leaf and the first healthy leaf (>80% of the leaf was green) from the bottom of the plant (designated "top" and "bottom" leaves hereafter) [19,21]. For each leaf, the number of wingless aphids was estimated using standard guidelines that had 7 density categories [22]. The categories included 0 wingless/leaf, 1–25 wingless/leaf, 26–50 wingless/leaf, 51–100 wingless/leaf, 101–500 wingless/leaf, 501–1000 wingless/leaf, and >1000 wingless/leaf. All these visual assessments were performed weekly, after treatment applications and prior to sorghum harvest. Manually infested plants were not subjected to visual assessments but served to increase aphid numbers across all plots to promote aphid establishment.

2.6. Cumulative Aphid Days and Sprayed Plants

We calculated the cumulative aphid days (CADs) per plot for the wingless using the procedure described by Lytle et al. [23]. CADs are used to calculate the area under the aphid population curve, and it provides a single number describing the aphid density or "pressure" over the entire growing season [23]. We also recorded the total number of sprayed plants in each plot under the ET, RS, PS, and TT treatments for every insecticide application. The number of sprayed plants was used to calculate the number of cumulative sprayed plants (CSP) in the overall evaluation phases.

2.7. Yield Sampling

At sorghum maturity, one middle row was randomly selected in each plot, and sorghum heads from 3 m sections of an inner row were harvested by hand and dried using rackets located in a dark room at 25 °C. We excluded the manually infested plants from all plots during the harvest process. A thresher (Almaco, SSBT-C2, Nevada, IA, USA) was used to remove the seeds from the panicles. The seeds were weighed in grams using an electronic balance (Scout™ Pro, SP202, Parsippany, NJ, USA), and the % moisture content was measured using a grain moisture tester (Agratronix, MT-16, Streetsboro, OH, USA) from each sample per plot.

2.8. Data Processing

The number of wingless was averaged over 24 leaves measured in each plot. Accordingly, the lower limit of detection for a plot is $1/24 \cong 0.042$. If a plot had no wingless aphids observed, its count was replaced by half of the detection limit, after which the plot underwent natural log transformation for statistical analysis.

2.9. Statistical Analysis

The natural log of counts post-treatment for each response variable was analyzed using the linear mixed model. The fixed effects of the model included block, treatment (Trt), evaluation phase (EP), sample day (D), and all two-way and three-way interactions between treatment, phase, and sample day (Table 2). The random effect of the model was plotted (an error term vector corresponding to repeated measurements over time). The natural log of the baseline counts one day prior to the 1st evaluation phase served as a covariate. The variance–covariance structure of the plot was taken as either unstructured, first-order autoregressive with heterogeneous variance, or compound symmetry depending on model convergence and fitness criteria. Model fitness was supported by the attributes of conditional studentized residuals. Treatment effects were assessed via back-transformed least squares means (LSMs), their standard errors (SEs), and mean differences. Disregarding the results of the test for three-way interactions, treatment groups were compared at each evaluation phase and sample day. Statistical analysis was performed via Statistical Analysis Software (SAS version 9.4; Cary, NC, USA) PROC MIXED with the option DDFM = KR in the MODEL statement.

Table 2. *p*-value for the type III test of the fixed effect of wingless aphid counts by season.

p-Value for Type III Test of Fixed Effect	2021 Season Wingless Aphids	2022 Season Wingless Aphids
Evaluation baseline	0.492	0.285
Block	0.018	0.452
Treatment (Trt)	0.068	<0.001
Evaluation phase (EP)	<0.001	<0.001
Sample day (D)	<0.001	<0.001
Trt × EP	<0.001	<0.001
Trt × D	0.098	0.113
EP × D	<0.001	<0.001
Trt × EP × D	<0.001	0.288

The yield and CAD variables were analyzed under the linear model, with fixed effects being block and treatment. Treatment effects were assessed via LSMs, SEs, and mean differences. Statistical analysis was performed via SAS PROC MIXED. All tests were conducted at the 0.05 significance level. Pairwise comparisons were carried out using two-sided tests. No multiplicity adjustment was applied.

3. Results

In the following sections, treatment abbreviations consist of untreated control (UC), economic threshold (ET), randomly sprayed (RS), plant specific (PS), and tally threshold (TT). Furthermore, the details of each treatment are outlined in Table 1. The *p* represents the *p*-value for the overall effect of treatment.

3.1. Aphid Abundance over Time in the 2021 Season

Initially, all treatments showed similar results one day after application in the first evaluation phase (p = 0.571). However, after six days, treatment differences occurred (p = 0.003). TT exhibited higher aphid counts 178 ± 32 (LSM ± SE), while RS had lower counts compared to the others. UC differed from ET, PS, and TT but was similar to RS during the initial evaluation.

In the subsequent evaluation phase, variations were observed one and six days after application (p = 0.027; <0.001). RS recorded lower counts one day after application, whereas ET had lower counts six days post-application. Conversely, TT showed higher counts (130 ± 18) one day after application, and UC (73 ± 16) had higher counts six days post-application. Significant differences were observed one day and six days after application in the third (p < 0.001; 0.025) and fourth evaluations (p = 0.018; 0.026), with UC exhibiting higher counts (≤37.7 ± 8.8) and ET lower counts, respectively. However, no significant differences were observed during the fifth and sixth evaluation phases. In summary, plots treated with ET, RS, and PS consistently exhibited lower aphid counts (≤176 ± 31) compared to UC and TT treatments across all evaluation phases and sample days (Figure 3).

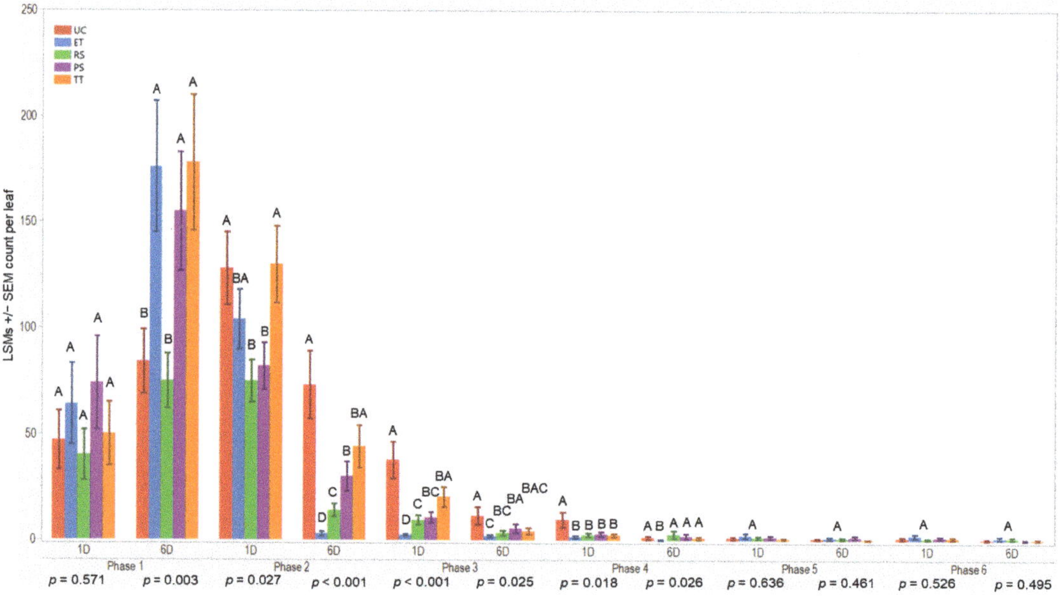

Figure 3. LSMs with SE counts per leaf of wingless in the 2021 season. LSMs with the same letter are not significantly different from each other (p >0.05).

3.2. Aphid Abundance over Time in the 2022 Season

All treatments initially displayed similar results on the first and sixth days after application in the first evaluation phase (p = 0.311; 0.068). However, variances in wingless aphid counts emerged by the sixth day after application in the second evaluation phase (p = 0.006). The UC treatment exhibited higher counts (81 ± 32) compared to ET. Throughout the third evaluation phase, ET consistently maintained lower counts compared to RS, while UC and TT showed higher counts (≤168 ± 71). These trends persisted through subsequent evaluation phases. To summarize, ET plots consistently recorded lower aphid counts, followed by PS and RS treatments (≤57 ± 15), whereas UC and TT treatments consistently displayed higher counts overall (Figure 4).

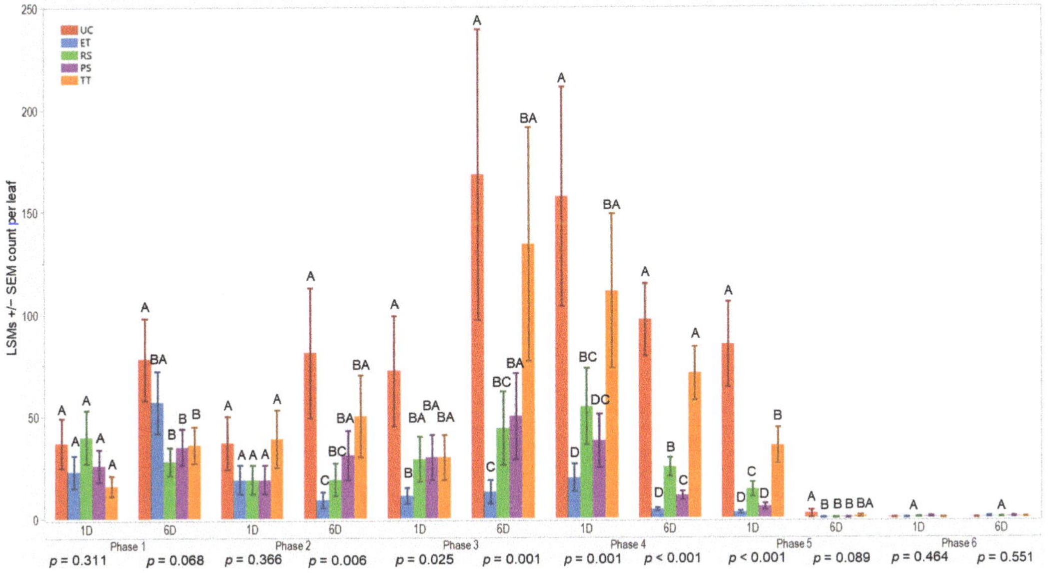

Figure 4. LSMs with SE counts per leaf of wingless in the 2022 season. LSMs with the same letter are not significantly different from each other ($p > 0.05$).

3.3. Cumulative Aphid Days

We observed significant differences in cumulative aphid days (CADs) across both years ($p = 0.014$; <0.001). In 2021, the separation of least-squares means (LSMs) for CADs indicated that the number of plots managed with the RS treatment was significantly lower (1173 ± 142) than those managed with the UC, PS, and TT treatments. In 2022, the CADs in the ET, RS, and PS treatments was significantly lower (≤1357 ± 200) compared to the UC and TT treatments (Figures 5 and 6).

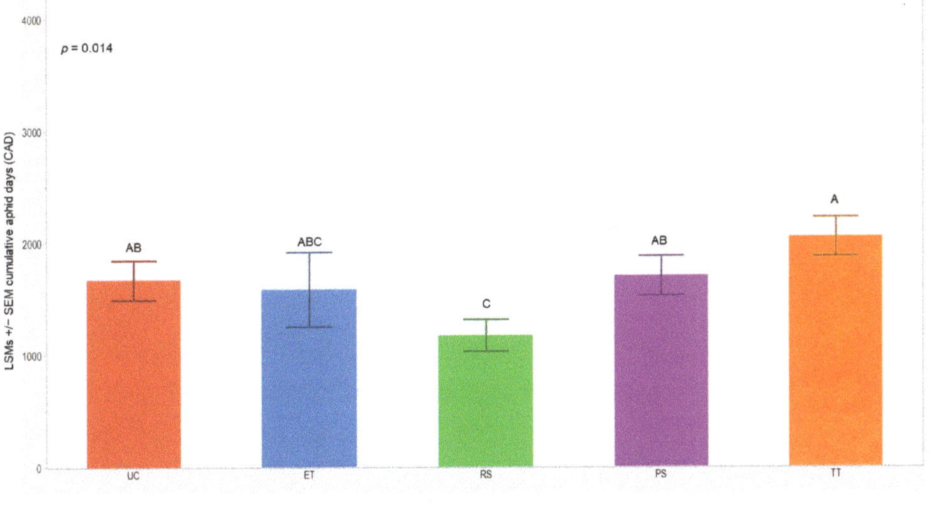

Figure 5. LSMs with SE of CADs: wingless in the 2021 season. LSMs with the same letter are not significantly different from each other ($p > 0.05$).

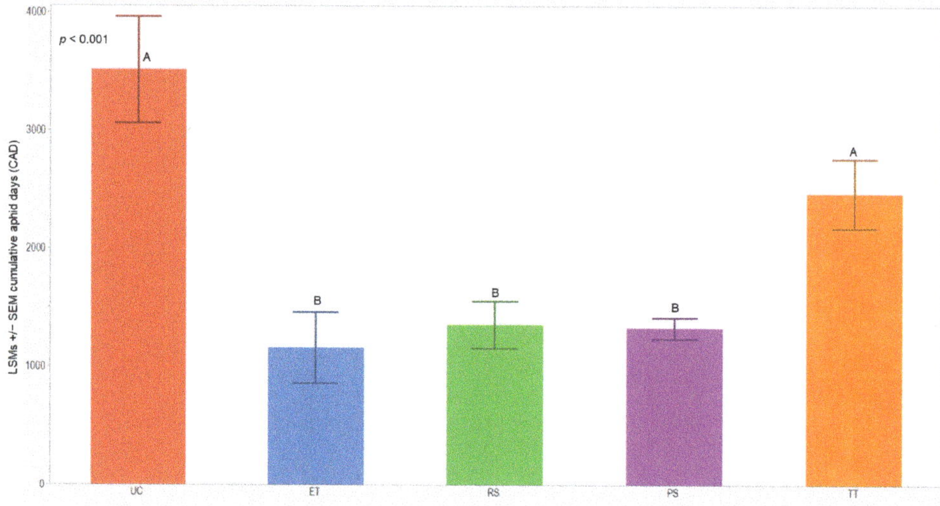

Figure 6. LSMs with SE of CADs: wingless in the 2022 season. LSMs with the same letter are not significantly different from each other ($p > 0.05$).

3.4. Cumulative Sprayed Plants and Grain Yield

In both seasons, we observed a higher number of cumulative sprayed plants in plots treated with ET (2588 plants) compared to those treated with RS (612 plants), PS (612 plants), and TT (214 plants). Notably, the number of cumulative sprayed plants in plots treated with RS, PS, and TT was sprayed 4 times lower than in the plots treated with ET in both seasons (Table 3). Regarding the yield response variable, measured by the dry weight of the grain standardized to 14% humidity, no significant differences were observed between treatments in either year of this study ($p = 0.246; 0.176$). However, the lowest yield was consistently observed in the UC treatment (1129 ± 84; 914 ± 62) in both years of this study (Figures 7 and 8).

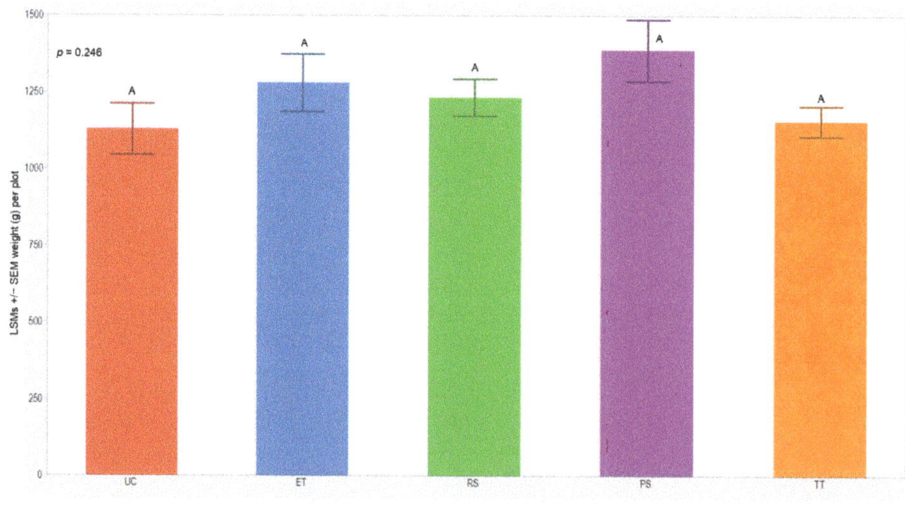

Figure 7. LSMs with SEs of yield in the 2021 season. LSMs with the same letter are not significantly different from each other ($p > 0.05$).

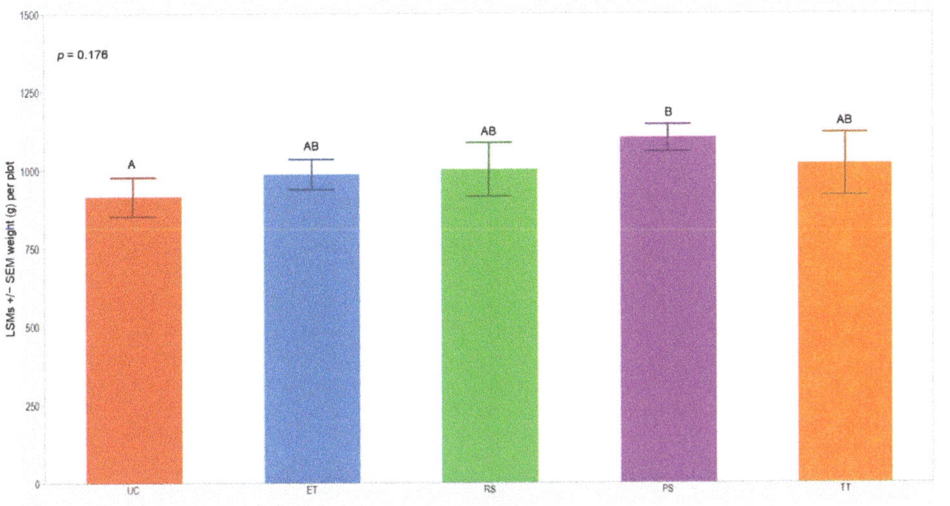

Figure 8. LSMs with SE of yield in the 2022 season. LSMs with the same letter are not significantly different from each other ($p > 0.05$).

Table 3. Cumulative number of sprayed plants (CSP) in the 2021 and 2022 seasons by treatment.

Treatment Abbreviation	Treatment Name	2021 Season CSP	2022 Season CSP
UC	Untreated control	-	-
ET	Economic threshold	1388	1200
RS	Randomly sprayed	280	332
PS	Plant-specific	280	332
TT	Tally threshold	92	122

4. Discussion

In this study, we demonstrated a reduction in the number of wingless aphids in plots where most plants were sprayed based on the established economic threshold (≥40 mean sorghum aphids per leaf), as well as in the RS and PS management strategies, where only individual plants were sprayed. Despite not observing aphid densities reaching an economic threshold level in 2020 and moderate numbers during the 2021 and 2022 growing seasons (i.e., CADs less than 5000), our results show that aphid suppression varied among sprayed treatments.

Overall, lower wingless counts were observed across the evaluation phases and sample days in plots managed with ET, followed by those managed with the RS and PS treatments. However, the number of cumulative sprayed plants found in plots managed with RS and PS was sprayed 4 times less than in plots managed with ET (Table 3). This suggests that compared with standard whole-plot applications based on an established economic thresholds (ET), spraying individual plants based on RS or PS management can reduce aphids and potentially mitigate the environmental effects of pesticides [14]. However, these results are based on plot studies, which further need to be evaluated at the field level.

Furthermore, we observed a considerable variation in wingless aphid density between the evaluation phases in both seasons (Figures 3 and 4). This suggests that the population dynamics of sorghum aphids changed over time [14]. According to Lytle et al. [23], interannual fluctuations in aphid populations are characteristic of many field crops where aphid populations are critical pests. For instance, economic infestations of sorghum aphids in Kansas have been less frequent and severe since 2017 [24]. The combination of different management practices, including host–plant resistance, planting date, and the adaptation of aphids' natural enemies, influences their reproduction and propagation in sorghum fields [17,24,25]. We consider the planting date to be a factor that influenced the aphid pressure at our field sites because early planting (May–June), similar to our planting dates, helps to reduce aphid infestations in Kansas [15]. Therefore, this can be attributed to our study's low aphid wingless pressure and CADs.

In both years of this study, the calculated CADs were lower in all the treated plots than in the UC treatment plots; however, the CADs in the UC treatment plots comprised less than 5000 CADs, resulting in no yield differences. This suggests that our pest pressure was not high enough to observe differences in yield. Similar studies evaluating the effect of insecticides and cultural management strategies on sorghum aphids revealed yield differences when the CADs were above 10,000 [23,26]. However, it also shows that we cannot affect yield when we manually infest aphid wingless, suggesting that the timing of infestation was too late into sorghum development, which is less susceptible to aphid injury than vegetative growth stages [19].

There is a need to identify alternatives for managing sorghum aphids to reduce further insecticide resistance and environmental effects [14,23]. Our current study provides information about the development of alternative management strategies for sorghum aphids. The RS and PS treatments based on individual plant applications could be alternatives for aphid management when the infestation intensity and pressure can injure sorghum plants. Developing and deploying these strategies will further improve the management options for this migratory and invasive pest.

5. Conclusions

Treating the entire sorghum field with insecticides could increase the risk of pest resistance and reduce the number of natural enemies contributing to aphid natural predation. Individual plant applications could be an alternative to common management strategies (i.e., treating the entire field) to reduce sprayed plants and improve aphid management. Our suggested approach revealed that RS and PS treatments based on individual plant applications can reduce aphid wingless populations. These results can inform decision-making processes for growers that can balance pest control and non-target environmental effects.

Author Contributions: I.G.: conceptualization, data curation, investigation, methodology, software, formal analysis, and writing—original draft preparation; Q.K.: conceptualization, data curation, formal analysis, investigation, methodology, software, and writing—original draft preparation; D.F.: conceptualization, investigation, methodology, funding acquisition, and writing—review and editing; A.S.: conceptualization, investigation, methodology, funding acquisition, and writing—review and editing; B.M.: conceptualization, investigation, methodology, resources, funding acquisition, and writing—original draft preparation. All authors have read and agreed to the published version of the manuscript.

Funding: The National Robotic Initiative supported this project with grant no. 2019-67021-28995. This is contribution no. 24-271-J from the Kansas Agricultural Experiment Station.

Data Availability Statement: The collected dataset used in this study will be available upon request from the main author.

Acknowledgments: We would like to thank the members of the Field Crops IPM Laboratory, including Kent Hampton, Nick Clark, Max Dunlap, Anna Keenan, Amanda Drouhard, Luke Carney, and Trevor Witt, for their support and data collection.

Conflicts of Interest: The authors declare no conflict of interest.

References

1. Mandava, N.B. *Handbook of Natural Pesticides: Methods: Volume I: Theory, Practice, and Detection*; CRC Press: Boca Raton, FL, USA, 2018; ISBN 978-1-351-08961-6.
2. Kogan, M. Integrated Pest Management: Historical Perspectives and Contemporary Developments. *Annu. Rev. Entomol.* **1998**, *43*, 243–270. [CrossRef]
3. Flint, M.L.; Van den Bosch, R. *Introduction to Integrated Pest Management*; Springer Science & Business Media: Berlin/Heidelberg, Germany, 2012; ISBN 1-4615-9212-7.
4. Pimentel, D.; Levitan, L. Pesticides: Amounts Applied and Amounts Reaching Pests. *BioScience* **1986**, *36*, 86–91. [CrossRef]
5. Krupke, C.; Long, E. Intersections between Neonicotinoid Seed Treatments and Honey Bees. *Curr. Opin. Insect Sci.* **2015**, *10*, 8–13. [CrossRef] [PubMed]
6. Sánchez-Bayo, F.; Wyckhuys, K.A.G. Worldwide Decline of the Entomofauna: A Review of Its Drivers. *Biol. Conserv.* **2019**, *232*, 8–27. [CrossRef]
7. Krupke, C.; Holland, J.D.; Long, E.Y.; Eitzer, B.D. Planting of Neonicotinoid-Treated Maize Poses Risks for Honey Bees and Other Non-Target Organisms over a Wide Area without Consistent Crop Yield Benefit. *J. Appl. Ecol.* **2017**, *54*, 1449–1458. [CrossRef]
8. Bonmatin, J.-M.; Giorio, C.; Girolami, V.; Goulson, D.; Kreutzweiser, D.P.; Krupke, C.; Liess, M.; Long, E.; Marzaro, M.; Mitchell, E.A.D.; et al. Environmental Fate and Exposure; Neonicotinoids and Fipronil. *Environ. Sci. Pollut. Res.* **2015**, *22*, 35–67. [CrossRef] [PubMed]
9. Jensen, S.E. Insecticide Resistance in the Western Flower Thrips, *Frankliniella occidentalis*. *Integr. Pest Manag. Rev.* **2000**, *5*, 131–146. [CrossRef]
10. Kranthi, K.R.; Jadhav, D.R.; Kranthi, S.; Wanjari, R.R.; Ali, S.S.; Russell, D.A. Insecticide Resistance in Five Major Insect Pests of Cotton in India. *Crop Prot.* **2002**, *21*, 449–460. [CrossRef]
11. Margni, M.; Rossier, D.; Crettaz, P.; Jolliet, O. Life Cycle Impact Assessment of Pesticides on Human Health and Ecosystems. *Agric. Ecosyst. Environ.* **2002**, *93*, 379–392. [CrossRef]
12. Van Lenteren, J.C. The State of Commercial Augmentative Biological Control: Plenty of Natural Enemies, but a Frustrating Lack of Uptake. *BioControl* **2012**, *57*, 1–20. [CrossRef]
13. Brewer, M.J.; Peairs, F.B.; Elliott, N.C. Invasive Cereal Aphids of North America: Ecology and Pest Management. *Annu. Rev. Entomol.* **2019**, *64*, 73–93. [CrossRef] [PubMed]
14. Haar, P.J.; Buntin, G.D.; Jacobson, A.; Pekarcik, A.; Way, M.O.; Zarrabi, A. Evaluation of Tactics for Management of Sugarcane Aphid (Hemiptera: Aphididae) in Grain Sorghum. *J. Econ. Entomol.* **2019**, *112*, 2719–2730. [CrossRef] [PubMed]
15. Ciampitti, I.; Ruiz, D.; Onofre, R.; Jardine, R.; Lancaster, S.; Whitworth, J.; Aguilar, J. Kansas Sorghum Management 2021. 2021. Available online: https://bookstore.ksre.ksu.edu/pubs/MF3046.pdf (accessed on 4 March 2024).
16. Singh, B.U.; Padmaja, P.G.; Seetharama, N. Biology and Management of the Sugarcane Aphid, *Melanaphis sacchari* (Zehntner) (Homoptera: Aphididae), in Sorghum: A Review. *Crop Prot.* **2004**, *23*, 739–755. [CrossRef]
17. Bowling, R.D.; Brewer, M.J.; Kerns, D.L.; Gordy, J.; Seiter, N.; Elliott, N.E.; Buntin, G.D.; Way, M.O.; Royer, T.A.; Biles, S.; et al. Sugarcane Aphid (Hemiptera: Aphididae): A New Pest on Sorghum in North America. *J. Integr. Pest Manag.* **2016**, *7*, 12. [CrossRef] [PubMed]
18. McCornack, B.; Zukoff, S.; Michaud, J.; Whitworth, J. Scouting Sugarcane Aphids. Available online: https://www.myfields.info/sites/default/files/page/ScoutCard%20KSU%20v05312017.pdf (accessed on 10 November 2023).
19. Gordy, J.W.; Brewer, M.J.; Bowling, R.D.; Buntin, G.D.; Seiter, N.J.; Kerns, D.L.; Reay-Jones, F.P.F.; Way, M.O. Development of Economic Thresholds for Sugarcane Aphid (Hemiptera: Aphididae) in Susceptible Grain Sorghum Hybrids. *J. Econ. Entomol.* **2019**, *112*, 1251–1259. [CrossRef] [PubMed]
20. Carroll, M.W.; Radcliffe, E.B.; MacRae, I.V.; Ragsdale, D.W.; Olson, K.D.; Badibanga, T. Border Treatment to Reduce Insecticide Use in Seed Potato Production: Biological, Economic, and Managerial Analysis. *Am. J. Potato Res.* **2008**, *86*, 31. [CrossRef]
21. Lindenmayer, J.C.; Giles, K.L.; Elliott, N.C.; Knutson, A.E.; Bowling, R.; Brewer, M.J.; Seiter, N.J.; McCornack, B.; Brown, S.A.; Catchot, A.L.; et al. Development of Binomial Sequential Sampling Plans for Sugarcane Aphid (Hemiptera: Aphididae) in Commercial Grain Sorghum. *J. Econ. Entomol.* **2020**, *113*, 1990–1998. [CrossRef]
22. Bowling, R.; Brewer, M.; Knutson, A.; Biles, S.; Way, M.; Sekula-Ortiz, D. *Scouting Sugarcane Aphids in South, Central, and West Texas*; Texas A&M AgriLife Research: Weslaco, TX, USA, 2021; p. 2. Available online: http://agrilife.org/sca/files/2016/06/Sugarcane-Aphid-Quick-Checker.pdf (accessed on 10 March 2024).
23. Lytle, A.J.; Huseth, A.S. Impact of Foliar Insecticide Sprays on *Melanaphis sacchari* (Hemiptera: Aphididae) and Natural Enemy Populations in Grain Sorghum. *Crop Prot.* **2021**, *149*, 105764. [CrossRef]
24. Zukoff, A.; McCornack, B.; Whitworth, R. Sorghum Insect Pest Management. Available online: https://bookstore.ksre.ksu.edu/pubs/MF742.pdf (accessed on 15 January 2024).

25. Brewer, M.J.; Gordy, J.W.; Kerns, D.L.; Woolley, J.B.; Rooney, W.L.; Bowling, R.D. Sugarcane Aphid Population Growth, Plant Injury, and Natural Enemies on Selected Grain Sorghum Hybrids in Texas and Louisiana. *J. Econ. Entomol.* **2017**, *110*, 2109–2118. [CrossRef]
26. Pekarcik, A.J.; Jacobson, A.L. Evaluating Sugarcane Aphid, *Melanaphis sacchari* (Hemiptera: Aphididae), Population Dynamics, Feeding Injury, and Grain Yield among Commercial Sorghum Varieties in Alabama. *J. Econ. Entomol.* **2021**, *114*, 757–768. [CrossRef] [PubMed]

insects

MDPI

Article

Prey Switching and Natural Pest Control Potential of Carabid Communities over the Winter Wheat Cropping Season

Ambre Sacco--Martret de Préville [1,2,*], Karin Staudacher [3], Michael Traugott [3], David A. Bohan [4], Manuel Plantegenest [5,*,†] and Elsa Canard [1,†]

1 IGEPP, INRAE, Institut Agro, Université de Rennes, 35653 Le Rheu, France; elsa.canard@inrae.fr
2 Functional Diversity in Agro-Ecosystems, Crop Research Institute, Drnovská 507, Ruzyně, 161 06 Praha 6, Czech Republic
3 Mountain Agriculture Research Unit, Institute of Ecology, University of Innsbruck, Technikerstraße 25, 6020 Innsbruck, Austria; karin.wastian@sinsoma.com (K.S.); michael.traugott@uibk.ac.at (M.T.)
4 Agroécologie, AgroSup Dijon, INRAE, Université de Bourgogne Franche-Comté, 17 rue Sully, BP 86510, 21065 Dijon CEDEX, France; david.bohan@inrae.fr
5 IGEPP, INRAE, Institut Agro, Université de Rennes, 35000 Rennes, France
* Correspondence: ambresacco@gmail.com (A.S.--M.d.P.); manuel.plantegenest@agrocampus-ouest.fr (M.P.)
† These authors contributed equally to this work.

Simple Summary: Carabids are abundant and diverse arthropods in agroecosystems and are well known for their generalist diet and therefore biocontrol potential. Most studies on carabids' diet have focused on periods of peak pest infestations but their feeding habits through the cropping season are poorly known. In this study, we explored how carabids' trophic interactions with pests (aphids and slugs) and alternative prey (spiders, earthworms and springtails) varied from the wheat seedling emergence to wheat ripening by analysing the prey DNA in their gut. Carabids seemed to switch between prey depending on their availability at the time, reaffirming carabids' opportunistic diet. Regarding biocontrol, carabid taxa contributed complementarily to pest predation over the season and per pest taxon, highlighting the importance of favouring diversity in agroecosystems.

Abstract: To date, evaluating the diets of natural enemies like carabids has largely been limited to spatially explicit and short-term sampling. This leaves a knowledge gap for the intra-annual dynamics of carabid diets, and the provision and timing of delivery of natural pest control services. Season-long pitfall trapping of adult carabids was conducted in conventional winter wheat fields, from November 2018 to June 2019, in five French departments. Diagnostic Multiplex PCR of carabid gut contents was used to determine the dynamics of carabid diets. The overall detection rate of target prey DNA was high across carabid individuals (80%) but varied with the prey group. The rate of detection was low for pests, at 8.1% for slugs and 9.6% for aphids. Detection of intraguild predation and predation on decomposers was higher, at 23.8% for spiders, 37.9% for earthworms and 64.6% for springtails. Prey switching was high at the carabid community level, with pest consumption and intraguild predation increasing through the cropping season as the availability of these prey increased in the environment, while the detection of decomposer DNA decreased. Variation in diet through the cropping season was characterized by: (i) complementary predation on slug and aphid pests; and (ii) temporal complementarity in the predominant carabid taxa feeding on each pest. We hypothesize that natural pest control services delivered by carabids are determined by complementary contributions to predation by the different carabid taxa over the season.

Keywords: food web; pest control; intraguild predation; alternative prey; molecular gut content analysis; multiplex PCR diagnostic

1. Introduction

Biological control theory posits that the efficiency of pest control depends on natural enemy abundance and diversity [1], as well as the complementarity of their contribution to control, including attacking different species and life stages of pests and facilitating prey capture by other natural enemies [2]. For instance, population growth of the weed *Alopecurus myosuroides* can be controlled by complementary seed predation by vertebrates and invertebrates in Swedish winter wheat fields [3]. The taxa contributing most to pest control may vary throughout the cropping season [2]. In addition, the presence of alternative prey to pests in their absence can be critical to maintaining large populations of natural enemies [4]. Therefore, the effectiveness of biocontrol services may depend on both the turnover of predator taxa and the presence of alternative prey in agricultural fields. Deciphering the temporal variation in trophic interactions between predators and their prey, both pests and non-pests, is essential to assess the switch in their diet over the agricultural season and evaluate the dynamics of pest control services in agroecosystems.

Carabids are among the most abundant and diverse families of arthropod generalist predators in temperate agricultural fields. It has been proposed that the polyphagous and opportunistic diet of carabids may make them good natural regulators of a wide range of pests [5,6]. Indeed, carabids have been observed preying on weed seeds [7] and a wide variety of invertebrate pests [8,9]. Carabids also consume alternative resources that could be beneficial organisms, including decomposers [7,10] and other natural enemies, causing intraguild predation [11,12]. Beyond these disservices caused by the direct predation on beneficial organisms, alternative prey might also divert carabids from pest control [13]. Conversely, alternative prey can ensure the presence and aggregation of carabids in agroecosystems by acting as alternative food resources and sustaining them during periods of low pest prey availability [13]. Our expectation is that the diversity of carabid species, when combined with the breadth, complexity and variation of their diet, makes carabids a model community of predators for studying the balance of ecosystem services and disservices provided by generalist natural enemies. In addition, it may highlight the indirect role played by non-pest prey in supporting biocontrol services as alternative food sources for generalist predators.

Carabid diet and feeding preferences have been studied in the laboratory using panels of prey in cafeteria or voracity tests [10,14]. These laboratory tests can provide repeatable information on predatory behaviour by simulating competition or risk of intraguild predation [15] but they do not mirror the frequency of prey–predator interactions under natural conditions [16]. In the field, diets have been inferred by observing the population dynamics of carabid prey [17,18] or indirectly by using predator exclusion cages [19]. Such methods provide only correlative data, however, and are frequently unable to pinpoint specific prey–predator interactions for which the test is evidence of feeding [20]. Molecular gut content analyses (MGCAs), and in particular diagnostic multiplex PCR, offer a direct test for unravelling specific predator–prey trophic interactions [12,20].

Diagnostic multiplex PCR has been successfully used to determine the diet of the most abundant carabid species and made it possible to sample for short-term, spatially explicit variation in carabid diets. This has included examining whether trophic interactions change with farming system [9,21], habitat heterogeneity [22] and pest abundance [12]. This sampling has focussed only on specific periods of the cropping season, such as periods of pest presence (but see [but [11]]), and therefore overlooked the important role of alternative prey in the carabid diet, particularly when pests are absent from fields. Indeed, as alternative food resources, alternative prey can provide a complementary indirect ecosystem service in maintaining natural enemy populations [23], which can only be evaluated by following diets through time. Furthermore, focusing on specific periods of the cropping season excludes sets of carabid species from consideration.

Our approach is to work at the community level, explicitly highlighting the dynamics of the trophic ecology of the entire carabid community and their predation on pests and alternative prey. Our aim is to determine which carabid species are the most effective predators of pests, when this predation take place and what other resources are involved in the population cycle.

In the present study, we determine carabid species' abundance and their diets over the entire cropping season of conventional winter wheat crops in France. Using multiplex PCR, we designed an approach to evaluate trophic interactions between carabids and five groups of potential prey, including problematic pests of winter wheat and alternative prey, natural enemies (intraguild predation) and decomposers.

The multiplex PCR is applied to the entire carabid community sampled to estimate the diet of carabids and thereby the services provided over the wheat cropping season. Following the expectation of Gray et al. [24], we assume that prey consumption will be primarily determined by the presence and abundance (availability) of prey resources in the field. Hence, we hypothesise that prey consumption increases with prey availability (i). As pest and spider abundance increases over the wheat cropping season [11,12], we expect their consumption by carabids will increase accordingly (ii). We also expect the predation rates on decomposers to remain more stable throughout the season, given their near-constant availability (iii). These contrasting dynamics lead us to predict switching in the prey eaten by the carabids over the cropping season, and especially around the time of peak pest availability.

2. Materials and Methods

2.1. Field Sampling

Across five regions of France, carabids were collected in ten conventional wheat fields, with two fields per region (Supplementary Material S1): Essonne (Île-de-France), Ille-et-Vilaine (Bretagne), Loire-Atlantique (Pays de la Loire), Vendée (Pays de la Loire) and Meurthe-et-Moselle (Lorraine). The regions of study lay on a climatic, landscape and farmland gradient. Ille-et-Vilaine, Loire-Atlantique and Vendée are oceanic regions with a dense hedgerow network across the farmland ("bocage" structure) dominated by a polyculture-livestock system. The Essonne and Meurthe-et-Moselle regions have an altered oceanic and semi-continental climate, respectively, and their farmland mainly exhibits cereals in open fields. The two selected fields per region were chosen randomly across the farmland. Field size ranged from 0.5 ha for the smallest field to 9.5 ha for the largest one.

Each field was sampled in each of five sampling sessions (November 2018, December 2018, April 2019, May 2019 and June 2019), using pitfall traps (9.5 cm diameter, 520 mL volume). Sampling sessions were designed depending on the growth stage of the wheat and periods of activity of invertebrates. Autumn sessions (November and December 2018) are periods of emergence and post-emergence of wheat seedlings when plants are sensitive to slugs. The session in early spring (April 2019) corresponds to arthropod activity resuming after the winter. The May session is linked to the main colonization period of cereal aphids. The last session in June is associated with wheat ripening, a period when aphid infestation reaches its peak. In each field, 25 dry pitfall traps were opened and emptied twice a day (morning and late afternoon) for three consecutive days. No trapping was carried out in the Meurthe-et-Moselle region in the November and December 2018 sampling sessions.

The pitfall traps were filled with clay pellets to provide cryptic refuges that can reduce predation among the captured animals. At emptying, all trapped carabids were placed on ice before being transferred to individual 2 mL Eppendorf tubes upon arrival in the laboratory. Carabids were stored at −20 °C as fast as possible to stop the digestion process, but a first rapid identification occurred on the live carabids to the genus level or morphotypes to have an idea of the community structure.

2.2. Carabids Selection for MGCA

The November and December 2018 sessions were pooled for analysis due to the low abundance of carabids trapped during this cold period. These pooled samples are henceforth referred to as the autumn session.

At each trapping point (i.e., a trapping session in a specific field), 48 individuals from among all the trapped carabids were sub-sampled for MGCA. Each species was sub-sampled according to its proportion in the entire community so that they constituted a representative subsample of the local trapped communities. Where fewer than 48 individuals were trapped (13 trapping points out of 38), all the individuals were retained for molecular analyses. Where possible, all beetles selected for MGCA were re-identified with Roger et al.'s identification key [25] to the species level, but were otherwise identified to genus, while making sure they did not thaw out. All subsampled beetles were then placed individually in individual 1.5 mL Eppendorf tubes on ice.

The wheat-field specialist, *Poecilus cupreus*, was overwhelmingly dominant at many sampling points. To ensure a specific diversity in the sample, a maximum of 30 individuals belonging to this species was included for each point. Where necessary, a weighting was used to account for this deviation from the pro rata sampling.

2.3. DNA Extraction

To reduce the risk of contamination, all extractions were carried out in a sterilised, pre-PCR laboratory. For those taxa smaller than 15 mm, the whole body of each carabid was processed for DNA extraction. For taxa longer than 15 mm, such as *Pterostichus melanarius* and *Carabus* spp., the gut was dissected and processed for DNA extraction.

To extract DNA, each carabid was transferred into an individual 2 mL Eppendorf tube filled with lysis buffer, 10 µL of Proteinase K (20 mg/mL) and a volume of TES adjusted to carabid size (0.1 M TRIS, 10 mM EDTA, 2% SDS, pH 8). TES volume was 290 µL for carabid species smaller than 4 mm, 630 µL for *P. melanarius*, 990 µL for *Carabus* spp. and 430 µL for all other species. Cuticles and soft tissues were crushed and homogenised with glass beads of various sizes (two cycles of 1 min at 5000 RPM in a Precellys® 24 Tissue Homogenizer, Bertin Technologies, Montigny-le-Bretonneux, France). The samples were centrifuged at 800 RPM for 1 min and incubated overnight at 58 °C. DNA was extracted using the BioSprint 96 DNA Blood Kit (Qiagen, Hilden, Germany) in a BioSprint 96 device (Qiagen) in accordance with manufacturer's instructions. An amount of 200 µL of DNA extract was obtained per sample and stored at −20 °C until PCR amplification.

Four negative controls per batch of 96 samples were included to check for DNA contamination (two at the lysis step and two at the extraction step).

2.4. Primer Design, Multiplex PCR Assay and PCR Amplification

A multiplex PCR assay was designed based on previously published group-specific primer pairs for earthworms, springtails, aphids and spiders, and a newly designed pair for slugs (Supplementary Material S2). Primer Premier 5 (PREMIER Biosoft International, Palo Alto, CA, USA) was used to design primers targeting slug 18S DNA. Primer sensitivity and specificity evaluation was performed for all primers pairs during the multiplex assay (Supplementary Material S3).

The new multiplex assay was conducted on a 10 µL volume containing 5 µL of QIAGEN Multiplex PCR Mastermix (Qiagen), 1 µL of 10× Primer mix (for final primers concentration in PCR see Supplementary Material S2), 1.5 µL of DNA extract, 0.5 µL of BSA (10 mg/mL, Qiagen) and 2 µL of PCR water to adjust the volume.

PCR cycling conditions included three steps: an initial cycle of 15 min at 95 °C, then 35 cycles of 30 s at 94 °C, 90 s at 63.5 °C and 60 s at 72 °Cand finally a 10 min-cycle at 72 °C. PCR products were displayed using the QIAxcel electrophoresis system (Qiagen) following the protocol of Staudacher et al. [12].

2.5. Statistical Analysis

All statistical analyses were carried out with the software R [26] version 3.6.1.

The DNA detection rate of a specific prey was calculated as the number of carabid individuals for which this prey DNA was detected in their gut divided by the total number of individuals that were positive for at least one of the five tested prey. The DNA detection rate was then modelled as a mixed generalized linear model with prey group (five levels: aphid, earthworm, slug, spider, springtail), session (four levels: autumn, April, May, June) and their interaction as fixed effects and with the field, nested in the region, and individual carabid codes as random effects. The model was fitted with the "glmer()" function from the "lme4" package (v. 1.1.35.1) [27], using a binomial distribution and the logit link function. Detection of overdispersion was tested using Pearson residual χ^2 statistic with the "simulateResiduals()" function from the "DHARMa" package (v. 0.4.6) [28]. Significance of the explanatory variables was checked with the "Anova()" function from the "car" package (v. 3.1.2) [29]. Pairwise comparisons between the levels of each factor were carried out using "emmeans()" and "pairs()" functions from the "emmeans" package (v. 1.10.0)to check the significance of detection rate differences between sessions [30].

As we limited the number of *P. cupreus* individuals retained for molecular analysis to a maximum of 30 at each sampling point, we created a corrected version of the dataset respecting the observed relative abundance of this species in the pitfall traps. This was achieved by randomly resampling the appropriate number of *P. cupreus* among the 30 tested individuals. For the sake of clarity, this dataset is henceforth termed the "community dataset" and it is used uniquely for analyses carried out at the community level.

The "plotweb()" function from the "bipartite" package (v. 2.19) was used to graph trophic interaction networks (food webs) of carabid communities and their dynamics over the wheat cropping season [31].

At the carabid genus level, patterns of temporal diet variation were drawn using individuals positive for at least one prey group. Seasonal predation profiles, defined as the list of predation rates for each combination of prey and session, were assessed for the most abundant genera only (>40 individuals). Consequently, only nine carabid genera were considered: *Amara, Anchomenus, Bembidion, Brachinus, Harpalus, Nebria, Notiophilus, Poecilus* and *Trechus*. Similarities and specificities between seasonal predation profiles of those genera were studied by Principal Component Analysis using the "PCA()" and "explor()" functions of the "FactoMineR" (v. 2.10) [32] and "explor" (v. 0.3.10) [33] packages. The procedure was followed by a hierarchical clustering using the "HCPC()" function to group selected genera in clusters of similar predation profiles and visualized with the "fviz_dend()" function, both from the "FactoMineR" package (v. 2.10) [32]. From the list of outputs of the "HCPC()" function, "$desc.var" provided the levels of factors that significantly contributed to describing each cluster.

3. Results

3.1. Carabid Communities

A total of 5843 carabids were trapped during sampling among which 1819 were processed for MGCA. A total of 57 taxa were identified, with abundances ranging between one and 625 individuals (details in Supplementary Material S4).

3.2. Detection Rates at Community Level

A total of 80% of processed carabid individuals tested positive for at least one prey in the multiplex PCR. Across all carabids examined, the detection rates for pests were 8.1% for slugs and 9.6% for aphids. Spider DNA was detected in 23.8% of carabids. Of all carabids, 37.9% contained decomposer DNA, 37.9% of earthworm and 64.6% of springtail. Detection rates for prey varied depending on region: from 7.1% in Meurthe-et-Moselle to 12.5% in Vendée for aphids, from 0.9% in Meurthe-et-Moselle to 10.4% in Ille-et-Vilaine for slugs, from 11.8% in Essonne to 45.5% in Vendée for spider, from 52% in Essonne to 72% in Loire-Atlantique for springtails and from 20.0% in Essonne to 48.0% in Meurthe-et-Moselle for earthworms.

The number of carabids which tested positive to pest and spider DNA increased through the season, while the number of carabids positive to decomposer DNA peaked in spring (April–May) (Figure 1).

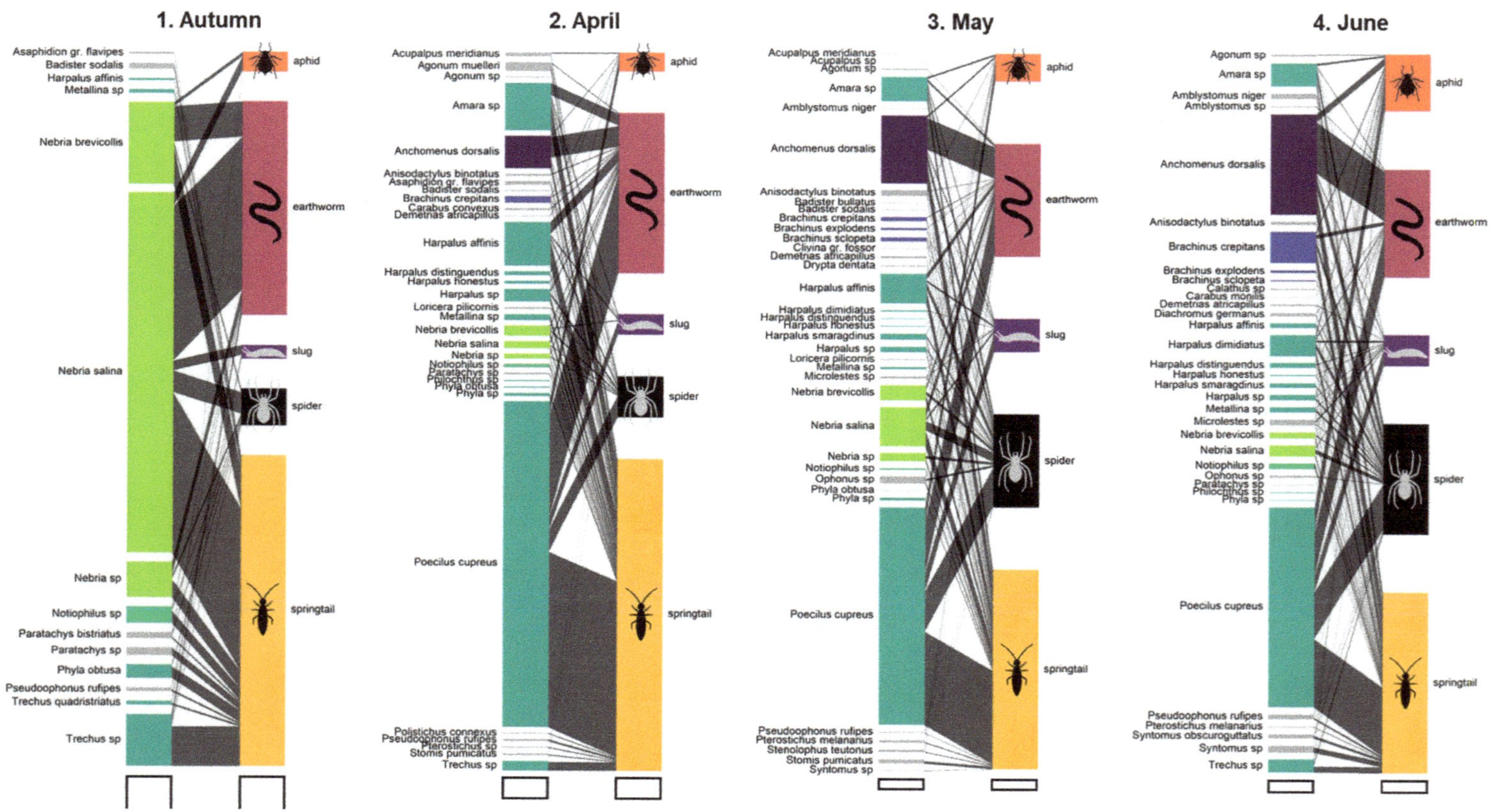

Figure 1. Bipartite food webs showing the number of individuals of each carabid taxon testing positive for DNA of each prey type over the wheat cropping season. The width of the left-hand blocks represents the number of DNA detections for each taxon of carabids. The width of the right-hand blocks represents the number of DNA detections for each prey group. The width of the link represents the number of DNA detections for each prey for each carabid taxon. The colours of the carabid taxa and prey groups are consistent from one figure to the next. White rectangles outlined in black under the food webs all represent 20 DNA detections.

3.3. Factors Influencing Prey DNA Detection Rates at Carabid Community Scale

Prey group, session and their interactions all have a significant effect on the probability of prey DNA detection in carabids (Table 1).

Table 1. Analysis of deviance table testing the significance of the effect of prey group, session and their interactions on the probability of detecting prey DNA in the gut of carabids. Chisq stands for the chi-square statistic, Df for the degree of freedom and Pr (>Chisq) for *p*-values associated with the chi-square statistic.

Fixed Effect	Chisq	Df	Pr (>Chisq)
(Intercept)	75.224	1	<0.001
Prey group	302.143	4	<0.001
Session	49.939	3	<0.001
Prey group: Session	236.196	12	<0.001

Overall, detection rates increased over sessions from autumn to summer, but not for all prey types. Significant increases in detection of pest DNA were observed (aphid: autumn < June, with $p < 0.001$, April < (May, June), with $p < 0.01$ and $p < 0.001$, respectively, and May < June, with $p < 0.001$; slug: autumn < (May, June), with $p < 0.01$ and $p < 0.05$, respectively, and April < (May, June), with $p < 0.001$ and $p < 0.05$, respectively) and of spider DNA (spider: (autumn, April) < (May, June) with $p < 0.001$ in all cases). Detection of earthworm DNA significantly decreased from autumn to summer (earthworm: autumn > (April, June) with $p < 0.05$ and $p < 0.001$, respectively). Detection of springtail DNA remained generally high, except for a significant drop in June (springtail: (autumn, April, May) > June, with $p < 0.001$ in all cases).

The overall detection rates and their seasonal changes are represented in Figure 2. Results from the "community dataset", obtained by resampling, were broadly similar.

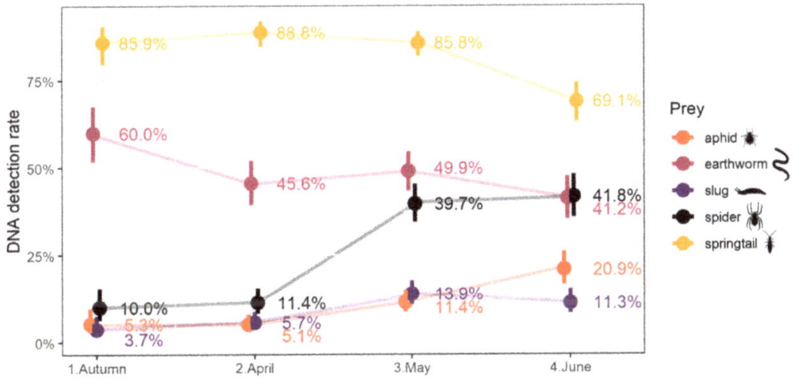

Figure 2. Variations throughout the cropping season of the DNA detection rate in carabid beetle communities for each prey group.

3.4. Seasonal Predation Profile of Predominant Carabid Genera

Using hierarchical clustering, the nine most frequent genera were grouped into five clusters based on the seasonal variation pattern of their diet (Figure 3). Cluster 1 was solely composed of the *Notiophilus* genus and its diet did not differ significantly from the average at any period. Significantly higher rates of DNA detection of the two pests in April characterised cluster 2 (aphid: $p < 0.01$ and slug: $p < 0.05$), which was also formed of a single genus, *Brachinus*. Cluster 3 was the only multi-genera cluster (*Amara*, *Bembidion*, *Harpalus*, *Poecilus*, *Trechus*) and was characterised by a significantly higher rate of slug DNA detection in May ($p < 0.05$). Cluster 4 included only the *Anchomenus* genus and was distinguished by significantly higher rates of earthworm DNA detection in April ($p < 0.05$), May ($p < 0.05$) and June ($p < 0.05$), and a lower springtail consumption rate in April ($p < 0.05$). Finally, the *Nebria* genus composed cluster 5 and had significantly higher rates of earthworm and aphid DNA detection in autumn ($p < 0.01$, $p < 0.05$), spider in April ($p < 0.05$) and aphid in June ($p < 0.05$).

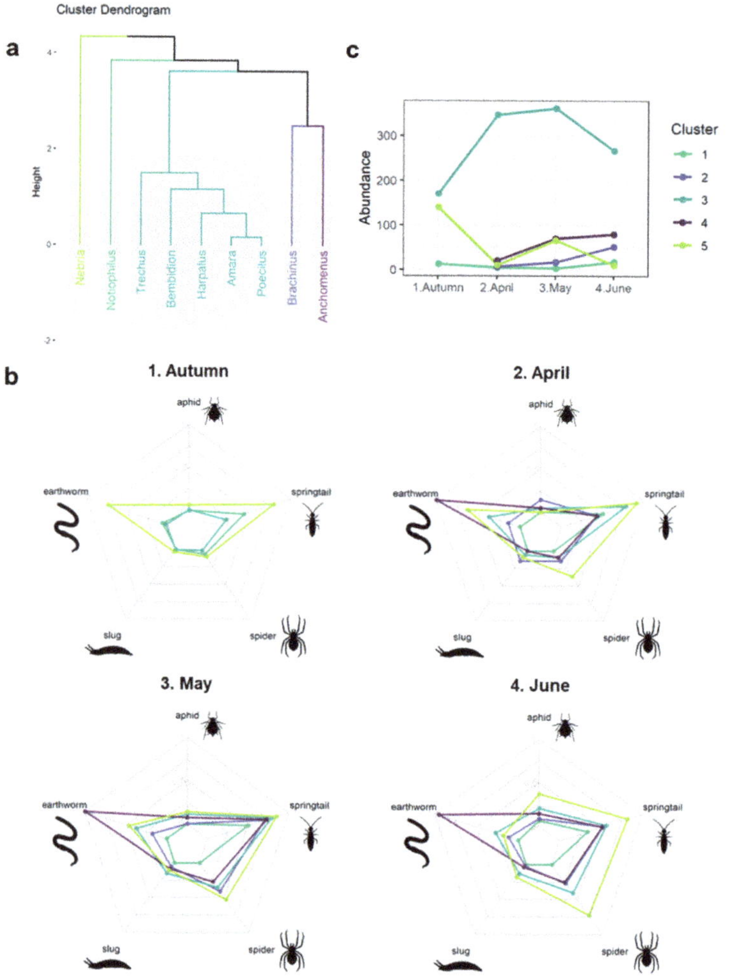

Figure 3. Cluster analyses of the most abundant carabid genera based on their predation profile, i.e., their seasonal predation patterns, (**a**), DNA detection rates for each combination of prey and cluster by session (**b**) and abundance variations of carabid genera grouped by clusters throughout the cropping season (**c**). The dendrogram presents the clustering of carabid genera based on the similarities in the seasonal pattern of prey DNA detection rates. Colours indicate clusters (green: cluster 1, indigo: cluster 2, blue: cluster 3, violet: cluster 4, light green: cluster 5). The absence of a specific colour on diamond plot (**b**) indicates that no individuals of the cluster associated with the colour were tested.

4. Discussion

The present study suggests a prey switching in carabids' diet consistent with prey availability in the environment. This relationship highlights the importance of alternative prey in sustaining carabid populations in the absence of pests. Temporal and complementary predation were also unveiled among carabid taxa through their dietary specificities.

This study corroborates previous findings that carabids prey upon two major wheat pests, aphids and slugs. However, detection rates of pest DNA were low compared to those of alternative prey DNA. We expected that pest detection rates would match the period of pest availability and be highest during peak abundance [12,21]. Indeed, aphid DNA detection increased sharply over the cropping season, to a peak in June, in a similar way to aphid infestation in the field (see Supplementary Material S5 for details), as has been found in previous studies [12,21]. The slug period of activity extends from November to June, with a peak in spring [9,34], and as expected slug DNA detection also peaked in April–May. However, this pest is a major crop pest in autumn, when they can attack wheat seedlings, while slug DNA detection remained low in autumn. This pattern might be due to a low slug infestation in 2018 (see Supplementary Material S5 for details) but also to the composition of carabid assemblages during the cold season. *Nebria* was the most abundant carabid genus during this period, but had low rates of slug consumption, in agreement with a previous result in the literature [9].

Despite their predation on a wide variety of pests, the generalist diet could result in an overall decrease in pest consumption due to intraguild predation. Indeed, in May and June, 40% of tested carabids were positive to spider DNA. As expected, this increase is positively correlated with spider availability in the environment (see Supplementary Material S5 for details). Intraguild predation rate on spiders was higher in this study than in most precedent works [12,21], but was consistent with the results of Davey et al. [11]. Interestingly, Davey et al. [11] showed that such high predation rates did not reduce the spider population, which continued to increase, potentially mitigating the risk of disservice resulting from intraguild predation by carabids. This is consistent with previous meta-analyses that have suggested that antagonistic interactions between natural enemies do not always interfere with biocontrol services [35,36].

Carabids were found consuming other service providers such as decomposers: DNA detection rates as high as 37.9% for earthworms and 64.6% for springtails were obtained during this analysis, which is at least double what was observed in previous studies [7,12,21]. Such differences might be due to the low infestation rates of pests in French fields in 2018–2019 resulting in carabids using decomposers as alternative prey [23]. Springtail DNA detection rates decreased slightly at the end of cropping season but remained high for most carabid species. Springtails are considered the basal prey of several carabid species [37], and our study corroborates these expectations. Decomposers are present and abundant all year round in French fields and, as alternative prey, might support carabid populations in seasons when pest availability is low. Settle [38] found that detritus-feeding and plankton-feeding insects were alternative prey for generalist predators and sustained their populations in rice fields early in the growing season. Yoo [4] observed that thrips were alternative prey of the generalist predator stinkbug, *Orius insidiosus*, facilitating its earlier colonization of fields. In both cases, the presence of alternative prey was believed to provide sufficient food resources to natural enemies to give them a head start in suppressing subsequent pest outbreaks [4,38]. In our study, earthworms and springtails are the predominant components of the carabid diet, but their detection rate declines over the cropping season in favour of pests and spiders. We argue, therefore, that alternative prey might contribute indirectly to pest consumption by sustaining carabid populations when pest prey is scarce. Carabids of the genus *Anchomenus* appeared as something of an outlier to this pattern. The detection rate of earthworm DNA in *Anchomenus* remained consistently above 95% throughout the year, suggesting a strong preference for earthworms. This preference has not previously been demonstrated. Kamenova [39] has shown that the carbon isotopic signature of *Anchomenus dorsalis* is linked to the isotopic signature of the crop, suggesting feeding on crop herbivores. This apparent difference may be an effect of prey communities differing between studies, but it is still consistent with our results because while high rates of earthworm DNA were detected in *Anchomenus* in our study, presumably due to their elevated availability, our results indicate that *Anchomenus* also eats all other prey tested and is not a strict earthworm specialist.

In addition, our results highlight the complementarity contribution of different carabid species to pest predation throughout the cropping season and the importance of considering their diversity to properly assess their potential. Indeed, for the two pest groups we examined, the carabid species providing the most predation changed over time. *Nebria* and *Brachinus* contributed to predation in a relay, while the taxa providing the greatest predation of slugs changed from *Brachinus* to a set of other genera (*Amara*, *Bembidion*, *Harpalus*, *Poecilus* and *Trechus*). *Nebria* outperformed other carabid genera in both the autumn and early summer, suggesting that this genus has the potential to predate a lot of aphids in wheat crops at two different times of the season, possibly autumn aphids and grain aphids, respectively. This is consistent with previous carbon isotopic analysis which suggested that *Nebria salina* consumed prey feeding on crop plants [39]. *Nebria* did not prey upon slugs in the autumn. Adult slugs protect themselves from predator attacks by secreting mucus, and *Nebria brevicollis* has been found to prey less on live slugs than other species of carabids, such as *Pterostichus madidus*, and with a success rate that decreases with slug size [40,41] but see also [42]. In the light of those results, we find no evidence from our fields that carabids importantly consume slugs in autumn, when they can cause damage in winter-sown crops. However, the increase in slug DNA detected in carabids in the spring, including for *Nebria*, suggests that carabids do consume slugs, but maybe at specific life stages. Indeed, other species of carabids, *Poecilus cupreus* and *Pterostichus melanarius*, have been observed consuming eggs of *Deroceras reticulatum* and *Arion lusitanicus* in laboratory conditions [41], suggesting that some control may be exerted by carabids on early life stages of slugs. Interestingly, in spring, genera of smaller carabids, including *Bembidion* and *Trechus*, had similar slug consumption rates to the larger-bodied genera, suggesting that they also contribute to slug predation. In addition to temporal complementarity of predation by the carabid community, our data suggest complementarities of pest predation. In spring, *Brachinus* showed evidence of complementary predation for both aphids and slugs. *Amara* and *Harpalus* have generally been considered to be granivores and to date studies have focused on their ability to control weeds by consuming their seeds [43]. Our results indicate that they also contribute to predation of slugs. However, the parasitic behaviour of *Brachinus* on the carabid genus *Amara* [44] could potentially interfere with its pest and weed consumption.

While the complementarity contribution of the different carabid taxa to pest predation in our results is exciting, it is important to remember that it is solely based on the data of pest DNA detection in the carabid gut. Many other factors impacting pest control remain, such as carabid abundance or metabolic rate, which bring nuances to the results. *Nebria* and *Brachinus* stood out for their higher detection rate of pests' DNA, *Nebria* for aphids in June and *Brachinus* for aphids and slugs in April, but these genera have a really low abundance at those specific times (Figure 3c), which suggest a low impact on pest control. Bigger species are also expected to have a higher impact than smaller ones due to their higher metabolic rates, and consequently their higher food intake. A lower contribution from small species is therefore suspected from *Bembidion* and *Trechus* genera on slug predation despite their similar detection rate in May to bigger genera like *Amara*, *Poecilus* or *Harpalus*. Another limitation of our study includes the presence of potentially important sources of variation in the detection of prey DNA. It is unlikely that carabids digest all types of prey at the same rate. Previous studies on rates of digestion by *P. melanarius* have indicated that the detectability half-life is highly prey-dependent, with lower rates for aphids than for other prey (between 5–20 h in most cases [45,46]), highly variable values for earthworms that were as high as 90 h [10,47] and intermediate values between 20–40 h for slugs and spiders [11,47]. Consequently, pest consumption by carabids might therefore be underestimated when compared to alternative prey consumption. The digestion rate also varies among predator species, with Greenstone et al. [8] showing that different predator taxa had different digestion rates after eating the same quantity of Colorado potato beetles, resulting in variation in post-predation detection times. Information on the variation in digestion rates among carabids is effectively non-existent and the study of this phenomenon has largely been limited to *P. melanarius*; it is therefore difficult to estimate variations between carabid species. Variability in detection rates also depends on amplicon length or primer specificity [45]. In our multiplex PCR, primer concentrations were adjusted so that detection was similar for each prey type with an equivalent amount of DNA, mitigating biases due to amplicon length. Lastly, mistaking secondary predation events (carabid–spider–aphid) for primary predation events (carabid–aphid) could lead to an overestimation of aphid predation by carabids. While the risk is high in situations where both predators did not have the time to digest their respective prey, detection of secondary predation is greatly reduced when they do [48,49]; misinterpretation of trophic links is therefore expected to be limited in our study. An uncertainty however remains when trying to assess carabids' predation impact due to our inability to distinguish predation events from scavenging ones with our methodology. Multiple carabid species have been shown to prefer dead prey to live ones in laboratory experiments [50–52], which could suggest a more saprophagous diet for carabids even in natural conditions and consequently a lower impact on prey populations. This could also explain the high detection rates of springtail DNA found in the gut of every carabid species, despite genera like *Notiophilus*, *Leistus* or *Loricera* having evolved specialized and crucial morphological adaptations for hunting and consuming springtails that are absent from other carabid species [53–55].

5. Conclusions

The current study contributed to a better understanding of carabids' diet dynamics and revealed prey switching consistent with prey availability in the environment. Our results highlight the importance of alternative prey, particularly decomposers, in sustaining carabid populations in the absence of pests and in maintaining their consumption of pests over the year. Adopting agricultural practices that promote decomposers, such as mulching or conservation practices, could therefore result in a positive bottom-up effect on the natural enemy communities and consequently be beneficial to the farmers through a top-down effect on pests. However, it remains to be shown, in practice, that carabid predation results in the regulation of pest populations and therefore in a biological control service.

Supplementary Materials: The following supporting information can be downloaded at: https://www.mdpi.com/article/10.3390/insects15080610/s1. Supplementary Material S1: Map of France showing the location of the five regions where carabid beetles were sampled. Supplementary Material S2: Primer pairs composing the multiplex PCR assay for assessing trophic interactions in wheat crops. Supplementary Material S3: Taxa used for testing primers' sensitivity and specificity. Supplementary Material S4: Distribution of the number of selected carabid beetles for MGCA per taxon and session. Supplementary Material S5: Correlation plots between prey abundance and their DNA detection rate in the gut of carabid beetles for aphids, slugs and spiders, and discussion.

Author Contributions: Conceptualization, E.C. and M.P.; methodology, E.C., M.P., K.S. and M.T.; formal analysis, A.S.--M.d.P., D.A.B., E.C. and M.P.; investigation, E.C., M.P. and A.S.--M.d.P.; resources, M.T.; data curation, A.S.--M.d.P.; writing—original draft preparation, A.S.--M.d.P.; writing—review and editing, A.S.--M.d.P., E.C., M.P., D.A.B., K.S. and M.T.; visualization, A.S.--M.d.P.; supervision, E.C. and M.P.; project administration, E.C.; funding acquisition, E.C. and M.P. All authors have read and agreed to the published version of the manuscript.

Funding: This work was funded by the Ministère de l'Agriculture et de la Souveraineté alimentaire, AAP CasDar project ARENA n°5653, "Anticiper les régulations naturelles" and by the Agence Nationale de la Recherche, ANR NGB (ANR-17-CE32-0011), "Next Generation Biomonitoring of change in ecosystem structure and function".

Data Availability Statement: Data will be available on the Recherche Data Gouv Repository.

Acknowledgments: The authors would like to thank the farmers and all the people of the ARENA network who participated in carabid sampling, as well as Franck Duval for his help in carabid identification. We thank the Sinsoma company and especially Christiane Zeisler for sharing her expertise in this molecular study.

Conflicts of Interest: The authors declare no conflicts of interest. The funders had no role in the design of the study; in the collection, analyses, or interpretation of data; in the writing of the manuscript; or in the decision to publish the results.

References

1. Letourneau, D.; Jedlicka, J.; Bothwell, S.; Moreno, C. Effects of Natural Enemy Biodiversity on the Suppression of Arthropod Herbivores in Terrestrial Ecosystems. *Annu. Rev. Ecol. Syst.* **2009**, *40*, 573–592. [CrossRef]
2. Snyder, W.E. Give predators a complement: Conserving natural enemy biodiversity to improve biocontrol. *Biol. Control* **2019**, *135*, 73–82. [CrossRef]
3. Daouti, E.; Jonsson, M.; Vico, G.; Menegat, A. Seed predation is key to preventing population growth of the weed Alopecurus myosuroides. *J. Appl. Ecol.* **2022**, *59*, 471–482. [CrossRef]
4. Yoo, H.J.S.; O'Neil, R.J. Temporal relationships between the generalist predator, Orius insidiosus, and its two major prey in soybean. *Biol. Control* **2009**, *48*, 168–180. [CrossRef]
5. Kromp, B. Carabid beetles in sustainable agriculture: A review on pest control efficacy, cultivation impacts and enhancement. *Agric. Ecosyst. Environ.* **1999**, *74*, 187–228. [CrossRef]
6. Lövei, G.L.; Sunderland, K.D. Ecology and Behavior of Ground Beetles (Coleoptera: Carabidae). *Annu. Rev. Entomol.* **1996**, *41*, 231–256. [CrossRef] [PubMed]
7. Frei, B.; Guenay, Y.; Bohan, D.A.; Traugott, M.; Wallinger, C. Molecular analysis indicates high levels of carabid weed seed consumption in cereal fields across Central Europe. *J. Pest. Sci.* **2019**, *92*, 935–942. [CrossRef] [PubMed]
8. Greenstone, M.H.; Szendrei, Z.; Payton, M.E.; Rowley, D.L.; Coudron, T.C.; Weber, D.C. Choosing natural enemies for conservation biological control: Use of the prey detectability half-life to rank key predators of Colorado potato beetle. *Entomol. Exp. Appl.* **2010**, *136*, 97–107. [CrossRef]
9. Reich, I.; Jessie, C.; Ahn, S.-J.; Choi, M.-Y.; Williams, C.; Gormally, M.; Mc Donnell, R. Assessment of the Biological Control Potential of Common Carabid Beetle Species for Autumn- and Winter-Active Pests (Gastropoda, Lepidoptera, Diptera: Tipulidae) in Annual Ryegrass in Western Oregon. *Insects* **2020**, *11*, 722. [CrossRef]
10. King, R.A.; Vaughan, I.P.; Bell, J.R.; Bohan, D.A.; Symondson, W.O.C. Prey choice by carabid beetles feeding on an earthworm community analysed using species- and lineage-specific PCR primers. *Mol. Ecol.* **2010**, *19*, 1721–1732. [CrossRef]
11. Davey, J.S.; Vaughan, I.P.; King, R.A.; Bell, J.R.; Bohan, D.A.; Bruford, M.W.; Holland, J.M.; Symondson, W.O.C. Intraguild predation in winter wheat: Prey choice by a common epigeal carabid consuming spiders. *J. Appl. Ecol.* **2013**, *50*, 271–279. [CrossRef]
12. Staudacher, K.; Jonsson, M.; Traugott, M. Diagnostic PCR assays to unravel food web interactions in cereal crops with focus on biological control of aphids. *J. Pest. Sci.* **2016**, *89*, 281–293. [CrossRef] [PubMed]
13. Carbonne, B.; Petit, S.; Neidel, V.; Foffova, H.; Daouti, E.; Frei, B.; Skuhrovec, J.; Řezáč, M.; Saska, P.; Wallinger, C.; et al. The resilience of weed seedbank regulation by carabid beetles, at continental scales, to alternative prey. *Sci. Rep.* **2020**, *10*, 19315. [CrossRef] [PubMed]
14. Honek, A.; Martinkova, Z.; Saska, P.; Pekar, S. Size and taxonomic constraints determine the seed preferences of Carabidae (Coleoptera). *Basic. Appl. Ecol.* **2007**, *8*, 343–353. [CrossRef]
15. Charalabidis, A.; Dechaume-Moncharmont, F.-X.; Carbonne, B.; Bohan, D.A.; Petit, S. Diversity of foraging strategies and responses to predator interference in seed-eating carabid beetles. *Basic. Appl. Ecol.* **2019**, *36*, 13–24. [CrossRef]
16. Loughridge, A.H.; Luff, M.L. Aphid Predation by Harpalus rufipes (Degeer) (Coleoptera: Carabidae) in the Laboratory and Field. *J. Appl. Ecol.* **1983**, *20*, 451–462. [CrossRef]
17. Hajek, A.E.; Hannam, J.J.; Nielsen, C.; Bell, A.J.; Liebherr, J.K. Distribution and Abundance of Carabidae (Coleoptera) Associated with Soybean Aphid (Hemiptera: Aphididae) Populations in Central New York. *Ann. Entomol. Soc. Am.* **2007**, *100*, 876–886. [CrossRef]
18. Symondson, W.O.C.; Sunderland, K.D.; Greenstone, M.H. Can generalist predators be effective biocontrol agents? *Annu. Rev. Entomol.* **2002**, *47*, 561–594. [CrossRef] [PubMed]
19. Lang, A.; Filser, J.; Henschel, J.R. Predation by ground beetles and wolf spiders on herbivorous insects in a maize crop. *Agric. Ecosyst. Environ.* **1999**, *72*, 189–199. [CrossRef]
20. Birkhofer, K.; Bylund, H.; Dalin, P.; Ferlian, O.; Gagic, V.; Hambäck, P.A.; Klapwijk, M.; Mestre, L.; Roubinet, E.; Schroeder, M.; et al. Methods to identify the prey of invertebrate predators in terrestrial field studies. *Ecol. Evol.* **2017**, *7*, 1942–1953. [CrossRef]
21. Roubinet, E.; Birkhofer, K.; Malsher, G.; Staudacher, K.; Ekbom, B.; Traugott, M.; Jonsson, M. Diet of generalist predators reflects effects of cropping period and farming system on extra- and intraguild prey. *Ecol. Appl.* **2017**, *27*, 1167–1177. [CrossRef] [PubMed]
22. Staudacher, K.; Rubbmark, O.R.; Birkhofer, K.; Malsher, G.; Sint, D.; Jonsson, M.; Traugott, M. Habitat heterogeneity induces rapid changes in the feeding behaviour of generalist arthropod predators. *Funct. Ecol.* **2018**, *32*, 809–819. [CrossRef] [PubMed]

23. Iuliano, B.; Gratton, C. Temporal Resource (Dis)continuity for Conservation Biological Control: From Field to Landscape Scales. *Front. Sustain. Food Syst.* **2020**, *4*, 127. [CrossRef]
24. Gray, C.; Ma, A.; McLaughlin, O.; Petit, S.; Woodward, G.; Bohan, D.A. Ecological plasticity governs ecosystem services in multilayer networks. *Commun. Biol.* **2021**, *4*, 75. [CrossRef] [PubMed]
25. Roger, J.-L.; Jambon, O.; Bouger, G. *Clé de Détermination Des Carabides—Paysages Agricoles du Nord Ouest de la France*; SAD-Paysage, Institut National de Recherche Agronomique: Rennes, France, 2013.
26. R Core Team. *R: A Language and Environment for Statistical Computing*; R Foundation for Statistical Computing: Vienna, Austria, 2019. Available online: https://www.R-project.org/ (accessed on 9 August 2024).
27. Bates, D.; Mächler, M.; Bolker, B.; Walker, S. Fitting Linear Mixed-Effects Models Using Lme4. *J. Stat. Softw.* **2015**, *67*, 1–48. [CrossRef]
28. Hartig, F.; Lohse, L. DHARMa: Residual Diagnostics for Hierarchical (Multi-Level/Mixed) Regression Models, R Package Version 0.4.6. 2022. Available online: https://CRAN.R-project.org/package=DHARMa (accessed on 9 August 2024).
29. Fox, J.; Weisberg, S.; Price, B.; Adler, D.; Bates, D.; Baud-Bovy, G.; Bolker, B.; Ellison, S.; Firth, D.; Friendly, M.; et al. Car: Companion to Applied Regression 2019, R Package Version 3.1.2. Available online: https://CRAN.R-project.org/package=car (accessed on 9 August 2024).
30. Lenth, R.V.; Buerkner, P.; Herve, M.; Love, J.; Riebl, H.; Singmann, H. Emmeans: Estimated Marginal Means, Aka Least-Squares Means 2020, R Package Version 1.10.0. Available online: https://CRAN.R-project.org/package=emmeans (accessed on 9 August 2024).
31. Dormann, C.; Gruber, B.; Fründ, J. Introducing the bipartite Package: Analysing Ecological Networks. *R News* **2008**, *8/2*, 8–11.
32. Husson, F.; Josse, J.; Le, S.; Mazet, J. FactoMineR: Multivariate Exploratory Data Analysis and Data Mining 2018, R Package Version 2.10. Available online: https://CRAN.R-project.org/package=FactoMineR (accessed on 9 August 2024).
33. Barnier, J. Explor: Interactive Interfaces for Results Exploration 2019, R Package Version 0.3.10. Available online: https://CRAN.R-project.org/package=explor (accessed on 9 August 2024).
34. Scaccini, D.; Panini, M.; Chiesa, O.; Nicoli Aldini, R.; Tabaglio, V.; Mazzoni, E. Slug Monitoring and Impacts on the Ground Beetle Community in the Frame of Sustainable Pest Control in Conventional and Conservation Agroecosystems. *Insects* **2020**, *11*, 380. [CrossRef] [PubMed]
35. Janssen, A.; Montserrat, M.; HilleRisLambers, R.; Roos AM de Pallini, A.; Sabelis, M.W. Intraguild Predation Usually does not Disrupt Biological Control. In *Trophic and Guild in Biological Interactions Control*; Brodeur, J., Boivin, G., Eds.; Springer Netherlands: Dordrecht, The Netherlands, 2006; pp. 21–44. [CrossRef]
36. Rosenheim, J.A.; Harmon, J.P. The Influence of Intraguild Predation on the Suppression of a Shared Prey Population: An Empirical Reassessment. In *Trophic and Guild in Biological Interactions Control*; Brodeur, J., Boivin, G., Eds.; Springer Netherlands: Dordrecht, The Netherlands, 2006; pp. 1–20. [CrossRef]
37. Sunderland, K.D. The Diet of some Predatory Arthropods in Cereal Crops. *J. Appl. Ecol.* **1975**, *12*, 507–515. [CrossRef]
38. Settle, W.H.; Ariawan, H.; Astuti, E.T.; Cahyana, W.; Hakim, A.L.; Hindayana, D.; Lestari, A.S. Managing Tropical Rice Pests Through Conservation of Generalist Natural Enemies and Alternative Prey. *Ecology* **1996**, *77*, 1975–1988. [CrossRef]
39. Kamenova, S.; Leroux, C.; Polin, S.E.; Plantegenest, M. Community-wide stable isotope analysis reveals two distinct trophic groups in a service-providing carabid community. *Bull. Entomol. Res.* **2018**, *108*, 130–139. [CrossRef]
40. McKemey, A.R.; Symondson, W.O.C.; Glen, D.M.; Brain, P. Effects of Slug Size on Predation by *Pterostichus melanarius* (Coleoptera: Carabidae). *Biocontrol Sci. Technol.* **2001**, *11*, 81–91. [CrossRef]
41. Oberholzer, F.; Frank, T. Predation by the Carabid Beetles *Pterostichus melanarius* and *Poecilus cupreus* on Slugs and Slug Eggs. *Biocontrol Sci. Technol.* **2003**, *13*, 99–110. [CrossRef]
42. McKemey, A.R.; Symondson, W.O.C.; Glen, D.M. Predation and prey size choice by the carabid beetle *Pterostichus melanarius* (Coleoptera: Carabidae): The dangers of extrapolating from laboratory to field. *Bull. Entomol. Res.* **2003**, *93*, 227–234. [CrossRef] [PubMed]
43. Honek, A.; Martinkova, Z.; Jarosik, V. Ground beetles (Carabidae) as seed predators. *EJE* **2003**, *100*, 531–544. [CrossRef]
44. Saska, P.; Honek, A. Development of the beetle parasitoids, *Brachinus explodens* and *B. crepitans* (Coleoptera: Carabidae). *J. Zool.* **2004**, *262*, 29–36. [CrossRef]
45. Berg K von Traugott, M.; Symondson, W.O.C.; Scheu, S. The effects of temperature on detection of prey DNA in two species of carabid beetle. *Bull. Entomol. Res.* **2008**, *98*, 263–269. [CrossRef]
46. Sheppard, S.K.; Bell, J.; Sunderland, K.D.; Fenlon, J.; Skervin, D.; Symondson, W.O.C. Detection of secondary predation by PCR analyses of the gut contents of invertebrate generalist predators. *Mol. Ecol.* **2005**, *14*, 4461–4468. [CrossRef]
47. Harper, G.L.; King, R.A.; Dodd, C.S.; Harwood, J.D.; Glen, D.M.; Bruford, M.W.; Symondson, W.O.C. Rapid screening of invertebrate predators for multiple prey DNA targets. *Mol. Ecol.* **2005**, *14*, 819–827. [CrossRef] [PubMed]
48. Guenay, Y.; Trager, H.; Glarcher, I.; Traugott, M.; Wallinger, C. Limited detection of secondarily consumed plant food by DNA-based diet analysis of omnivorous carabid beetles. *Environ. DNA* **2021**, *3*, 426–434. [CrossRef]
49. Harwood, J.D.; Phillips, S.W.; Sunderland, K.D.; Symondson, W.O.C. Secondary predation: Quantification of food chain errors in an aphid–spider–carabid system using monoclonal antibodies. *Mol. Ecol.* **2001**, *10*, 2049–2057. [CrossRef]
50. Ferrante, M.; Barone, G.; Lövei, G.L. The carabid Pterostichus melanarius uses chemical cues for opportunistic predation and saprophagy but not for finding healthy prey. *BioControl* **2017**, *62*, 741–747. [CrossRef]

51. von Berg, K.; Traugott, M.; Scheu, S. Scavenging and active predation in generalist predators: A mesocosm study employing DNA-based gut content analysis. *Pedobiologia* **2012**, *55*, 1–5. [CrossRef]
52. Mair, J.; Port, G.R. Predation by the carabid beetles Pterostichus madidus and Nebria brevicollis is affected by size and condition of the prey slug Deroceras reticulatum. *Agric. For. Entomol.* **2001**, *3*, 99–106. [CrossRef]
53. Baulechner, D.; Jauker, F.; Neubauer, T.A.; Wolters, V. Convergent evolution of specialized generalists: Implications for phylogenetic and functional diversity of carabid feeding groups. *Ecol. Evol.* **2020**, *10*, 11100–11110. [CrossRef] [PubMed]
54. Baulechner, D.; Jauker, F.; Wolters, V. Carabid adaptation to a collembolan diet: Hunting efficiency and nutritional value. *Ecol. Entomol.* **2022**, *47*, 242–248. [CrossRef]
55. Hintzpeter, U.; Bauer, T. The antennal setal trap of the Ground beetle Loricera pilicornis: A specialization for feeding on Collembola. *J. Zool.* **1986**, *208*, 615–630. [CrossRef]

insects

MDPI

Article

A New Isolate *Beauveria bassiana* GxABT-1: Efficacy against *Myzus persicae* and Promising Impact on the Beet Mild Yellow Virus-Aphid Association

Kenza Dessauvages [1], Mathilde Scheifler [2], Frédéric Francis [1,*] and Ibtissem Ben Fekih [1,*]

[1] Functional and Evolutionary Entomology, Terra, Gembloux Agro-Bio Tech, University of Liege, Passage des Déportés 2, 5030 Gembloux, Belgium; kenza.dessauvages@uliege.be

[2] Evolution and Ecophysiology Group, Functional and Evolutionary Entomology, Terra, Gembloux Agro-Bio Tech, University of Liege, 5030 Gembloux, Belgium; mathilde.scheifler@uliege.be

* Correspondence: frederic.francis@uliege.be (F.F.); fekih.ibtissem@gmail.com (I.B.F.)

Simple Summary: Investigating new microbial control agents to overcome the use of chemical insecticides is of utmost importance to control *Myzus persicae*, the main vector of Beet Mild Yellow Virus (BMYV). Our study explored the efficiency of two *Beauveria bassiana* isolates (GHA and GxABT-1) against *M. persicae* and on the transmission of BMYV. After 8 days (post fungal spray), a mortality rate of more than 90% was registered among treated *M. persicae*. The impact of these fungi as endophytes on the BMYV-*M. persicae* association was assessed by (1) treating sugar beet seeds with fungal suspensions, (2) proving the ability of the fungi to colonize the plant, (3) studying the impact on the aphid's life cycle and its attraction towards plants, and (4) evaluating the virus load in treated plants. The fungi were able to colonize all parts of the plant, which led to alterations in the aphids' life cycle and their attractiveness to the plant. Although the fungi were not able to prevent virus transmission, the viral load appears to be reduced. We suggest performing the experiments on a larger scale and using different methods to inoculate the plants to explore whether the efficiency of the tested fungi could be enhanced.

Abstract: Within the context of ecofriendly alternatives to neonicotinoids, we explored the direct and endophytic potential of two *Beauveria bassiana* isolates, GHA from BotaniGard and the new endemic isolate GxABT-1, against the *Sugar Beet Mild Yellow Virus* (*BMYV*)-*Myzus persicae* pathosystem. A mortality rate of 96 and 91% was registered after 8 days of treatment with GHA and Gx-ABT-1, respectively. To assess the endophytic impact, sugar beet seeds were treated, and the ability of the fungi to colonize the plant was assessed and correlated with the aphids' (1) life cycle, (2) attraction towards the plants, and (3) ability to transmit BMYV. Both fungi colonized the plants, and the GxABT-1 isolate impaired the aphids' life cycle. *Myzus persicae* were more attracted to leaf discs from non-treated plants than to the fungal-treated ones. Interestingly, when the choice test dealt only with the fungal treatments, aphids were more attracted to leaves from plants harboring Gx-ABT-1 than those with GHA. Moreover, no significant impact was observed for BMYV transmission despite the slight decrease in the viral load in GxABT-1 isolate-treated plants. Our findings constitute a baseline to delve more into the performance of the new endemic isolate *B. bassiana* in other pathosystems using different treatment methods.

Keywords: aphid-borne virus; beet yellowing; endophyte; hypocrealean fungi; microbial control; multitrophic interactions

1. Introduction

The green peach aphid, *Myzus persicae* (Sulzer, 1776), is a highly polyphagous aphid species responsible for the transmission of numerous viruses in various economically important cash crops [1]. *Myzus persicae* is also one of the major pests of sugar beet

(*Beta vulgaris* ssp. *vulgaris* L.), being the main vector of sugar beet-associated viruses [2–5]. The 'beet polerovirus' beet mild yellow virus (BYMV) is one of the economically important yellowing viruses causing yield losses of up to 30% [2,6,7]. For many years, neonicotinoid insecticides (i.e., clothianidin, imidacloprid, and thiamethoxam) have been the major chemicals used in seed treatments or foliar applications against aphids and associated beet yellow viruses [4,8]. However, due to the adverse effects on pollinators and the environment, the European Commission took a series of measures since 2018 to restrict the use of some of these chemicals [9,10] and then to completely ban their use in 19 January 2023 (Case C-162/21) [11]. Following these decisions, the beet yellowing disease is likely to become a major threat to the sugar beet industry. Therefore, there is an urgent need to explore innovative and more eco-friendly, sustainable alternative methods of pest control [4,12].

Entomopathogenic fungi (EPF) are among the most promising alternatives to neonics, offering multiple ecosystem services. For example, they have a direct entomotoxic effect on insect pests, indirectly promote plant growth, and induce systemic resistance against pathogens and herbivores [13]. Globally distributed, EPF show great potential as effective biocontrol agents [13,14]. However, some of the EPF, including *Beauveria bassiana* (Hypocreales: Cordycipitaceae) and *Metarhizium brunneum* (Hypocreales: Clavicipitaceae), exhibit an endophytic lifestyle by colonizing and living asymptomatically within plant tissues without causing disease in their host [15,16]. Different studies have reported the beneficial impact of their endophytic presence at various levels, including enhancing plant growth, altering the attractiveness of plants to both bio-aggressors and natural enemies, and influencing the development and reproduction of pests, among other factors [17,18]. Finally, both *M. brunneum* and *B. bassiana* exhibit a direct or endophytic effect against insect herbivores [13].

Endophytic EPF may also be effective in the management of plant diseases [19]. A few studies have examined their impact on virus transmission by insect vectors, revealing the potential to reduce this transmission, which could be linked to the insect feeding behavior or to systemic resistance to the virus induced by the endophytic EPF [20–22]. Research into understanding the mechanisms behind these phenomena is expanding, and many aspects remain poorly understood. To date, no studies have been conducted on the colonization of sugar beet by endophytic EPF and their potential impacts on aphids and the viruses they transmit. In this study, we explored the direct and endophytic effects of an endemic new isolate of *B. bassiana* in the sugar beet-BMYV infected *M. persicae* pathosystem. This exploratory study serves as a foundation for directing subsequent research efforts towards finding more effective eco-friendly microbial agents to decrease beet yellow disease.

2. Materials and Methods

2.1. Plants, Insects and Virus

Sugar beet (*Beta vulgaris* ssp. *vulgaris* L.) seeds coated with hymexazol (active ingredient in Tachigaren 70 WS) were provided by the Royal Belgian Institute for Beet Improvement (IRBAB). The treated seeds were sown in seed trays before being transplanted individually into 10 cm diameter pots upon reaching the 2-leaf stage, using a universal potting soil (reference: TERS50, La Plaine Chassart, Fleurus, Belgium). Sugar beet plants were maintained in a growth chamber under controlled conditions at 23 ± 1 °C, 60% RH, and LD 16:8 h.

A BMYV-infected *M. persicae* clone was provided by SESVanderHave (Tienen, Belgium). Aphid rearing was maintained on sugar beet plants in rearing cages (BugDorm 4M4545, MegaView Science Co. Ltd., Taichung, Taiwan). Four- to six-leaf-stage sugar beet plants were added twice per month to maintain the rearing of viruliferous *M. persicae* and for the infection of new plants with BMYV.

2.2. *Beauveria bassiana* Isolates

2.2.1. Origin

Two isolates of *B. bassiana* were used in our study: (i) *B. bassiana* GHA (the active ingredient of the commercial mycoinsecticide BotaniGard ® 22WP (Certis Europe, Brussels, Belgium)) was generated from culture stock kept in a 10% glycerol solution and stored at −80 °C in cryovial tubes at the Laboratory of Functional and Evolutionary Entomology (Gembloux Agro-Bio Tech, Gembloux, Belgium). The isolate was grown on Sabouraud Dextrose Agar (SDA; Darmstadt, Germany) media and incubated at 23 °C for 14 days until fully sporulated; and ii) *B. bassiana* GxABT-1 from soil collected from the hedgerow of a sugar beet field site (50°33′48.9′′ N, 4°40′15.2′′ E) in Gembloux (Belgium). Three soil cores (around 10 cm of depth) from the plants' (weeds and/or flowers) rhizosphere were taken randomly using a hand shovel. After each sampling, the tools were rinsed thoroughly with water, followed by 70% ethanol, and then rinsed again with water. The collected samples were stored in small, sealed plastic bags and carried out to the laboratory for fungal isolation following the bait method by Zimmermann [23]. The different samples were sieved (2 mm mesh), and around 20 mL were added to plastic cups (30 mL), with up to 1 mL of tap water added to balance the water content of the samples. Subsequently, three larvae of *Tenebrio molitor* L. (Coleoptera: Tenebrionidae) were added to each cup, which was sealed with perforated lids. The different cups were then incubated in the dark in closed boxes at 21 °C, and a moist paper towel was added to each box to maintain high humidity. The cups were inverted every 2 days over a period of four–five weeks and assessed daily for the presence of dead larvae. Once dead, the collected cadavers were surface-sterilized with 70% ethanol for 10 s, washed three times with ddH_2O sterile water, and placed on filter paper to absorb the remaining water droplets. The different specimens were incubated in the dark on SDA media supplemented with streptomycin (0.5 mL of 0.6 g mL^{-1}), tetracycline (0.5 mL of 0.05 g mL^{-1}), and cyclohexamide (1 mL of 0.05 g mL^{-1}) at 23 °C and checked for fungal outgrowth for up to 5–8 days. From a total of 27 incubated specimens, only one cadaver showed a fungal outgrowth.

2.2.2. Morphological and Molecular Identification

Fungal features, such as colony color and conidia shape, were used to identify the fungal isolate at the genus level. The conidia shape was observed under a light microscope at 400x magnification. Morphological identification was conducted using the key provided by Humber [24].

To pursue fungal identification at the species level, mycelia and conidia cultured on SDA media for 14 days were harvested by scraping off a portion of the fungal culture using a sterile cell spreader. The harvested fungal structures (mycelia and conidia) were then transferred to Eppendorf tubes with sterile steel beads and ground into a fine powder using Retsch-MM 400 (Verder scientific, Haan, Germany). A second grinding was performed after freezing the samples in liquid nitrogen for 30 s. Genomic DNA extraction was performed using a DNeasy Plant Mini kit (Qiagen, Hilden, Germany) following the manufacturer's instructions. PCR amplification of the ITS1–5.8S–ITS2 region of the rRNA gene cluster [25] was performed using the primer pairs ITS5 (5′-GGAAGTAAAAGTCGTAACAAGG-3′)/ITS4 (5′-TCCTCCGCTTATTGATATGC-3′) [26]. PCR amplification was performed in reaction volumes of 50 µL containing 25 µL of Q5 High-Fidelity 2X Master Mix (New England Biolabs, Hitchin, UK), 2.5 µL of each primer (10 µM), and 5 µL genomic DNA (10 ng/µL). Cycling conditions were denaturation at 98 °C for 3 min, followed by 35 cycles of denaturation at 98 °C for 1 min, annealing at 55 °C for 1 min, and extension at 72 °C for 1 min 30 s, with a final extension at 72 °C for 10 min. PCR amplifications were run on 1% agarose gels (1× TAE buffer) at 110 V for 30 min by electrophoresis, and the products visualized with SYBR Safe (Invitrogen, Carlsbad, CA, USA) in a Biorad Universal Hood II Gel Doc Imaging System. Prior to sequencing, the PCR products were purified with NucleoSpin Gel and a PCR Clean-up Kit (Macherey-Nagel, Düren, Germany) according to the manufacturer's instructions. Sanger sequencing was performed by Eurofins Genomics (Ebersberg, Germany), and a Basic Local Alignment Search Tool (BLAST) on the National Center for Biotechnology Information (NCBI) database was performed. Afterwards, the gene sequence was aligned with reference sequences obtained from Genbank (Table S1) using MEGA X [27]. The phylogenetic tree of ITS sequences was constructed with the IQ-TREE program [28]. *Beauveria amorpha* was used as the outgroup. Kimura's two-parameter model (estimated via Akaike Information Criterion using jModelTest v.2.0 [29]) was applied, and maximum likelihood analysis was run with 10,000 ultrafast bootstrap replicates.

2.3. Fungal Inoculum Preparation

Fungal inoculum was prepared by harvesting conidia using a sterilized spatula in sterilized 0.03% Tween 80. The fungal solution was then filtered through sterile cheesecloth to remove hyphal debris. Conidia counts was determined using a Burker hemocytometer (Marienfeld, Germany), and the concentration was adjusted to 1×10^8 conidia mL^{-1}. Conidial viability was checked by plating 100 µL of 1×10^5 conidia mL^{-1} dilution on SDA plates for 24 h at 23 °C. Germinated and non-germinated conidia were counted, and only those showing >90% viability were used for the bioassays either against aphids or as seed treatments.

2.4. Direct Effect of Fungal Treatment on *Myzus persicae*

A group of ten young adult (fourth instar) aphids was picked up from infested sugar beet plants, placed in a Petri dish (90 mm diameter), and sprayed with 1 mL of a 1×10^8 conidia mL^{-1} solution from a distance of 10 cm. Control aphids were sprayed with 1 mL of a 0.03% Tween 80 solution following the same procedure as the fungal-treated aphids. Each group of treated aphids was then transferred jointly to a sugar beet leaf in a plastic cup containing 1.5% water agar, as described in Ben Fekih et al. [30]. The sugar beet leaf secured in the water agar served as a food source for the aphids during the bioassay. A total of five replicates for each fungal isolate were used, and the different cups were maintained at 23 °C for LD 16:8 h. Dead aphids were recorded daily over a period of 8 days. The cadavers collected daily were immediately surface-sterilized by dipping in 70% ethanol for 10 s, followed by three rinses in sterile distilled water, after which the cadavers were incubated at 23 °C in a Petri dish containing a filter paper moistened with 600 µL of sterile distilled water. The incubated cadavers were checked daily for up to five days for fungal outgrowth.

2.5. Seed Sterilization and Treatments

Non-coated sugar beet seed sterilization was conducted following the methods of Rasool et al. [31]. The seeds were immersed in 70% ethanol for 3 min, then in 2% sodium hypochlorite (NaClO) for 10 min, followed by seven rinses with sterile distilled water. Surface-sterilized seeds were air-dried on sterile absorbent filter paper under a laminar flow hood for 40 min. The efficacy of the seed sterilization was assessed by spreading 100 µL of the final rinse water on SDA medium and incubating it for 10 days in darkness at 23 °C. Once dry, the seeds were immersed in a 15 mL solution of 1×10^8 conidia mL^{-1} suspensions of each of the fungal isolates. For the control, surface-sterilized seeds were immersed in a 0.03% Tween 80 solution. The seeds were immersed for each of the treatments and agitated at 100 rpm at 25 ± 2 °C for 24 h. Following the incubation period, the seeds were sown in seed trays and transplanted into individual pots of 10 cm diameter after 2 weeks. Six-leaf stage plants (40 days post inoculation (DPI)) were then used to proceed with the experiments.

2.6. Assessment of Endophytic Colonization

Samples of roots, hypocotyls, leaves, and petioles were collected from ten randomly chosen treated plants to assess the endophytic colonization of both fungi. From each plant part, six pieces of respective tissues with different sizes were cut with sterile scissors (roots,

1.5 cm; hypocotyls, 2 cm; petioles, 2 cm; leaves, 3 × 3 cm). The various collected plant pieces were subsequently surface-sterilized by immersion for 2 min in 70% and 2% NaClO, followed by three rinses in ddH_2O sterile water, following Parsa et al. [32]. Sterilization efficacy was assessed as described above. The edges of the surface-sterilized plant tissues were trimmed using sterile scalpels and cut into small pieces (roots, 1 cm; hypocotyls, 0.5 cm; petioles, 0.5 cm; leaves, 0.5 cm). Then, the different plant pieces were partially inserted into SDA media supplemented with the antibiotics mentioned above (Figure 1). The different Petri dishes were incubated at 25 °C in darkness for a period of 14 days. The monitoring of outgrowth mycelia from plant tissues was conducted daily.

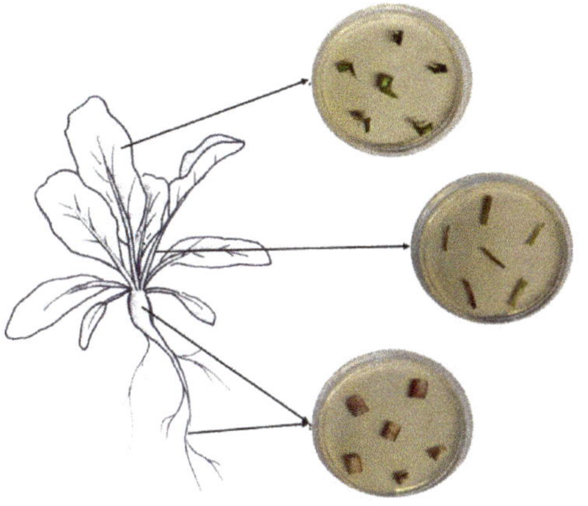

Figure 1. Assessment of fungal endophytic colonization in sugar beet plants.

2.7. Aphid Life Cycle Monitoring on Whole Treated Sugar Beet Plants

Adult aphids were collected from infested sugar beet plants and placed on a sugar beet leaf in a plastic cup containing 1.5% water agar, as described above, to allow for nymph production. The different cups were incubated under controlled conditions at 23 °C for LD 16:8 h. Afterwards, a total of 16 nymphs (24 h-old first instar) produced by these adults were picked up and transferred individually onto the first leaf of treated and non-treated sugar beet plants (of the same age) to assess the effects of the different treatments on their life cycle. The nymphs were confined in clip-cages made from polyurethane foam tubes (internal diameter of 3.2 cm). The clip-cage preparation procedure was inspired by the protocol proposed in Haas et al. [33]. A 2 cm section was cut from the polyurethane foam tube and then incised in the middle to form two discs. A disk of insect-proof film (Nortene®) was glued onto one surface of each disc with non-toxic glue. The insect-proof film was made to prevent aphids from escaping while allowing ventilation. Three alligator hair clips were used to bind the two discs of the clip-cage to the leaf. Nymph survival was checked daily for the first 3 days and then every 2 days to monitor their development until adult form and subsequent nymph production. Once the leaves hosting the aphids showed signs of yellowing, the aphids were moved individually to a younger leaf of the same host plant. We ensured a continuous fresh food source for *M. persicae*. An aphid is considered an adult when it produces its first nymph. For this experiment, different parameters were recorded, such as the nymphal development time (time covering all instars period), nymphal mortality (all instars included) occurred during the nymphal development stage, adult fecundity (first instar nymph production), adult longevity (duration of adult stage), and aphid longevity (overall aphid lifespan).

2.8. Choice Test

Leaf discs from six-leaf sugar beet plants (treated and non-treated) were used for the choice assays following Fingu-Mabola et al. [34]. One leaf disc was cut from leaves of the same age per plant and placed on filter paper moistened with ddH_2O in Petri dishes (90 mm in diameter) for 10 min. The filter papers were used to keep the leaf pieces fresh and to reduce volatile emissions due to the cutting [34]. Afterwards, two discs from each treatment (fungal-treated and control) were placed 6 cm apart in a Petri dish filled with moistened filter papers. A total of ten young adult aphids, kept for 2 h without food, were carefully placed in the center of the arena (Petri dish) to assess feeding preference. The different arenas were kept in a ventilated incubator at 23 °C under LD 16:8 h. The total number of aphids on each of the two discs was estimated after 2, 4, and 6 h. Each pairwise treatment was repeated 10 times, with new leaf discs and new aphids (10 arenas).

2.9. Virus Transmission

Four viruliferous aphids carrying BMYV were used to infest six-leaf stage plants (treated and non-treated). The different infested plants were, afterwards, covered individually with microperforated plastic to prevent aphid escape. Four days later, the infested plants were sprayed with an acetamiprid-based systemic insecticide (KB® Multisect) to eliminate the aphids. Aphid-free plants were kept in the growth chamber at 23 °C under LD 16:8 h for a period of 4 weeks before proceeding with the virus transmission check experiment.

After the virus incubation period, leaf pieces from the 36 plants for each treatment were collected. Four leaf discs per plant were collected using a 5 mL Eppendorf tube from leaves of different ages to ensure a similar amount of tissue was gathered from all the plants. These samples were then analyzed with a double antibody sandwich ELISA following the manufacturer's instructions (Loewe Biochemica GmbH, Sauerlach, Germany). Plates were first coated with 1:250 diluted IgG in coating buffer (pH 9.6) and incubated for 4 h at 37 °C. Leaf samples were ground in grinding bags, diluted to 1:20 in sample buffer (pH 7.4), and duplicates of each sample were placed into 96-well ELISA microplates. The plates were then washed three times with wash buffer (pH 7.2–7.4) before adding 100 μL of the samples. Each sample was duplicated. The plates were incubated overnight at 4 °C. Then, the plates were washed and incubated with AP-conjugate diluted to 1:250 in conjugate buffer (pH 7.4) at 37 °C for a period of 4 h. The plates were washed and incubated with 4-nitrophenylphosphate-di-Na-Salt in substrate buffer (pH 9.8) and left at room temperature for 1–2 h in the dark. Color development was recorded at 415 nm in an ELISA reader (Bio-Rad, iMark Microplate Reader). The absorbances of the two replicates per sample were averaged. Plants were considered virus-infected if the absorbance value was at least twice that of the OD405.

2.10. Statistical Analysis

Statistical analyses of the infection bioassays were conducted as follows: The 'surv' function from the 'survival' [35] package was used to combine survival times and observed events (event: death). Within the same package, the 'survfit' function was employed to fit the Kaplan–Meier survival model to these data, estimating the survival function of aphids over time. The 'ggsurvplot' function from the 'survminer' [36] package was used to graphically visualize the results with confidence intervals.

Data related to plant colonization were analyzed using a binomial generalized linear mixed model (GLMM), where fungal isolates (GHA and GxABT-1) and plant parts (root, petiole, and leaf) were treated as fixed effects. Random effects included plant number and pieces of each plant part.

Analyses of the aphid life cycle were conducted using a GLMM with various distributions tailored to the nature of each variable. Gamma distribution was employed to model the duration of life stages and aphid longevity, while nymphal mortality data were analyzed using binomial distribution (binary data). Additionally, a Poisson distribution was applied to model the number of first instar nymphs produced per aphid over its lifetime. Considering experiments were conducted at two different times, a "time" variable was introduced to the model as a random factor.

Concerning choice test, a GLMM with a Poisson distribution was employed using the "arena" variable as a random factor for repeated measures over time.

For the BMYV transmission experiments, two complementary approaches were used. Binary data (i.e., infected/non-infected plants) were analyzed using a GLMM, while absorbance data were analyzed using a linear mixed-effects model (LMM). This analysis was conducted to investigate potential differences in viral load. Following the modeling, analyses of variance were conducted to further explore the effects of treatments, and Tukey's pairwise comparisons were performed when significant differences were observed. All analyses and visualizations of the data were performed using R (version 4.1.3; RStudio Team, 2022).

3. Results

3.1. Fungal Identification

From the 27 total *T. molitor* used to bait nine soil samples, only one cadaver showed a typical mycosis caused by Hypocrealean fungi. Conidial dimensions of the endemic fungal isolate were $2.5 \pm 0.1 \times 2.1 \pm 0.2$ µm (Figure 2), which is in the range of fungi within the genus *Beauveria* sp. (Hypocreales, Cordycipitaceae). DNA sequencing of the fungal isolate and the high similarity search using NCBI BLAST confirmed it as *B. bassiana*. Phylogenetic analyses constructed with a total of seven ITS sequences of *Beauveria* sp. downloaded from GenBank strongly supported that our fungal isolate was included in the *B. bassiana* clade (Figure 3).

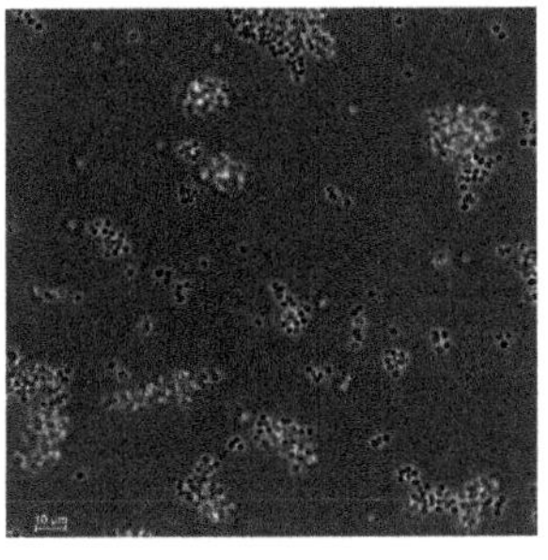

Figure 2. Conidia of the soil-borne fungi *Beauveria bassiana* isolated from the plant rhizosphere in the hedgerow of a sugar beet field. Photos were taken using a Nikon Eclipse Ti2-E inverted automated microscope with a Nikon camera DSQi2.

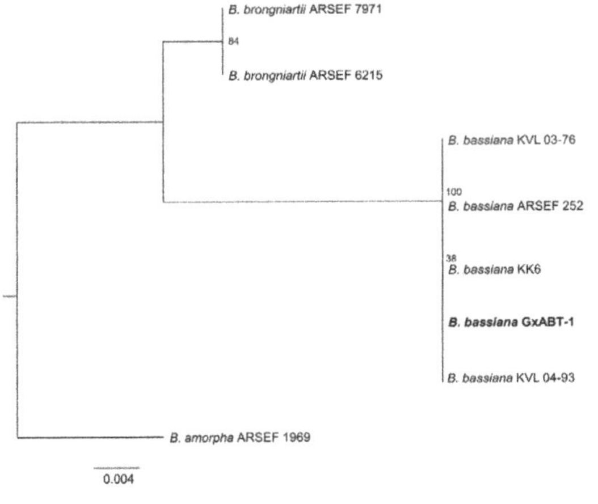

Figure 3. Phylogenetic tree of the ITS1–5.8S–ITS2 region of *Beauveria bassiana* isolates and reference sequences from Genbank (Table S1) analyzed by Maximum Likelihood. Numbers at branches indicate support values, based on 10,000 ultrafast bootstrap replicates. The endemic new isolate of *B. bassiana* isolate is in bold.

3.2. Effect of *Beauveria bassiana* on Aphid Survival

Treatments with both *B. bassiana* isolates significantly reduced the survival time of *M. persicae* individuals (Log-Rank test: $\chi^2 = 25$, df = 1, $p < 0.001$) compared to those in the control (Figure 4). No significant difference in survival was observed between aphids exposed to both fungal isolates. On average, aphids exposed to *B. bassiana* GHA and GxABT-1 survived 6 and 5 days, respectively. At 8 DPI, on average, 6 ± 4% of aphids treated with GHA isolate survived, 9 ± 5% with GxABT-1 isolate, compared to 56 ± 7% for the control.

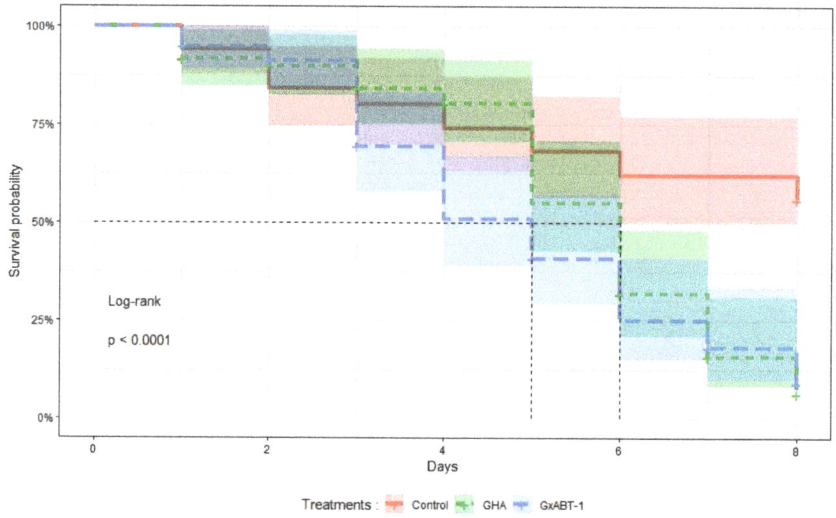

Figure 4. Kaplan–Meier survival curves showing the proportion of surviving treated aphids over time after exposure to two isolates of *Beauveria bassiana*, GHA and GxABT-1. The mean survival time is indicated by the vertical dashed lines for each isolate.

3.3. Endophytic Colonization of Sugar Beet Plants

Inoculation of seeds with *B. bassiana* GHA and GxABT-1 led to the colonization of all tested tissues (i.e., roots, petioles, and leaves) (Figure 5), although we observed varying levels of colonization in the different parts of the sugar beet plants. Since no EPF growth was observed in the control group, it was excluded from the statistical analyses. A significant difference in colonization levels was found between the fungal isolates (GLMM: $\chi^2 = 7.51$, df = 1, $p = 0.006$), with the GHA isolate being more prevalent in the tissues in general (Table 1). The interaction between plant tissues and fungal isolates was not significant (GLMM: $\chi^2 = 5.13$, df = 2, $p = 0.07$), suggesting the same colonization pattern for both fungi. Both fungal isolates colonized the roots, petioles, and leaves in the same way, which means that EPF were more commonly found in roots, followed by petioles, and least often in leaves for both isolates.

Table 1. Percentage (95% confidence intervals, n = 60) of colonized sugar beet tissue pieces (root, petiole, and leaf) at 45 days post seed inoculation with EPF isolates *B. bassiana* GHA and GxABT-1. The significance letters (a, b, c, d, e) indicate statistically different groups determined by a post-hoc test using the estimated marginal means method (emmeans).

Treatments	Root	Petiole	Leaf
B. bassiana GHA	60.0% (46.5–72.2) e	30.0% (19.2–43.4) cd	18.3% (9.9–30.9) ab
B. bassiana GxABT-1	38.3% (26.4–51.8) de	28.3% (17.8–41.6) bc	13.3% (6.3–25.1) a

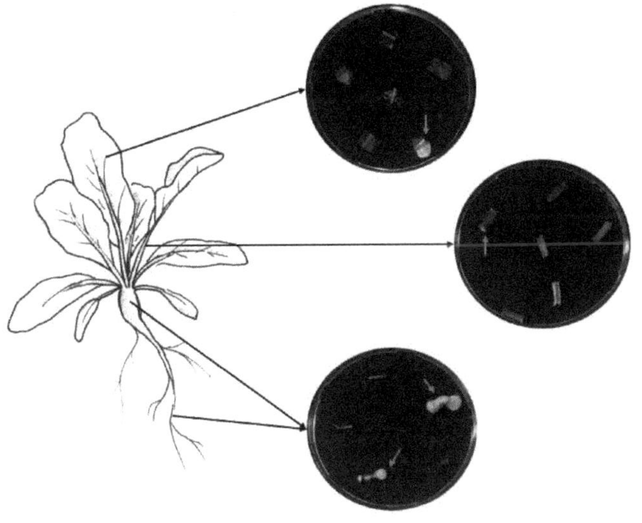

Figure 5. Fungal growth on sugar beet tissues in Petri dishes as a confirmation of endophyte presence.

3.4. Indirect Effect on Aphid's Development and Reproduction

Significant effects of fungal treatment on both nymphal development time (GLMM: $\chi^2 = 9.22$, df = 2, $p = 0.009$) and aphid fecundity (GLMM: $\chi^2 = 7.41$, df = 2, $p = 0.02$) were observed (Table 2). Subsequent post-hoc analysis using Tukey's test showed that only *B. bassiana* GxABT-1 had a significant impact on both previously mentioned parameters (Table S2). Only *B. bassiana* GxABT-1-treated plants significantly increased the duration of nymphal development stage ($p = 0.003$) by 5.8 ± 3.9 days (from 12.2 ± 1.2 to 18 ± 2.7 days on average) (Figure 6a, Table 2). In addition, the same fungal isolate negatively impacted aphid fecundity by significantly decreasing nymph production ($p = 0.02$): on average, the aphids produced 5.0 ± 4.6 fewer offspring compared to the control treatment (from 17.6 ± 2.5 for the control to 12.6 ± 2.1 for the GxABT-1 isolate) (Figure 6b).

Furthermore, an increased mortality rate of nymphs (all instars included) was observed for both GxABT-1 and GHA (53.3% and 28.6% against 14.3% for control), although the difference was only marginally significant (GLMM: $\chi^2 = 5.30$, df = 2, $p = 0.07$). The endophytic EPF did not impact adult stage or longevity (Table 2).

Table 2. Life cycle parameters (mean ± SE) of *Myzus persicae* fed on sugar beet plants inoculated with *Beauveria bassiana* GHA or GxABT-1 compared to non-inoculated plants. The significance letters (a, b) indicate statistically different groups determined by a post-hoc Tukey test, and asterisks indicate significant differences between the treatments: * $p < 0.05$, ** $p < 0.01$.

Parameters	Control	*B. bassiana* GHA	*B. bassiana* GxABT-1	*p*-Value
	Mean ± SE			
Nymphal mortality (all instars included) (%)	14.3 ± 9.7 a	28.6 ± 12.5 a	53.3 ± 13.3 a	0.071
Nymphal development time (days)	12.2 ± 1.2 b	15.2 ± 1.6 ab	18.0 ± 2.7 a	0.010 **
Adult stage (days)	12.5 ± 1.3 a	12.8 ± 1.6 a	10.0 ± 1.4 a	0.377
Aphid longevity (days)	23.6 ± 1.4 a	23.9 ± 2.3 a	21.6 ± 2.4 a	0.654
Adult fecundity (total)	17.6 ± 2.5 b	15.7 ± 3.3 ab	12.6 ± 2.1 a	0.025 *

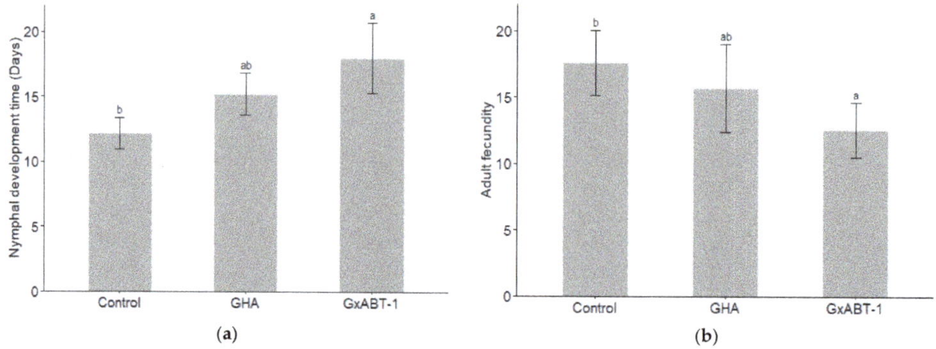

Figure 6. Impact of fungal endophytic colonization of sugar beet plants by *Beauveria bassiana* GHA or GxABT-1 isolates on aphid (**a**) nymphal development time and (**b**) adult fecundity (number of produced first instar nymphs). The significance letters (a, b) indicate statistically different treatments determined by a post-hoc Tukey test.

3.5. Host-Choice Tests

The preliminary test involving aphid preference between the two control plants showed no bias. The aphids were distributed evenly across the two leaf discs. Aphids showed preferences towards leaf discs from control over those treated with fungi (for both fungal isolates) (Figure 7). They were found more frequently on control discs compared to those treated with *B. bassiana* GHA (GLMM: $\chi^2 = 37.20$, df = 1, $p < 0.001$) and *B. bassiana* GxABT-1 (GLMM: $\chi^2 = 9.10$, df = 1, $p = 0.002$). However, when given the choice between leaf discs from both *B. bassiana*-treated plants, aphids showed a preference for the ones treated with *B. bassiana* GxABT-1 over the ones with *B. bassiana* GHA (GLMM: $\chi^2 = 6.66$, df = 1, $p = 0.010$). The time factor did not influence the choice of aphids among the different treatments (p-values > 0.05) (Figure 7).

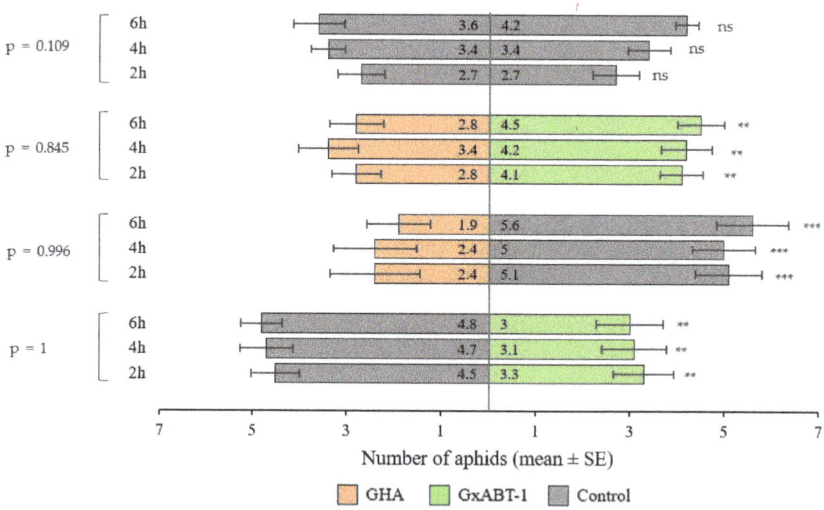

Figure 7. Plant selection by aphids between control and fungal inoculated plants (*Beauveria bassiana* GHA or GxABT-1) during dual-choice tests. Each test involved the release of 10 aphids and observation after 2, 4, and 6 h. The numbers included in the horizontal bars represent the mean number of aphids visting each of the treatements. To the left, the p-values represent tests comparing aphid distribution over time. ns, not significant. Asterisks indicate significance levels: ** $p < 0.01$, *** $p < 0.001$.

3.6. *Impact on Beet Mild Yellow Virus Transmission*

Neither of the two *B. bassiana* isolates showed an effect on the transmission of BMYV in terms of infected plants (Figure 8a) (GLMM: $\chi^2 = 0.17$, df = 2, $p = 0.918$). However, a slight decrease but not significant in the registered absorbance—which reflect the viral load in the tested samples—was observed for the treated plants over those of the controls (LMM: $\chi^2 = 5.61$, df = 2, $p = 0.060$). Delving deeper with a post-hoc test reveals that this difference lies between the control and the endemic isolate GxABT-1, where a slight decrease in viral load was observed (Tukey: $p = 0.0595$).

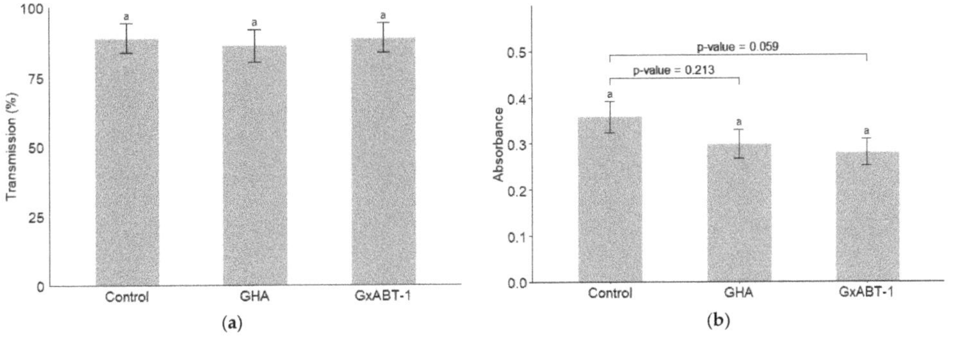

Figure 8. Impact of entomopathogenic fungi colonization on Beet Mild Yellow Virus transmission by *Myzus persicae*. (**a**) Percentage of BMYV-infected plants; (**b**) comparison of the absorbance for inoculated vs. control plants. The significance letter (a) indicates no statistical difference between treatments (post-hoc Tukey test).

4. Discussion

Hypocrealean EPF such as *B. bassiana* are extensively studied due to their great interest as microbial control agents against numerous insect pests [14]. The endophytic potential of these EPF has been reported in recent years [37]. For example, *B. bassiana* has shown a direct and endophytic impact against other aphid species, such as *Aphis gossypii*, in addition to *M. persicae* [13]. Furthermore, our study investigated the efficiency of the new endemic isolate *B. bassiana* GxABT-1 against *M. persicae*, both directly and endophytically, and its potential impact on BMYV transmission in sugar beet plants. Investigations exploring the potential of endemic isolates of hypocrealean EPF are of great interest, since by being part of the local biodiversity, they are more adapted than other introduced fungal isolates to control local pests and reduce risks to non-target organisms [38]. In addition, introduced exotic organisms as biocontrol agents might face challenges to adapt to local abiotic and biotic factors, while native fungal isolates can circumvent these difficulties [39].

The direct potential of the two *B. bassiana* fungal isolates, GxABT-1 and GHA, has been assessed in vitro using the spray method. Both fungal isolates were equally virulent against *M. persicae*, with mortality rates of 94 and 91% at 8 DPI and an LT50 of 5 and 6 days, respectively. These results are consistent with other studies evaluating the pathogenicity of *B. bassiana* against *M. persicae* [40,41].

Besides their direct efficacy, both fungal isolates showed the ability to colonize, after 40 DPI, all sugar beet plant tissues, with *B. bassiana* GHA being significantly more prevalent in the tissues than GxABT-1. However, interestingly, only plants harboring the new isolate GxABT-1 impacted the aphid life cycle by extending the duration of the nymphal development stage while reducing adult fecundity. Our results are in accordance with Jaber and Araj [42], who reported that the endophytic colonization of *Capsicum annuum* by *B. bassiana* ATCC 74040 caused an extended development time and reduced fecundity of *M. persicae*. However, another study [20] reported an increase in the mortality rate and a reduced fecundity of *M. persicae* feeding on *B. bassiana* GHA-treated tobacco plants. The reduced fecundity was also noted in cotton aphids feeding on *B. bassiana* GHA-inoculated cotton plants [18]. However, the study by Wilberts et al. [43] reported that the treatment of *C. annuum* by *B. bassiana* ARSEF 3097 (ATCC 74040) did not affect *M. persicae* life cycle parameters [43], which contradicts the results by Jaber and Araj [42]. Both studies applied the fungal treatment by root inoculation on the same plant species and using the same *B. bassiana* isolate, but in two different ways, which may explain the differences in efficacy. In the study by Wilberts et al. [43], the roots were rinsed and then soaked in a spore solution, whereas in the study by Jaber and Araj [42], the plants—still in their substrate—were watered with the inoculum. These different results reporting the indirect effect of *B. bassiana* against aphids suggest that the efficiency of the tested endophytic EPF is likely to be dependent on the fungal isolate, plant host, and inoculation method. Following the fungal colonization of plant tissues, the indirect entomotoxic effect might be explained by the production of secondary metabolites, which would impair insect pest fitness [44]. Similarly, the endophytic potential of EPF might involve an upregulation of the genes responsible for the production of phytohormones, which play a role in plant immune responses against herbivores [45].

Concerning the choice test experiments, leaf discs from fungal-treated plants were visited significantly less by *M. persicae* than those in the control. Interestingly, aphids facing choices between fungal-treated leaf discs showed that they preferred leaves from GxABT-1-treated plants than those treated with GHA. Our results contradict some other studies that have shown either no effect of the endophytic colonization on the aphids' choice or a higher attractiveness of colonized plants. For example, Fingu-Mabola et al. [34] reported that *M. persicae* colonies were more attracted to tobacco plant leaf discs inoculated with *B. bassiana* GHA than non-inoculated discs. The same study reported that this attractiveness might be related to the high production of aldehydes by *B. bassiana*-treated plants. It has previously been shown that aldehydes attract aphids [46]. However, in a "whole-plant attractiveness" study, the inoculation with *B. bassiana* ATCC 74040 of *C. annuum* did not impact the choice of *M. persicae* [43]. Similarly, in this context, the endophytic colonization of melon plants by *B. bassiana* (EABb 04/01-Tip and EABb 01/33-Su isolates) did not influence host plant selection [47]. Previous studies have shown that some insect pests other than aphids could detect the presence of EPF in plants and avoid them. For example, the black vine weevil, *Otiorhynchus sulcatus*, avoided grape vines colonized by *B. bassiana* ATCC 74040 when released into a two-choice olfactometer set-up [48]. Similarly, the western tarnished plant bug (*Lygus hesperus*) and the southern green stink bug (*Nezara viridula*) were deterred by cotton plants also colonized by *B. bassiana* GHA under both no-choice and choice assays [49]. These interesting results push towards investigating whether these choices are specific to fungal isolate, host plant, and herbivore associations or whether they are more likely related to the experimental set-up, i.e., whether the test was carried out on plant parts or on whole plants.

Regarding the virus transmission assay, irrespective of the fungal isolate, the endophytic colonization did not significantly impact the transmission rate of BMYV. Nevertheless, treatment with the GXABT-1 isolate reduced the viral load by 22% compared to the control. However, the difference in viral load between both treatments was only marginally significant (p = 0.059), indicating a potential negative effect of *B. bassiana* GxABT-1 on viral transmission. To elucidate the accurate potential of our endemic isolate, it would be interesting to investigate its performance by adopting different methods to treat sugar beet plants (seed treatments, spray, or watering) while ensuring we have heavily BMYV-infected plants that can be used as a source of inoculum for large-scale experiments. Monitoring the evolution of viral load on fungal-treated plants over time and correlating the dynamics of both microbes (virus and fungi) would also help assess the real effect of the endophytic fungi on limiting the virus' establishment and maintenance in the host. In addition, it would be important to explore the potential of these treatments on non-persistently transmitted potyvirus, one of the most widely distributed pathogens in economically important crops [50]. Investigating the potential of these fungal treatments as well as other microbial agents on different plant pathogens would be beneficial to promote their use in organic [51] and sustainable farming practice.

5. Conclusions

The performance of introduced microbial control agents in a new area is continuously jeopardized by the local (biotic and abiotic) factors of the host environment. Therefore, exploring the agricultural and ecological benefits of endemic microbial control agents is important for overcoming compatibility-related issues. Within this context, we performed our study in which we provided baseline information on the promising potential of the new endemic isolate *B. bassiana*, isolated from a sugar beet agricultural field, against the beet yellowing virus. Interesting results have been reported, but more thorough experiments are still needed at the biological, chemical, and molecular levels to elucidate the performance of this new fungal isolate.

Supplementary Materials: The following supporting information can be downloaded at: https://www.mdpi.com/article/10.3390/insects15090697/s1, Table S1: Reference sequences from Genbank used to construct *Beauveria* spp. phylogenetic tree; Table S2: Pairwise comparison *p*-values (Tukey) for the two life-cycle variables impacted by the endophytic presence of *B. bassiana* isolates GHA and GxABT-1 (nymphal development time and adult fecundity).

Author Contributions: Conceptualization, I.B.F. and F.F.; methodology, K.D. and I.B.F.; software, K.D. and M.S.; validation, I.B.F. and F.F.; formal analysis, K.D., M.S. and I.B.F.; investigation, K.D. and I.B.F.; resources, I.B.F. and F.F.; data curation, K.D., M.S. and I.B.F.; writing—original draft preparation, K.D.; writing—review and editing, M.S., I.B.F. and F.F.; supervision, I.B.F. and F.F.; project administration, I.B.F. and F.F.; funding acquisition, I.B.F. All authors have read and agreed to the published version of the manuscript.

Funding: I.B.F. was funded by the National Fund for Scientific Research F.R.S.-FNRS: PDR T.0088.21.

Data Availability Statement: All data and scripts used for the analysis will be provided upon request.

Acknowledgments: We are grateful to Jeannine Bortels and Arnaud Segers for the technical support provided for the field work, to Joachim Carpentier for his support regarding the molecular experiments, and to Grégoire Noël for the advice regarding the statistical analysis.

Conflicts of Interest: The authors declare no conflicts of interest.

References

1. Ali, J.; Bayram, A.; Mukarram, M.; Zhou, F.; Karim, M.F.; Hafez, M.M.A.; Mahamood, M.; Yusuf, A.A.; King, P.J.H.; Adil, M.F.; et al. Peach–Potato Aphid Myzus Persicae: Current Management Strategies, Challenges, and Proposed Solutions. *Sustainability* **2023**, *15*, 11150. [CrossRef]
2. Kozłowska-Makulska, A.; Beuve, M.; Syller, J.; Szyndel, M.S.; Lemaire, O.; Bouzoubaa, S.; Herrbach, E. Aphid Transmissibility of Different European Beet Polerovirus Isolates. *Eur. J. Plant Pathol.* **2009**, *125*, 337–341. [CrossRef]
3. Schliephake, E.; Graichen, K.; Rabenstein, F. Investigations on the Vector Transmission of the Beet Mild Yellowing Virus (BMYV) and the Turnip Yellows Virus (TuYV)/Untersuchungen Zur Vektorübertragung Des Milden Rübenvergilbungsvirus (Beet Mild Yellowing Virus) Und Des Wasserrübenvergilbungsvirus (Turnip Yellows Virus). *Z. Für Pflanzenkrankh. Pflanzenschutz J. Plant Dis. Prot.* **2000**, *107*, 81–87.
4. Francis, F.; Then, C.; Francis, A.; Gbangbo, Y.A.C.; Iannello, L.; Ben Fekih, I. Complementary Strategies for Biological Control of Aphids and Related Virus Transmission in Sugar Beet to Replace Neonicotinoids. *Agriculture* **2022**, *12*, 1663. [CrossRef]
5. Zhu, Y.; Stahl, A.; Rostás, M.; Will, T. Temporal and Species-Specific Resistance of Sugar Beet to Green Peach Aphid and Black Bean Aphid: Mechanisms and Implications for Breeding. *Pest Manag. Sci.* **2024**, *80*, 404–413. [CrossRef]
6. Hossain, R.; Menzel, W.; Lachmann, C.; Varrelmann, M. New Insights into Virus Yellows Distribution in Europe and Effects of Beet Yellows Virus, Beet Mild Yellowing Virus, and Beet Chlorosis Virus on Sugar Beet Yield Following Field Inoculation. *Plant Pathol.* **2021**, *70*, 584–593. [CrossRef]
7. Stevens, M.; Patron, N.J.; Dolby, C.A.; Weekes, R.; Hallsworth, P.B.; Lemaire, O.; Smith, H.G. Distribution and Properties of Geographically Distinct Isolates of Sugar Beet Yellowing Viruses. *Plant Pathol.* **2005**, *54*, 100–107. [CrossRef]
8. Hauer, M.; Hansen, A.L.; Manderyck, B.; Olsson, Å.; Raaijmakers, E.; Hanse, B.; Stockfisch, N.; Märländer, B. Neonicotinoids in Sugar Beet Cultivation in Central and Northern Europe: Efficacy and Environmental Impact of Neonicotinoid Seed Treatments and Alternative Measures. *Crop Prot.* **2017**, *93*, 132–142. [CrossRef]

9. European Commission Commission Implementing Regulation (EU) 2018/783 of 29 May 2018 Amending Implementing Regulation (EU) No 540/2011 as Regards the Conditions of Approval of the Active Substance Imidacloprid (Text with EEA Relevance). Available online: https://eur-lex.europa.eu/eli/reg_impl/2018/783/oj (accessed on 21 July 2024).
10. European Commission Commission Implementing Regulation (EU) 2018/784 of 29 May 2018 Amending Implementing Regulation (EU) No 540/2011 as Regards the Conditions of Approval of the Active Substance Clothianidin (Text with EEA Relevance). Available online: https://eur-lex.europa.eu/eli/reg_impl/2018/784/oj (accessed on 21 July 2024).
11. InfoCuria Judgment—19/01/2023—Pesticide Action Network Europe and Others Case C-162/21. Available online: https://curia.europa.eu/juris/document/document.jsf?text=&docid=269405&pageIndex=0&doclang=EN&mode=lst&dir=&occ=first&part=1&cid=7397235Acknowledgments (accessed on 24 July 2024).
12. Ballivian, E. Concerning: Bees at Risk: Abuses on Derogations for Neonicotinoids | Policy Commons. 2023. Available online: https://coilink.org/20.500.12592/tp7h5v (accessed on 9 September 2024).
13. Francis, F.; Fingu-Mabola, J.C.; Ben Fekih, I. Direct and Endophytic Effects of Fungal Entomopathogens for Sustainable Aphid Control: A Review. *Agriculture* **2022**, *12*, 2081. [CrossRef]
14. Meyling, N.V.; Eilenberg, J. Ecology of the Entomopathogenic Fungi *Beauveria Bassiana* and *Metarhizium Anisopliae* in Temperate Agroecosystems: Potential for Conservation Biological Control. *Biol. Control* **2007**, *43*, 145–155. [CrossRef]
15. Berbee, M.L. The Phylogeny of Plant and Animal Pathogens in the Ascomycota. *Physiol. Mol. Plant Pathol.* **2001**, *59*, 165–187. [CrossRef]
16. Vega, F.E.; Posada, F.; Catherine Aime, M.; Pava-Ripoll, M.; Infante, F.; Rehner, S.A. Entomopathogenic Fungal Endophytes. *Biol. Control* **2008**, *46*, 72–82. [CrossRef]
17. Hu, S.; Bidochka, M.J. Root Colonization by Endophytic Insect-pathogenic Fungi. *J. Appl. Microbiol.* **2021**, *130*, 570–581. [CrossRef] [PubMed]
18. Lopez, D.C.; Zhu-Salzman, K.; Ek-Ramos, M.J.; Sword, G.A. The Entomopathogenic Fungal Endophytes Purpureocillium Lilacinum (Formerly Paecilomyces Lilacinus) and Beauveria Bassiana Negatively Affect Cotton Aphid Reproduction under Both Greenhouse and Field Conditions. *PLoS ONE* **2014**, *9*, e103891. [CrossRef]
19. Ownley, B.H.; Griffin, M.R.; Klingeman, W.E.; Gwinn, K.D.; Moulton, J.K.; Pereira, R.M. *Beauveria Bassiana*: Endophytic Colonization and Plant Disease Control. *J. Invertebr. Pathol.* **2008**, *98*, 267–270. [CrossRef] [PubMed]
20. Fingu-Mabola, J.C.; Bawin, T.; Francis, F. Direct and Indirect Effect via Endophytism of Entomopathogenic Fungi on the Fitness of Myzus Persicae and Its Ability to Spread PLRV on Tobacco. *Insects* **2021**, *12*, 89. [CrossRef] [PubMed]
21. González-Mas, N.; Quesada-Moraga, E.; Plaza, M.; Fereres, A.; Moreno, A. Changes in Feeding Behaviour Are Not Related to the Reduction in the Transmission Rate of Plant Viruses by *Aphis Gossypii* (Homoptera: Aphididae) to Melon Plants Colonized by *Beauveria Bassiana* (Ascomycota: Hypocreales). *Biol. Control* **2019**, *130*, 95–103. [CrossRef]
22. Jaber, L.R.; Salem, N.M. Endophytic Colonisation of Squash by the Fungal Entomopathogen Beauveria Bassiana (Ascomycota: Hypocreales) for Managing Zucchini Yellow Mosaic Virus in Cucurbits. *Biocontrol Sci. Technol.* **2014**, *24*, 1096–1109. [CrossRef]
23. Zimmermann, G. The 'Galleria Bait Method' for Detection of Entomopathogenic Fungi in Soil. *J. Appl. Entomol.* **1986**, *102*, 213–215. [CrossRef]
24. Humber, R.A. Identification of Entomopathogenic Fungi. *Man. Tech. Invertebr. Pathol.* **2012**, *2*, 151–187.
25. White, T.J.; Bruns, T.; Lee, S.; Taylor, J. Amplification and Direct Sequencing of Fungal Ribosomal RNA Genes for Phylogenetics. In *PCR Protocols*; Innis, M.A., Gelfand, D.H., Sninsky, J.J., White, T.J., Eds.; Academic Press: San Diego, CA, USA, 1990; pp. 315–322, ISBN 978-0-12-372180-8.
26. Meyling, N.V.; Pilz, C.; Keller, S.; Widmer, F.; Enkerli, J. Diversity of Beauveria Spp. Isolates from Pollen Beetles Meligethes Aeneus in Switzerland. *J. Invertebr. Pathol.* **2012**, *109*, 76–82. [CrossRef] [PubMed]
27. Kumar, S.; Stecher, G.; Li, M.; Knyaz, C.; Tamura, K. MEGA X: Molecular Evolutionary Genetics Analysis across Computing Platforms. *Mol. Biol. Evol.* **2018**, *35*, 1547–1549. [CrossRef] [PubMed]
28. Nguyen, L.-T.; Schmidt, H.A.; von Haeseler, A.; Minh, B.Q. IQ-TREE: A Fast and Effective Stochastic Algorithm for Estimating Maximum-Likelihood Phylogenies. *Mol. Biol. Evol.* **2015**, *32*, 268–274. [CrossRef] [PubMed]
29. Darriba, D.; Taboada, G.L.; Doallo, R.; Posada, D. jModelTest 2: More Models, New Heuristics and High-Performance Computing. *Nat. Methods* **2012**, *9*, 772. [CrossRef] [PubMed]
30. Ben Fekih, I.; Jensen, A.B.; Boukhris-Bouhachem, S.; Pozsgai, G.; Rezgui, S.; Rensing, C.; Eilenberg, J. Virulence of Two Entomophthoralean Fungi, Pandora Neoaphidis and Entomophthora Planchoniana, to Their Conspecific (Sitobion Avenae) and Heterospecific (Rhopalosiphum Padi) Aphid Hosts. *Insects* **2019**, *10*, 54. [CrossRef]
31. Rasool, S.; Cárdenas, P.D.; Pattison, D.I.; Jensen, B.; Meyling, N.V. Isolate-Specific Effect of Entomopathogenic Endophytic Fungi on Population Growth of Two-Spotted Spider Mite (Tetranychus Urticae Koch) and Levels of Steroidal Glycoalkaloids in Tomato. *J. Chem. Ecol.* **2021**, *47*, 476–488. [CrossRef]
32. Parsa, S.; Ortiz, V.; Vega, F.E. Establishing Fungal Entomopathogens as Endophytes: Towards Endophytic Biological Control. *JoVE J. Vis. Exp.* **2013**, e50360. [CrossRef]
33. Haas, J.; Lozano, E.R.; Poppy, G.M. A Simple, Light Clip-Cage for Experiments with Aphids. *Agric. For. Entomol.* **2018**, *20*, 589–592. [CrossRef]
34. Fingu-Mabola, J.C.; Martin, C.; Bawin, T.; Verheggen, F.J.; Francis, F. Does the Infectious Status of Aphids Influence Their Preference Towards Healthy, Virus-Infected and Endophytically Colonized Plants? *Insects* **2020**, *11*, 435. [CrossRef]

35. Therneau, T.M.; Lumley, T.; Elizabeth, A.; Cynthia, C. A Package for Survival Analysis in R. Available online: https://cran.r-project.org/web/packages/survival/index.html (accessed on 9 September 2024).
36. Kassambara, A.; Kosinski, M.; Biecek, P.; Fabian, S. Survminer: Drawing Survival Curves Using "Ggplot2" 2021. Available online: https://cran.r-project.org/web/packages/survminer/survminer.pdf (accessed on 9 September 2024).
37. Mantzoukas, S.; Eliopoulos, P.A. Endophytic Entomopathogenic Fungi: A Valuable Biological Control Tool against Plant Pests. *Appl. Sci.* **2020**, *10*, 360. [CrossRef]
38. Dalbon, V.A.; Acevedo, J.P.M.; Ribeiro Junior, K.A.L.; Silva, J.M.d.; Ferro, M.M.d.M.; Negrisoli Júnior, A.S.; Fonseca, H.G.; Santana, A.E.G.; Porcelli, F. Native Entomopathogenic Fungi Isolated from Rhynchophorus Palmarum (Linnaeus, 1758) in Northeast Brazil. *Insects* **2024**, *15*, 159. [CrossRef] [PubMed]
39. Sánchez-Peña, S.R.; Lara, J.S.-J.; Medina, R.F. Occurrence of Entomopathogenic Fungi from Agricultural and Natural Ecosystems in Saltillo, México, and Their Virulence towards Thrips and Whiteflies. *J. Insect Sci.* **2011**, *11*, 1. [CrossRef] [PubMed]
40. Biryol, S.; Demirbağ, Z.; Erdoğan, P.; Demir, I. Development of Beauveria Bassiana (Ascomycota: Hypocreales) as a Mycoinsecticide to Control Green Peach Aphid, Myzus Persicae (Homoptera: Aphididae) and Investigation of Its Biocontrol Potential. *J. Asia-Pac. Entomol.* **2022**, *25*, 101878. [CrossRef]
41. Javed, K.; Javed, H.; Mukhtar, T.; Qiu, D. Pathogenicity of Some Entomopathogenic Fungal Strains to Green Peach Aphid, Myzus Persicae Sulzer (Homoptera: Aphididae). *Egypt. J. Biol. Pest Control* **2019**, *29*, 92. [CrossRef]
42. Jaber, L.R.; Araj, S.-E. Interactions among Endophytic Fungal Entomopathogens (Ascomycota: Hypocreales), the Green Peach Aphid *Myzus Persicae* Sulzer (Homoptera: Aphididae), and the Aphid Endoparasitoid *Aphidius Colemani* Viereck (Hymenoptera: Braconidae). *Biol. Control* **2018**, *116*, 53–61. [CrossRef]
43. Wilberts, L.; Vuts, J.; Caulfield, J.C.; Thomas, G.; Birkett, M.A.; Herrera-Malaver, B.; Verstrepen, K.J.; Sobhy, I.S.; Jacquemyn, H.; Lievens, B. Impact of Endophytic Colonization by Entomopathogenic Fungi on the Behavior and Life History of the Tobacco Peach Aphid Myzus Persicae Var. Nicotianae. *PLoS ONE* **2022**, *17*, e0273791. [CrossRef]
44. Vega, F.E. The Use of Fungal Entomopathogens as Endophytes in Biological Control: A Review. *Mycologia* **2018**, *110*, 4–30. [CrossRef]
45. Bamisile, B.S.; Afolabi, O.G.; Siddiqui, J.A.; Xu, Y. Endophytic Insect Pathogenic Fungi-Host Plant-Herbivore Mutualism: Elucidating the Mechanisms Involved in the Tripartite Interactions. *World J. Microbiol. Biotechnol.* **2023**, *39*, 326. [CrossRef]
46. Chehab, E.W.; Kaspi, R.; Savchenko, T.; Rowe, H.; Negre-Zakharov, F.; Kliebenstein, D.; Dehesh, K. Distinct Roles of Jasmonates and Aldehydes in Plant-Defense Responses. *PLoS ONE* **2008**, *3*, e1904. [CrossRef]
47. González-Mas, N.; Sánchez-Ortiz, A.; Valverde-García, P.; Quesada-Moraga, E. Effects of Endophytic Entomopathogenic Ascomycetes on the Life-History Traits of Aphis Gossypii Glover and Its Interactions with Melon Plants. *Insects* **2019**, *10*, 165. [CrossRef]
48. Rondot, Y.; Reineke, A. Association of Beauveria Bassiana with Grapevine Plants Deters Adult Black Vine Weevils, Otiorhynchus Sulcatus. *Biocontrol Sci. Technol.* **2017**, *27*, 811–820. [CrossRef]
49. Sword, G.A.; Tessnow, A.; Ek-Ramos, M.J. Endophytic Fungi Alter Sucking Bug Responses to Cotton Reproductive Structures. *Insect Sci.* **2017**, *24*, 1003–1014. [CrossRef] [PubMed]
50. Bera, S.; Arena, G.D.; Ray, S.; Flannigan, S.; Casteel, C.L. The Potyviral Protein 6K1 Reduces Plant Proteases Activity during Turnip Mosaic Virus Infection. *Viruses* **2022**, *14*, 1341. [CrossRef] [PubMed]
51. Parizad, S.; Bera, S. The Effect of Organic Farming on Water Reusability, Sustainable Ecosystem, and Food Toxicity. *Environ. Sci. Pollut. Res.* **2023**, *30*, 71665–71676. [CrossRef]

insects

MDPI

Article

Can a Mixture of Farnesene Isomers Avert the Infestation of Aphids in Sugar Beet Crops?

Denise Kuhn [1,2,*], Nils Nägele [2], Till Tolasch [1], Georg Petschenka [2] and Johannes L. M. Steidle [1,3,*]

[1] Department of Chemical Ecology 190t, Institute of Biology, University of Hohenheim, 70599 Stuttgart, Germany; tolasch@uni-hohenheim.de

[2] Department of Applied Entomology 360c, Institute of Phytomedicine, University of Hohenheim, 70599 Stuttgart, Germany; nils.naegele@uni-hohenheim.de (N.N.); georg.petschenka@uni-hohenheim.de (G.P.)

[3] KomBioTa—Center of Biodiversity and Integrative Taxonomy, University of Hohenheim, 70599 Stuttgart, Germany

* Correspondence: denise.kuhn@uni-hohenheim.de (D.K.); jsteidle@uni-hohenheim.de (J.L.M.S.)

Simple Summary: Many studies have shown that pesticide use has negative environmental effects. Therefore, alternative measures to manage insect pests and protect plant health are needed. Potential tools include chemicals that cause behavioral changes in pests and are already a part of the integrated pest management toolbox. Aphids are a significant pest in sugar beet (*Beta vulgaris*). They cause visual damage and act as vectors for various phytopathogens that reduce the yield. When attacked by predators, aphids emit an alarm pheromone, which causes other aphids to stop feeding, move away, and sometimes even abandon the host plant. We tested whether the artificially release of this pheromone could be an environmentally friendly alternative to prevent aphid infestation in sugar beet. Application of the pheromone resulted in reduced aphid densities in experimental sugar beet patches at two out of three experimental sites. This indicates the potential of this pheromone as an aphid repellent in sugar beet production. However, more research is required to ensure the reliability of this method.

Abstract: The negative impact of pesticides on the environment and the potential of pest species to develop pesticide resistance make it necessary to explore new methods of pest control. Pheromones and other behavior-modifying semiochemicals are already important in integrated pest management (IPM). (E)-ß-farnesene (EBF) is a semiochemical that acts as an alarm pheromone in aphids. Upon perception of EBF, aphids stop feeding, move away, and sometimes even abandon the host plant. The aphids *Myzus persicae* and *Aphis fabae* are significant crop pests and vectors of many harmful phytopathogens affecting sugar beet (*Beta vulgaris*). Field trials were conducted at different locations in Germany to test whether dispensers containing a mixture of farnesene isomers (FIMs) including EBF were able to reduce the infestation of these species on sugar beet. Our results showed a reduction in aphid abundance in the FIM-treated patches in two out of three sites. Therefore, we hypothesize that FIM dispensers could prevent aphid infestation and could be used in combination with other IPM measures. However, more research is required to increase the effect and ensure the reliability of this method.

Keywords: pest control; insecticide replacement; (E)-ß-farnesene; alarm pheromone; aphids; sugar beet

1. Introduction

There is general agreement that intensified agriculture contributes considerably to the decline in insect diversity and abundance [1,2]. Insecticides are very likely to play an important role in this process, affecting not only insect pests, but also non-target insects such as pollinators and predators [3–7]. Thus, there is a pressing need for alternative methods to replace or reduce insecticides. Synthetic semiochemicals have been found to be effective tools for pest control in various agricultural fields [8], especially in viticulture, fruit cultivation, forestry, and horticulture [9]. As natural compounds, they mediate communication between individuals within the same species (pheromones [10]) or of different species (allelochemicals [11]). Pheromones possess the capability of inducing long-lasting physiological or behavioral reactions in conspecifics [10], such as sexual attraction, mating, dispersion, oviposition, aggregation, and alarm behavior [10,12,13]. In pest management, synthetic pheromones are used mainly for (a) mating disruption, (b) monitoring, and (c) mass trapping (attract and kill). For instance, two important tortricid pests, *Eupoecilia ambiguella* (Hübner, 1796) (Lepidoptera: Tortricidae) in vineyards and *Cydia pomonella* (L., 1758) (Lepidoptera: Tortricidae) in orchards, are mainly controlled with pheromones [12]. Regarding allelochemicals, various studies have focused on the use of synthetic volatile secondary plant metabolites that are released by infested plants to defend themselves against herbivores, such as methyl salicylate (MeSA), (Z)-3-hexenol, dodecanoic acid (DDA), and (E)-ß-farnesene (EBF), as part of integrated pest management (IPM) [13–15]. These volatiles act either as repellents of pest insects [16], or to attract natural enemies to improve biological pest control [17], or in a 'push and pull' strategy, where pests (i.e., aphids) are lured away from the crop by repellents or deterrents (push) to establish themselves elsewhere where attractive stimuli are provided (pull) [14].

Many aphids use the sesquiterpene (E)-ß-farnesene (EBF) as an alarm pheromone, which they emit through their cornicles in the presence of natural enemies [18–21]. Upon detection, it causes short-term behavioral changes in other individual aphids. They stop feeding on their host plant or even withdraw their stylet from the plant tissue and move away from the volatile source, which could include dropping off the host plant [22–25]. Moreover, prolonged exposure to high levels of EBF can lead to the production of winged aphids in the next generation that are able to migrate from the "danger zone" to colonize another host plant [26–28]. Also, the normal population growth rate is disrupted and the rate of infestation is reduced [29,30]. In addition, under laboratory conditions, EBF has been shown to serve as a foraging kairomone [31] and attracts various natural enemies of aphids, such as ladybirds [31,32], syrphid flies [33], and parasitoid wasps [34]. Results from field experiments on EBF as an attractant for natural enemies to control aphids, however, are inconsistent [34–37]. Some studies have demonstrated the effect of EBF as a kairomone on various natural enemies under field conditions [34], but others showed that the observed effect was due to the exceptionally high concentrations of pheromone used [36].

Sugar beet *Beta vulgaris* subsp. *vulgaris* (Amaranthaceae) is mainly infested by two aphid species [38], *Aphis fabae* (Scopoli 1763) (Hemiptera: Aphididae) and *Myzus persicae* (Sulzer 1776) (Sternorrhyncha: Aphididae). While sucking the phloem, these aphids deprive the plants of important nutrients, resulting in retarded growth of sugar beet, loss of sugar content up to 50%, water stress, and leaf deformities [39,40]. In addition, both aphid species act as vector of a variety of viruses such as beet chlorosis virus (BCHV), beet mosaic virus (BtMV), and beet yellow virus (BYV). While the first two viruses induce orange-yellow hues on the surface of the leaves [41], BYV causes leaf blade discoloration with necrosis in the form of dots or dashes. BYV is the most prevalent virus of sugar beet in Central Europe, and frequently results in significant reductions in yield and sugar content [42].

In this study, we investigated the potential use of a commercially available mixture of farnesene isomers (FIMs) to control aphids. The mixture contained low levels of the aphid alarm pheromone EBF and was therefore suitable for use as a preventative method to avoid aphid infestation of sugar beets and virus transmission.

2. Materials and Methods

2.1. Experimental Sites

Experiments were performed on conventionally managed *B. vulgaris* fields in Mannheim (49°35′05.7″ N 8°27′18.0″ E), Heidelberg (49°22′25.8″ N 8°39′39.6″ E), and Dettenheim (49°10′59.2″ N 8°24′07.2″ E) (Southern Germany). No insecticides were applied to the crop during the experimental period. Sugar beet is usually sown at 110.000 seeds/ha, with about 80.000 seeds/ha emerging, giving 8 plants/m^2.

2.2. Dispenser Preparation and Application

We used Eppendorf tubes (1.5 mL) as FIM dispensers, which were filled with 0.7 mL of a commercially available farnesene isomer mixture (FIM, Sigma Aldrich, St. Louis, MO, USA, product number: W383902), containing 15.1% EBF as well as 0.1% alpha-tocopherol as stabilizer. The mixture was used instead of pure EBF because it was more cost effective. We applied dispensers filled with FIMs onto conventionally managed *B. vulgaris* fields in three locations in Southern Germany and monitored the numbers of aphids on the plants. Immediately before application, the tubes were perforated with a needle four times just above the fill level (diameter of the perforations: 0.1 mm) to facilitate evaporation. FIM dispensers were placed in the field in spring as soon as sugar beets had emerged, and the first winged female aphids were detected in yellow pan traps. Dispensers were installed on a plastic stick 5 cm above the ground downwind (i.e., away from the control patches to avoid potential contamination, see arrows in Figure 1A,C,E) and checked weekly for potential drying out.

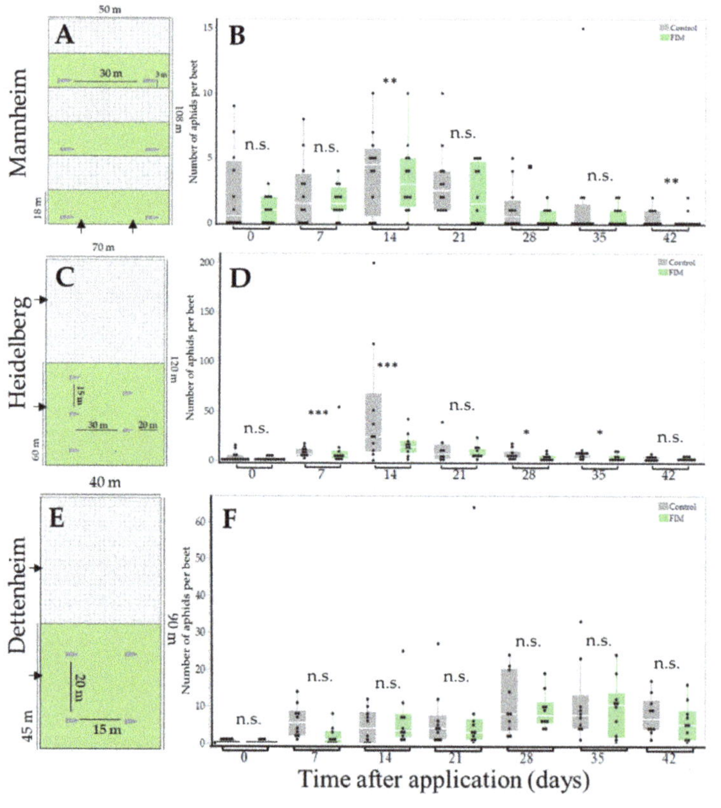

Figure 1. The use of FIM dispensers in sugar beet fields as a potential strategy to prevent aphid colonization and to inhibit colony growth. An overview of the experimental design (field patches with dispenser: green; without dispenser: gray) is shown on the left for Mannheim (**A**), Heidelberg (**C**), and Dettenheim (**E**) and the number of aphids found per sugar beet sampled are shown on the right for Mannheim (**B**), Heidelberg (**D**) and Dettenheim (**F**). Black arrows indicate typical wind

direction. The graphs on the right show the number of aphids at specific sampling times over a period of 42 days. The data are presented as box–whisker plots. Boxes indicate interquartile ranges and whiskers indicate minimum and maximum data points. Black dots represent raw data, and white lines indicate median values. Note that for better visualization of the Heidelberg dataset, the outliers are excluded. The y-axis of each location is adjusted according to the number of individuals found, resulting in different y-axes for Mannheim, Heidelberg, and Dettenheim. "***" significant at $p < 0.001$, "**" significant at $p < 0.01$, "*" tendency with $0.05 > p < 0.06$, "n.s." not significant at $p > 0.06$.

2.3. Sampling

The first sampling took place before FIM dispensers were placed in the field (labeled "0 days" in Figure 1A,C,E), to determine the initial population density of aphids. After installation of the dispensers, weekly samples were taken for seven consecutive weeks (i.e., 7, 14, 21, 28, 35, and 42 days after application). Sampling was performed by cutting all leaf material from randomly selected beet plants within each respective patch (see below) directly above the soil and preserving it in a Whirl-Pak™ (125 mm × 380 mm, volume: 1065 mL, Sigma Aldrich, St. Louis, MO, USA) filled with 70% ethanol. Care was taken to ensure that the leaf material was completely covered with ethanol.

2.4. Experimental Setups

Different experimental designs were used at the three sites. The fact that no insecticide was allowed to be applied to these patches for the duration of the trial limited the areas that the farmers could make available for the trials. Therefore, the numbers and distribution of the dispensers needed to be adapted to the available patches. As a result, different sizes and shapes of patches were available. In addition, sowing times could not be standardized, due to the responsibilities of the individual farmers and varying weather conditions. The experimental and the control patches were not separated in the field.

In Mannheim, *B. vulgaris* was sown at the beginning of March 2023. Within the cultivated field, six patches in a row (W × L: 18 m × 50 m, 900 m^2 each) were selected and staked (Figure 1A). The patches alternated between experimental patches with dispensers and control patches without dispensers. In each experimental patch, two FIM dispensers were placed directly at the edge, 30 m apart and 3 m from each end of the patch. During the experiment, 4 out of 7200 beet plants in each patch were sampled 7, 14, 21, 28, 35, and 42 days after FIM application. given that the sampled patches were smaller than the other patches (i.e., $n = 12$ for each treatment, 84 beet plants in total).

In Heidelberg, *B. vulgaris* was sown in April 2023 in a field measuring 120 m × 70 m (8400 m^2). The field was divided into two equal patches of 120 m × 35 m (4200 m^2) each (Figure 1C). In the experimental patches, five FIM dispensers were evenly spaced 20 m from each end of the field, 30 m apart horizontally, 15 m apart vertically. At each sampling time, 10 out of 33.600 beet plants per patch were randomly collected 7, 14, 21, 28, 35, and 42 days after FIM application.

In Dettenheim, we used a similar design as in Heidelberg. *B. vulgaris* was sown at the end of March 2023 in a field measuring 260 m × 150 m (39.000 m^2). An area at the edge of the field, measuring 40 m × 90 m (3600 m^2), was selected and divided into two equal patches (45 m × 40 m, each 1800 m^2) (Figure 1E). Four FIM dispensers were evenly spaced in the experimental patch (distance between dispensers: 20 m vertical, 15 m horizontal). At each sampling time, 10 out of 14.400 beets per patch were randomly collected 7, 14, 21, 28, 35, and 42 days after FIM application.

2.5. Preparation of Samples

To obtain all arthropods from the beet leaves in the Whirl-Paks™ (Thermo Fisher Scientific, Waltham, MA, USA), the ethanol was poured through a sieve (mesh size 0.5 mm) and the sample bag was rinsed using ethanol. Then, the insects were counted and identified to species level using a stereomicroscope (VWR® VisiScope® 360 stereo zoom microscope, Avantor®, Ismaning, Germany). We aimed to differentiate the found aphids into "alates" and "nymphs". However, the number of alates on the plants was sometimes too small to make a statistical analysis. So, we combined nymphs and alates.

2.6. Statistical Analyses

Data were analyzed using R software (version 4.0.3) [43,44] with pre-installed packages as well as the packages multcomp [45], car [46], and glmm [47]. Significance was assumed at $p < 0.05$, a tendency was assumed at $0.05 < p > 0.06$. In the Heidelberg dataset, outliers were identified based on the four-sigma rule [48] and excluded from the dataset. An analysis of their influence revealed that the results did not depend on the presence of the outliers.

For comparing initial aphid numbers between the control and experimental patches (i.e., before the dispenser application took place, labeled as "0"), Wilcox rank tests were conducted, since the data were not normally distributed. The rest of the data from the three locations (Mannheim, Heidelberg, and Dettenheim) were analyzed separately using a generalized linear mixed model with negative binomial distribution (Figures 1 and 2). Note that seasonal patterns were not statistically analyzed. Data from time point "0" were excluded from this analysis. The occurrence of aphids on individual sugar beet plants (yes or no) was analyzed separately for each location using a generalized mixed model with binomial distribution. Models were followed by Tukey tests [49]. The presence and number of natural enemies on the sugar beet could not be statistically analyzed due to the low number of individuals. Graphs were created using JMP Pro 17 (JMP Pro 17, 2023 JMP Statistical Discovery LLC, Cary, NC, USA).

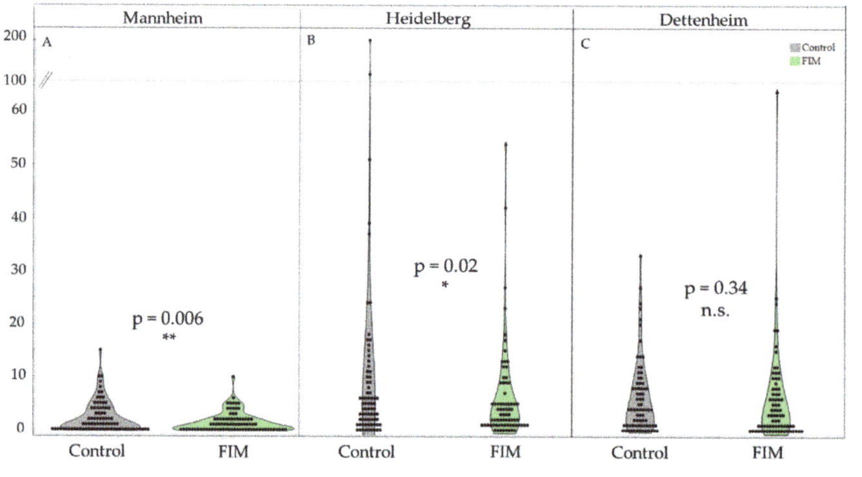

Figure 2. Total number of aphids found on *B. vulgaris* in the control patches (gray) and the patches with FIM dispensers (green) in Mannheim (**A**), Heidelberg (**B**), and Dettenheim (**C**). Data are presented as violin graphs to show the distribution of the raw data. Note that for better visualization of the Heidelberg dataset, the outliers are excluded. Furthermore, the *y*-axis has been interrupted to improve the visibility of individual data points. Dots represent raw data. "**" significant at $p < 0.01$, "*" significant at $p < 0.05$, "n.s." not significant at $p > 0.06$.

3. Results

Only two aphid species, *M. persicae* and *A. fabae*, were found infesting the sugar beets. The numbers of natural enemies on the beet plants were too small for statistical analysis. The total numbers of natural enemies in Mannheim were as follows: hymenopteran parasitoids: 8; parasitized aphids: 4; coccinellidae (larvae and adults): 11; syrphid larvae: 1. Those for Heidelberg were as follows: hymenopteran parasitoids: 11; parasitized aphids: 21; coccinellidae (larvae and adults): 56; syrphid larvae: 2. In Dettenheim, the numbers were as follows: hymenopteran parasitoids: 17; parasitized aphids: 2; coccinellidae (larvae and adults): 21; syrphid larvae: 4 (for further information, please see Supplementary Materials Table S1).

At the beginning of the experiment, before placing the dispensers in the field (i.e., at day 0), we did not observe any differences in the numbers of aphids between the two treatments at the different locations (Wilcox test, Mannheim: W = 79, $p > 0.69$; Heidelberg: W = 119, $p = 0.79$; Dettenheim: W = 127, $p = 0.43$) (Figure 1B,D,F). After we distributed dispensers containing FIMs across the experimental fields, we observed a significant reduction in aphid numbers in the FIM-treated patches in Mannheim compared with the control patches throughout the season (Figure 2A). This was also the case in Heidelberg, where aphid abundance was generally highest (Figure 2B), regardless of whether extreme outliers in the control treatments were included or excluded. No differences were found between the control and FIM-treated patches in Dettenheim (Figure 2C).

When analyzing the data in more detail, there was a general increase in the aphid population 14 d after application in Mannheim, followed by a decrease 28 d after application (Figure 1B). We documented significantly more aphids in the control patches than in the experimental patches with FIM dispensers at 14 d ($p = 0.039$ *), 28 d ($p = 0.011$ *), and 42 d ($p = 0.0002$ ***) after application, but not after 7 d, 21 d, or 35 d. In Heidelberg, we also observed an overall increase in aphid abundance starting at 14 days post-application in both the control and FIM-treated patches (Figure 1D). Twenty-eight days after application and at subsequent time points, the aphid populations had almost completely collapsed, regardless of treatment. As in Mannheim, we found significantly more aphids in the control than in the FIM-treated patches at 7 d ($p < 0.000$ ***), 14 d ($p = 0.000$ **), 28 ($p = 0.015$ *), 35 d ($p = 0.010$ *), and 42 d ($p = 0.000$) after application. In contrast to the other locations, we observed an increase in aphid abundance over the whole season in Dettenheim, and no differences between controls and FIM-treated patches (Figure 1F).

4. Discussion

So far, many studies have shown that the aphid alarm pheromone EBF repels aphids from colonizing plants, reduces aphid population growth, stimulates the production of alates (winged aphids) to leave their fully infested host plant [27], and attracts their natural enemies [50]. These effects indicate that EBF could be used for the control of aphid pests in crop plants. We studied the use of a mixture of commercial farnesene isomers (FIMs) containing synthetic EBF to control the aphid species *M. persicae* and *A. fabae* on sugar beet fields in three locations in Southern Germany.

While we did not find significant effects at one of the sites, at the two other sites, sugar beet plants in patches with FIM dispensers were found to be significantly less infected with aphids than those in control patches when the data for the entire season were analyzed together. Comparing experimental and control patches on the individual sampling dates, the number of aphids on the plants was reduced significantly in the FIM patches for at least half of the sample dates at these sites. This indicates that their is possible to achieve a reduction in aphid numbers by using FIM dispensers in sugar beet.

While it is unclear whether this reduction was due to a repellence effect on the aphids, a reduction in population growth, or the desertion of the host plants by alates, the effect of natural enemies can be excluded [48]. Their numbers were very low in general and did not increase in the presence of EBF. Possibly, the attraction of natural enemies with the use of EBF does not work in intensively managed agricultural landscapes with low biodiversity. Alternatively, it might well be that EBF does not act as a kairomone in aphid–natural enemy interaction under natural conditions and in natural concentrations. This was indicated by a detailed field study with pea aphids *Acyrthosiphon pisum* [36].

Generally, the observed reduction in the number of aphids in the presence of EBF was relatively small, and there was no effect on one of the locations. This agrees with earlier studies, which also found only small effects of EBF on aphid population size in crop fields of peas [51], wheat [51], cabbage [17], and iceberg lettuce [52]. Thus, EBF seems to have only a limited effect when studied under field conditions. Potential explanations for this reduced efficiency in the field in contrast to the laboratory include (1) the chemical instability of EBF, which oxidizes over time due to certain double bonds in the molecule, rendering the alarm pheromone ineffective [27,53], (2) the large distance between the points of release, (3) the low concentration of the EBF used in the experiments, and (4) the environmental conditions at different sites, which can lead to different microclimates and can make the FIM dispenser either more or less effective.

In addition, as field trials are difficult to standardize, there are certain aspects that could have influenced the numbers of aphids found on the plants. Firstly, the main wind direction was considered, but there is still the possibility of a change in wind direction. As a result, FIM could have been evaporated and delivered in the wrong direction, leaving the experimental patches unaffected, which may have affected aphid infestation. In addition, if the wind speed was too high or too low, the evaporated FIMs may have been carried either too far or not far enough to cover the entire expected area. Furthermore, due to the different responsibilities of the different farmers and the different weather and soil conditions at the different sites, sowing took place at different times (March to April), which could have influenced aphid infestation.

From these explanations, chemical instability is unlikely to have been involved in our experiment, because FIMs contain alpha-tocopherol as a stabilizer and we observed the effects of EBF in our experiments over several weeks. In contrast, the distance between the release points and the sampled plants could well have been too large (Mannheim: 21.1 m; Heidelberg: 36.1 m; Dettenheim: 17.7 m), especially in the absence of wind. However, small effects or no effects of EBF on aphids were also observed in studies in which the compound was released by the plants themselves [53,54], indicating that distance to the release point was not the only important factor. The EBF concentration necessary to induce a behavioral or physiological response by aphids is unclear. Dose–response studies are scarce and the amount of EBF required to stimulate a response in aphids has rarely been studied in terms of absolute concentrations in field experiments. For *M. persicae*, which we have also studied, van Emden et al. found identical behavioral responses to EBF when applied at 1, 10, and 100 $\mu g\ mL^{-1}$ [55]. Jing-Gong et al. showed that the response of *M. persicae* was stronger towards 1 $\mu g\ mL^{-1}$ EBF compared with 0.1 $\mu g\ mL^{-1}$, although the responses of related aphids in the genus *Aphis* were mostly higher at lower concentrations [56]. In the present study, we applied 0.7 mL of an undiluted mixture containing 15.1% EBF in each dispenser, firstly, because we had carried out preliminary studies on the exposure of FIMs for seven days under laboratory conditions and found that the evaporation was low, and secondly, because we wanted to achieve long-term exposure under field conditions. Possibly, this concentration was too high to induce a response. More studies are required to identify the optimal field concentration of the mixture of farnesene isomers used.

5. Conclusions

In summary, our study revealed that aphid infestation in sugar beet fields can be reduced using a cost-efficient mixture of farnesene isomers (FIMs). Obviously, FIM dispensers could be an affordable alternative to insecticides. However, there is a strong need to understand the basics to optimize the current system and to increase the effect and make it more reliable. For instance, the optimal concentration of FIMs for field application is unclear. Further research is also required to elucidate the appropriate density of dispensers in the field, their optimal distance, and their position within the field, depending on the wind direction. Considering their high potential to control aphid populations in the field without harmful insecticides, more efforts to optimize the use of FIMs are certainly justified.

Supplementary Materials: The following supporting information can be downloaded at: https://www.mdpi.com/article/10.3390/insects15100736/s1, Table S1: Total number of natural enemies found on the sugar beet plants.

Author Contributions: Conceptualization, D.K.; methodology, D.K. and N.N.; software, T.T.; validation, D.K., G.P. and J.L.M.S.; formal analysis, D.K. and N.N.; investigation, D.K.; resources, G.P.; data curation, D.K. and N.N.; writing—original draft preparation, D.K.; writing—review and editing, J.L.M.S., G.P. and D.K.; visualization, D.K. and J.L.M.S.; supervision, J.L.M.S., G.P. and D.K.; project administration, D.K.; funding acquisition, G.P. All authors have read and agreed to the published version of the manuscript.

Funding: Südzucker AG funded the travel costs to visit the fields.

Data Availability Statement: All data underlying our study have been deposited as Supplementary Material. There are no restrictions on data availability.

Acknowledgments: We thank Südzucker AG for providing access to their sugar-beet fields and for the financial support of the travel expenses of DK.

Conflicts of Interest: The study was financially supported by Südzucker AG but did not cause any conflict of interests between the authors and the company.

References

1. Tilman, D.; Fargione, J.; Wolff, B.; D'Antonio, C.; Dobson, A.; Schindler, D.; Schlesinger, W.H.; Simberloff, D.; Swackhamer, D. Forecasting Agriculturally Driven Global Environmental Change. *Science* **2001**, *292*, 281–284. [CrossRef] [PubMed]
2. Seibold, S.; Gossner, M.M.; Simons, N.K.; Blüthgen, N.; Müller, J.; Ambarlı, D.; Ammer, C.; Bauhus, J.; Fischer, M.; Habel, J.C.; et al. Arthropod Decline in Grasslands and Forests Is Associated with Landscape-Level Drivers. *Nature* **2019**, *574*, 671–674. [CrossRef] [PubMed]
3. Sánchez-Bayo, F.; Baskaran, S.; Kennedy, I.R. Ecological Relative Risk (EcoRR): Another Approach for Risk Assessment of Pesticides in Agriculture. *Agric. Ecosyst. Environ.* **2002**, *91*, 37–57. [CrossRef]
4. Desneux, N.; Decourtye, A.; Delpuech, J.-M. The Sublethal Effects of Pesticides on Beneficial Arthropods. *Annu. Rev. Entomol.* **2007**, *52*, 81–106. [CrossRef]
5. Sánchez-Bayo, F. Insecticides Mode of Action in Relation to Their Toxicity to Non-Target Organisms. *J. Environ. Anal. Toxicol.* **2012**, *s4*, 2. [CrossRef]
6. Vivekanandhan, P.; Swathy, K.; Thomas, A.; Kweka, E.J.; Rahman, A.; Pittarate, S.; Krutmuang, P. Insecticidal Efficacy of Microbial-Mediated Synthesized Copper Nano-Pesticide against Insect Pests and Non-Target Organisms. *Int. J. Environ. Res. Public Health* **2021**, *18*, 10536. [CrossRef]
7. Serrão, J.E.; Plata-Rueda, A.; Martínez, L.C.; Zanuncio, J.C. Side-Effects of Pesticides on Non-Target Insects in Agriculture: A Mini-Review. *Sci. Nat.* **2022**, *109*, 17. [CrossRef]
8. Bruce, T.J.A.; Aradottir, G.I.; Smart, L.E.; Martin, J.L.; Caulfield, J.C.; Doherty, A.; Sparks, C.A.; Woodcock, C.M.; Birkett, M.A.; Napier, J.A.; et al. The First Crop Plant Genetically Engineered to Release an Insect Pheromone for Defence. *Sci. Rep.* **2015**, *5*, 11183. [CrossRef]
9. Norin, T. Semiochemicals for Insect Pest Management. *Pure Appl. Chem.* **2007**, *79*, 2129–2136. [CrossRef]
10. Karlson, P.; Butenandt, A. Pheromones (Ectohormones) in Insects. *Annu. Rev. Entomol.* **1959**, *4*, 39–58. [CrossRef]
11. Whittaker, R.H. *Communities and Ecosystems*; Macmillan: London, UK; University of California: Los Angeles, CA, USA, 1970; ISBN 978-0-02-427400-7.
12. Hoffmann, C.; Doye, E. How to Measure Mating Disruption Efficacy of Pheromone Products for Registration Purposes in the Field? *Pheromones Semiochem. Integr. Prod.* **2017**, *126*, 59–70.
13. Park, S.-W.; Kaimoyo, E.; Kumar, D.; Mosher, S.; Klessig, D.F. Methyl Salicylate Is a Critical Mobile Signal for Plant Systemic Acquired Resistance. *Science* **2007**, *318*, 113–116. [CrossRef] [PubMed]
14. Zhou, H.; Chen, L.; Liu, Y.; Chen, J.; Francis, F. Use of Slow-Release Plant Infochemicals to Control Aphids: A First Investigation in a Belgian Wheat Field. *Sci. Rep.* **2016**, *6*, 31552. [CrossRef] [PubMed]
15. Herrbach, E. Effect of Dodecanoic Acid on the Colonisation of Sugar Beet by Aphids and the Secondary Spread of Virus Yellows. *Ann. Appl. Biol.* **1987**, *111*, 477–482. [CrossRef]
16. Ninkovic, V.; Glinwood, R.; Ünlü, A.G.; Ganji, S.; Unelius, C.R. Effects of Methyl Salicylate on Host Plant Acceptance and Feeding by the Aphid *Rhopalosiphum padi*. *Front. Plant Sci.* **2021**, *12*. [CrossRef]
17. Cui, L.-L.; Francis, F.; Heuskin, S.; Lognay, G.; Liu, Y.-J.; Dong, J.; Chen, J.-L.; Song, X.-M.; Liu, Y. The Functional Significance of E-β-Farnesene: Does It Influence the Populations of Aphid Natural Enemies in the Fields? *Biol. Control* **2012**, *60*, 108–112. [CrossRef]
18. Bowers, W.S.; Nault, L.R.; Webb, R.E.; Dutky, S.R. Aphid Alarm Pheromone: Isolation, Identification, Synthesis. *Science* **1972**, *177*, 1121–1122. [CrossRef]
19. Nault, L.R.; Edwards, L.J.; Styer, W.E. Aphid Alarm Pheromones: Secretion and Reception. *Environ. Entomol.* **1973**, *2*, 101–105. [CrossRef]
20. Francis, F.; Vandermoten, S.; Verheggen, F.; Lognay, G.; Haubruge, E. Is the (E)-β-Farnesene Only Volatile Terpenoid in Aphids? *J. Appl. Entomol.* **2004**, *129*, 6–11. [CrossRef]

21. Vandermoten, S.; Mescher, M.C.; Francis, F.; Haubruge, E.; Verheggen, F.J. Aphid Alarm Pheromone: An Overview of Current Knowledge on Biosynthesis and Functions. *Insect Biochem. Mol. Biol.* **2012**, *42*, 155–163. [CrossRef]
22. Edwards, J.S. Defence by Smear: Supercooling in the Cornicle Wax of Aphids. *Nature* **1966**, *211*, 73–74. [CrossRef]
23. Strong, F.E. Observations on Aphid Cornicle Secretions. *Ann. Entomol. Soc. Am.* **1967**, *60*, 668–673. [CrossRef]
24. Montgomery, M.E.; Nault, L.R. Comparative Response of Aphids to the Alarm Pheromone, (E)-ß-Farnsene. *Entomol. Exp. Appl.* **1977**, *22*, 236–242. [CrossRef]
25. Wohlers, P. Escape responses of pea aphids *Acyrthosiphon pisum* to alarm pheromones and additional stimuli. *Entomol. Exp. Appl.* **1980**, *27*, 156–168. [CrossRef]
26. Sloggett, J.J.; Weisser, W.W. Parasitoids Induce Production of the Dispersal Morph of the Pea Aphid, *Acyrthosiphon pisum*. *Oikos* **2002**, *98*, 323–333. [CrossRef]
27. Kunert, G.; Otto, S.; Röse, U.S.R.; Gershenzon, J.; Weisser, W.W. Alarm Pheromone Mediates Production of Winged Dispersal Morphs in Aphids. *Ecol. Lett.* **2005**, *8*, 596–603. [CrossRef]
28. Podjasek, J.O.; Bosnjak, L.M.; Brooker, D.J.; Mondor, E.B. Alarm Pheromone Induces a Transgenerational Wing Polyphenism in the Pea Aphid, *Acyrthosiphon pisum*. *Can. J. Zool.* **2005**, *83*, 1138–1141. [CrossRef]
29. Su, J.; Zhu, S.; Zhang, Z.; Ge, F. Effect of Synthetic Aphid Alarm Pheromone (E)-β-Farnesene on Development and Reproduction of *Aphis gossypii* (Homoptera: Aphididae). *J. Econ. Entomol.* **2006**, *99*, 1636–1640. [CrossRef]
30. de Vos, M.; Cheng, W.Y.; Summers, H.E.; Raguso, R.A.; Jander, G. Alarm Pheromone Habituation in *Myzus persicae* Has Fitness Consequences and Causes Extensive Gene Expression Changes. *Proc. Natl. Acad. Sci. USA* **2010**, *107*, 14673–14678. [CrossRef]
31. Francis, F.; Lognay, G.; Haubruge, E. Olfactory Responses to Aphid and Host Plant Volatile Releases: (E)-β-Farnesene an Effective Kairomone for the Predator *Adalia bipunctata*. *J. Chem. Ecol.* **2004**, *30*, 741–755. [CrossRef]
32. Al Abassi, S.; Birkett, M.A.; Pettersson, J.; Pickett, J.A.; Wadhams, L.J.; Woodcock, C.M. Response of the Seven-Spot Ladybird to an Aphid Alarm Pheromone and an Alarm Pheromone Inhibitor Is Mediated by Paired Olfactory Cells. *J. Chem. Ecol.* **2000**, *26*, 1765–1771. [CrossRef]
33. Francis, F.; Martin, T.; Lognay, G.; Haubruge, E. Role of (E)-ß-Farnesene in Systematic Aphid Prey Location by *Episyrphus balteatus* Larvae (Diptera: Syrphidae). *Eur. J. Entomol.* **2005**, *102*. [CrossRef]
34. Micha, S.G.; Wyss, U. Aphid Alarm Pheromone (E)-β-Farnesene: A Host Finding Kairomone for the Aphid Primary Parasitoid *Aphidius uzbekistanicus* (Hymenoptera: Aphidiinae). *Chemoecology* **1996**, *7*, 132–139. [CrossRef]
35. Joachim, C.; Vosteen, I.; Weisser, W.W. The Aphid Alarm Pheromone (E)-β-Farnesene Does Not Act as a Cue for Predators Searching on a Plant. *Chemoecology* **2015**, *25*, 105–113. [CrossRef]
36. Joachim, C.; Weisser, W.W. Does the Aphid Alarm Pheromone (E)-β-Farnesene Act as a Kairomone under Field Conditions? *J. Chem. Ecol.* **2015**, *41*, 267–275. [CrossRef] [PubMed]
37. Vosteen, I.; Weisser, W.W.; Kunert, G. Is There Any Evidence That Aphid Alarm Pheromones Work as Prey and Host Finding Kairomones for Natural Enemies? *Ecol. Entomol.* **2016**, *41*, 1–12. [CrossRef]
38. Kennedy, J.S.; Ibbotson, A.; Booth, C.O. The Distribution of Aphid Infestation in Relation to the Leaf Age: *Myzus persicae* (Sulz.) and *Aphis fabae* (Scop.) on Spindle Trees and Sugar Beet Plants. *Ann. Appl. Biol.* **1950**, *37*, 651–679. [CrossRef]
39. Sémétey, O.; Bressan, A.; Richard-Molard, M.; Boudon-Padieu, E. Monitoring of Proteobacteria and Phytoplasma in Sugar Beets Naturally or Experimentally Affected by the Disease Syndrome 'Basses Richesses'. *Eur. J. Plant Pathol.* **2007**, *117*, 187–196. [CrossRef]
40. Smith, H.G.; Hallsworth, P.B. The Effects of Yellowing Viruses on Yield of Sugar Beet in Field Trials, 1985 and 1987. *Ann. Appl. Biol.* **1990**, *116*, 503–511. [CrossRef]
41. Dusi, A.N.; Peters, D. Beet Mosaic Virus: Its Vector and Host Relationships. *J. Phytopathol.* **1999**, *147*, 293–298. [CrossRef]
42. Hossain, R.; Menzel, W.; Lachmann, C.; Varrelmann, M. New Insights into Virus Yellows Distribution in Europe and Effects of Beet Yellows Virus, Beet Mild Yellowing Virus, and Beet Chlorosis Virus on Sugar Beet Yield Following Field Inoculation. *Plant Pathol.* **2021**, *70*, 584–593. [CrossRef]
43. R Core Team. *R: A Language and Environment for Statistical Computing;* R Foundation: Vienna, Austria, 2020.
44. RStudio Team. *RStudio: Integrated Development for R;* RStudio PBC: Boston, MA, USA, 2020.
45. Hothorn, T.; Bretz, F.; Westfall, P. Simultaneous Inference in General Parametric Models. *Biom. J* **2008**, *50*, 346–363. [CrossRef]
46. Fox, J.; Weisberg, S. *An {R} Companion to Applied Regression;* Sage Publications: Thousand Oaks, CA, USA, 2019.
47. Knudson, C.; Geyer, C.J.; Benson, S. Glmm: Generalized Linear Mixed Models via Monte Carlo Likelihood Approximation. 2022.
48. Sachs, L. *Angewandte Statistik. Anwendung Statistischer Methoden*, 7th ed.; Springer: Berlin/Heidelberg, Germany, 1992.
49. Tukey, J.W. Comparing Individual Means in the Analysis of Variance. *Biometrics* **1949**, *5*, 99–114. [CrossRef] [PubMed]
50. Bushra, S.; Tariq, M. How Aphid Alarm Pheromone Can Control Aphids: A Review. *Arch. Phytopathol. Plant Prot.* **2014**, *47*, 1563–1573. [CrossRef]
51. Xu, Q.; Hatt, S.; Lopes, T.; Zhang, Y.; Bodson, B.; Chen, J.; Francis, F. A Push–Pull Strategy to Control Aphids Combines Intercropping with Semiochemical Releases. *J. Pest Sci.* **2018**, *91*, 93–103. [CrossRef]
52. Ester, A.; Gut, J.; Van Oosten, A.M.; Pijnenburg, H.C.H. Controlling Aphids in Iceberg Lettuce by Alarm Pheromone in Combination with an Insecticide. *J. Appl. Entomol.* **1993**, *115*, 432–440. [CrossRef]

53. Beale, M.H.; Birkett, M.A.; Bruce, T.J.A.; Chamberlain, K.; Field, L.M.; Huttly, A.K.; Martin, J.L.; Parker, R.; Phillips, A.L.; Pickett, J.A.; et al. Aphid Alarm Pheromone Produced by Transgenic Plants Affects Aphid and Parasitoid Behavior. *Proc. Natl. Acad. Sci. USA* **2006**, *103*, 10509–10513. [CrossRef]
54. Agelopoulos, N.G.; Chamberlain, K.; Pickett, J.A. Factors Affecting Volatile Emissions of Intact Potato Plants, *Solanum Tuberosum*: Variability of Quantities and Stability of Ratios. *J. Chem. Ecol.* **2000**, *26*, 497–511. [CrossRef]
55. van Emden, H.F.; Dingley, J.; Dewhirst, S.Y.; Pickett, J.A.; Woodcock, C.M.; Wadhams, L.J. The Effect of Artificial Diet on the Production of Alarm Pheromone by *Myzus persicae*. *Physiol. Entomol.* **2014**, *39*, 285–291. [CrossRef]
56. Jing-Gong, X.; Feng, Z.; Yu-Ling, F.; Wei, K.; Guang-Xue, Z.; Zhong-Ning, Z. Behavioural Response of Aphids to the Alarm Pheromone Component (E)-β-Farnesene in the Field. *Physiol. Entomol.* **2002**, *27*, 307–311. [CrossRef]

insects

MDPI

Article

Factors to Male-Female Sex Approaches and the Identification of Volatiles and Compounds from the Terminalia of *Proholopterus chilensis* (Blanchard) (Coleoptera: Cerambycidae) Females in *Nothofagus obliqua* (Mirb.) Oerst. (Nothofagaceae) Forests in Chile

Diego Arraztio [1], Amanda Huerta [2], Andrés Quiroz [3,4], Washington Aniñir [5], Ramón Rebolledo [6] and Tomislav Curkovic [7,*]

[1] Doctorado en Ciencias Silvoagropecuarias y Veterinarias, Facultad de Cs. Agronómicas, Universidad de Chile, Santiago P.O. Box 1004, Chile; dsarrazt@uchile.cl
[2] Facultad de Ciencias Forestales y de la Conservación de la Naturaleza, Universidad de Chile, Santiago P.O. Box 9206, Chile; ahuerta@uchile.cl
[3] Departamento de Ciencias Químicas y Recursos Naturales, Universidad de La Frontera, Temuco 4811230, Chile; andres.quiroz@ufrontera.cl
[4] Centro de Investigación Biotecnológica Aplicada al Medio Ambiente (CIBAMA), Universidad de La Frontera, Temuco 4811230, Chile
[5] Doctorado en Ciencias de Recursos Naturales, Universidad de La Frontera, Temuco 4811230, Chile; w.aninir01@ufromail.cl
[6] Facultad de Ciencias Agropecuarias y Medio Ambiente, Universidad de La Frontera, Temuco 4811230, Chile; ramon.rebolledo@ufrontera.cl
[7] Facultad de Ciencias Agronómicas, Universidad de Chile, Santiago 8820808, Chile
* Correspondence: tcurkovi@uchile.cl

Simple Summary: *Proholopterus chilensis* (Cerambycidae: Cerambycinae) is a native species and an important damage agent in the Nothofagaceae forests of southern Chile, significantly affecting timber production in some areas, especially production of *Nothofagus obliqua*, with no sanitary management measures currently available. Therefore, the present research determined the predisposing factors for male–female sex approaches and identified, for the first time, some compounds that appear to mediate sexual behaviors that could be used to develop specific, safe, and effective control strategies. Interestingly, the chemical communication for mating purposes in *P. chilensis* preliminarily seems to include both possible "volatile or airborne" pheromones (eventually long-range ones) and short-range pheromonal compounds produced only by females, something, as far as we know, unique in Cerambycinae.

Abstract: During the spring–summer seasons between 2019 and 2023, in the localities of Maquehue (La Araucanía Region) and Llifén (Los Ríos Region), we collected 262 virgin *Proholopterus chilensis* (1 female/2.3 males) specimens emerging from the live trunks of *N. obliqua* trees, an atypical sex ratio in Cerambycidae, suggesting high male competition for females. Virgin specimens of both sexes were individually placed in panel traps in the field, capturing only males (n = 184) over the field study seasons and only in traps baited with females. This fact preliminarily suggests the *P. chilensis* females emit possible "volatile or airborne" pheromones (eventually being a long-range sex pheromone), something unusual in Cerambycinae, the subfamily to which it currently belongs, although the taxonomic status of the species is under debate. In Llifén and Santiago (Metropolitan Region), behavioral observations were conducted, which allowed us to define the conditions necessary for male–female encounters that were replicated when carrying out volatile captures (Head Space Dynamic = HSD) and collections of compounds from terminalias excised from females. In field trials, virgin females less than ten days old were more attractive than older ones and attracted males during the night, i.e., between 23:00 and 5:00 AM, when the ambient temperature exceeded 11.6 °C. The aeration of females under the conditions described above and subsequent analysis of extracts by

GC-MS allowed the identification of compounds absent in males and the control, including two oxygenated sesquiterpenes, a nitrogenous compound (C_{20}), and a long-chain hydrocarbon (C_{26}). From the terminalia extracts, hentriacontane, heptacosane, and heneicosyl, heptacosyl and docosyl acetates were identified by GC-MS, and their roles are unknown in the development of short-distance sexual behaviors, but they could serve to mark a trail leading the male towards the female in the final stage of approach for courtship/mating. Thus, we proposed the hypothesis that *P. chilensis* females emit both a long-range and a trail-pheromone, which, if confirmed, would be a rare case in this family. The specific identity of the compounds obtained by HSD, as well as the activities of these chemicals and those obtained from the terminalias, should be evaluated in future behavioral studies, as well as regarding their potential to attract males under field conditions. The current document is the first report on volatiles obtained from aerations and compounds extracted from female terminalia in *P. chilensis*.

Keywords: Cerambycinae female-produced sex pheromone; Cerambycidae sex ratio; Cerambycinae trail-pheromones

1. Introduction

Proholopterus chilensis (Blanchard) (Cerambycidae) is a long-horned beetle currently placed in the subfamily Cerambycinae, tribe Proholopterini, which includes species present only in Chile and Argentina [1]. In Chile, it is present between the regions of Maule (−35.4333, −71.6666) and Los Lagos (−41.4716, −72.9366) [2,3]. Its larval stage develops only in live trees of *Nothofagus obliqua* (Mirb.) Oerst., *Nothofagus alpina* Poepp. et Endl., and *Nothofagus dombeyi* (Mirb.) Oerst. (Nothofagaceae) [2]. The female oviposits on the bark, the eggs hatch in about a month, and their larvae develop in two years, boring extensive galleries in the trunks without killing the tree but severely affecting the quality of the valuable wood in commercial farms [4–7]. The last instar larva builds a pupal chamber [3] connected to the outside by an orifice plugged with wood chips, which the adult later removes upon emergence. The pupae are found between July and December, and adults emerge between November and February [8]. In Valdivia Province (Los Ríos Region), infestation (% of attacked trees) in *N. obliqua* ranged from 14% to ca. 44%; in the provinces of Osorno and Llanquihue (Los Lagos Region), it reached almost 10%, while in La Araucanía Region, it ranged between 7% and 29%. The damage can range between 40 and 50% of the total volume of the log. Consequently, the attacked wood can only be used for wood chips or as fuel. Because of that, currently, *P. chilensis* is considered one of the most important pests in renewal (secondary) *N. obliqua* forests [9–11].

The control of these xylophagous as juveniles is usually not feasible using conventional insecticides, biological control agents, or cultural or physical measures, and no sanitary management measures are currently available against *P. chilensis* [12,13]. On the other hand, in the last decade, the ethological control of adults has been studied for the management of Cerambycidae in the northern hemisphere, with promissory reports of mass trapping and mating disruption. These specific, innocuous, and effective control techniques are based on the use of intraspecific attractant compounds produced by the insects [14,15] and include aggregation pheromones (emitted by males to attract both sexes, frequently occurring in most Cerambycidae) or sex pheromones produced by the female to attract only the male (uncommon in cerambycids) [16–18]. In Cerambycidae, long-range sex pheromones have been described for the orientation of conspecifics from relatively long distances to the emitting source. Short-range pheromones have also been reported, usually long-chain alkane compounds [19,20], allowing both the meeting of sexes and the specific recognition required to achieve successful mating [21].

Insects produce and emit volatile pheromones into the environment during the calling phase (i.e., when releasing long range chemical cues), the first behavior conducted in male–female encounters for mating purposes [21]. Calling occurs under specific environmental (e.g., temperature) and endogenous (e.g., age) conditions, which can be determined and then reproduced in the laboratory to collect pheromones from individuals maintained in controlled environments and identified with various analytical techniques [19]. Calling has been studied in males of some cerambycids [22,23] but very rarely in females [24,25]. After calling, during the approach and the courtship for specific recognition [21], short-range pheromones emitted by females play a role in several cerambycids [20]. Regarding *P. chilensis*, however, the sexual behaviors and chemical signals used to achieve effective encounters and mating are entirely unknown. Therefore, the objectives of our study were (a) to study the emergence and sexual ratio of adults of *P. chilensis* in the field, (b) to characterize the conditions during the approaches between both sexes for courtship/mating purposes, and (c) to capture and identify the respective volatiles emitted by virgin females and compounds of their terminalia, which will allow for laying the groundwork for technological developments for *P. chilensis* management.

2. Methodology

2.1. Localities, Study Seasons and Insect ID

Field activities were carried out during the spring–summer seasons between 2019 and 2023 at sites with *Proholopterus chilensis*-infested *N. obliqua* trees in Maquehue (−38.8364, −72.6941) (La Araucanía Region), at a "Universidad de La Frontera" (UF) experimental station located in the central valley, and at a private property at Llifén (−40.1978, −72.2598) (Los Ríos Region), a colder locality located 200 km further south in the foothills of the Andes. In addition, during the COVID-19 pandemic in 2019–2020, observations of the behaviors of *P. chilensis* males and females were conducted in Santiago (−33.447487, −70.673676), Metropolitan Region. The collection of volatiles, extraction of terminalia components, and identification of chemical compounds from females were carried out at the Chemical Ecology Laboratory of the UF in Temuco (−38.73965 −72.59842), La Araucanía Region. Insect identification (ID) was performed by the entomologist authors using only external morphological characters and confirmed by comparison with previously identified material at the MEUCh collection, Entomological Museum, the University of Chile, where two vouchers (males) were deposited. The determination of sexes considered the following features: female antennae shorter than the specimen's body length, with antennae being longer in males; females with abdomens at least 25% wider than their elytras, as they are narrower or the same width in males; and males with two small spines merging from the margins of the last tergite, which is absent in females.

2.2. Emergence of *P. chilensis* from Trunks, Their Sex Ratio, and the Determination of the Attractive Sex and Attractiveness According to Age

To collect wild individuals from live *N. obliqua* trees, we fixed a prism (=cage, Figure 1) of ~10 × 20 cm made of wire (0.15 mm diameter) mesh (1.5 × 1.5 mm weft) over *P. chilensis* emergence orifices (~20 mm in greatest diameter) detected on the trunk (Figure 1). Approximately 800 cages (1–14 per tree) were installed between 2019 and 2023 (November to January each year) in Maquehue and Llifén, placed from ~0.4 to ~2 m in height. The cages were checked twice weekly (Maquehue) or daily (Llifén). When the cages contained specimens, these were taken to the laboratory and kept isolated in their respective cages at room temperature, with a cotton wick soaked in a 5% sugar solution used to feed the adults [25]. In this way, virgin *P. chilensis* of known age were sexed (as described by [3,7]) and kept for further use in indoor and field tests.

Virgin females or males of *P. chilensis* (aged 0–7 days from emergence) were individually placed in cages, as described above (one specimen/cage, one cage/trap). Each cage was stapled in the center of an interception trap (Alpha Scent, CA, USA, Figure 2). Traps have been successfully used in previous studies for adult cerambycid collections [26] and were hung in the field on live *N. obliqua* trees at a height of ~2 m, separated by ~20 m from each other, during the adult flight period (December to January). Three treatments were contrasted synchronously: female, male, and a control (trap without insects). After seven days in the field, the cumulative catches of *P. chilensis* were counted and the individuals sexed. A completely randomized block design (n = 8 replicates/treatment) was considered over three seasons between 2019 and 2023 (during the 2020–2021 season, no field work was possible because of the COVID-19 mandatory lock out). Given the structure of these results, no statistical analysis was conducted because in 2 out of 3 treatments catches were zero for all replicates, not providing variance for contrast purposes.

Figure 1. Wire mesh cages (n = 14) placed on a *N. obliqua* tree for capturing virgin *P. chilensis* adults during emergence. Sawdust produced by the larval activity can be seen accumulated at the base of the trunk (**left**); a *P. chilensis* emergence hole blocked with wood chips (**right**).

Figure 2. A *Proholopterus chilensis* female placed in a cage stapled to a flight interception trap (**left**); males captured in the trap's collecting cup (**right**).

To determine the effect of a female's age on her ability to attract males, we tested three treatments synchronously: individuals from 0 to 4, 5 to 9, or 10 to 14 days post-emergence from the trunk. Each replicate consisted of a trap arranged as described above, containing one female, which was kept for three days in the field during the peak of adult activity (January/2023). The cumulative captured congeners in three days were then counted and sexed. A completely randomized block design (n = 3) was considered, and the catches, trap, and day (CTD) were contrasted with the Kruskal–Wallis non-parametric and Tukey tests for mean comparisons ($p = 0.05$) [27].

2.3. Successful Behavioral Observations, Environmental Conditions, and Hours during P. chilensis Female–Male Sex Approaches

In the 2019–2020 season, at Santiago, transparent plastic cylinders ~70 cm long × ~13 cm section were used as indoor observation arenas. On the bottom, a mesh cotton strip was placed to allow the beetles to move and avoid the stress that they showed when trying to walk on the bare surface, as determined by preliminary observations. Another piece of mesh was arranged as a barrier blocking the section in the center of the cylinder, which allowed airflow (~0.5 m/s in the middle) generated by a fan located at one end, but not physical contact between the female (n = 24), placed upwind of the barrier) and the male (downwind). In addition, from 2021 to 2023, virgin females from 0 to 4 days old were individually tested in the field (Llifén) using (a) cages placed on interception traps (n = 12), as described in the previous section, and (b) free (not confined) specimens placed directly on the base of the trunk (n = 10). Only data for when males flew and actively approached the active (moving upward and downward and grooming themselves) females were considered for the analysis. Measurements included time (measured every 30 min), temperature, relative humidity (Cima control data logger OM-EL-USB-2, Omega brand, Deckenpfronn, Germany), light intensity (VWR International luxmeter, Radnor, PA, USA), and wind speed (EXTECH Instrument anemometer, Knoxville, TN, USA). A flashlight with a red filter was used during night observations (19:00 and 6:00 AM), allowing for the registering of the activity of males approaching females. No observations on windy or rainy nights were conducted in Llifén.

2.4. Capture of Volatiles and Collection of Compounds from the P. chilensis Terminalia of Females

Four head space dynamic (HSD) runs were made on *P. chilensis* females (n = 8) in the laboratory for five hours, i.e., between 23:00 and 4:00 AM, in darkness, under temperature and RH favorable for the approaching/encountering of both sexes (see results below). The HSD runs were conducted on individual females (n = 2) or groups of 2 or 4 (n = 1 ea.) specimens. They were confined to a 1325 mL Pyrex glass chamber, together with a fresh bark base obtained from live *N. obliqua* trunks, to facilitate insect displacement and avoid stress observed preliminarily in bare flasks, while the respective controls included either only bark or a male and bark (n = 4 in both). The capture of low-molecular-weight compounds was performed using a positive/negative pressure system inside the glass chamber, as described by [28], by pumping previously purified air over the insect and then draining it with another suction pump (flow rate of 1 L/min in both) into a Porapak Q column Water Corporation, Milford, MA, USA (100 mg; 80 Å 100 mesh). The compounds were then desorbed from the Porapak with hexane (GC grade; Merck, Darmstadt, Germany) (2 mL) or dichloromethane (GC grade; Merck, Darmstadt, Germany) (2 mL), obtaining two different extracts (not mixed). Before gas chromatography–mass spectrometry analysis (GC-MS), each extract was independently concentrated under a flow of nitrogen gas (70 mL/min) to 100 μL and analyzed [28,29].

The terminalia [30] exposed by four females (not observed in males) during the putative calling were excised with dissecting scissors and immediately fixed in a vial with hexane (GC grade). Then, before the filtration of the hexane containing the terminalia, Na_2SO_4 was added to remove water up to an extracted volume of 2000 μL and kept in an amber vial at −4 °C until GC-MS. In this study, the solvent alone was used as the control.

2.5. Identification of Volatiles and Terminalia Compounds from P. chilensis Females

The identification of compounds obtained from aerations and terminalia extracts was performed by GC-MS using a Thermo Electron Model Trace 1300 coupled to an ISQ 7000 Thermo Electron quadrupole mass spectrometer, with an integrated Xcalibur 4.1 data system (Thermo Fisher Scientific Inc., Waltham, MA, USA). In total, 1 μL of each extract was injected. An RTx-5MS column (30 m × 0.25 mm × 0.25 μm, Restek Corporation, Bellefonte, PA, USA) with helium as a carrier gas (1.0 mL/min) was used. The GC-MS procedure was performed with an initial temperature of 40 °C, which was maintained for 2 min and then increased at a rate of 5 °C/minute until 250 °C. The injector, transfer line, and ion trap temperatures were maintained at 250 °C. The mass detector used ionization energy of 70 eV. The mass range was from 30 to 500 amu. Compounds were identified by comparison with those in the NIST database ver. 2.3 (National Institute of Standards and Technology, Gaithersburg, MD, USA), and their retention indices were compared with those reported in the literature. Retention indices were calculated using an alkane series standard, as described by [31]. When identification by library could not be carried out, identification by mass/charge analysis was performed. A series of alkanes ranging from C_8 to C_{18} was used and then extrapolated based on the seriality of the alkanes. The formula used was the following:

$$K = 100(Z-1) + \frac{\mathrm{LogRt}(X) - \mathrm{LogRt}(Z-1)}{\mathrm{LogRt}(Z+1) - \mathrm{LogRt}(Z-1)}$$

where Rt is the retention time, X is the unknown compound, Z − 1 is the alkane preceding the unknown compound, and Z + 1 is the alkane following the unknown compound.

3. Results

3.1. Emergence of Virgin P. chilensis from Trunks, Their Sex Ratio, and the Determination of the Attractive Sex and Attractiveness According to Age

During the four study seasons (2019–2020, 2021–2022, 2022–2023, and 2023–2024), 262 adults of *P. chilensis* emerging from trunks were collected from late November to late December (Maquehue) and from mid-December to mid-January (Llifén). A total of 79 females and 183 males were obtained, with a 1:~2.3 sex ratio. In interception traps baited with *P. chilensis* virgin females, 184 conspecific adult males were captured in total over the course of the study (Table 1), without capturing a single female. In contrast, in traps baited with live males and control traps, there were no captures of this species.

Table 1. Male *Proholopterus chilensis* specimens captured in "traps per treatment" (n/season) during the 2019–2023 period.

	2019–2020 n = 4	2021–2022 n = 3	2022–2023 n = 4	Total Captures
Control	0	0	0	0
Male	0	0	0	0
Female	58	80	46	184

Females of *P. chilensis* attracted males throughout the tested age range, but significantly fewer (p = 0.036) for individuals older than nine days (2.7 ± 0.6 CTD of males) compared to those of 5–9 days of age (8.3 ± 1.5 CTD), suggesting an optimum of attraction in young adult specimens, since females of 0–4 days of age captured 6.7 ± 1.5 CTD, with no differences compared with the other two age categories. For this reason, we used individuals of less than ten days of age in subsequent studies.

3.2. Successful Behavioral Observations, Environmental Conditions, and Hours during P. chilensis Female–Male Sex Approaches

All females placed freely on trunks, in cages, or in the plastic cylinders developed characteristic behaviors, including the projection of the terminalia, not observed among males. While stationary, females placed on trunks projected the terminalia and dragged it on the bark. After a few seconds doing so, the *P. chilensis* females retracted the terminalia and walked on the substrate for short distances (cm), stopping and projecting the terminalia again. This sequence was repeated many times by the females while climbing up and down the trunk. These behaviors were only observed during the period when males were actively flying (or walking, n = 24, in cylinders) during night hours, sometimes even around the box where the cages with females were taken to the field to set trials. Given the nocturnal sexual activity, only one approaching male was observed in detail when flying toward the female, showing casting flight, as described in [25], but males were observed when flying to a light source; in this case, the male flew in a straight line. Arriving males (n = 11) landed on the trunk and walked following the same route apparently marked by the female with her terminalia, whereas they (n = 10) were apparently disoriented after landing near the cages. Oriented approaches from males to putative calling females of *P. chilensis* occurred at 0–1 m/s (range of wind speed during observations), 30–50% RH, 0.1–0.3 lux, and 14.5–25.5 °C and between 22:30 and 5:00 AM in 2019 in Santiago, as well as at 0–1 m/s, 60–86% RH, 0–4 lux, and 11.6–21.3 °C and between 23:00 and 5: 00 AM in the period 2021–2023 in Llifén (Figure 3: pooled data obtained during observations of females placed freely on trunks, in cages, and in plastic cylinders).

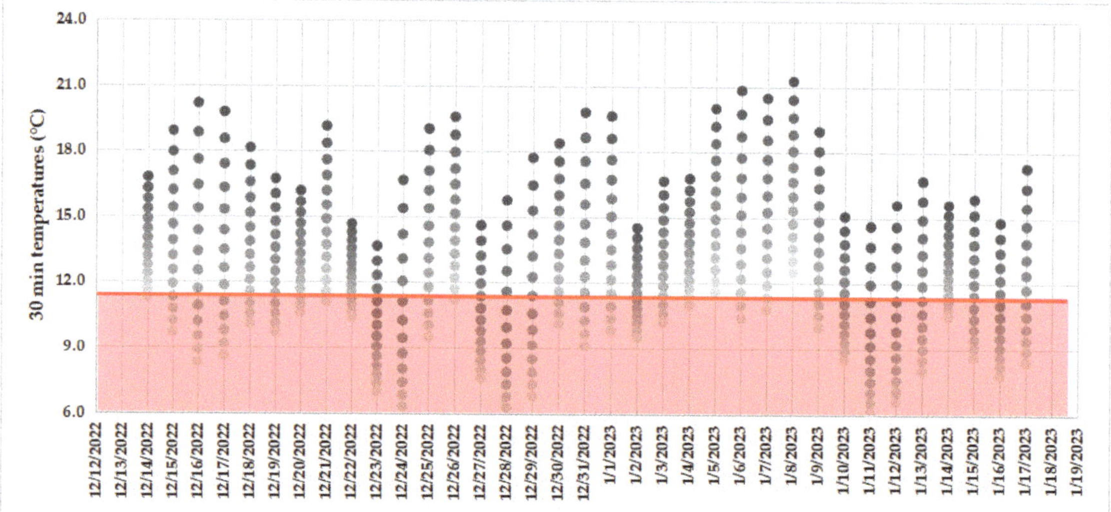

Figure 3. Pooled data (measured in Santiago—females in cylinders—and Llifén—field trials with females in cages and free on the trunk) on temperatures during the oriented approaches of males to females of *P. chilensis* taken every 30 min (dots in each column inside the graph) for each night of observation. The grey tone of the dot varies from light to dark as the night hour progresses. The pink area delimited by the red line represents the temperatures in which no male–female approaches were observed.

3.3. *Identification of Volatiles and Terminalia Compounds from P. chilensis Females*

The analysis of chromatograms obtained from aeration extracts showed that the following compounds (Table 2) were present in females and absent in the respective controls: two oxygenated sesquiterpenes, a nitrogenous compound (C_{20}), and a long-chain hydrocarbon (C_{26}), plus an unidentified compound, all present at a 24.1 min retention time. In one female aerated individually, all five compounds were detected, and only three were found in the other one (those eluted at 24.1, 26.57, and 41.21 min); in the group of four females, only two compounds were found (at 24.1 and 26.57 min), and none were detected in the other group (a pair).

The fragments that explain this diagnosis for the family compound with a retention time of 24.87, i.e., an oxygenated sesquiterpene, were C_4H_9 for the fragment 57 m/z, and $C_{10}H_{13}O_2$ for the fragment 165 m/z. For the family compound with a retention time of 26.57, the representative fragments were C_5H_9O for the 85 m/z fragment, C_5H_9 for the fragment 69 m/z, C_6H_{11} for the fragment 83 m/z, and C_4H_9 for the fragment 57 m/z. In the case of the nitrogenated compound with a retention time of 41.21, the representative fragments correspond to $C_{15}H_{16}N$ for the fragment 210 m/z and $C_{20}H_{27}N$ for the fragment

281 *m/z*. We were not able to identify a more precise result because the parameters for a specific ID were too low.

Table 2. Analytical features and compounds identified at the generic level by GC-MS in aeration extracts obtained by HSD from virgin females of *Proholopterus chilensis*.

Retention Time (min)	Family Compound Name [1]	Area [2] (%)	Main Mass Fragments (*m/z*)
24.1	Not identified	6.23	57 (100), 93 (76), 81 (70), 96 (63), 73 (56), 95 (53), 107 (48), 79 (46), 41 (46), 91 (40)
24.87	Oxigenated sesquiterpene 1	53.41	57 (100), 161 (79), 71 (59), 45 (54), 41 (39), 203 (37), 85 (36), 133 (32), 77 (31)
26.57	Oxigenated sesquiterpene 2	4.30	57 (100), 69 (89), 205 (83), 71 (57), 83 (40), 111 (37), 85 (100), 70 (35), 41 (30)
41.21	Nitrogenated compound (C_{20})	33.17	210 (100), 211 (17), 281 (94)
43.77	Long-chain hydrocarbon (C_{26})	2.88	43 (100), 57 (700), 71 (680), 85 (600), 41 (418), 71 (144)

[1]: Due to the fragmentation pattern, only the chemical family can be suggested. [2]: The peak chromatogram relative area.

An analysis of the extracts obtained from the terminalias of virgin *P. chilensis* females is presented in Table 3. The molecular weights, formulas, and retention indices of the compounds found are reported as fitting parameters > 90%. We detected high-molecular-weight linear alkanes, representing 67.48% (relative areas), and long-chain esters (29.6%). This identification was based on the characteristic fragmentation patterns of both classes of compounds but was not achieved for the compound whose retention time was 44.21 min.

Table 3. Analytical features and compounds identified by GC-MS in terminalia extracts obtained from virgin females of *Proholopterus chilensis*.

Retention Time (min)	Compound Name	Area [1] (%)	MW [2] (g/mol)	Chemical Composition	RI [3]	Main Mass Fragments (*m/z*)
40.03	Heneicosyl acetate	14.12	354	$C_{23}H_{46}O_2$	2515	43 (100), 83 (88), 57 (86), 97 (81), 55 (10), 69 (62), 71 (54), 61 (49), 41 (41), 70 (39)
43.29	Docosyl acetate	8.78	368	$C_{24}H_{48}O_2$	2618	43 (100), 57 (64), 97 (59), 83 (58), 55 (57), 61 (50), 69 (46), 111 (34), 41 (32), 71 (31)
44.21	Not identified	2.90	-	-	-	130 (100), 117 (59), 131 (54), 95 (36), 55 (30), 57 (29), 43 (28), 81 (27), 71 (26)
44.82	Heptacosane	22.69	380	$C_{27}H_{56}$	2700	57 (100), 71 (85), 85 (63), 43 (50), 99 (26), 55 (24), 41 (19), 69 (19), 113 (16), 83 (15)
48.21	Heptacosyl acetate	6.71	438	$C_{29}H_{58}O_2$	3071	97 (100), 57 (94), 83 (91), 69 (77), 55 (75), 43 (61), 71 (59), 111 (54), 41 (40), 85 (39)
48.87	Hentriacontane	44.79	436	$C_{31}H_{64}$	3100	57 (100), 71 (86), 85 (64), 43 (48), 99 (27), 55 (23), 69 (18), 41 (17), 113 (15), 83 (13)

[1]: peak chromatogram relative area; [2]: molecular weight; [3]: retention index.

4. Discussion

The collection of *P. chilensis* adults from trunks occurred in periods shorter than those reported in the literature [6,8]. The sex ratio found (1:2.3) among them is uncommon in Cerambycidae breeding, but no previous records exist for *P. chilensis*. Studies of other Cerambycidae species [32,33] mostly show parity in sexes (about 1:1) or sex ratios favorable to females. However, there are some reports of male-favorable sex ratios (1:2.2) obtained in cerambicinaes collected on flowers or at feeding or mating sites [34] and somewhat lower (1:1.5) ratios in species characterized by showing aggressive male competition for females, performing pre-mating walks (something observed in *P. chilensis*), or post-mating guarding [35]—behavior described by [36], as well as in another Chilean Cerambycidae species. We cannot rule out the possibility that some sampling effects affected our sex ratio results, but at least no sampling bias was expected regarding emergence hole size (similar between sexes) and the hole position along the trunk (random) used as criteria for installing cages. Moreover, cage placement started long before the onset and until the end of the flight and presence of adults in the field, discharging some possible temporal bias for emergence. Among the emerged adults, all were alive, and no mortality attributed to malformation, diseases or arthropod parasitoids, or predators was observed that might affect one sex more than the other one.

The fact that only males were captured in traps baited with virgin females, as well as the absence of *P. chilensis* captures in male-baited and control traps, preliminarily suggests that the females emit possible "volatile or airborne" pheromones to attract only males, a characteristic reported to date only in the subfamilies Prioninae and Lepturinae [20,30], but practically unknown in the subfamily Cerambycinae [18]. It is, however, important to point out that the taxonomic status of *P. chilensis* is uncertain: it has been classified as a Cerambycine [37], a Lepturinae [38], a Cerambycinae [1], and an *incertae sedis* [39] and mentioned as a case of misplaced taxa [40]. Nevertheless, in the last decade, there have been reports suggesting a deviation from this pattern in a few cerambicinaes. For instance, a compound collected in aerations of *Xylotrechus quadripes* (Chevrolat) females attracted only males, although it showed electroantennographic responses in both sexes [41]. Furthermore, a compound emitted by flowers of the orchid *Disa forficaria* Bolus attracted only males of *Chorothyse hessei* Quentin and Villiers that pollinate this plant [42], and two studies of cerambycid pest species in Asia, *Aromia bungii* Faldermann [43] and *Rhytidodera simulans* White [44], with extracts in the laboratory, did the same, suggesting a sex pheromone produced by the female. Thus, to the best of our knowledge, our preliminary findings in *P. chilensis* correspond to one of the first studies in Cerambycinae in which behavioral evidence (like also proposed for other species [44,45]) preliminarily suggests the existence of female-produced "volatile or airborne" pheromones, eventually acting as long-range sex pheromones (Table 2).

The mechanisms that could explain the *P. chilensis* female age effect on attractiveness to males are not known. However, a study of a Japanese cerambycid suggests that sexual maturity is reached in females after five days of adult life [46], but it should be considered that the *P. chilensis* females were prevented from mating (most were confined), which may extend the calling period in individuals that remain virgins. Additionally, in most studies with methodologies equivalent to the one used in ours, females up to 15 days from emergence were used, a period in which most maintained attractive capacity [45,47]. The literature points to calling behaviors occurring that start at one week of life in some Cerambycidae females: in *Anoplophora glabripennis* (Motschulsky) (Lamiinae), age did not affect calling in specimens aged 7–34 days old [48], and this behavior occurred between 7 and 24 days in *Glycobius speciosus* (Say) (Cerambycinae) [49].

Behaviors like those observed in *P. chilensis* females during the putative calling (terminalia projection and contact with the bark) have been reported in Cerambycidae [50,51].

The ultimate causes of these behaviors are unknown, but the fact that at least one male was observed approaching a (putative) calling female by casting flight in the field [zigzagging in 25] suggests responses to airborne semiochemicals, eventually including responses to female volatiles, host cues, or both. These responses did not occur below 11.6 °C, a factor that has also been reported to be restrictive in other cerambycids [52]. However, it is possible that the temperature during field tests did not exceed the minimum for either male-oriented flight (or walking) and/or for the performance of calling in females, so we cannot conclude that this reached either threshold. Methodologically similar studies [25], with males and females of a diurnal Cerambycidae species, also established a similar range of hours per day (5 h, with females prevented from mating during the hours of study).

Regarding the C_{20} and the C_{26} unspecifically identified compounds, these do not seem to be very promising as long-range chemicals since their molecular weight suggests low volatility (especially C_{26}), making sesquiterpenes better candidates. With regard to the terpene compounds collected by HSD and identified generically, it should be noted that these types of chemicals have been reported as one of the main structural types of the pheromones present in cerambycid species [20] or as minor components of their pheromonal blends [53,54], but unlike our results, all were emitted by males. For instance, among monoterpenes, in *Megacyllene caryae* (Gahan), (S)-(−)-limonene has been identified as a component of the male-produced odor blend [55]. In *Megacyllene antennata* (White), (S)-α-terpineol, (S)-limonene, and terpinolene have been reported as well [56]. On the other hand, sesquiterpene hydrocarbons such as β-elemene, β-caryophyllene, α-humulene, and α-farnesene have been found on the surfaces of male and female elytras acting as contact pheromones, being them attractive to both sexes [57]. Furthermore, sesquiterpenes as fuscumol acetate ((E)-6,10-Dimethyl-5,9-undecadien-2-yl acetate; vapor Pressure: 1.61×10^{-2} mm Hg at 25 °C) have been reported to be attractants produced by males of many cerambycid species [58], mainly in Cerambycinae, and this particular compound was significantly attractive and identified by our group as probable components of a long-range aggregation pheromone for the South American cerambycid *Eryphus laetus* (Bl.) (Cerambycinae) [26]. The same article notes that under field conditions, this compound and two other sesquiterpenes (fuscumol and geranyl acetate) tested in the field individually did not attract *P. chilensis* despite conducting the experiment in a heavily infested site during their flying time, suggesting that these chemicals have no biological activity on this species. There are also some studies of nitrogen-containing semiochemicals (e.g., amides) for some compounds identified in arthropods [59,60] and at least one report of long-range pheromones produced by the female in the cerambicinae *Vesperus xatarti* Mulsant [61]. Finally, several C_{26} hydrocarbons (e.g., hexacosene) have been reported as cuticular pheromonal compounds (i.e., with contact but not long-range activity), but unlike in our study, they were produced by males of the species *Pidonia* spp. (Lepturinae) [62]. This is the first report of possible volatile or airborne chemicals and possible contact compounds obtained from *P. chilensis* females by HSD. It is now necessary to advance with identifying and evaluating these compounds in terms of their behavioral role and, eventually, as potential long- or short-range attractants.

Among the chemicals found in the terminalia, the following are considered relatively more volatile [63–65]: heneicosyl acetate, also reported in the profile of pods of *Vigna radiata* (L.) R. Wilczek (Fabaceae) and in the pre-anal glands of the lizard *Hemidactylus flaviviridis* Ruppell (Chordata: Squamata: Gekkonidae), and heptacosyl acetate, reported as an indicator compound for nematode infestation (*Anguina funesta* Price, Fisher and Kerr, Tylenchida: Anguinidae) in Poaceae [66]. As far as we know, no insect-associated pheromonal activity has been reported for these two chemicals [58]. On the other hand, docosyl acetate has been identified in labial glands of male bumblebees of the genus *Bombus* (Hymenoptera: Apidae), from which the sex pheromone is emitted [67], and has been listed as a minor component of the sex pheromone of the lepidopteran *Myelois cribrella* Hubner (Pyralidae), without electroantennographic activity [68]. The other compounds identified from terminalias have lower volatility: hentriacontane (PV: 1.40×10^{-11} mm Hg at 25 °C) has been reported in low proportions in the cuticles of adults of both sexes of *Cerambyx wellensi* Kuster (Cerambycidae: Lamiinae) [69], whose mixture of components evidenced roles in sexual behavior, but at a short range; heptacosane (PV: 2.81×10^{-7} mm Hg at

25 °C) has been detected in the elytra and thorax of females of two species of *Tetropium* (Spondylinidae) [70] and is recognized as a main pheromonal (contact) chemical, as it stimulates mating. Compounds of this type, with short-range activity, could serve as marking pheromones (trail-pheromones: widely described for ants and termites [71]) to guide males of *P. chilensis* walking on the substrate presumably previously marked by the female during calls, as observed in our field evaluations. This is like the behavior described in *Nadezhdiella cantori* (Hope) (Cerambycinae, [72]). Mixtures of long-chain alkanes (similar to those reported in Table 3) have also been described with short-range activity in females of several Cerambycidae subfamilies [20,51], which are detected by the antennae, maxillary palps, or tarsi of males in contact with female bodies or the marked substrate, something observed in males during our field tests. Together with long-range pheromones, these types of compounds complement a chemical communication system that enables specific encounters and recognition in some cerambycids [20,51]. These compounds might function during courtship, but no reports were found for Cerambycidae. Our study constitutes the first identification of compounds from the terminalia of *P. chilensis*, some of which could be part of the biosynthetic pathway of final compounds, and others could be degradation products [20] or have direct roles in the respective sexual behaviors and male orientation [51], which should be tested and elucidated in future research. All these compounds should be evaluated for their potential utility for *P. chilensis* management as well. A similar complex mate-finding system has been reported in cerambycids for *A. glabripennis* by [50,73], where males first attract females using a long-range pheromone, and then the male follows a trail-pheromone released by the female on the substrate. Interestingly, the trail marked with the latter compound is avoided by conspecific females and might act as a spacing pheromone to avoid intraspecific competition [50]. The putative trail-pheromone compound(s) produced by *P. chilensis* would help the arriving male to follow the female that first walks up and then down on the trunk. According to our findings, it is possible that chemical communication in the male–female approach/encounter in *P. chilensis* could be a hybrid. No reports like that were found for cerambycids, but a similar system has been described in some *Callosobruchus* beetles (Chrysomelidae) where females produce both types of pheromones [74], involving, first, a long- (as informed in compounds similar to those presented in Table 2) and then a short-range pheromone serving for recognition and, eventually, for courtship. In addition, both types of compounds are produced by the *P. chilensis* female, a mechanism that needs to be confirmed and, as far as we know, is not reported in the literature for cerambycids.

5. Conclusions

Proholopterus chilensis that emerged from live *Nothofagus obliqua* trunks showed a sex ratio highly favorable to males, suggesting intense competition for females.

Only females of *P. chilensis* are attractive to males, which were not attractive to either sex, something unusual in Cerambycinae.

Females of *P. chilensis* attract males during the night, a period that can extend for up to 5 h in the absence of mating, and their attractiveness depends, at least, on their age and environmental temperature.

The analysis of extracts from aerations of *P. chilensis* females allowed the generic identification of four volatile compounds, while five compounds were specifically identified from their terminalias. These compounds formed a complex and complementary chemical communication system in which the first ones would attract males from relatively long distances. Then, the compounds found in the terminalia helped the female mark the trail (on the trunk) towards her location during the putative calling/courtship phase so that the encounter with males could occur. This hypothesis should be further tested, as should the compounds found for their possible role in sexual behavior and as potential attractants for *P. chilensis* management.

Author Contributions: Conceptualization, T.C., A.H., and D.A.; Methodology, T.C., A.Q., D.A., and W.A.; Statistical analysis, T.C. and D.A.; Chemical procedures and analysis, A.Q., W.A., and D.A.; insect ID, T.C., R.R., and D.A.; Supervision, T.C., D.A., R.R., and A.H.; Resources, T.C., A.H., R.R., and A.Q.; Writing—original draft preparation, T.C., D.A., A.H., and R.R.; Writing—review and editing, T.C., D.A., A.H., A.Q., and W.A.; Project administration, T.C. and D.A.; Funding acquisition, T.C., A.H., R.R., and A.Q. All authors have read and agreed to the published version of the manuscript.

Funding: "Fondo de Investigación del Bosque Nativo" from "Corporación Nacional Forestal" (FIBN-CONAF); grant "Manejo sostenible de *Proholopterus chilensis* mediante la identificación y uso de semioquímicos y técnicas silvícolas en Roble-Raulí-Coihue" FIBN 021/2018 (T.C.).

Informed Consent Statement: Informed consent was obtained from the owner of private property involved in the study.

Data Availability Statement: The chromatograms obtained in this study are available on request from the corresponding author.

Conflicts of Interest: The authors declare no conflicts of interest. The funders had no role in the design of this study; the collection, analysis, or interpretation of data; the writing of the manuscript, or the decision to publish this research.

References

1. Monne, M. Catalogue of the Cerambycidae (Coleoptera) of the Neotropical Region. Part I. Subfamily Cerambycinae. *Zootaxa* **2005**, *946*, 1–765. [CrossRef]
2. Barriga, J.E.; Curkovic, T.; Fichet, T.; Henríquez, J.; Macaya, J. Nuevos antecedentes de coleópteros xilófagos y plantas hospederas en Chile, con una recopilación de citas previas. *Rev. Chil. De Entomol.* **1993**, *20*, 65–91.
3. Artigas, J.N. Familia Cerambycidae. In *Entomología Económica: Insectos de Interés Agrícola, Forestal, Médico y Veterinario*, 1st ed.; Ediciones Universidad de Concepción: Concepción, Chile, 1994; Chapter 27.6, pp. 120–160.
4. Gara, R.; Cerda, L.; Krahmer, E. Sobre la emergencia y vuelo de dos cerambícidos chilenos: Holopterus Chil. Y Cheloderus childreni. *Turrialba* **1978**, *18*, 117–121.
5. Cabrera, P. Impacto de *Holopterus chilensis* (Coleoptera: Cerambycidae) en renovales de *Nothofagus obliqua* en la provincia de Valdivia, Chile: Avances hacia la evaluación del daño. *Bosque* **1997**, *18*, 9–19. [CrossRef]
6. Baldini, A.; Alvarado, A. *Manual de Plagas y Enfermedades del Bosque Nativo en Chile*, 1st ed.; FAO: Santiago, Chile, 2008; 240p.
7. Diaz, A. Gusanera del Roble: *Holopterus chilensis*. In *Entomología Forestal en Chile*, 1st ed.; Lanfranco, D., Ruiz, C., Eds.; Ediciones Universidad Austral de Chile: Valdivia, Chile, 2010; pp. 337–348.
8. Baldini, A.; Pancel, L. *Agentes de Daño en el Bosque Nativo*; Editorial Universitaria: Santiago, Chile, 2002; 412p.
9. Ruiz, C.; Montalva, C.; González, M. Native Forest Health in Chile: Toward a Strategy of Sustainable Management. In *Forest Pest and Disease Management in Latin America*; Estay, S., Ed.; Springer International Publishing: Cham, Switzerland, 2020; Chapter 7, pp. 89–103. [CrossRef]
10. Alzamora, R.M.; Apiolaza, L.A.; Ruiz, C.; Lanfranco, D. Site, Tree and Silvicultural Factors Influencing the Infestation of Xylophagous Insects on Nothofagus Forests. *J. For. Res.* **2020**, *9*, 226. [CrossRef]
11. Alvarado, T. Análisis de Factores de Predisposición de *Proholopterus chilensis* (Coleoptera: Cerambycidae) en el Tipo Forestal Roble-Raulí-coihue, Región de la Araucanía. Bachelor Thesis, Universidad de Chile, Santiago, Chile, 2020; 48p.
12. Curkovic, T. La sierra del manzano: Antecedentes biológicos y bases para el desarrollo del control de adultos. *Aconex* **2008**, *98*, 10–14.
13. Rodríguez-González, A.; Pelaez, H.J.; González-Nuñez, M.; Casquero, P.A. Control of egg and neonate larvae of *Xylotrechus arvicola* (Coleoptera: Cerambycidae), a new vineyard pest, under laboratory conditions. *Aust. J. Grape Wine Res.* **2017**, *23*, 112–119. [CrossRef]
14. Sánchez-Husillos, E.; Etxebeste, I.; Pajares, J. Effectiveness of mass trapping in the reduction of *Monochamus galloprovincialis* Olivier (Col.: Cerambycidae) populations. *J. Appl. Entomol.* **2015**, *139*, 747–758. [CrossRef]
15. Sweeney, J.; Silk, P.J.; Rhainds, M.; MacKay, W.; Hughes, C.; Van Rooyen, K.; MacKinnon, W.; Leclair, G.; Holmes, S.; Kettela, E.G. First report of mating disruption with an aggregation pheromone: A case study with *Tetropium fuscum* (Coleoptera. Cerambycidae). *J. Econ. Entomol.* **2017**, *110*, 1078–1086. [CrossRef]
16. Mitchell, R.F.; Graham, E.E.; Wong, J.C.; Reagel, P.F.; Striman, B.L.; Hughes, G.P.; Hanks, L.M. Fuscumol and fuscumol acetate are general attractants for many species of cerambycid beetles in the subfamily Lamiinae. *Entomol. Exp. Appl.* **2011**, *141*, 71–77. [CrossRef]
17. Mitchell, R.F.; Reagel, P.F.; Wong, J.H.C.; Meier, L.R.; Diaz Silva, W.; Mongold-Diers, J.; Millar, J.G.; Hanks, L.M. Cerambycid beetle species with similar pheromones are segregated by phenology and minor pheromone components. *J. Chem. Ecol.* **2015**, *41*, 431–440. [CrossRef] [PubMed]

18. Hanks, L.M.; Millar, J.G. Sex and aggregation-sex pheromones of cerambycid beetles: Basic science and practical applications. *J. Chem. Ecol.* **2016**, *42*, 631–654. [CrossRef] [PubMed]
19. Millar, J.G.; Haynes, K.F. *Methods in Chemical Ecology: Chemical Methods*; Springer: New York, NY, USA, 1998; Volume 1, p. 390.
20. Millar, J.G.; Hanks, L.M. Chemical Ecology of cerambycids. In *Cerambycidae of the World*; Wang, Q., Ed.; CRC Press: Boca Raton, FL, USA, 2017; pp. 161–208.
21. Matthews, R.W.; Matthews, J.R. *Insect Behavior*; Springer: New York, NY, USA, 2010; 520p.
22. Lacey, E.S.; Ray, A.M.; Hanks, L.M. Calling behavior of the cerambycid beetle *Neoclytus acuminatus acuminatus* (F.). *J. Insect Behav.* **2007**, *20*, 117–128. [CrossRef]
23. Lemay, M.A.; Silk, P.J.; Sweeney, J. Calling behavior of *Tetropium fuscum* (Coleoptera: Cerambycidae: Spondylidinae). *Can. Entomol.* **2010**, *142*, 256–260. [CrossRef]
24. Barbour, J.; Cervantes, E.; Lacey, E.; Hanks, L. Calling behavior in the primitive longhorned beetle *Prionus californicus* Mots. *J. Insect Behav.* **2006**, *19*, 623–629. [CrossRef]
25. Curkovic, T.; Ferrera, C. Female calling and male flight orientation and searching behaviors in *Callisphyris apicicornis*: Evidence for a female-produced sex attractant pheromone. *Int. J. Agric. Nat. Resour.* **2012**, *39*, 147–158. [CrossRef]
26. Curkovic, T.; Arraztio, D.; Huerta, A.; Rebolledo, R.; Cheuquel, A.; Contreras, A.; Millar, J.G. Generic Pheromones Identified from Northern Hemisphere Cerambycidae (Coleoptera) Are Attractive to Native Longhorn Beetles from Central-Southern Chile. *Insects* **2022**, *13*, 1067. [CrossRef]
27. Conover, W.J. *Practical Nonparametric Statistics*; John Wiley and Sons, Inc.: Hoboken, NJ, USA, 1999; 608p.
28. Parra, L.; Mutis, A.; Ceballos, R.; Lizama, M.; Pardo, F.; Perich, F.; Quiroz, A. Volatiles Released from *Vaccinium corymbosum* Were Attractive to *Aegorhinus superciliosus* (Coleoptera: Curculionidae) in an Olfactometric Bioassay. *Environ. Entomol.* **2009**, *38*, 781–789. [CrossRef]
29. Mutis, A.; Parra, L.; Manosalva, L.; Palma, R.; Candia, O.; Lizama, M.; Pardo, F.; Perich, F.; Quiroz, A. Electroantennographic and Behavioral Responses of Adults of Raspberry Weevil *Aegorhinus superciliosus* (Coleoptera: Curculionidae) to Odors Released from Conspecific Females. *Environ. Entomol.* **2010**, *39*, 1276–1282. [CrossRef]
30. Monné, M.L.; Monne, M.A.; Wang, Q. General Morphology, Classification and Biology of Cerambycidae. In *Cerambycidae of the World*; Wang, Q., Ed.; CRC Press: Boca Raton, FL, USA, 2017; pp. 2–70.
31. Kovats, E.; Keuleman, A. The Kovats retention index system. *Rep. Anal. Chem.* **1964**, *36*, 31–41.
32. Akessé, E.N.; Ouali-N'goran, S.W.M. Reproductive Parameters of *Diastocera trifasciata* (Fabricius, 1775) (Coleoptera: Cerambycidae: Lamiinae), Cashew Branches Girdler in Côte d'Ivoire, Under Semi-Natural Conditions. *Int. J. Sci. Res. Publ.* **2018**, *8*, 686–698. [CrossRef]
33. Torres-Vila, L.M.; Mendiola-Díaz, F.J.; Sánchez-González, Á. Adult size and sex ratio variation of *Cerambyx welensii* (Coleoptera: Cerambycidae) in Mediterranean oak (Fagaceae) woodlands. *Can. Entomol.* **2018**, *150*, 334–346. [CrossRef]
34. Wang, Q.; Chen, L.Y. Mating behavior of a flower-visiting longhorn beetle *Zorion guttigerum* (Westwood) (Coleoptera: Cerambycidae: Cerambycinae). *Naturwissenschaften* **2005**, *92*, 237–241. [CrossRef]
35. Hanks, L.M.; Millar, J.G.; Paine, T.D. Mating behavior of the eucalyptus longhorned borer (Coleoptera: Cerambycidae) and the adaptive significance of long "horns". *J. Insect Behav.* **1996**, *9*, 383–393. [CrossRef]
36. Curkovic, T.; Muñoz, J. Caracterización del cortejo y cópula en *Callisphyris apicicornis: Herramienta* para definir la viabilidad para desarrollar estrategias de manejo. *Agrociencia* **2011**, *45*, 453–464.
37. Cerda, M. Lista sistemática de los cerambícidos chilenos. *Rev. Chil. Entomol.* **1986**, *14*, 29–39.
38. Vitali, F. Versetzung der Tribus Holopterini Lacordaire, 1869 zur Unterfamilie Lepturinae (Coleoptera, Cerambycidae). *Entomofauna* **2002**, *23*, 29–33.
39. Monne, M. Catalogue of the type-species of the genera of the Cerambycidae, Disteniidae, Oxypeltidae and Vesperidae (Coleoptera) of the Neotropical Region. *Zootaxa* **2012**, *3213*, 1–183. [CrossRef]
40. Švácha, P.; Lawrence, J.F. Chapter 2.4. Cerambycidae Latreille, 1802. In *Handbook of Zoology: Arthropoda: Insecta: Coleoptera, Beetles*; Volume 3: Morphology and Systematics (Phytophaga); Leschen, R.A.B., Beutel, R.G., Eds.; Walter de Gruyter: Berlin, Germany, 2014; pp. 77–177.
41. Mangalgikar, P.; Madhura Bhanu, K.R.; Belavadi, V.; Muniyappa, C.; Ammagarahalli, B. 1-octadecene, A Female Produced Aggregation Pheromone of the Coffee White Stem Borer (*Xylotrechus quadripes*). *Horticulturae* **2023**, *9*, 173. [CrossRef]
42. Cohen, C.; Liltved, W.R.; Colville, J.F.; Shuttleworth, A.; Weissflog, J.; Svatoš, A.; Bytebier, B.; Johnson, S.D. Sexual deception of a beetle pollinator through floral mimicry. *Curr. Biol.* **2021**, *31*, 1962–1969. [CrossRef]
43. Wang, W.C.; Cao, D.D.; Men, J.; Wei, J.R. (R)-(+)-citronellal identified as a female-produced sex pheromone of *Aromia bungii* Faldermann (Coleoptera: Cerambycidae). *Egypt. J. Biol. Pest Control.* **2018**, *28*, 1–6. [CrossRef]
44. Sabri, M.S.; Abdullah, F. Mating behaviour and evidence of a female sex pheromone in *Rytidodera simulans* White (Coleoptera: Cerambycidae). *J. Entomol. Res.* **2016**, *40*, 313–326. [CrossRef]
45. Curkovic, T.; Rodríguez, D.; Huerta, A.; Bergmann, J.; Ceballos, R. Behavioral and physiological response of male *Callisphyris apicicornis* (Coleoptera: Cerambycidae) to virgin con-specific females' extracts. *Chil. J. Agric. Res.* **2018**, *78*, 470–477. [CrossRef]

46. Nakayama, Y.; Jikumaru, S.; Togashi, K. Reproductive traits and diel activity of adult *Monochamus saltuarius* (Coleoptera: Cerambycidae) at two different temperatures. *J. For. Res.* **1998**, *3*, 55–61. [CrossRef]
47. Cervantes, D.E.; Hanks, L.M.; Lacey, E.S.; Barbour, J.D. First documentation of a volatile sex pheromone in a longhorned beetle (Coleoptera: Cerambycidae) of the primitive subfamily Prioninae. *Ann. Entomol. Soc. Am.* **2006**, *99*, 718–722. [CrossRef]
48. Xu, T.; Hansen, L.; Teale, S.A. Female calling behaviour in the Asian longhorned beetle (Coleoptera: Cerambycidae). *Can. Entomol.* **2019**, *151*, 600–607. [CrossRef]
49. Adams, K.B. The Bionomics of a Monophagous, Native Wood Borer, Glycobius speciosus (Say) (Coleoptera: Cerambycidae). Ph.D. Thesis, State University of New York College of Environmental Science and Forestry, New York, NY, USA, 2017; 142p.
50. Hoover, K.; Keena, M.; Nehme, M.; Wang, S.; Meng, P.; Zhang, A. Sex-specific trail pheromone mediates complex mate finding behavior in *Anoplophora glabripennis*. *J. Chem. Ecol.* **2014**, *40*, 169–180. [CrossRef]
51. Fukaya, M.; Yasui, H. Pheromones in Longhorn Beetles with a Special Focus on Contact Pheromones. In *Insect Sex Pheromone Research and Beyond: From Molecules to Robots*; Ishikawa, Y., Ed.; Springer: Singapore, 2020; pp. 77–107.
52. Kiriyama, S.; Iwata, R.; Fukaya, M.; Hoshino, Y.; Yamanaka, Y. Mating behavior of *Rosalia batesi* (Coleoptera: Cerambycidae) is mediated by male-produced sex pheromones. *Insects* **2018**, *9*, 48. [CrossRef]
53. Fonseca, M.G.; Vidal, D.M.; Zarbin, P.H. Male-produced sex pheromone of the cerambycid beetle *Hedypathes betulinus*: Chemical identification and biological activity. *J. Chem. Ecol.* **2010**, *36*, 1132–1139. [CrossRef]
54. Xu, T.; Hansen, L.; Teale, S.A. Mating and adult feeding behaviour influence pheromone production in female Asian longhorn beetle *Anoplophora glabripennis* (Coleoptera: Cerambycidae). *Agric. For. Entomol.* **2021**, *23*, 276–286. [CrossRef]
55. Lacey, E.S.; Moreira, J.A.; Millar, J.G.; Hanks, L.M. A male-produced aggregation pheromone blend consisting of alkanediols, terpenoids, and an aromatic alcohol from the cerambycid beetle *Megacyllene caryae*. *J. Chem. Ecol.* **2008**, *34*, 408–417. [CrossRef] [PubMed]
56. Mitchell, R.F.; Ray, A.M.; Hanks, L.M.; Millar, J.G. The common natural products (S)-α-terpineol and (E)-2-hexenol are important pheromone components of *Megacyllene antennata* (Coleoptera: Cerambycidae). *Environ. Entomol.* **2018**, *47*, 1547–1552. [CrossRef] [PubMed]
57. Yasui, H.; Yasuda, T.; Fukaya, M.; Akino, T.; Wakamura, S.; Hirai, Y.; Fukuda, T. Host plant chemicals serve intraspecific communication in the white-spotted longicorn beetle, *Anoplophora malasiaca* (Thomson) (Coleoptera: Cerambycidae). *Appl. Entomol. Zool.* **2007**, *42*, 255–268. [CrossRef]
58. Pherobase. 2024. Available online: www.pherobase.com (accessed on 20 July 2024).
59. Renthal, R.; Lohmeyer, K.; Borges, L.M.F.; Pérez de León, A.A. Surface lipidome of the lone star tick, *Amblyomma americanum*, provides leads on semiochemicals and lipid metabolism. *Ticks Tick-Borne Diseases* **2019**, *10*, 138–145. [CrossRef] [PubMed]
60. Sánchez-Hernández, E.; Martín-Ramos, P.; Niño-Sánchez, J.; Diez-Hermano, S.; Álvarez-Taboada, F.; Pérez-García, R.; Santiago-Aliste, A.; Martín-Gil, J.; Diez-Casero, J.J. Characterization of *Leptoglossus occidentalis* Eggs and Egg Glue. *Insects* **2023**, *14*, 396. [CrossRef]
61. Boyer, F.D.; Malosse, C.; Zagatti, P.; Einhorn, J. Identification and synthesis of vesperal, the female sex pheromone of the longhorn beetle *Vesperus xatarti*. *Bull. Société Chim. Fr.* **1997**, *134*, 757–764. [CrossRef]
62. Tanigaki, T.; Yamaoka, R.; Sota, T. The role of cuticular hydrocarbons in mating and conspecific recognition in the closely related longicorn beetles *Pidonia grallatrix* and *P. takechii*. *Zool. Sci.* **2007**, *24*, 39–45. [CrossRef]
63. Khannoon, E.R. Secretions of pre-anal glands of house-dwelling geckos (Family: Gekkonidae) contain monoglycerides and 1,3-alkanediol. A comparative chemical ecology study. *Biochem. Syst. Ecol.* **2012**, *44*, 341–346. [CrossRef]
64. Wang, H.M.; Bai, P.H.; Zhang, J.; Zhang, X.M.; Qin, H.U.I.; Zheng, H.X.; Zhang, X.H. Attraction of bruchid beetles *Callosobruchus chinensis* (L.) (Coleoptera: Bruchidae) to host plant volatiles. *J. Integr. Agric.* **2020**, *19*, 3035–3044. [CrossRef]
65. Boevé, J.L.; Giot, R. Chemical composition: Hearing insect defensive volatiles. *Patterns* **2011**, *2*, 100352. [CrossRef]
66. Koli, P.; Agarwal, M.; Kessell, D.; Mahawar, S.; Du, X.; Ren, Y.; McKirdy, S.J. Metabolite Variation between Nematode and Bacterial Seed Galls in Comparison to Healthy Seeds of Ryegrass Using Direct Immersion Solid-Phase Microextraction (DI-SPME) Coupled with GC-MS. *Molecules* **2023**, *28*, 828. [CrossRef] [PubMed]
67. Coppée, A.; Terzo, M.; Valterova, I.; Rasmont, P. Intraspecific variation of the cephalic labial gland secretions in *Bombus terrestris* (L.) (Hymenoptera: Apidae). *Chem. Biodivers.* **2008**, *5*, 2654–2661. [CrossRef] [PubMed]
68. Attygalle, A.B.; Cai-Hong, W.U.; Schwarz, J.; Vostrowsky, O.; Hasenfuss, I.; Bestmann, H.J. Sex pheromone of female *Myelois cribrella* Hübner (Lepidoptera: Pyralidae) Chemical identification, electrophysiological evaluation, and field attractancy tests. *J. Chem. Ecol.* **1998**, *14*, 485–494. [CrossRef] [PubMed]
69. Sánchez-Osorio, I.; López-Pantoja, G.; Domínguez, L.; López-Manzano, M.R.; Rosell, G.; Guerrero, Á. Contact chemoreceptive mate recognition in *Cerambyx welensii* Küster (Coleoptera: Cerambycidae). *Agric. For. Entomol.* **2023**, *25*, 622–636. [CrossRef]
70. Silk, P.J.; Sweeney, J.; Wu, J.; Sopow, S.; Mayo, P.D.; Magee, D. Contact Sex Pheromones Identified for Two Species of Longhorned Beetles (Coleoptera: Cerambycidae) *Tetropium fuscum* and *T. cinnamopterum* in the Subfamily Spondylidinae. *Environ. Entomol.* **2011**, *40*, 714–726. [CrossRef]
71. Howse, P.; Stevens, I.; Jones, O. *Insect Pheromones and Their Use in Pest Management*; Chapman & Hall: London, UK, 1998; 369p.
72. Wang, Q.; Zeng, W.; Chen, L.; Li, J.; Yin, X. Circadian reproductive rhythms, pair-bonding, and evidence for sex-specific pheromones in *Nadezhdiella cantori* (Coleoptera: Cerambycidae). *J. Insect Behav.* **2002**, *15*, 527–539. [CrossRef]

73. Graves, F.; Baker, T.C.; Zhang, A.; Keena, M.; Hoover, K. Sensory aspects of trail-following behaviors in the Asian longhorned beetle, *Anoplophora glabripennis*. *J. Insect Behav.* **2016**, *29*, 615–628. [CrossRef]
74. Shimomura, K.; Ohsawa, K. Hybrid Sex Pheromone Communication Systems in Seed Beetles. In *Insect Sex Pheromone Research and Beyond: From Molecules to Robots*; Ishikawa, Y., Ed.; Springer: Singapore, 2020; pp. 61–76.

insects

Article

Native Japanese Polygonaceae Species as Potential Native Insectary Plants in Conserving Indigenous Natural Enemies

David Wari [1,*], Junichiro Abe [2] and Toshio Kitamura [1]

[1] Western Region Agricultural Research Center (WARC), National Agriculture and Food Research Organization (NARO), Fukuyama 721-8514, Hiroshima, Japan
[2] Institute of Plant Protection (NIPP), National Agriculture and Food Research Organization (NARO), Tsukuba 305-8666, Ibaraki, Japan
* Correspondence: warid692@affrc.go.jp; Tel.: +81-(0)84-923-4150

Simple Summary: Manipulating a natural vegetation to regulate pest densities is a crucial strategy within conservation biological control (CBC). Much of the CBC has revolved around the non-native invasive species that, in many ways, are causing concerns among local farmers, hence the need for alternate but noninvasive insectary plants. In this study, we bio-prospected potential native plants that conserve indigenous natural enemies and their potential role in regulating pests on vegetable crops. Field surveys revealed Japanese native Polygonaceae species as native plants of interest. Interplanting Polygonaceae plant species with eggplants resulted in reduced pest populations compared to eggplant without Polygonaceae plants. Reduced pest numbers could be related to an increased number of natural enemies, possibly supported by the native Polygonaceae plants. Combined with other results, preliminary results show that Polygonaceae species may serve as a native insectary plant species in supporting indigenous natural enemies.

Abstract: Conservation biological control (CBC) is the application of agricultural practices that utilize insectary plants to conserve and enhance natural enemies, thereby increasing their efficiency to suppress pests. Most of the insectary plants used in CBC are non-native invasive insectary plants, which are costly and pose threats to the local ecosystems and biodiversity. Alternative to non-native insectary plants, the use of native plants is proposed. Hence, the aim of this study is to identify native plant species that can be used as alternatives to non-native insectary plants to conserve and promote indigenous natural enemies (INEs) for sustainable pest management. To achieve this, first, we bio-prospected the surrounding habitats of organic fields in the western region of Japan (i.e., Hiroshima Prefecture) to identify native plant species as prospective native insectary plants. As a result, among various Japanese native plants surveyed, Polygonaceae plant species seem to host a variety of INEs, showing potential as a native insectary plant. We then conducted open field experiments to test the role of Polygonaceae plants in promoting INEs, thereby indirectly suppressing pest densities on vegetable crops such as eggplants. Results show that significantly high densities of INEs (green lacewing, $p = 0.024$; *Orius* spp., $p = 0.001$: GLM) were observed on eggplants with Polygonaceae plants compared to eggplants without Polygonaceae plants, leading to a significant reduction in pest densities (thrips, $p = 0.000$; whiteflies, $p = 0.002$: GLM) on the eggplants with Polygonaceae plants. Furthermore, molecular analysis revealed that *Orius* spp., as a representative INE in this study, migrated from Polygonaceae plants to eggplants, suggesting that Polygonaceae plants may conserve and promote INEs to vegetable crops, resulting in pest suppression. Here, we discuss the roles of Polygonaceae plants (and other native plants) in regulating pest densities on crops.

 https://doi.org/10.3390/insects16020232

Keywords: conservation biological control (CBC); natural enemies; invasive plants; organic farming; *Orius* spp.; *Persicaria lapathifolia*; *Persicaria longiseta*

1. Introduction

To manage crop damage and losses by insect pests, farmers mostly rely on pesticides. However, excessive use of pesticides induces pest resistance and, with climate change at the helm, evolutionary changes could potentially trigger the movement of destructive pest species globally threatening the current pest control efforts in place [1]. To minimize the usage of pesticides and its negative impact on the environment, organic farming has been suggested and promoted by the government and research organizations. Organic farming is a type of agriculture system that uses agricultural farming techniques that causes minimal to no environmental impact on the local ecosystems and biodiversity [2]. One of the farming techniques adopted to control pests in organic farming is biological control. A biological control method used in organic farming that is not new but trending of late is the engineering of agroecosystems. The classical term herein is conservation biological control (hereafter referred to as CBC), which involves engineering of habitats with the aim of promoting natural enemies to regulate pest occurrences [3].

By and large, mostly non-native horticultural insectary plants have been used throughout CBC as hosts for natural enemies. While non-native horticultural insectary plants are well recognized and documented to enhance biological pest control, non-native insectary plants are somewhat causing concerns among locals via threats to native ecosystems and biodiversity. Nonetheless, excessive use of non-native insectary plants can be minimized by utilizing native plants. Essentially, native plant species are critical in maintaining biodiversity and a healthy ecosystem. From the agronomic and conservation perspective, native plants can be beneficial in many ways; they are adapted to the local environmental conditions [4], require little maintenance costs [5], have no risk of becoming invasive [4], and are readily available to natural enemies, providing food resources, shelter, and breeding grounds [6]. Unlike non-native insectary plants, research on utilizing native plants to conserve and promote natural enemies in CBC for pest control is still very patchy.

In Japan, the following flowering plants have showed the potential to harbor and/or promote natural enemies: white clover (*Trifolium repens* L. (Fabaceae)) and *Gramineae* plants [7], *Rudbeckia hirta* L. (Asteraceae) [8], *Salvia farinacea* Benth. (Lamiaceae) [9], *Verbena* × *hybrida* Voss. (Verbenaceae) [10], *Scaevola aemula* R.Br. (Goodeniaceae) [11], *Cleome hassleriana* Chod. (Cleomaceae), *Sesamum indicum* L. (Pedaliaceae) [12], and *Setaria viridis* Beauv (Poaceae) [13]. However, most, if not all, of these plants and/or grasses reported so far are non-native horticultural insectary plants. Little is known or reported on the use of native plants as ecologically sound alternatives to supporting indigenous natural enemies (hereafter referred to as INEs). In this study, we surveyed the agroecological habitats surrounding organic fields to identify native plant species that have the potential to conserves INEs. We identified native plant species as potential native insectary plants and tested their role in conserving and promoting INEs that can regulate pest occurrences on vegetable crops. Our preliminary data show that native plants have the potential to conserve and promote INEs. Here, we discuss the role of native plants as better alternative to non-native invasive insectary plant species in supporting diversity of INEs and their potential role in delivery of function in terms of pest suppression. Furthermore, from these preliminary results, we propose further surveys and experiments to build towards advancing the concept of sustainable CBC in organic farming in Japan through native insectary plants.

2. Materials and Methods

2.1. Study Site

The native plants surveys were conducted in organic fields in Jinsekikogen-town (34°51′14″, 133°12′59″), Higashi-Hiroshima (34°29′12″, 132°39′6″), and the National Agriculture and Food Research Organization (NARO), Western Region Agricultural Research Center (WARC) (NARO-WARC) research field (34°30′10″, 133°23′18″) in Fukuyama City, Hiroshima Prefecture, (Western) Japan. For the organic fields in Jinsekikogen-town and the Higashi-Hiroshima, the vegetation is almost surrounded by bushes and shrubs, with no applications of any synthesized chemical compounds such as pesticides, herbicides, and fertilizers. On the other hand, while no synthesized chemical compounds were used in NARO-WARC fields, since the fields are used in rotation between research groups, pesticides have been used to control other pests and diseases.

2.2. Surveying of Native Japanese Plants

Native plants inhabiting the field margins, hedgerows, and weedy areas surrounding the organic farms in Western Japan were surveyed from July to October between the years 2023 and 2024. The surveying months were selected to coincide with the peak cultivation periods of agriculturally important vegetables such as eggplants (*Solanum melongena* L. Solanaceae) and spring onions (*Allium fistulosum* L. Amaryllidaceae) in Western Japan. Essentially, plants at their floral stages were identified and the floral parts tapped five–ten times onto a plastic tub (40 cm by 20 cm) and the number of arthropods that fell onto the white tub were quantified and recorded. Furthermore, the number of branches, leaves per branch, as well as flower per plant were recoded for shrubby plants, while, for creeping plants with a web of branches that could not be quantified, a 50 cm by 50 cm quadrat was used to measure the area and quantify the number of branches, number of leaves, and number of flowers to calculate the density of the plants and correlate it with the number of arthropod species. However, since a variety of native plants with different growth patterns were observed, the arthropod trends on native plants are presented as observed or absent for each sampling month. The detailed data can be produced upon request.

Similarly, Japanese native plant seeds (native plants that have been cultivated in controlled settings for preservation of the native flowering plants of Japan) that were identified from the 2023 field surveys were ordered from ESPEC Mic Corp. (Osaka, Japan), germinated, and transplanted onto the NARO-WARC research field to confirm if the native plants harbored INEs. Additionally, *Ocimum basilicum* L. (Lamiaceae), *O. tenuiflorum* L. (Lamiaceae), and *Trifolium pratense* L. (Fabaceae), non-native insectary plants known to conserve and promote natural enemies, were also geminated and planted in the fields to compare with the native plant species. Arthropod sampling was as discussed above. Moreover, *Orius* spp. specimens were also sampled and stored in a glass container filled with 99% EtOH for further experiments.

2.3. Semi-Field Studies on the Effect of Japanese Native Polygonaceae Plant Strips on Arthropod Species on Eggplants

To determine the role of Japanese native plants on conserving and promoting INEs resulting in suppression of pest (eggplant pests of agricultural importance such as aphids and thrips) populations, semi-field studies were performed at NARO-WARC research field. Japanese native Polygonaceae plant (*Persicaria longiseta* (Bruijn) Kitagawa 1937 and *P. lapathifolia* (L.) Delarbre 1800) strips were planted surrounding the eggplant beds (see Figure A1). Two treatments, eggplants with Polygonaceae plant strips and eggplants without Polygonaceae plant strips with three biological replicates each, were conducted. Each replicate had 24 eggplant plants. Since the field was limited to carry out the experiment, each treatment with their replicates were spaced 10 m to each other. After field ploughing, prior to bed preparations, S604 (NPK with 16:10:14 ratio (Zen-noh West Co., Ltd., Hiroshima, Japan)), a readily available fertilizer, was applied with rates according to the recommendations by the manufacturer. To mimic the eggplant cultivations in Western Japan, eggplant seeds were sowed on 8 March and transplanted on 21 May 2024. The eggplant cultivar used in this study was the medium in length eggplant cv. Kuromasari LF. Polygonaceae seeds were sowed in early April and transplanted when the plantlets were about 8 weeks old following the transplantation of eggplants on 29 May 2024. Arthropod surveys (both pests and INEs) were performed on at least a fortnight basis from 19 June through to 29 October. Eggplant seedlings were 15 weeks old (from sowing to the start (19 June) of the survey) when the survey was started. Six leaves, two each from the base of the plant, the mid-section, and the canopy of the eggplants, were surveyed to determine the population trends of arthropods. The six leaves were used as representative of the whole plant. Due to the smaller land area used in this study (see Figure A1), all 24 eggplants for each replicate per treatment were surveyed to clearly capture the population trends of pests and INEs on eggplants.

Native Polygonaceae plants were also surveyed to determine the population dynamics of pests and INEs. Since the whole plant was laborious to survey, one branch each with its flower buds for each plant for *P. longiseta* and *P. lapathifolia* were tapped five–ten times onto a plastic tub (40 cm by 20 cm) and the arthropods that fell onto the plastic tub were quantified. The arthropod numbers on the Polygonaceae plants are expressed as number of arthropods per branch per plant surveyed.

Pests (such as aphids, thrips, and whiteflies) and INEs (e.g., coccinellids and green lacewings) were all collectively surveyed at their family level. For instance, there are multiple species of aphids that use eggplants as hosts, such as *Aphis gossypii* [14], *Aulacorthum solani* [15], and *Macrosiphum euphorbiae* [16], to name a few. Furthermore, thrips species such as *Thrips palmi* and *Frankliniella intonsa* [13] are also reported to infest eggplants. To identify each aphid species in the field and, at the same time, identify other pests and natural enemies in the field is time-consuming and laborious. Therefore, pests such as aphids, thrips, and whiteflies were generally identified and quantified as aphid species, thrips species, and whitefly species, respectively. Likewise for coccinellids and green lacewing surveys. However, in the case of *Orius* species, since increased densities of *Orius* spp. were observed on Polygonaceae plants, species identification at the species level was essential. Therefore, *Orius* spp. Adults, as the INE of interest in this study from both eggplants and *P. lapathifolia*, were collected and stored separately in 99% EtOH for identification and species composition analysis, as well as detection of *P. lapathifolia* DNA using molecular tools. *Orius* spp. from *P. lapathifolia* were chosen because of their higher densities compared to *P. longiseta* (See Figure A2).

2.4. *Orius spp.* DNA Extraction

DNA from the adult *Orius* spp. specimens collected from eggplants and *P. lapathifolia* was extracted using 2× STE buffer (200 mM NaCl, 20 mM Tris-Cl, pH 8.0, and 2 mM EDTA) and Proteinase K (TaKaRa Bio Inc., Shiga, Japan). In brief, *Orius* spp. specimen (1 individual each) was homogenized in a sterile 1.5 mL tube. STE buffer and Proteinase K were then added, followed by heat treatment, once at 55 °C for 30 min followed by 95 °C for 15 min and a subsequent centrifugation at 20,000× *g* for 3 min. The resultant supernatant was then used as the DNA template for PCR.

2.5. Molecular Identification of Orius spp. Using Multiplex PCR

PCR was performed as described in Hinomoto et al. [17] with minor modifications. Essentially, 5× KAPA2G Robust HotStart ReadMix with Dye (KAPA Biosystems, Wilmington, MA, USA) was used as the polymerase with primer sets described in Hinomoto et al. [17]. Amplification was performed in a TP350 PCR Thermal Cycler Dice Touch (TaKaRa Bio Inc.) with the following conditions: 1 cycle at 94 °C for 3 min, followed by 30 cycles of 94 °C for 15 s, 56 °C for 15 s, and 72 °C for 1 min, with a subsequent final extension at 72 °C for 7 min. The PCR products were electrophoresed on a 1% agarose gel in TAE (40 mM Tris, 40 mM acetic acid, and 1 mM ethylenediaminetetraacetic acid) stained with SYBR Safe DNA Gel Strain (Invitrogen, San Francisco, CA, USA) and observed under WSE-5400-UP Printgraph Classic (ATTO Corporation, Tokyo, Japan) mounted on a Safe Imager Blue Light Transilluminator (Invitrogen, San Francisco, CA, USA). *Orius* spp. were estimated from the DNA bands on the 1% agarose gel as reported in Hinomoto et al. [17]. Ideally, *O. minutus* (L.) should show an approximate PCR fragment length of 910 bp, *O. sauteri* (L.) with 660 bp and 320 bp, *O. strigicollis* (Poppius) with 680 bp and 490 bp, and *O. nagaii* Yasunaga with 610 bp and 190 bp. *Orius tantillus* (Motchulxky, 1863) does not occur in the main island of Japan so *O. tantillus* amplification is not expected.

2.6. Detection of P. lapathifolia DNA in Adult Orius spp. Individuals

Orius spp. DNA samples used in identification and species composition were also used in the detection of *P. lapathifolia* DNA. PCR was performed using the 5× KAPA2G Robust HotStart ReadMix with Dye using the primer sets Pl_F 5′ CTTTCCAAGGCCCACCTCAC 3′ and Pl_R 5′ GAACATGATCCCCACCGGAC 3′. Amplification was performed in a TP350 PCR Thermal Cycler Dice Touch with the following conditions: 1 cycle at 95 °C for 3 min, followed by 40 cycles of 95 °C for 15 s, 55 °C for 15 s, and 72 °C for 30 s, with a subsequent final extension at 72 °C for 7 min. The PCR products were electrophoresed on a 1% agarose gel in TAE stained with SYBR safe and observed under WSE-5400-UP Printgraph Classic mounted on a Safe Imager Blue Light Transilluminator. *Persicaria lapathifolia* DNA was estimated from the DNA bands on the 1% agarose gel. Ideally, *P. lapathifolia* DNA in the gut of *Orius* spp. should show an approximate PCR fragment length of 500 bp.

2.7. Statistical Analysis

Statistical analyses were performed only for the arthropod pests (aphids, whiteflies, and thrips) and INEs (coccinellids, green lacewings, and *Orius* spp.) observed on eggplants. The pests and INE counts were tested for normality (Shapiro–Wilk test and Lilliefors test) and homogeneity (Bartlett's Test) using open-source software OpenStat (https://smerser.github.io/OpenStat/OpenStatMain.htm accessed on 21 December 2024). The total number of pests and INEs were then transformed using 'x + 1' where necessary. The effects of 'treatments' (eggplants with Polygonaceae plant strips and without Polygonaceae plant strips) and 'sampling interval' (weeks) factors and their interaction on the densities of pests and natural enemies on eggplants were analyzed using Sum of Squares by Regression, a partial Generalized Linear Model (GLM) by OpenStat software ver. 30.0. Chi-square test analysis for detection of *P. longiseta* DNA in the gut of *Orius* spp. was performed using the Microsoft excel software.

3. Results

3.1. Japanese Native Plants Survey

In the search for native plants that conserve and promote INEs, we surveyed the agroecosystems surrounding the organic farms in Western Japan. A two-year (2022–2023) survey revealed more than 20 flowering wild plant species (both native and non-native plant species) (Table 1). The collected flowering plants were cross-referenced with the wild plants listed in [18,19] to confirm that the plants were of the correct species. While most of the flowering plants surveyed showed frequent occurrences of known pests such as aphids, thrips, and spider mites, several native plants seem to harbor INEs. Native plants from the Polygonaceae family such as *P. filiformis* (Thunb.) Nakai, *P. lapathifolia*, *P. longiseta*, *P. orientalis* (L.) Spach (a non-native but neutralized ornamental flowering plant in Japan), and *P. pubescens* (Blume) H.Hara seem to harbor INEs such as *Orius* spp., coccinellids, green lacewings, phytoseiid mites, hoverfly species from the Syrphidae family, and the occasional occurrence of parasitoids, which are important natural enemies of agriculturally important pests such as aphids, thrips, and spider mites. Other native plants such as *Commelina communis* (L.) (Commelinaceae), *Eclipta prostrata* (L.) L. (Asteraceae), *Eupatorium japonicum* Thunberg ex Murray. (Asteraceae), *Impatiens textori* Miq. (Balsaminaceae), *Paederia foetida* L. (Rubiaceae), *Patrinia scabiosifolia* Link (Caprifoliaceae), and *Salvia japonica* Thund (Lamiaceae) were also some of the native plants observed to harbor INEs and could be considered as potential candidates in future studies (Table 1).

Table 1. List of native plants surveyed (between 2023 to 2024) from the surrounding habitats of organic fields in Western Japan.

Study Site	**Plants Observed**		**July**	**August**	**September**	**October**
	Family	**Species**				
Fukuyama	Asteraceae	*Cirsium japonicum* DC. (1838)	■	▲	▲	■
	Asteraceae	*Eupatorium japonicum* Thunberg ex Murray.	▲•	▲•	▲•	▲•
	Caryophyllaceae	*Dianthus superbus* L.	■	■	■	■
	Caprifoliaceae	*Patrinia scabiosifolia* Link	▲•	▲•	▲•	▲•
	Commelinaceae	*Commelina communis* L.	▲•	▲•	▲•	▲•
	Fabaceae	*Trifolium pratense* L. †	▲•	▲	▲•	□
	Lamiaceae	*Ocimum basilicum* L. †	▲•	▲•	▲•	▲•
	Lamiaceae	*Ocimum tenuiflorum* L. †	□	▲•	▲•	▲•
	Lamiaceae	*Salvia japonica* Thunb.	▲•	□	□	□
	Papaveraceae	*Macleaya cordata* (Willd.) R. Br.	□	□	□	□
	Polygonaceae	*Persicaria longiseta* (Bruijn) Kitagawa 1937	▲•	▲•	▲•	▲•
	Polygonaceae	*Persicaria lapathifolia* (L.) Delarbre 1800	▲•	▲•	▲•	▲•
Higashi-Hiroshima	Acanthaceae	*Rostellularia procumbens* (L.) Nees (1847)	□	□	•	□
	Commelinaceae	*Commelina communis* L.	□	■	▲•	▲
	Onagraceae	*Oenothera biennis* L. *	□	□	▲•	□
	Polygonaceae	*Persicaria longiseta* (Bruijn) Kitagawa 1937	□	▲	▲•	■
	Polygonaceae	*Persicaria pubescens* (Blume) H. Hara	□	□	▲	•
	Polygonaceae	*Polygonum thunbergii* Sieb. & Zucc.	□	□	□	▲

Table 1. *Cont.*

Study Site	Plants Observed		July	August	September	October
	Family	Species				
Jinsekikogen-town	Apiaceae	*Cnidium officinale* Makino *	□	□	▲	□
	Apiaceae	*Ostericum sieboldii* (Miq.) Nakai	□	□	□	▲
	Asteraceae	*Aster yomena* (Kitam.) Honda	□	□	□	▲
	Asteraceae	*Eclipta prostrata* (L.) L.	□	▲●	▲●	□
	Asteraceae	*Sigesbeckia pubescens* (Makino) Makino	□	□	□	▲
	Balsaminaceae	*Impatiens textori* Miq.	□	□	▲●	▲●
	Brassicaceae	*Barbarea vulgaris* W.T. Aiton	□	▲	▲	□
	Commelinaceae	*Commelina communis* L.	□	▲●	▲●	▲●
	Fabaceae	*Lotus corniculatus* L.	□	▲	□	□
	Lamiaceae	*Clinopodium multicaule* (Maxim.) O. Kuntze. var. multicaule	□	□	■	□
	Lamiaceae	*Isodon inflexus* (Thunb.) Kudo	□	□	□	▲
	Lamiaceae	*Mosla scabra* (Thunb.) C.Y. Wu et H.W. Li	□	□	□	▲
	Lamiaceae	*Salvia japonica* Thunb.	□	▲●	▲●	□
	Onagraceae	*Circaea mollis* Siebold & Zucc.	□	▲●	□	□
	Rosaceae	*Agrimonia pilosa* Ledab. var. japonica	□	▲	▲	□
	Rubiaceae	*Paederia foetida* L.	□	▲●	□	□
	Polygonaceae	*Persicaria filiformis* (Thunb.) Nakai	□	□	▲	□
	Polygonaceae	*Persicaria longiseta* (Bruijn) Kitagawa 1937	□	▲●	▲●	▲●
	Polygonaceae	*Persicaria orientalis* (L.) Spach *	□	▲●	▲●	▲●
	Polygonaceae	*Polygonum thunbergii* Sieb. & Zucc.	□	□	□	▲

▲: Herbivores observed, ●: natural enemies present, ■: no arthropod (including natural enemies and pests) species observed, □: plants not available during survey, *: non-native but neutralized in Japan, †: non-native insectary plants.

Persicaria longiseta and *P. lapathifolia* showed a relatively increased number of INEs, therefore, both plants were selected as representatives of the Polygonaceae species and the population trends of pests (aphids and thrips) and INEs (*Orius* spp., coccinellids, and green lacewings) were compared to a known non-native insectary plant such as *O. basilicum*. Results show that thrips were mostly dominant on the *P. longiseta* and *P. lapathifolia*, with few to no occurrences of aphids, while both thrips and aphids were observed to occur almost simultaneously throughout the survey on *O. basilicum* (Figure 1). As for the INEs, while sporadic but low numbers of coccinellids and green lacewing occurrences were observed on *P. longiseta*, *P. lapathifolia*, and *O. basilicum*, a large number of *Orius* spp. was observed on the three plants but substantially high on Polygonaceae plants, especially *P. lapathifolia*. As is a known phenomenon, the population trends of *Orius* spp. somewhat followed that of the thrips on Polygonaceae plants, suggesting that Polygonaceae plants could be an important native host plant of *Orius* spp. with thrips as the food source. From these results, we selected *P. longiseta* and *P. lapathifolia* as potential native insectary plants that can conserve and promote *Orius* spp. and possibly other INEs.

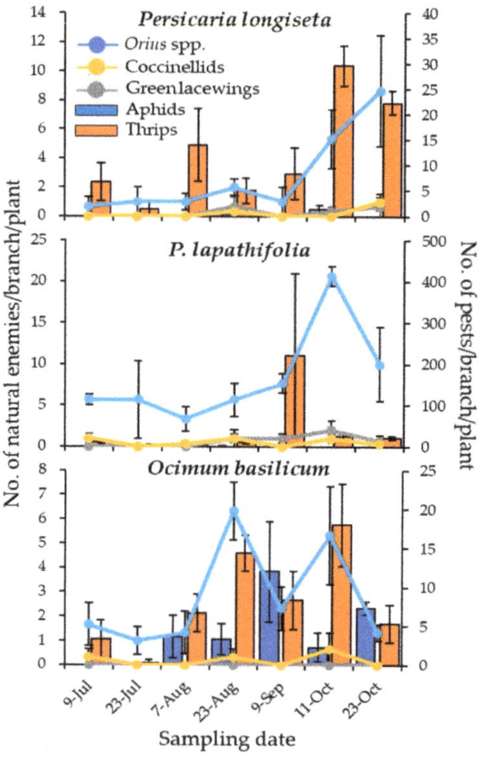

Figure 1. Population dynamics of pests (blue bar graphs for aphids and orange bars for thrips) and INEs (blue lines for *Orius* spp., orange lines for coccinellid beetles, and grey lines for green lacewings) on *Persicaria longiseta*, *P. lapathifolia*, and *Ocimum basilicum*. (n = 3, error bars: SE.).

3.2. Preliminary Field Studies on the Potential Role of Polygonaceae Plants as Insectary Plants

To determine the role of Polygonaceae plants as potential Japanese native insectary plants in conserving and promoting INEs, we performed field studies. Population trends of major pests such as aphids, thrips, and whiteflies were determined together with INEs such as coccinellids, green lacewings, and *Orius* spp. Results shows that, while there was no significant difference between the treatments (eggplants with and without Polygonaceae plant strips) for the aphids ($p = 0.4675$), some marginal differences could be observed when taking into consideration the effect of sample intervals ($p = 0.0717$) and the interaction between treatments and sampling intervals ($p = 0.0627$). On the other hand, whiteflies ($p = 0.0011$) and thrips ($p = 0.0000$) were significantly reduced in eggplants with Polygonaceae plant strips, as with the effects of sampling intervals ($p = 0.0155$ for whiteflies and $p = 0.0000$ for thrips) as well as the effect of the interaction between the treatment and sampling intervals ($p = 0.0037$ for whiteflies and $p = 0.0000$ for thrips) (Figure 2, Table 2). Similarly, INEs such as green lacewing and *Orius* spp. were observed to be significantly higher on eggplants with Polygonaceae plant strips compared to eggplants without Polygonaceae plant strips ($p = 0.0241$, $p = 0.0241$ for green lacewings and *Orius* spp., respectively) (Figure 2, Table 2), suggesting that *Orius* spp. and green lacewings could be involved in suppressing the thrips and whitefly densities on eggplants. Furthermore, to determine if pests, as well as INEs were present on the Polygonaceae plants, we also surveyed the pests and INE trends on the Polygonaceae plant strips. As expected, aphids and thrips (pests) and coccinellids, green lacewings, and *Orius* spp. (INEs) were present and seemed to increase on *P. lapathifolia* as the survey progressed (Figure A2). Whether the higher densities of *Orius* spp. on *P. lapathifolia* migrated to eggplants remains unknown; therefore, we performed laboratory experiments to confirm their species composition and the migration pattern.

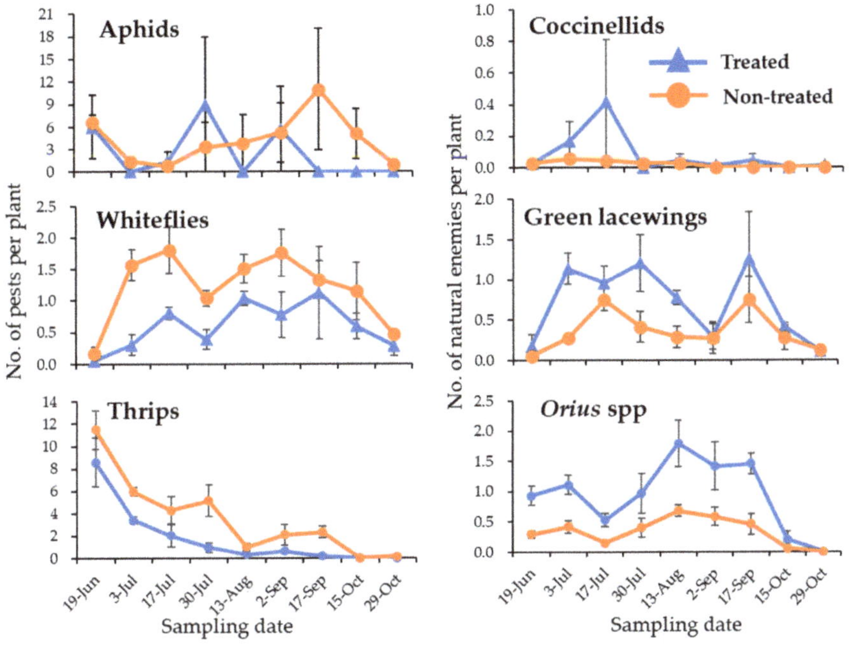

Figure 2. Population dynamics of pests (aphids, whiteflies, and thrips) and INEs (coccinellid, green lacewings, and *Orius* spp.) on eggplants. Blue lines (treated) represent eggplants with Polygonaceae plant strips and orange lines (non-treated) are eggplants without Polygonaceae plant strips (n = 3, error bars: SE).

Table 2. Results of the partial GLM analysis showing the interaction effects between treatments and sampling intervals on the pests (aphids, whiteflies, and thrips) and INEs (coccinellids, green lacewings, and *Orius* spp.) on eggplants.

Arthropod Species Observed		Response								
		Treatments			Sampling Intervals			Treatment × Sampling Intervals		
		d.f	*F*	*p*	*d.f*	*F*	*p*	*d.f*	*F*	*p*
Pests	Aphids	1	0.934	0.4675	8	2.094	0.0717	8	2.170	0.0627
	Whiteflies	1	4.847	0.0011	8	2.964	0.0155	8	3.791	0.0037
	Thrips	1	19.464	0.0000	8	13.049	0.0000	8	14.978	0.0000
INEs	Coccinellids	1	1.321	0.2711	8	1.198	0.3241	8	1.244	0.3015
	Green lacewings	1	2.868	0.0241	8	2.089	0.0723	8	2.288	0.0510
	Orius spp.	1	5.406	0.0005	8	3.170	0.0108	8	4.952	0.0010

3.3. Identification and Species Composition of Orius spp. on Eggplants and P. lapathifolia

Identifying the species that inhabits the given agroecosystem is essentially fundamental in designing a successful CBC. Adult *Orius* spp. individuals sampled during the survey were identified using the multiplex PCR method [17]. There are five known species of *Orius* species distributed in Japan, four on the main island of Honshu, *O. sauteri*, *O. strigicollis*, *O. minutus*, and *O. nagaii* [17], and *O. tantillus* on the Okinawa islands [11]. The species composition of the *Orius* spp. on eggplants and *P. lapathifolia* is shown in Figure 3. *Orius sauteri* seems to be the dominant species at the earlier sampling dates (June and July) on the eggplants (Figure 3A) and on *P. lapathifolia* (late June to mid-July) (Figure 3B). *Orius minutus* sporadically occurred in the earlier dates and seems to increase in its occurrence as the survey progressed. *Orius nagaii* only appeared in surveys between June and July, while *O. strigicollis* was only observed once on eggplants on 13 August. Coincidently, the occurrence pattern of *Orius* spp. on *P. lapathifolia* resembled that of those on eggplants, indicating that *Orius* spp. could be migrating from *P. lapathifolia* to eggplants. Overall, the dominant occurrence of *O. sauteri* on *P. lapathifolia* (and eggplants) in Fukuyama or at least in Hiroshima Pref. in Western Japan suggests that *P. lapathifolia* could be an important host plant species for *O. sauteri*. This information could be critical in designing biological control strategies that involve specific natural enemy and plant species.

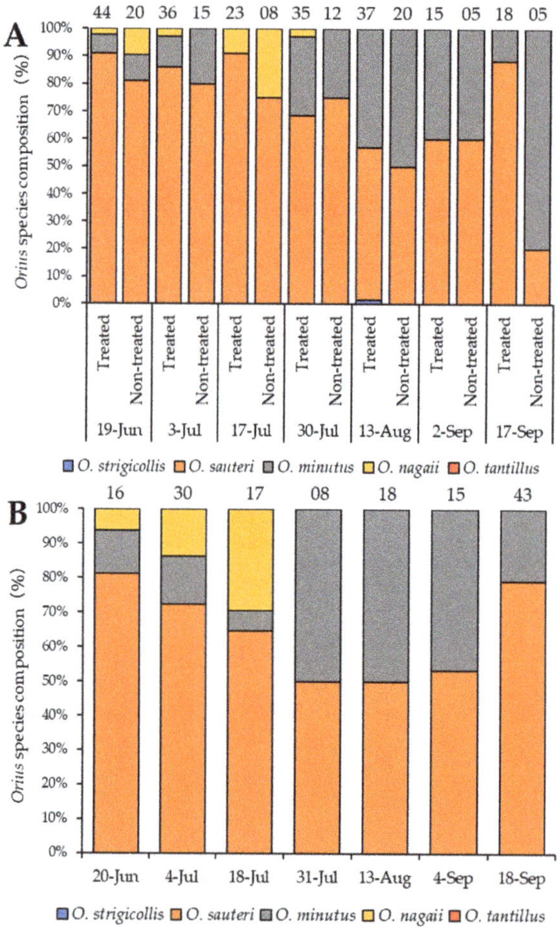

Figure 3. Species composition of *Orius* spp. collected on eggplants (**A**) and on *Persicaria lapathifolia* (**B**). The numbers at the top end of the bar graphs represent the number of *Orius* spp. individuals used for a specified sampling date. *Orius strigicollis* is represented with blue bars, *O. sauteri* with orange bars, *O. minutus* with grey bars, *O. nagaii* with yellow bars, and *O. tantillus* with red bars. Treated and non-treated in '**A**' im-plies eggplants with Polygonaceae plant strips and eggplants without Polygonaceae plant strips, respectively.

3.4. Detection of P. lapathifolia DNA in Orius spp.

To confirm if the *Orius* spp. (candidate INE used in this set of experiments) migrates from *P. lapathifolia* to eggplants, we designed *P.-lapathifolia*-specific primers and performed PCR to detect *P. lapathifolia* DNA in the gut of adult *Orius* spp. Upon confirming the amplification of *P. lapathifolia* DNA, we performed PCR on *Orius* spp. adult individuals collected from eggplants as well as from *P. lapathifolia*. The results are presented in Tables 3 and 4. Results show that *P. lapathifolia* DNA was amplified in the gut of adult *Orius* spp. collected from eggplants (Table 3). In an unexpected turn of events, while *P. lapathifolia* DNA was detected in the gut of adult *Orius* spp., collected from eggplants with and without Polygonaceae plant strips, the detection frequency in *Orius* spp. collected from eggplants without Polygonaceae plant strips was higher than that of the eggplants with Polygonaceae plant strips between June and mid-July surveys (Table 3, $p = 0.03556$ for 19 June, $p = 0.00001$ for 3 July, and $p = 0.00004$ for 17 July), suggesting that adult *Orius* spp. could migrate as far as 10 m or beyond in search for prey materials. Adult *Orius* spp. collected on *P. lapathifolia* also confirmed the detection of *P. lapathifolia* DNA in their guts (Table 4). How far laterally the *Orius* spp. can migrate or, rather, for how long before the plant material is completely metabolized and degraded in the gut of the *Orius* spp. remains for further elucidation. Nonetheless, all things considered, although the detection frequencies were low, the fact that *P. lapathifolia* DNA was detected in the gut of an adult *Orius* spp. individual collected on eggplants suggests that *Orius* spp. did migrate from *P. lapathifolia* to eggplants. These results, together with the population trends of INEs and pests on eggplants, suggest that native Japanese Polygonaceae plants can support and promote INEs when interplanted with vegetable crops, resulting in suppression of pest densities.

Table 3. Detection frequency of *Persicaria lapathifolia* DNA in the guts of adult *Orius* spp. collected on eggplants.

	19-June		03-July		17-July		30-July		13-August		02-September		17-September	
	Treated *	Non-Treated #	Treated	Non-Treated	Treated	Non-Treated	Treated	Non-Treated	Treated	Non-Treated	Treated	Non-Treated	Treated	Non-Treated
Tot. No. of *Orius* individuals observed	44	20	36	15	23	8	35	12	37	20	15	5	18	5
No. of *Orius* spp. that showed *P. lapathifolia* DNA amplification	3	4	1	4	4	2	0	1	3	0	0	0	1	0
Detection frequency (%)	6.8	20.0	2.8	26.7	17.4	25.0	0.0	8.3	8.1	0.0	0.0	0.0	5.6	0.0
$\chi2$ (p-value) at $\alpha = 0.05$	0.03556		0.00001		0.00004		0.12469		0.35644		1.00000		0.03868	

*: Adult *Orius* spp. collected from the eggplants with Polygonaceae plant strips, #: adult *Orius* spp. collected from the eggplants without Polygonaceae plant strips.

Table 4. Detection frequency of *Persicaria lapathifolia* DNA in the guts of adult *Orius* spp. collected on *P. lapathifolia*.

	20-June	04-July	18-July	31-July	13-August	04-September	18-September
Tot. No. of *Orius* individuals observed	16	30	17	8	18	15	43
No. of *Orius* spp. that showed *P. lapathifolia* DNA amplification	3	5	3	0	1	4	4
Detection frequency (%)	18.8	16.7	17.6	0.0	5.6	26.7	9.3

4. Discussion

The introduction of non-native insectary plants to conserve and promote introduced natural enemies seems to be functioning well in managing pest populations. However, the introduced non-native flora and fauna are causing loss of biodiversity and threatening local ecosystems [20–23]. Native plants that are adapted to the local environmental conditions with minimal risk of becoming invasive are potential alternatives as substitutes to non-native plants. Studies on native plants and/or managing local habitats for conservation of biological control is not a new concept. As native plants vary between geographical locations, so does the plant species used. A range of studies have reported the use of native plants in supporting natural enemies [24–28]. Furthermore, several of these studies have shown that indigenous arthropods have higher tendencies to visit and use native plants to shelter and breed compared to the non-native plants [24,26]. Thus, native plants do have the potential to conserve and promote indigenous arthropod species, especially INEs, which could be significant in CBC in regulating pest populations in vegetable productions.

In Japan, mostly non-native horticultural insectary plants have been used extensively in biological control of pests. As a substitute for these non-native insectary plants, we surveyed the surrounding habitats of organic fields in Western Japan to identify native plants as potential insectary plants. Among the native plants surveyed, we identified Polygonaceae plants as a potential insectary plant. Polygonaceae, especially *P. longiseta* and *P. lapathifolia,* were previously reported to harbor INEs such as phytoseiid mites, important natural enemies of spider mites in fruits (peach) orchards [29]. In North America, *Polygonum* (Polygonaceae) plants are semi-aquatic weeds and are considered occasional pests [30]. There are very few reports demonstrating the role of Polygonaceae plants in promoting and conserving beneficial insects [29,31]. While Polygonaceae plants are mostly referred to as plants of many seeds [32], their somewhat dense flowering could provide floral resources to support arthropod communities, including natural enemies.

Floral resources are essentially vital in supporting diversity of natural enemies, which, in turn, delivers the desired function in pest reduction [33–35]. The relatively large number of *Orius* spp. with other INEs observed on *P. longiseta* and *P. lapathifolia* in NARO-WARC fields when planted and managed in a controlled manner suggests that Polygonaceae plants could serve as a useful reservoir, providing refugia, alternative foods, and places to breed, supporting a diversity of arthropod species, including natural enemies and predators. Diversity in natural enemies and predators have, in theory, been linked to delivery of function, i.e., suppression or reduction in pest populations [33,34]. Polygonaceae plants, in part, do support a variety of INEs but does this equate to supporting the delivery of function? Our results, although preliminary, confirmed the movement of adult *Orius* spp. as the model INE between insectary plants and vegetable crop. We verified by detection of *P. lapathifolia* DNA in the gut of an adult *Orius* spp. collected on eggplants, and the almost similar species composition pattern between *Orius* spp. on eggplants and the native plant *P. lapathifolia* suggests that natural enemies do move from insectary plants and vegetable crops. Whether the movement of natural enemies from insectary plants to vegetable crops results in pest reduction, achieving delivery of function, is still debatable as we could only observe significant reduction in two pest species but not all of them. Further studies are needed to comprehensively and conclusively prove that Polygonaceae plants can indeed support diversity of natural enemies, resulting in delivery of function in pest suppression. Nonetheless, we can only infer that Polygonaceae plants may, in part, when integrated with other biological control techniques, support the diversity of natural enemies and the delivery of function in suppression of pests.

Providing an assortment of habitat and resources can support diversity of natural enemies and predators, as a result, enhancing biological control services. This is a known concept that is widely reviewed [36–38] and proven in a variety of field experiments [35,39–42]. Replacing the already known and proven non-native invasive insectary plants is a task that will take time and effort. However, for the sake of preserving the biodiversity and maintaining a healthy ecosystem, native plants are the way forward. Furthermore, to achieve a very effective CBC approach with native plants, one native plant may not be enough. Predictive studies have shown that plant diversity promotes pest reduction through movement patterns, host interaction, and predation from natural enemies [34]. In addition, studies have shown that INE species distribution and abundance can be influenced by the host plant chemistry and features (flower/leaf morphology and architecture), plant origin (native), in addition to the prey and the vegetative sites it provides for mating, breeding, shelter, and hibernation [43–47]. This study with its preliminary results has set the basis that INE species can breed and proliferate on native (flowering) plants. Here, we suggest further studies on searching and screening for potential native insectary plants that can support a diversity of INEs. In this digital age where mundane man-hours in ecological surveys can be approached with revolutionary sequencing technology, i.e., Next-Generation Sequencing (NGS), native plant search and screening using the environmental DNA (eDNA) technology can be a useful approach to identifying potential native insectary plants. Furthermore, to maximize delivery of function in pest suppression, multiple annual field studies and laboratory experiments in tracing natural enemy movements, pest consumption rates, pest–natural enemy–native insectary plant interaction studies are a prerequisite.

5. Conclusions

In this era where climate crisis and movement of invasive species between borders are threatening local biodiversity and ecosystems, urgent actions are a prerequisite. In conservation biological pest control, flowering plants are utilized to provide food (pollen and nectar), shelter, and breeding grounds to support natural enemies/predators. However, most of the flowering plants are non-native invasive plants. Ecologically sound yet noninvasive substitutes in the likes of native plants are suggested. To replace the already available non-native invasive insectary plants will be a task too tedious to achieve in the given timeframe. To realize this urgency is already a step forward and progress nonetheless. Our field studies have identified Polygonaceae plants with a plethora of other native plants as potential native plants that can support INEs. Results are still preliminary and, therefore, further experiments are needed to elucidate the role of Polygonaceae plants as well as other native plants in supporting the abundance and diversity of INEs that can yield the desired delivery of function through pest suppression.

Author Contributions: Conceptualization, D.W., J.A. and T.K.; methodology, D.W. and T.K.; formal analysis, D.W.; investigation, D.W., J.A. and T.K.; data curation, D.W.; writing—original draft preparation, D.W.; writing—review and editing, J.A. and T.K.; supervision, J.A. and T.K.; funding acquisition, D.W., J.A. and T.K. All authors have read and agreed to the published version of the manuscript.

Funding: This research received no external funding.

Data Availability Statement: The data presented in this study are available upon request from the corresponding author.

Acknowledgments: The authors wish to extend their gratitude towards the organic farmers in Jinsekikogen-town and Higashi-Hiroshima for allowing us to survey wild plants inhabiting the surroundings of their farms. The authors also express their gratitude to Akihide Fushimi for identifying the wild plants. And, finally, the authors also acknowledge Kitamura Tomoko (technical assistant) for her assistance throughout the course of the experiments.

Conflicts of Interest: The authors declare no conflicts of interest.

Appendix A

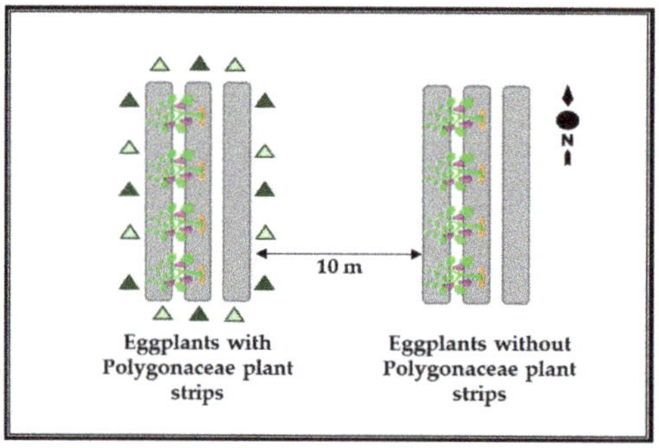

Figure A1. Schematic layout of the field experiments carried out in WARC-NARO experimental fields. Two treatments (eggplants with Polygonaceae plant strips and another without Polygonaceae plant strips) replicated three times. Three beds per treatment with a total of 24 plants (8 plants for each bed). Eggplants were spaced 1 m apart from each other on the beds. Each bed was 1 m by 10 m and the beds were spaced 1.5 m apart. Each treatment therefore covered a total crop area of 70 m^2. For eggplant with Polygonaceae plant strips, an additional 1 m from the edge beds was used for planting Polygonaceae plantings, therefore covering a total crop+Polygonaceae area of roughly 100 m^2. Treatments and replicates were separated by 10 m (a 100 m^2 in terms of area) bare zone so that interaction between the treatments and replicates was minimized. Covering both the crop area and the non-crop area, a total of 1620 m^2 of land area was used in this study. A randomized block design was used to assess the impact of native plant on regulating pest densities on eggplants. For the eggplants with Polygonaceae plant strips, *P. longiseta* and *P. lapathifolia* were interplanted around the eggplant beds as shown above with light green and dark green triangles, respectively.

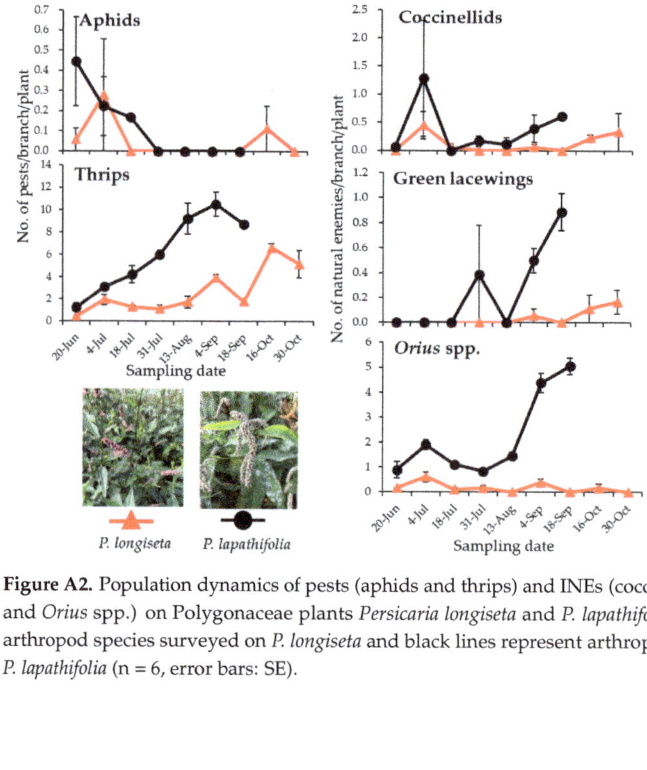

Figure A2. Population dynamics of pests (aphids and thrips) and INEs (coccinellid, green lacewings, and *Orius* spp.) on Polygonaceae plants *Persicaria longiseta* and *P. lapathifolia*. Red lines represent arthropod species surveyed on *P. longiseta* and black lines represent arthropod species surveyed on *P. lapathifolia* (n = 6, error bars: SE).

References

1. Ma, C.S.; Zhang, W.; Peng, Y.; Zhao, F.; Chang, X.Q.; Xing, K.; Zhu, L.; Ma, G.; Yang, H.P.; Rudolf, V.H.W. Climate warming promotes pesticide resistance through expanding overwintering range of a global pest. *Nat. Commun.* **2021**, *12*, 5351. [CrossRef] [PubMed]
2. Rigby, D.; Càceres, D. Organic farming and the sustainability of agricultural systems. *Agric. Syst.* **2001**, *68*, 21–40. [CrossRef]
3. Begg, G.S.; Cook, S.M.; Dye, R.; Ferrante, M.; Franck, P.; Lavigne, C.; Lovei, G.L.; Mansion-Vaquie, A.; Pell, J.K.; Petit, S.; et al. A functional overview of conservation biological control. *Crop Prot.* **2017**, *97*, 145–158. [CrossRef]
4. Pandey, S.; Gurr, G.M. Conservation biological control using Australian native plants in a brassica crop system: Seeking complementary ecosystem services. *Agric. Ecosyst. Environ.* **2019**, *280*, 77–84. [CrossRef]
5. Isaacs, R.; Tuell, J.; Fiedler, A.; Gardiner, M.; Landis, D. Maximizing arthropod-mediated ecosystem services in agricultural landscapes: The role of native plants. *Front. Ecol. Environ.* **2009**, *7*, 196–203. [CrossRef]
6. Zaviezo, T.; Munoz, A.E. Conservation biological control of arthropod pests using native plants. *Curr. Opin. Insect Sci.* **2023**, *56*, 101022. [CrossRef]
7. Ohno, K.; Takemoto, H. Species Composition and seasonal occurrence of *Orius* spp. (Heteroptera: Anthocoridae), predacious natural enemies of *Thrips palmi* (Thysanoptera: Thripidae), in eggplant fields and surrounding habitats. *Appl. Entomol. Zool.* **1997**, *32*, 27–35. [CrossRef]
8. Nagai, K.; Hikawa, M. Evaluation of Black-eyed Susan *Rudbeckia hirta* L. (Asterales: Asteraceae) as an insectary plant for a predacious natural enemy *Orius sauteri* (Poppius) (Heteroptera: Anthocoridae). *Jpn. J. Appl. Entomol. Zool.* **2012**, *56*, 57–64. [CrossRef]
9. Toshikazu, K.; Yoshimitsu, H.; Yoshiyuki, H.; Miwa, I. Establishment of an integrated pest management system for major insect pests in open field eggplants using insectary plants. *Bull. Yamaguchi Prefect. Agri. For. Gen. Tech. Cen.* **2014**, *5*, 55–63.
10. Hinomoto, N.; Abe, J.; Nagasaka, K. Control of thrips in cucumber by *Nesidiocoris tenuis*-banker plant system. In Proceedings of the IOBC/WPRS Working Group "Integrated Control in Protected Crops, Temperate Climate", Niagara Falls, ON, Canada, 4–8 June 2017; Volume 124, pp. 207–213.
11. Seko, T.; Abe, J.; Miura, K.; Hikawa, M. The contribution of a beneficial insectary plant *Scaevola aemula* to survival and long-term establishment of flightless *Harmonia axyridis* in greenhouses. *BioControl* **2017**, *62*, 221–231. [CrossRef]
12. Nakano, R.; Hinomoto, N. Tracking the movement of *Nesidiocoris tenuis* among banker plants and crops in a tomato greenhouse by DNA markers of host plants. *BioControl* **2021**, *66*, 659–671. [CrossRef]
13. Sasaki, M.; Kawamoto, K.; Nakaima, H.; Uchida, N.; Onuma, M. Distribution and occurrence of *Orius tantillus* (Motschulsky) (Hemiptera: Anthocoridae) on *Setaria viridis* Beauv in Okinawa Island. *Res. Bull. Plant Prot. Serv. Jap.* **1999**, *35*, 139–142.
14. Hayashi, M.; Abe, J.; Owashi, Y.; Miura, K. Estimation of movement from insectary plants to crop plants in *Orius* bugs (Heteroptera: Anthocoridae) by molecular gut content analysis. *Appl. Entomol. Zool.* **2020**, *55*, 361–365. [CrossRef]
15. Ohta, I.; Takeda, M. Rapid suppressing of the foxglove aphid, *Aulacorthum solani* (Kaltenbach) on green peppers and eggplants by releasing an aphid parasitoid, *Aphidius gifuensis* Ashmead. *Proc. Kansai Plant Prot. Soc.* **2016**, *58*, 17–21. [CrossRef]
16. Yukawa, J.; Yamaguchi, D.; Mizota, K.; Setokuchi, O. Distribution and host range of an Aphidophagous species of Cecidomyiidae, *Aphidoletes aphidimyza* (Diptera), in Japan. *Appl. Entomol. Zool.* **1998**, *33*, 185–193. [CrossRef]
17. Hinomoto, N.; Muraji, M.; Noda, T.; Shimizu, T.; Kawasaki, K. Identification of five *Orius* species in Japan by multiplex polymerase chain reaction. *Biol. Control* **2004**, *31*, 276–279. [CrossRef]
18. Ito, Y.; Tachikake, K.; Nakamura, S.; Hamada, N.; Watanabe, Y. *Wild Plants of Hiroshima Prefecture in Spring and Early Summer*; The Chugoku Shinbun: Hiroshima, Japan, 1994; ISBN 4-88517-186-5. (In Japanese)
19. Ito, Y.; Tachikake, K.; Nakamura, S.; Hamada, N.; Watanabe, Y. *Wild Plants of Hiroshima Prefecture in Summer & Autumn*; The Chugoku Shinbun: Hiroshima, Japan, 1994; ISBN 4-88517-197-0. (In Japanese)
20. Celesti-Grapow, L.; Alessandrini, A.; Arrigoni, P.V.; Assini, S.; Banfi, E.; Barni, E.; Bovio, M.; Brundu, G.; Cagiotti, M.; Camarda, I.; et al. Non-native flora of Italy: Species distribution and threats. *Plant Biosyst.* **2010**, *144*, 12–28. [CrossRef]
21. Jonsson, M.; Wratten, S.D.; Landis, D.A.; Tompkins, J.-M.L.; Cullen, R. Habitat manipulation to mitigate the impacts of invasive arthropod pests. *Biol. Invasions* **2010**, *12*, 2933–2945. [CrossRef]
22. Tallamy, D.W.; Narango, D.L.; Mitchell, A.B. Do non-native plants contribute to insect declines? *Ecol. Entomol.* **2021**, *46*, 729–742. [CrossRef]
23. Rai, P.K.; Singh, J.S. Invasive alien plant species: Their impact on environment, ecosystem services and human health. *Ecol. Indic.* **2020**, *111*, 106020. [CrossRef]
24. Baisden, E.C.; Tallamy, D.W.; Narango, D.L.; Boyle, E. Do Cultivars of Native Plants Support Insect Herbivores? *Horttechnology* **2018**, *28*, 596–606. [CrossRef]
25. Burkett, C.A.; Corwin, R.; Lautenbach, J.; Serrani-Gallego, I.; Joseph, S.A.; Burkett, F.; Hendrickson, B.J.; Jackson, D.A.; McCool, M.; Molina, C.M.; et al. An Introduction to the Native and Non-Native Plant-Insect Interactions and Potential Pollinators of Puerto Williams and Yendegaia, Cabo de Hornos, Chile. *microPubl. Biol.* **2024**, *2024*. [CrossRef]

26. Lerch, D.; Blüthgen, N.; Mody, K. Home sweet home: Evaluation of native versus exotic plants as resources for insects in urban green spaces. *Ecol. Solut. Evid.* **2024**, *5*, e12380. [CrossRef]
27. Retallack, M.J.; Thomson, L.J.; Keller, M.A. Predatory arthropods associated with potential native insectary plants for Australian vineyards. *Aust. J. Grape Wine Res.* **2019**, *25*, 233–242. [CrossRef]
28. Howard, S.R.; Symonds, M.R.E. Complex preference relationships between native and non-native angiosperms and foraging insect visitors in a suburban greenspace under field and laboratory conditions. *Sci. Nat.* **2023**, *110*, 16. [CrossRef]
29. Wari, D.; Yamashita, J.; Kataoka, Y.; Kohara, Y.; Hinomoto, N.; Kishimoto, H.; Toyoshima, S.; Sonoda, S. Population survey of phytoseiid mites and spider mites on peach leaves and wild plants in Japanese peach orchard. *Exp. Appl. Acarol.* **2014**, *63*, 313–332. [CrossRef]
30. Heppner, J.B.; Habeck, D.H. Insects Associated with Polygonum (Polygonaceae) in North Central Florida. I. Introduction and Lepidoptera. *Fla. Entomol.* **1976**, *59*, 231–239. [CrossRef]
31. Koner, A.; Das, S.; Karmakar, A.; Barik, A. Attraction of the biocontrol agent, *Galerucella placida* towards volatile blends of two Polygonaceae weeds, *Rumex dentatus* and *Polygonum glabrum*. *J. Chem. Ecol.* **2022**, *48*, 165–178. [CrossRef]
32. Costea, M.; Tardif, F.J.; Hinds, H.R. Polygonum. In *Flora of North America (Online)*; Flora of North America Editorial Committee, Ed.; Oxford University Press: Oxford, UK, 1993. Available online: www.eFloras.org (accessed on 24 December 2024).
33. Letourneau, D.K.; Jedlicka, J.A.; Bothwell, S.G.; Moreno, C.R. Effects of natural enemy biodiversity on the suppression of arthropod herbivores in terrestrial ecosystems. *Ecol. Evol. Syst.* **2009**, *40*, 573–592. [CrossRef]
34. Letourneau, D.K.; Armbrecht, I.; Rivera, B.S.; Lerma, J.M.; Carmona, E.J.; Daza, M.C.; Escobar, S.; Galindo, V.; Gutiérrez, C.; López, S.D.; et al. Does plant diversity benefit agroecosystems? A synthetic review. *Ecol. Appl.* **2011**, *21*, 9–21. [CrossRef]
35. Blaauw, B.R.; Isaacs, R. Wildflower plantings enhance the abundance of natural enemies and their services in adjacent blueberry fields. *Biol. Control* **2015**, *91*, 94–103. [CrossRef]
36. Pickett, C.H.; Bugg, R.L. Enhancing biological control: Habitat management to promote natural enemies of agricultural pests. *J. Appl. Entomol.* **1998**, *126*, 204–205. [CrossRef]
37. Landis, D.A.; Wratten, S.D.; Gurr, G.M. Habitat management to conserve natural enemies of arthropod pests in agriculture. *Annu. Rev. Entomol.* **2000**, *45*, 175–201. [CrossRef] [PubMed]
38. Wäckers, F.L.; Romeis, J.; van Rijn, P. Nectar and pollen feeding by insect herbivores and implications for multitrophic interactions. *Annu. Rev. Entomol.* **2007**, *52*, 301–323. [CrossRef]
39. Wanner, H.; Gu, H.; Dorn, S. Nutritional value of floral nectar sources for flight in the parasitoid wasp, *Cotesia glomerata*. *Physiol. Entomol.* **2006**, *31*, 127–133. [CrossRef]
40. Büchi, R. Mortality of pollen beetle (*Meligethes* spp.) larvae due to predators and parasitoids in rape fields and the effect of conservation strips. *Agric. Ecosyst. Environ.* **2002**, *90*, 255–263. [CrossRef]
41. Blaauw, B.R.; Isaacs, R. Larger wildflower plantings increase natural enemy density, diversity, and biological control of sentinel prey, without increasing herbivore density. *Ecol. Entomol.* **2012**, *37*, 386–394. [CrossRef]
42. Blaauw, B.R.; Isaacs, R. Larger patches of diverse floral resources increase insect pollinator density, diversity, and their pollination of native wildflowers. *Basic Appl. Ecol.* **2014**, *15*, 701–711. [CrossRef]
43. Lundgren, J.G.; Fergen, J.K. The oviposition behavior of the predator *Orius insidiosus*: Acceptability and preference for different plants. *BioControl* **2006**, *51*, 217–227. [CrossRef]
44. Lundgren, J.G.; Fergen, J.K.; Riedell, W.E. The influence of plant anatomy on oviposition and reproductive success of the omnivorous bug, *Orius insidiosus*. *Anim. Behav.* **2008**, *75*, 1495–1502. [CrossRef]
45. Lundgren, J.G.; Wyckhuys, K.A.G.; Desneux, N. Population responses by *Orius insidiosus* to vegetational diversity. *BioControl* **2009**, *54*, 135–142. [CrossRef]
46. Pumariño, L.; Alomar, O.; Lundgren, J.G. Effects of floral and extrafloral resource diversity on the fitness of an omnivorous bug, *Orius insidiosus*. *Entomol. Exp. Appl.* **2012**, *145*, 181–190. [CrossRef]
47. Bosco, L.; Tavella, L. Distribution and abundance of species of the genus *Orius* in horticultural ecosystems of northwestern Italy. *Bull. Insectology* **2013**, *66*, 297–307.

MDPI AG
Grosspeteranlage 5
4052 Basel
Switzerland
Tel.: +41 61 683 77 34

Insects Editorial Office
E-mail: insects@mdpi.com
www.mdpi.com/journal/insects

www.ingramcontent.com/pod-product-compliance
Lightning Source LLC
Chambersburg PA
CBHW061830210926
51850CB00006B/846
* 9 7 8 3 7 2 5 8 4 7 6 7 9 *